MOLECULAR STRUCTURE AND ENERGETICS

Volume 3

Studies of Organic Molecules

MOLECULAR STRUCTURE AND ENERGETICS

Volume 3

Series Editors

Joel F. Liebman
University of Maryland Baltimore County

Arthur Greenberg
New Jersey Institute of Technology

Advisory Board

Distribution: VCH Verlagsgesellschaft mbH, P.O. Box 1260/1280. D-6940 Weinheim.
Federal Republic of Germany

USA and Canada: VCH Publishers, Inc., 303 N.W. 12th Avenue. Deerfield Beach, FL 33442–1705. USA

MOLECULAR STRUCTURE AND ENERGETICS

Volume 3

Studies of Organic Molecules

Edited by

Joel F. Liebman
Arthur Greenberg

Joel F. Liebman
Department of Chemistry
University of Maryland Baltimore County
Catonsville, Maryland 21228

Arthur Greenberg
Division of Chemistry
New Jersey Institute of Technology
Newark, New Jersey 07102

Library of Congress Cataloging-in-Publication Data

Main entry under title:

Studies of organic molecules.

(Molecular structure and energetics; v. 3)
Bibliography: p.
Includes index.
1. Chemistry, Physical organic. 2. Molecular structure. I. Liebman, Joel F. II. Greenberg, Arthur. III. Series.
QD476.S766 1986 547.1′22 85-22655
ISBN 0-89573-141-X

Printed in the United States of America.

ISBN 0-89573-141-X VCH Publishers
ISBN 3-527-26477-9 VCH Verlagsgesellschaft

SERIES FOREWORD

Molecular structure and energetics are two of the most ubiquitous, fundamental and, therefore, important concepts in chemistry. The concept of molecular structure arises as soon as even two atoms are said to be bound together since one naturally thinks of the binding in terms of bond length and interatomic separation. The addition of a third atom introduces the concept of bond angles. These concepts of bond length and bond angle remain useful in describing molecular phenomena in more complex species, whether it be the degree of pyramidality of a nitrogen in a hydrazine, the twisting of an olefin, the planarity of a benzene ring, or the orientation of a bioactive substance when binding to an enzyme. The concept of energetics arises as soon as one considers nuclei and electrons and their assemblages, atoms and molecules. Indeed, knowledge of some of the simplest processes, eg, the loss of an electron or the gain of a proton, has proven useful for the understanding of atomic and molecular hydrogen, of amino acids in solution, and of the activation of aromatic hydrocarbons on airborne particulates.

Molecular structure and energetics have been studied by a variety of methods ranging from rigorous theory to precise experiment, from intuitive models to casual observation. Some theorists and experimentalists will talk about bond distances measured to an accuracy of 0.001 Å, bond angles to 0.1°, and energies to 0.1 kcal/mol and will emphasize the necessity of such precision for their understanding. Yet other theorists and experimentalists will make equally active and valid use of such seemingly ill-defined sources of information as relative yields of products, vapor pressures, and toxicity. The various chapters in this book series use as their theme "Molecular Structure and Energetics", and it has been the individual authors' choice as to the mix of theory and of experiment, of rigor and of intuition that they have wished to combine.

As editors, we have asked the authors to explain not only "what" they know but "how" they know it and explicitly encouraged a thorough blending of data and of concepts in each chapter. Many of the authors have told us that writing their chapters have provided them with a useful and enjoyable (re)education. The chapters have had much the same effect on us and we trust readers will share our enthusiasm. Each chapter stands autonomously as a combined review and tutorial of a major research area. Yet clearly there are interrelations between them and to emphasize this coherence we have tried to have a single theme in each volume. Indeed the first four volumes of this series were written in parallel, and so for these there is an even higher degree of unity. It is this underlying unity of molecular structure and energetics with all of chemistry that marks the series and our efforts.

Another underlying unity we wish to emphasize is that of the emotions and of the intellect. We thus enthusiastically thank Alan Marchand for the opportunity to write a volume for his book series, which grew first to multiple volumes, and then became the current, autonomous series for which this essay is the foreword. We likewise thank Marie Stilkind of VCH Publishers for her versatility, cooperation and humor. We also wish to emphasize the support, the counsel, the tolerance and the encouragement we have long received from our respective parents, Murray and Lucille, Murray and Bella; spouses, Deborah and Susan; parents-in-law, Jo and Van, Wilbert and Rena; and children, David and Rachel. Indeed, it is this latter unity, that of the intellect and of emotions, that provides the motivation for the dedication for this series:

"To Life, to Love, and to Learning".

Joel F. Liebman
Baltimore, Maryland

Arthur Greenberg
Newark, New Jersey

PREFACE

The present volume examines seven topics in organic chemistry in which molecular structure and energetics are centers of focus. All the essays explicitly compare and interweave the results from experiment and computational theory.

Chapter 1, by Nelsen, examines hydrazines and their associated radical cations. It employs dynamic NMR, ESR, ultraviolet absorbence and photoelectron spectroscopies for the conformational analysis of these species. Quantum mechanical calculations are used in the analysis of photoelectron spectroscopy results to explore the origins and magnitudes of structural and electronic effects.

Chapter 2, by Stevenson, considers the issue of aromaticity and antiaromaticity in hydrocarbons. He employs calorimetric, gas and condensed phase, electron binding, and conformational data to derive quantitative measures of stabilization and destabilization. Model calculations are used to allow the factoring of these quantitative energies into conceptually more intelligible and, therefore, useful components.

Chapter 3, by Johnson, presents the current status of the investigation of small cyclic alkenes, alkynes and cumulenes. Many of these compounds are so strained that they are on the borderline of existence and isolation appears unlikely. Other members of these classes of compounds can be directly studied and the experimental results (eg, from photoelectron spectroscopy) compared with theory. Some examples of quantum mechanical calculations employed to suggest relative thermodynamic and kinetic stabilities are also discussed.

Chapter 4, by Smart, summarizes experimental and theoretical studies of the structural and thermochemical features of highly fluorinated organic molecules. These fluorocarbons represent extreme cases—substituent effects are often large, experimental accuracy is exceedingly difficult to achieve, and numerous pitfalls await the unsuspecting calculationist. The results are often striking. Moreover, conclusions about stability drawn from thermochemistry and kinetics often contradict each other.

Chapter 5, by Greenberg and Stevenson, considers the effects of substituents on the structure and energetics of strained and unsaturated organic molecules. It includes a large compilation of calculated and experimental data. Correlation analyses of the stabilization energies are introduced and applied here in order to understand the relative sensitivity of the various strained and unsaturated framework to a wide range of electronic effects.

Chapter 6, by Liebman, summarizes and exemplifies his macroincrementation reactions approach to the understanding of properties of organic molecules. Numerous stabilization/destabilization reactions presented elsewhere in these volumes are seen as examples of this approach. This method is demanding in that it requires the use of chemical intuition as to the important components and interactions of the molecule. However, it is also flexible in that any reviewer may assess or attempt to better the intuition of an original user.

Chapter 7, by Ōsawa and Kanematsu, examines novel organic molecules that have abnormally long carbon-carbon bonds. Their approach entails the complementary use of molecular mechanical and quantum mechanical calculations. For example, if there are significant differences between bond lengths so calculated, this disparity suggests there are special conjugative effects and other meaningful electronic factors ignored in the conventional molecular mechanics model derived from more normal species.

CONTRIBUTORS

Arthur Greenberg
Division of Chemistry
New Jersey Institute of Technology
Newark, New Jersey 07102

Richard P. Johnson
Department of Chemistry
University of New Hampshire
Durham, New Hampshire 03824

Ken Kanematsu
Institute of Synthetic Organic Chemistry
Faculty of Pharmaceutical Sciences
Kyushu University
62 Fukuoka 812
Japan

Joel F. Liebman
Department of Chemistry
University of Maryland Baltimore County
Catonsville, Maryland 21228

Stephen Nelsen
Department of Chemistry
University of Wisconsin
Madison, Wisconsin 53706

Eiji Ōsawa
Department of Chemistry
Faculty of Science
Hokkaido University
Sapporo 060
Japan

Bruce Smart
Central Research & Development Department
E.I. duPont de Nemours & Company
Experimental Station
Wilmington, Delaware 19898

Gerald R. Stevenson
Department of Chemistry
Illinois State University
Normal, Illinois 61761

Tyler A. Stevenson
Division of Chemistry
New Jersey Institute of Technology
Newark, New Jersey 07102

CONTENTS

CHAPTER 1

Hydrazine–Hydrazine Cation Electron Transfer Reactions

Stephen F. Nelsen

University of Wisconsin, Madison, Wisconsin

CONTENTS

1. INTRODUCTION

The hydrazine unit (R_4N_2 in Equation 1-1) has been important in understanding how bond rotations and angle changes affect electronic interactions. The nitrogens of unstrained examples of R_4N_2 are approximately tetrahedral (lone pairs about sp^3 hybridized), and there is a modest electronic preference for the lone pair–lone pair dihedral angle θ (see **I**) to be near 90°, which is the geometry for many acyclic and *N,N*-cycloalkyl hydrazines. Because the $\theta = 180°$ conformation is sterically less hindered than the 90° conformations if the nitrogens are tetrahedral, there apparently is an electronic preference for the 90° conformations.

MOLECULAR STRUCTURE AND ENERGETICS, Vol. 3

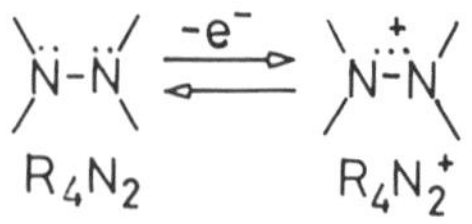

$$R_4N_2 \underset{}{\overset{-e^-}{\rightleftharpoons}} R_4N_2^{+\cdot}$$

A variety of angles and distances have been used to describe the amount of pyramidality at nitrogens. Most theoretical discussions of bending at nitrogen employ the angle the NN bond makes with the RNR plane, $\beta(N)$ (see **II**). The $\beta(N)$ is 0° for a planar nitrogen (pure p lone pair orbital), and 54.7° for a tetrahedral nitrogen (sp^3 lone pair orbital for an R_3N nitrogen atom). Use of β is misleading for many purposes because a plot of energy functions versus $\beta(N)$ is very curved. The calculated ionization potential of the lone pair dominated molecular orbital (MO) for simple amines is almost linear, with both the fractional p character at N, f_p, and the average of the three bond angles at nitrogen, $\alpha(av)$, which makes it convenient to use $\alpha(av)$

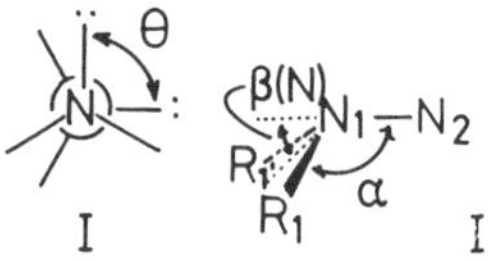

(1–1)

I II

values in discussing how pyramidal the unsymmetrical nitrogens of a hydrazine are.

The four sites for attachment of alkyl groups to the nitrogens of R_4N_2, allow building the hydrazine unit into *N,N′*-cyclic and bicyclic frameworks, and hydrazines with equilibrium θ values varying from 0 to 180° have been studied. The highest occupied molecular orbital (HOMO) of R_4N_2 is that dominated by the antibonding combination of lone pair atomic orbitals, while the second highest MO (SHOMO) is often that dominated by the bonding lone pair combination. The energy separation of these orbitals may be measured by photoelectron spectroscopy (PES) and provides one of the clearest examples of conformational analysis by this method. The photoelectron experiment measures the gas phase vertical energy gap from ground state neutral R_4N_2 to various states of its radical cation $R_4N_2^{+}$ in the same geometry as the neutral. This geometry is far from that of relaxed, ground state $R_4N_2^{+}$. The electronic preference for the cation is to have the lone pair axes coplanar ($\theta = 0$ or 180°), in contrast to the $\theta = 90°$ preference of neutral R_4N_2. The nitrogens are also greatly flattened in $R_4N_2^{+}$, so the lone pair hybridization is considerably richer in p character than is that of neutral R_4N_2.

Despite a formal spin and charge density of 0.5 at each nitrogen, most examples of $R_4N_2^{+}$ are long-lived in solution. This is in contrast to the radical cations of most saturated compounds, which exhibit rapid bond cleavage reactions following electron loss, prohibiting study of the electron loss as an isolated step. The long lifetime of $R_4N_2^{+}$ is of great experimental importance. It makes the thermodynamics for the electron transfer of Equation 1-1 easily available by cyclic voltammetry (CV). Changes in the free energy for electron loss, ΔG°_e, may be obtained by CV measurement of the formal oxidation potentials $E^{\circ\prime}$ for two examples of R_4N_2 by using Equation 1-2. The electrochemical measurement gives the relative adiabatic energy gap between R_4N_2 and $R_4N_2^+$ in solution.

$$\Delta(\Delta G_e^\circ),\ \text{kcal/mol} = 23.06\ (\Delta E^{\circ\prime}),\ \text{V} \tag{1-2}$$

Differences between vertical and adiabatic electron loss are especially large for $R_4N_2^+$ because of the large geometry difference between the two oxidation states, as shown in Figure 1-1. The energy gap labeled $E^{\circ\prime}$ is not measured in absolute size, but only relative to a standard. Hydrazines have been important in establishing how large bond rotation and pyramidality effects on neutral and cationic forms can be. This chapter reviews work in this area.

2. CONFORMATIONS OF NEUTRAL HYDRAZINES

A. Electron Diffraction and Calculational Studies

The literature through 1973 is well surveyed by Shvo.[1] A new gas phase electron diffraction (GED) study on H_4N_2 by Kohata, Fukuyima, and Kuchitsu[2] was analyzed without assuming equal HNN angles for the two different NH bonds (there are two types of NH bond because θ is 91 $\pm$ 2°; see **I**) and the outer HNN angle was determined as 106 $\pm$ 2°, the inner angle 112 $\pm$ 2°, along with r_g (NN) = 1.449(2), r_g (NH) 1.021(2) Å. A new and conservative analysis of the GED data for 1,2-dimethylhydrazine[3] employed the results of 4–31G level ab initio calculations to constrain the GED calculations. The authors concluded that although the conformation with CNNC angle of about 88° predominates, and the best fit to the GED data was obtained by including about 22% of a conformation with CNNC angle of about 151°, one could not rule out a conformation with CNNC angle of about 50°, which had been inferred as the minor conformation in a conventional GED analysis.[4] Indeed, even only the approximately 88° conformation being present could not be completely ruled out.

A general discussion of rotational barriers of single bonds by Brunck and Weinhold[5] has emphasized the importance of mixing occupied MOs with unoccupied σ^* MOs in determining the rotational barrier. The difficulty of doing accurate enough ab initio calculations on hydrazine for estimating rotational and inversion barriers has been emphasized by Cowley and co-workers,[6] who found that 4–31G calculations at STO-3G energy-minimized geometries make the conformation in which θ is about 90° only 0.4 kcal/mol stabler than the 180° conformation, which is far too small an energy difference. They present a perturbation molecular orbital (PMO) energy analysis. Voronkov and co-workers[7] have published a STO-4LGF analysis of the θ = 0 and 180° rotational barriers for H_4N_2. They obtain θ = 95° for the stablest conformation and rotational barriers of 4.2 and 10.8 kcal/mol, which are compared to those of previous calculations. They also consider torsional

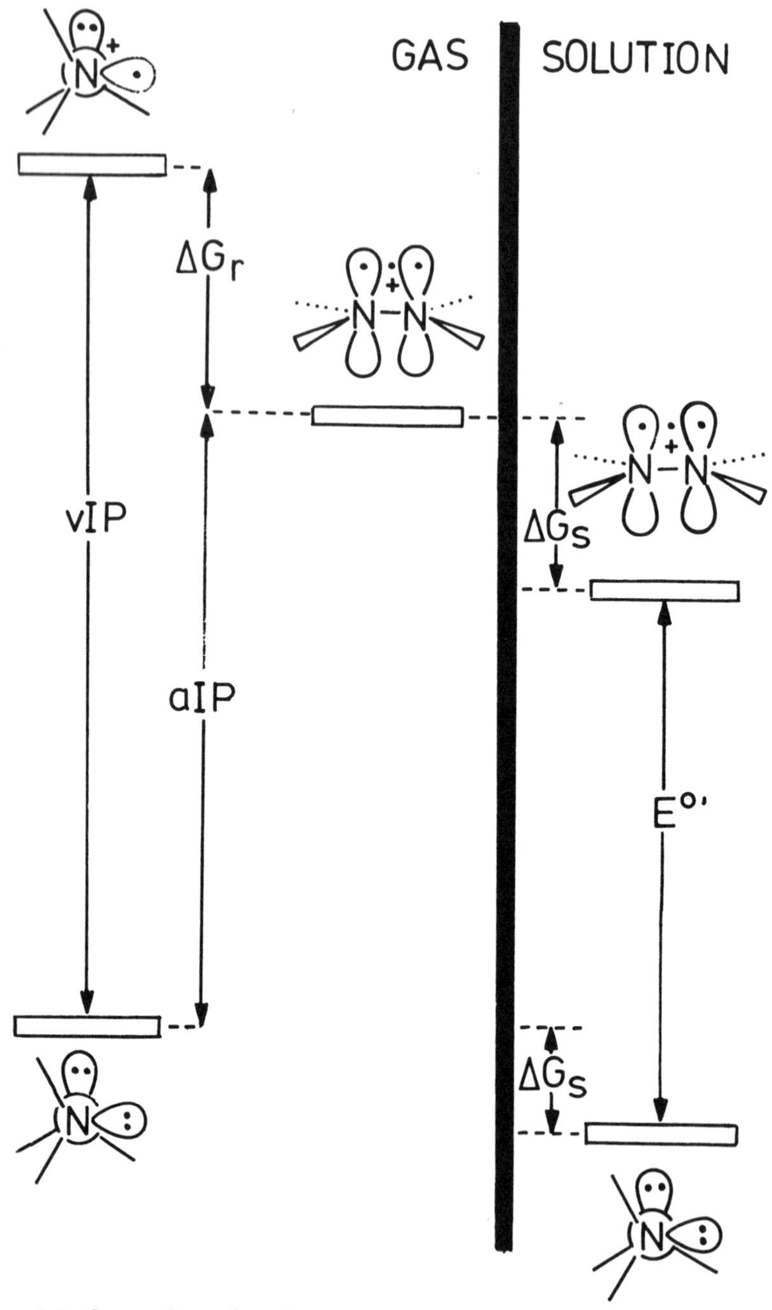

Figure 1-1. Comparison of vertical (v) and adiabatic (a) electron loss in solution and the vapor phase for an acyclic hydrazine.

vibronic structure at the noninteracting anharmonic oscillator approximation level, concluding that the population of the first five levels at 300 K are 78, 17, 3.6, 0.8, and 0.2%, emphasizing the ease of distortion from the equilibrium geometry.

B. Conformational Studies by Photoelectron Spectroscopy

The groups of Rademacher[8–11] and of Nelsen[12–16] pointed out the utility of PES for determining the most important feature of hydrazine conformations, the lone pair–lone pair dihedral angle θ. Because there is a large geometry difference between R_4N_2 and $R_4N_2^+$, the vertical ionization of PES gives a highly excited cation, where the vibrational levels are essentially a continuum; the lone pair ionization bands are broad and well fit by Gaussian curves. Because the bands typically overlap, deconvolution is necessary for accurate determination of the vertical ionization potential (vIP). Figure 1-2 shows the calculated behavior of the lone pair combination orbital energies for sp^3 orbitals held at the hydrazine distance as a function of θ, and illustrates the sensitivity of the energy difference between the lone pair MOs to dihedral angle θ.

It was already known from spectroscopic studies of other sorts that H_2NNH_2 and methylated hydrazines have equilibrium θ values near 90°, so it was expected that $IP_2 - IP_1 = \Delta$ would be small. For simple alkyl substitution, Δ depends on substitution pattern in the order R_2NNH_2 ($\Delta \simeq 1.25$ eV) > $RHNNH_2$ ($\simeq 0.95$) > H_2NNH_2 ($\simeq 0.85$) > RHNNHR ($\simeq 0.75$) > R_2NNHR ($\simeq 0.65$) > R_2NNR_2 ($\simeq 0.54$).[14] The "lone pairs" of amino nitrogens are not really nonbonding and do mix with the σ orbitals to a small extent, tending to decrease their mixing with each other when substituents are changed from hydrogen to alkyl. The lone pairs are not of equal energy in the unsymmetrical compounds, and part of the observed Δ reflects this energy difference in the (hypothetical) absence of mixing. The orbital crossing of Figure 1-2 is also obtained by semiempirical MO calculations on H_2NNH_2 and Me_2NNMe_2, but the crossing point moves to slightly lower θ values [83° for intermediate neglect of differential orbital (INDO) results on Me_2NNMe_2 with tetrahedral nitrogens[14]]. We have suggested,[17] however, that the fact that many acyclic and 1,1-cycloalkyl tetraalkylhydrazines give the same Δ value of 0.54 ± 0.03 eV requires an avoided crossing. An avoided crossing would make Δ insensitive to θ near the crossover point, as is observed experimentally; the sensitivity of Δ to θ near 90° surely cannot approach the 0.03_5 eV/degree given by semiempirical calculations. The STO-4LGF calculation of Voronkov and co-workers[7] gives an avoided crossing, but sharper cusps in the curves for IP_1 and IP_2 versus θ near the crossover point than we suspect is occurring experimentally. Their calculations were not geometry optimized. A summary of some photoelectron data for selected examples of R_4N_2 is included in Table 1-1.

Inclusion of rings sometimes causes large Δ values to be observed, and

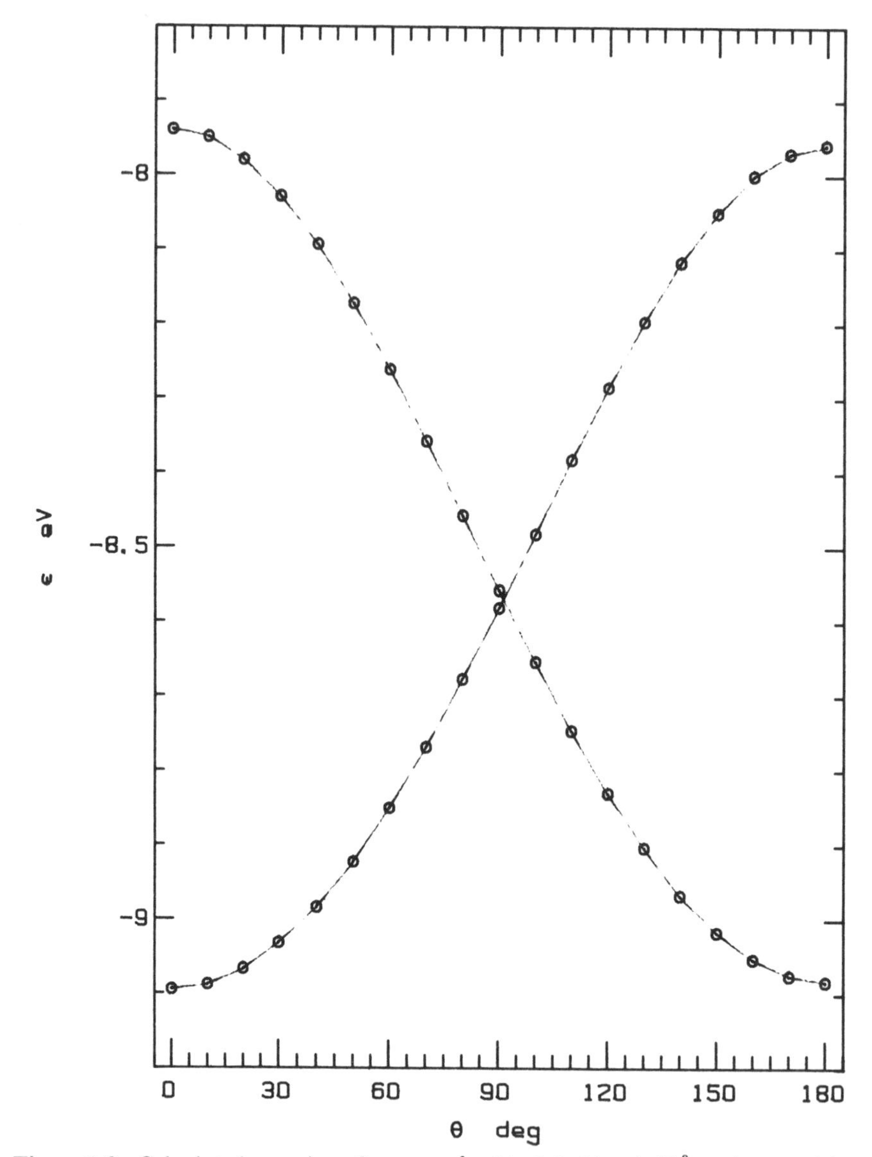

Figure 1-2. Calculated energies ε for two sp^3 orbitals held at 1.45Å and rotated from a dihedral angle of 0° to 180°.

Rademacher[8,18] showed by vibrational and GED spectroscopy that biaziridine (**6**) exists in the 180° conformation both in the gas phase and in solution, and as expected, its PE Δ value is large. Photoelectron spectroscopy has a very short time scale, so that a superposition of spectra for occupied conformations is observed. Diazetidine (**7**) clearly shows a peak that is caused by a large Δ anti conformation, but a small Δ-gauche conformation predominates[19] [although the anti conformation predominates for 1(1-azetidinyl)piperidine[19a]], and only gauche conformations are observed for dipyrrolidine (**8**)[17] and dipiperidine (**9**). The bending at nitrogen accentuated by decreasing the CNC angle in **6** appears to destabilize the gauche conformation by nonbonded interaction enough to cause its anti conformation to be populated. The PE spectrum of 9,9′-bibicyclo[3.3.1]nonane (**10**)[20] shows it to also exist in the 180° anti conformation, as it does in the crystal,[21] and examination of models is convincing that serious steric interactions result from gauche conformations with the bicyclic alkyl groups attached.

Dimethyldiaziridine (**11**)[10] has only a modest splitting despite its known trans geometry and inability to twist the ring, but dimethyldiazetidine (**12**) has a large Δ.[19a] Two conformations are observed for dimethylpyrazolidine (**13**).[9,11,16] The major one is assigned to the $\mathbf{T_{ee}}$ (*trans*-dipseudoequatorial) conformations, and the minor to the ring reversal form, the $\mathbf{T_{aa}}$ (*trans*-dipseudoaxial) conformation. The RNNR steric interaction in $\mathbf{T_{ee}}$ appears to be larger than the $RNNCH_2$ steric interactions in $\mathbf{T_{aa}}$ when R is larger than methyl, because the diethyl and diisopropyl compounds exist only in the $\mathbf{T_{aa}}$ conformations, as do all of the R = Me, Et, *i*-Pr, and *t*-Bu oxadiazolidine analogues (**23**).[22] A delicate interplay between steric destabilizations is also exhibited by derivatives of hexahydropyridazine.[11,13,19]

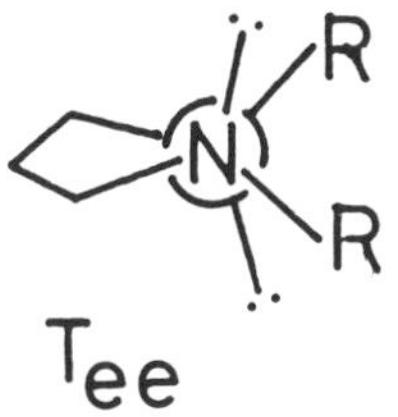

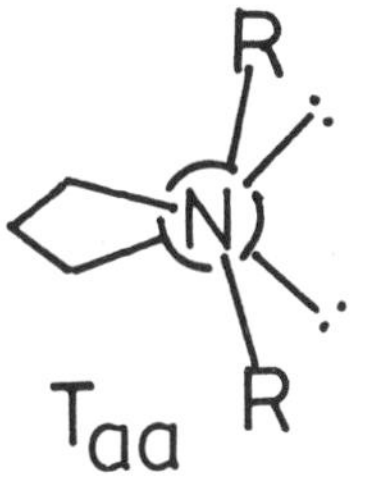

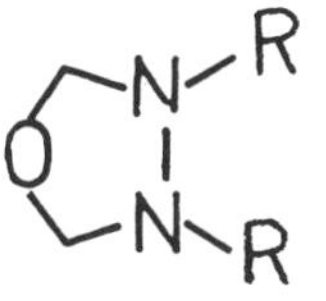

The dimethyl compound **14** exists predominantly in the **ee** form, for which X-ray crystallography on a model system has shown has θ very close to 180°. Although the gauche conformation seen by PES could in principle be either the **ae** or **aa** form, NMR work has demonstrated that only **ae** is actually occupied (see below). The hydropyridazine ring system is remarkably sensitive to substitution changes, and only minor amounts of **ee** forms could be detected by PES for **24** or **25**. The ability of PES to yield conformational information on these compounds should be noted, because their lack of symmetry proved to be disastrous for NMR studies. Methyl substitution at both C_3 and C_6 and ethyl substitution of both nitrogens essentially completely eliminates the **ee** conformation, but closure of a second ring as in **15** causes only the **ee** conformation to be occupied. We will discuss the

Table 1-1. Photoelectron Data for Some Tetraalkylhydrazines

	Compound		IP_1, (eV)	Δ (eV)	θ	Dev (eV)[a]	Ref.
1	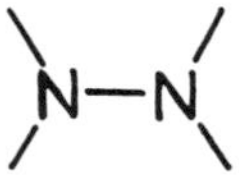 $Me_2N{-}NMe_2$		8.27	0.55	≃90°	+0.01	17
2	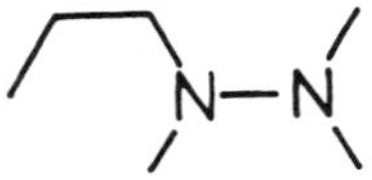		8.14	0.51	≃90°	−0.01	17
3	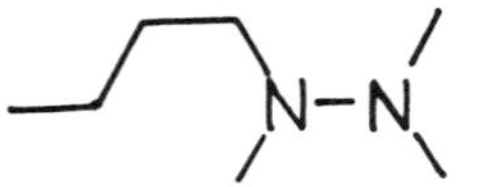		8.12	0.54	≃90°	−0.07	17
4	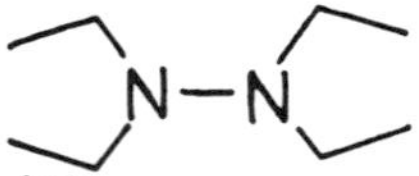		7.94	0.51	≃90°	+0.01	17
5	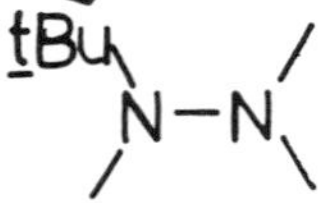 tBu		7.89	0.59	Gauche	−0.12	17
6	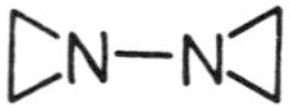		8.59	2.57	180°	+1.23	8
7	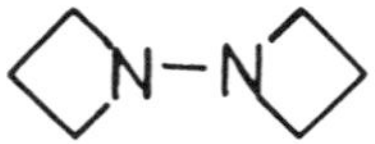	Major	8.25	0.73	Gauche	+0.26	19a
		Minor	≃8.2	≃1.9	180°	+0.8	
8	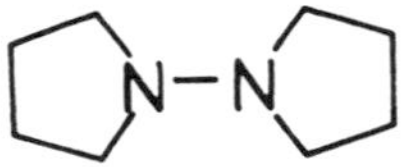		7.91	0.56	≃90°	0.00	17

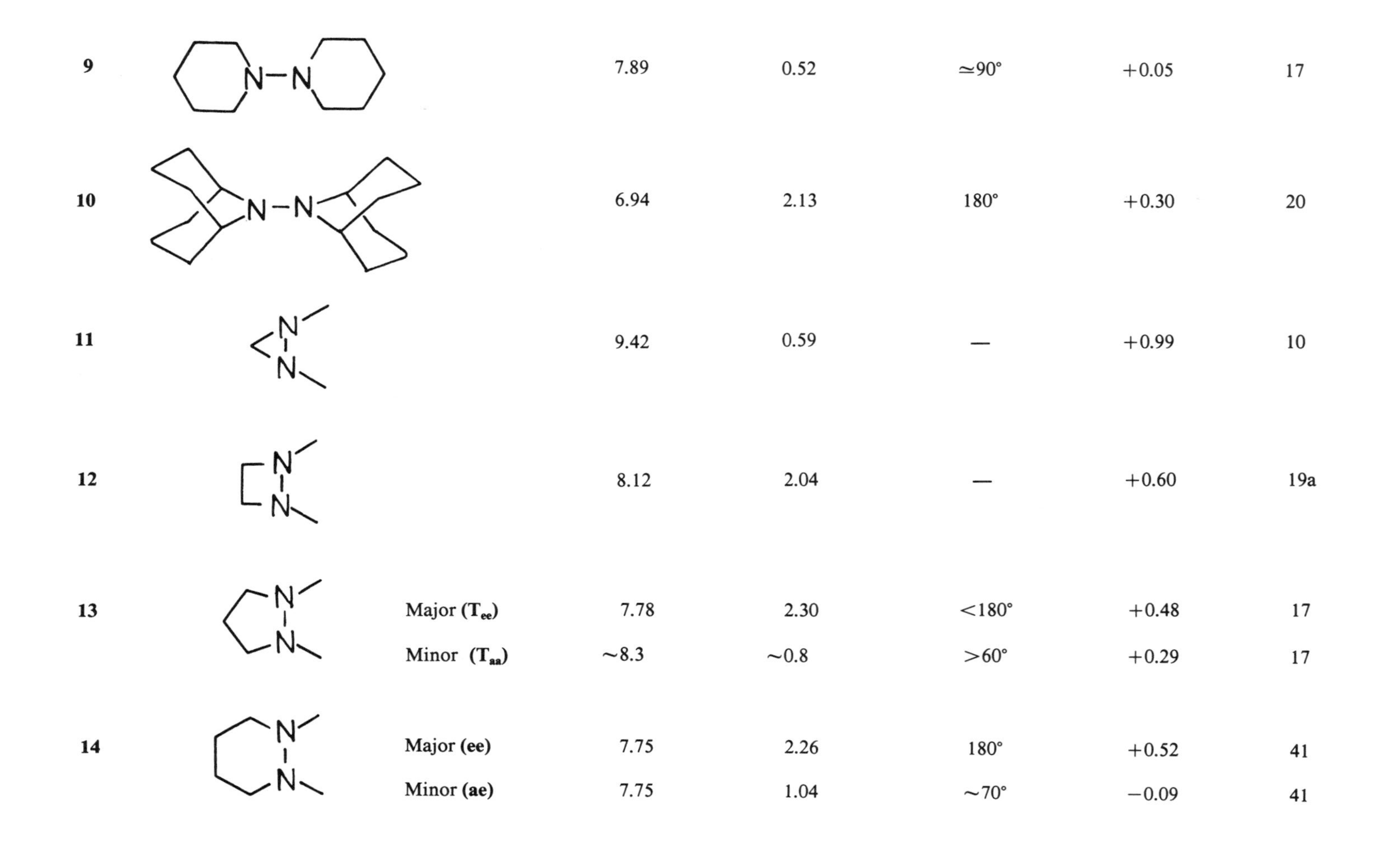

9			7.89	0.52	≃90°	+0.05	17
10			6.94	2.13	180°	+0.30	20
11			9.42	0.59	—	+0.99	10
12			8.12	2.04	—	+0.60	19a
13		Major (T_{ee})	7.78	2.30	<180°	+0.48	17
		Minor (T_{aa})	~8.3	~0.8	>60°	+0.29	17
14		Major (**ee**)	7.75	2.26	180°	+0.52	41
		Minor (**ae**)	7.75	1.04	~70°	−0.09	41

Table 1-1. (continued)

	Compound	IP_1, (eV)	Δ (eV)	θ	Dev (eV)[a]	Ref.
15	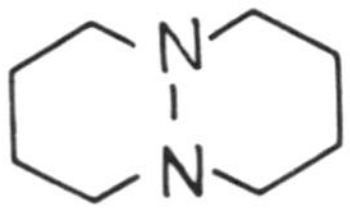	7.61	2.31	180°	+0.57	17
16	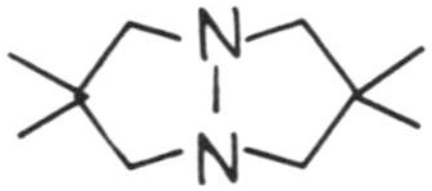	7.53	2.21	—	+0.44	17
17	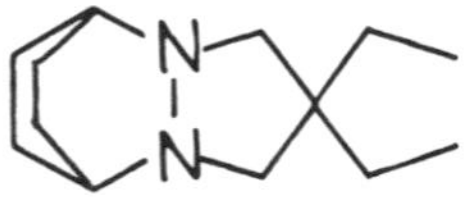	6.92	2.32	0°	+0.08	17
18	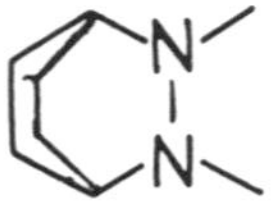	7.46	1.82	—	+0.16	17
19	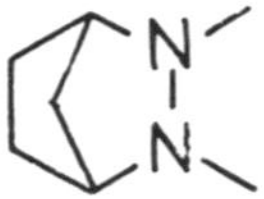	7.66	1.78	~120°	+0.26	17

20	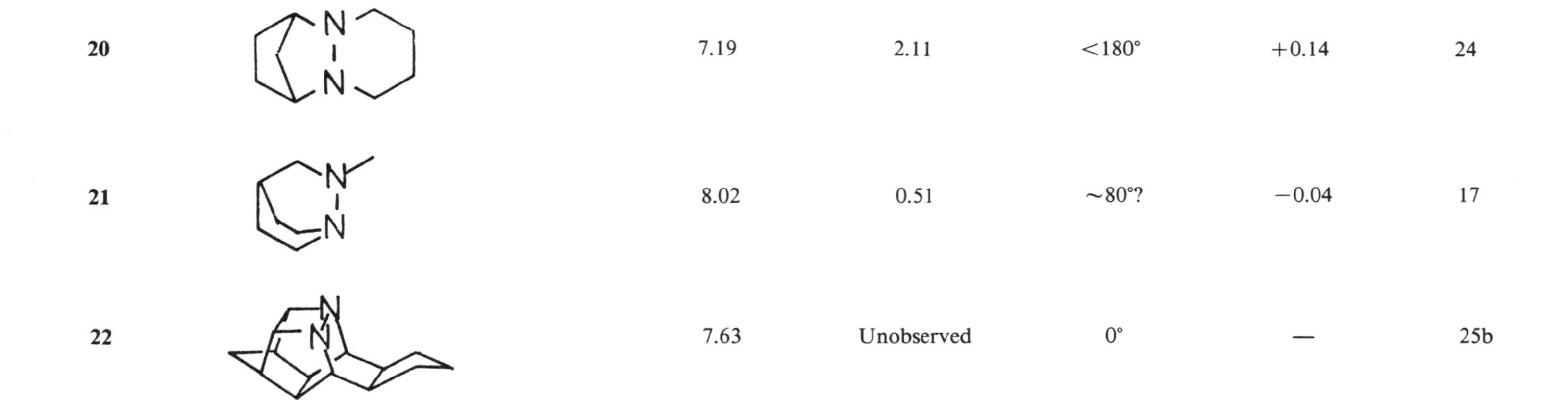	7.19	2.11	<180°	+0.14	24
21		8.02	0.51	~80°?	−0.04	17
22		7.63	Unobserved	0°	—	25b

[a]Dev is the difference between the average lone pair ionization energy $IP_{av} = (IP_1 + IP_2)/2$ and that estimated from the "effective number of carbons" in the alkyl groups (see the discussion after Table 1-2). A positive Dev means that the observed IP_{av} value is higher than that estimated.

thermodynamics of these equilibrium shifts after the NMR data have been presented.

14ee **14ae** **24** **25**

In contrast to the trans ring fusion of 1,6-diazabicyclo[4.4.0]decane **(15)** and 1,6-diazabicyclo[4.3.0]nonane,[17,23] 1,5-diazabicyclo[3.3.0]octane **(26)** has a cis ring junction, but the Δ value of 1.57 for its major conformation requires twisting about the NN bond, showing that it is in conformation **26a.** Tetramethyl substitution in **16** raises Δ to 2.21 eV, presumably because the *gem*-dimethyls force the pyrazolidine rings to assume envelope conformations, eclipsing the lone pairs (see **16a**). A similar 0° cis-fused conformation is adopted by **17.** The Δ values are similar for **18** and **19,** but we doubt that the θ values are close. The bicyclo[2.2.2]octyl system is probably much more twisted than the bicyclo[2.2.1]heptyl system. Compound **20** shares with **17** a fused 6,5-hydrazine unit, and their Δ values are rather close, but NMR work has shown that **20** is trans fused.[24] The 1,2-diazabicyclo[2.2.2]octane **21** has as low a Δ value as an acyclic compound; θ is near 60° only if there is no torsion in the bicyclooctyl system, and torsion to increase θ (which relieves interaction of the methyl with the methylene groups) clearly occurs. Our inability to measure Δ for Hünig's diazetidine **(22)**[25] points out a problem; when Δ is large, the bonding lone pair combination gets buried in the σ envelope, which moves to lower IP as the size of the alkyl groups increases.

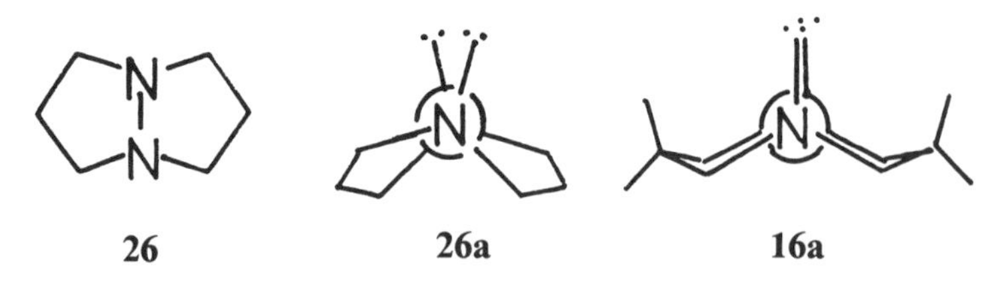

26 **26a** **16a**

We argue that the Δ values of Table 1-1 show that a single $\Delta = A \cos \theta + B$ curve[11] will not give reasonable θ values for R_4N_2. Not only is there a region of insensitivity of Δ to θ near the calculated crossover point of Figure 1-2, and examples with θ unquestionably very near 180° **(6, 10, 15)** have a Δ range of 0.44 eV (22% of the total range of Δ), but also application of such a curve leads to geometrically unreasonable θ values in several examples **(11, 12, 13T$_{ee}$, 20).** We believe this happens because of changes in lone pair

TABLE 1-2. n(eff) Values for Simple Alkyl Groups

Group	Number of carbons	n(eff)	Notes	
Me	1	1.00	Reference alkyl group	
Et	2	2.00	First β carbon:	1.00
i-Pr	3	2.87	Second β carbon:	0.87
t-Bu	4	3.64	Third β carbon:	0.77
n-Pr	3	2.48	First γ carbon:	0.48
t-Pe	5	4.13	First γ carbon:	0.49
i-Bu	4	2.94	Second γ carbon:	0.46
n-Bu	4	2.73	First δ carbon:	0.25
i-Pe	5	2.90	Second δ carbon:	0.17
n-Pe	5	2.79	First ϵ carbon:	0.06

hybridization as the alkyl groups are changed. Addressing this question requires consideration of the size of IP_1 and IP_2, as well as their difference.

$$\text{vIP(RX)} = \text{vIP(MeX)} + m'\mu(\text{R}) \qquad (1\text{-}3)$$

Lengthening alkyl groups lowers the IP values of RX compounds, and Danby and co-workers[26] showed that the effect is quantitatively described using empirical parameters μ(R), through Equation 1-3. The size of μ(Me) and μ(Et) are arbitrary, and we find it convenient to recast the Danby parameters as "effective number of carbons" $n(\text{eff}) = 1 + \mu(\text{R})/\mu(\text{Et})$. Examination of the n(eff) values of Table 1-2 shows that one is simply counting carbons, allowing for a decrease in effectiveness at lowering vIP as the distance from the heteroatom increases and allowing for "saturation" when more than one substituent is present.[27] The fall-off in n(eff) for the first β, γ, δ, and ϵ carbons runs 1.00, 0.48, 0.25, and 0.06, which is remarkably close to Taft's 0.45 fall-off in "inductive effect" per methylene group,[28] which would lead to numbers of 1.00, 0.45, 0.20, and 0.09 for the series above. The advantage of the numbers as n(eff) values is that one can easily include cyclic groups. For ethers, both cyclic and bicyclic substituents fit vIP versus n(eff) plots well.[27] For amines, deviations are clearly seen in vIP versus n(eff) plots when the alkyl groups force CNC angles away from those of *n*-alkylamines. This is expected, because the "lone pair" at nitrogen mixes with the σ framework, leading to changes in lone pair hybridization as the CNC angles are changed. Raising s character raises IP, and lowering it lowers IP. The effect is large, amounting to 0.88 eV for IP of manxine (**M**),[29] which has a flat nitrogen ($f_p \cong 1.0$, $\alpha_{av} \cong 120°$).

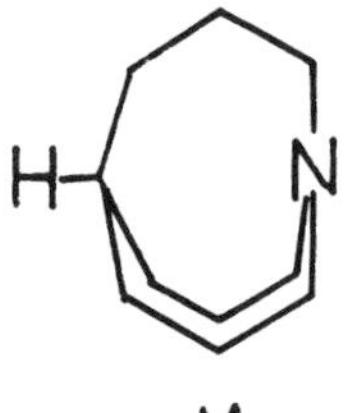

M

To apply this idea to hydrazines, we use $IP_{av} = (IP_1 + IP_2)/2$, attempting to extract the lone pair energy in the (hypothetical) absence of the bonding and antibonding interaction between the lone pair dominated MOs, which cause two peaks to be observed by PES. A plot of IP_{av} versus n(eff) for methyl- and ethyl-substituted hydrazines gives $I_c(R_2NNR_2) = 8.99_1 - 0.08_8$ Σn(eff), where n(eff) values of all the carbons are simply added to give Σn(eff), with a |Dev| (deviation of the observed IP_{av} from I_c) of 0.03 eV for 15 n-alkylhydrazines, which all show the same Δ values within experimental error (which is estimated to be on the order of 0.03 eV for our data). Dev values are also included in Table 1-1, and make the general point that there is a lot of scatter in an IP_{av} versus n(eff) plot for hydrazines. The most instructive case to consider, as always for conformational effects, is the six-membered ring case, **14.**

Figure 1-2 suggests that the large Δ **14ee** conformation should have a considerably lower IP_1 than the small Δ **14ae.** This is not the case, however; instead IP_1 for the two conformations is not resolved. This can happen only if the lone pair energies of **14ee** and **14ae** differ substantially; **14ee** apparently has a lone pair energy (2.26 − 1.04)/2 = 0.61 eV, or 14 kcal/mol stabilized relative to **14ae.** We contend that the major reason for this is increased s character, hence increased bending at N in **14ee** relative to **14ae.** Figure 1-2 does *not* represent what happens as θ is changed in a hydrazine properly at all, because the hybridization of the orbitals was held constant. In fact, the nitrogen atoms rehybridize as θ is changed. Although this effect is seen in geometry-optimized semiempirical calculations, such calculations are of little use for quantitatively considering the effect because they get the wrong NN distance for hydrazines (considerably shorter than expected), get the wrong dihedral angle for the minimum energy conformation [modified neglect of diatomic overlap (MNDO) "thinks" H_2NNH_2 has a $\theta = 180°$ minimum energy conformation], and give poor N inversion barriers. The STO-4LGF calculations of Voronkov and co-workers[7] do estimate IP_{av} for θ = 180° hydrazine to be 10.5 kcal/mol higher than that for gauche (θ = 95°) hydrazine, but these calculations were not geometry optimized.

X-ray crystallographic structures of **27** (which is **ee** in the crystal, with θ = 178°) and of **28** (which is **ae** in the crystal, θ = 70°)[30] demonstrate that there is a significant rehybridization at N when θ is forced to assume a 180° value. The unfavorable electronic interaction in **27** (θ being held at 180°) causes NN bond lengthening and increased bending at nitrogen, increasing the fractional s character in the lone pair MOs about 50%. Because of differences in their structures, the effect of flattening at nitrogen on Dev ought to be different for hydrazines and amines, but ΔDev calculated for going from α(av) = 107.5 to 112.2 for the admittedly imperfectly calibrated amine case is 0.44 eV, 0.72 times as large as the ΔDev for **14ee** and **14ae.** Table 1-1 indicates that other compounds with large Δ show positive increments in Dev also, so the rehybridization when electronically destabilized large Δ conformations are forced to be occupied by steric effects is not restricted to hexahydropyridazines. It is clear that if CNC and CNN angles are restricted by small rings, there should also be a positive increment in Dev, and biaziridine (**6**) shows the largest Dev yet found. Similarly diaziridine (**11**) shows almost as large a Dev value, despite its small Δ value, and the four-membered ring compounds **7** and **12** also have positive increments in Dev compared to compounds with less angle restriction. We also note that steric interactions are expected to cause flattening at N in compounds **16** and especially **17** (for which $\theta \sim 0°$) and probably are responsible for the modest Dev values despite large Δ. The same effect is seen for compound **20** versus **19.** More structural data will clearly be required to see whether useful estimates of α_{av} from Dev values can actually be made, but with the limited data available, it appears that this might be the case.

27

$\alpha(av) = 107.5$ ($f_p = 0.69$)

$d(NN)$:eq 1.48_6 Å

28

$\alpha(av) = 111.6$ (110.9 at NMe_{ax}, $f_p = 0.79$)
(112.2 at NMe_{eq}, $f_p = 0.82$)

$d(NN) = 1.45_0$ Å

Surprisingly little thermochemical information appears to be available on substituted hydrazines. The only heats of formation we have seen for alkylated hydrazines (thanks to Professors A. Greenberg and J.F. Liebman) are listed in Table 1-3. All sources for this table include ΔH_f values for mono-, *N,N'*-di-, and tetraphenylhydrazine, and the reference given in note *c* includes heats of combustion for six 2-propinylhydrazine derivatives.

Table 1-3. Heats of Formation for Three Alkylated Hydrazines

Compound	Heat of formation, 298 K (kJ/mol)	
	Liquid phase	Gas phase
H_2NNH_2	50.63[a]	95.40[a]
CH_3HNNH_2	54.0,[a] 54.2,[b] 54.8,[c] 54.1[c]	94.35[a], 94.6[b]
$CH_3HNNHCH_3$	48.9,[b] 51.63[c]	83.9[b]

[a]Wagman, D.D.; Evans, W.H.; Parker, V.B.; Schumm, R.H.; Halow, I.; Bailey, S.M.; Churney, K.L.; Nuttall, R.L. *J. Phys. Chem. Ref. Data* **1982,** *11,* Suppl. 2.
[b]Pedley, J.B.; Rylance, J.; Sussex–N.P.L. Computer-Analysed Thermochemical Data: Organic and Organometallic Compounds", University of Sussex: Brighton, UK, 1977.
[c]Schmidt, E.W. "Hydrazine and Its Derivatives", Wiley-Interscience: New York, p. 221. **1983**

C. NMR Studies of Hydrazine Conformations: Barriers to Conformational Change

The considerable work through about 1973 has been reviewed in detail by Shvo,[1] and little of this material is repeated. Most conformational information from NMR work has come from study of highly symmetrical compounds in which two atoms or groups of atoms are different when conformational interconversion of enantiomeric forms is slow, but interconvert when enantiomerization becomes fast, at higher temperature. The great power of NMR is that the barrier to conformational interconversion can be determined by dynamic NMR (DNMR) when the barrier is between about 6.5 and 22 kcal/mol. The principal problems are a necessity for high symmetry or a small number of atoms so the spectra are not too complex, and the difficulty of distinguishing between rotational and nitrogen inversion processes as the barriers measured. The lone pair–lone pair dihedral angle θ is not directly available from NMR studies. Comparison of ^{15}N shifts for some hydrazines with ^{13}C shifts for analogous compounds with both N atoms substituted by CH led to the claim[31] that nitrogen chemical shift is sensitive to θ, presumably because of the rehybridization discussed above.

The work of Lehn and co-workers, who studied hydrazines as part of a general program of nitrogen inversion barriers,[32] deserves special mention. They showed that **16** is cis fused, whereas **18** and **19** are trans fused, and measured many inversion barriers by DNMR. Shvo has extensive discussion of the complex problem of separating NN rotation from N inversion in acyclic compounds. Dewar and Jennings[34] pointed out that N inversion should be fastest when θ is 90°.

The most instructive molecule for consideration of NN bond rotational effects on rates of conformational interconversion is of course dimethylhexahydropyridazine **(14),** which has well-defined axial and equatorial positions for the methyl groups and lone pairs because of its chair six-membered ring. Despite considerable work by a variety of methods, the conformational picture for **14** was not clarified until low-temperature ^{13}C NMR provided the ability to not only observe separate peaks for **ee** and **ae** conformations, but to assign their structures properly, using the upfield shift for ^{13}C signals in the presence of axial substituents.

Katritzky and co-workers[35] were the first to explicitly state the complexity of the conformational problem for **14,** which has three types of conformation of different energy (there are two nitrogens that can invert, and a six-membered ring to reverse, so there are $2^3 = 8$ conformations, which occur in three groups of equal energy, four **ae,** two **ee,** and two **aa**), connected by four different transition states. There are two types of nitrogen inversion and two types of ring reversal, depending on whether the methyl groups and lone pairs must pass each other near the transition state (which will be described as a "passing" barrier). How these conformational equilibria were sorted out has been discussed in detail[36] and need not be repeated here, except to note that the interconversion barrier for the $NC\underline{H}_2$ protons in the ^1H NMR pro-

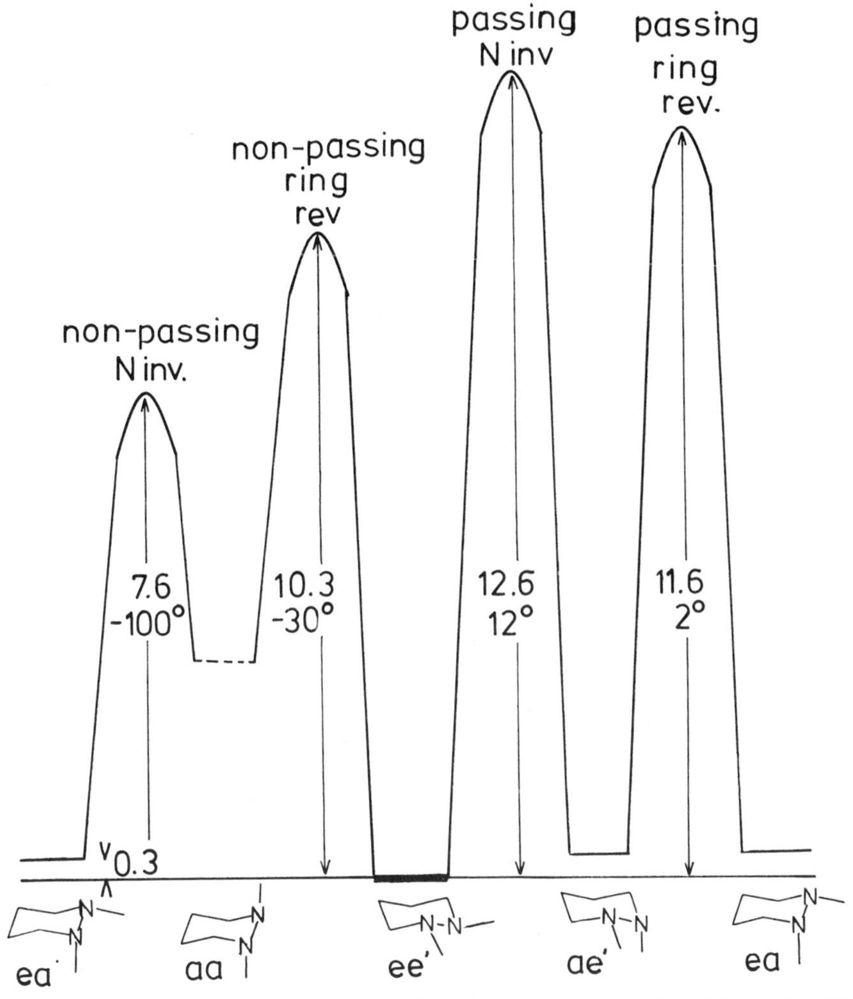

Figure 1-3. Comparison of NMR-derived barriers for conformational change in 1,2-dimethylhexahydropyridazine **(14)**.

vides the lowest of the "passing" barriers. The "passing" N inversion barrier is measurable by ^{13}C NMR for **27**, and **27** was argued to be a good enough model for that of **14**, so that because its "passing" N inversion barrier is higher than the ^{1}H NMR barrier of **14**, that "passing" ring reversal is faster than "passing" nitrogen inversion. The two non-"passing" barriers of **14** are directly observable by ^{13}C NMR and differ enough to allow their determination.

The results are shown in the energy diagram of Figure 1-3. It is seen that the "passing" N inversion (inversion of the axial N of **ae**) requires the lone pairs and methyl groups to pass each other (shown in Newmann projection in Figure 1-4), while the non-"passing" N inversion (inversion of the equa-

non-passing N inversion passing N inversion

aa θ~90° ae θ~30° ee

Figure 1-4. Newmann projections illustrating passing and nonpassing nitrogen inversion processes in 1,2-dimethylhexahydropyridazine **(14).**

torial N of **ae**) does not require such passing. Study of model compounds shows that nitrogen inversion that requires passing of a methyl group on nitrogen to a vicinal one attached to carbon causes only a small barrier increase, and it was estimated that about 4 kcal/mol of the 5-kcal/mol increase in N inversion barrier for "passing" compared to non-"passing" is caused by electronic interaction. The lone pair–lone pair dihedral angle is formally near 30° for the "passing" barrier, and 90° in non-"passing" N inversion, as shown in Figure 1-4. We note that the non-"passing" barrier for **ea** → **aa** (7.6 kcal/mol at −100°C) is close to that observed for **21** (7.8 at −113°C), which also has lone pairs formally perpendicular at the transition state. For comparison, **29,** the bicyclic amine analogue of **21,** has a nitrogen inversion barrier of 6.5 kcal/mol at −127°C, verifying that α-nitrogen substitution raises N inversion barriers significantly, even when the transition state has the most favorable θ value for N inversion. A large effect of θ on nitrogen inversion barrier has also been obtained computationally; Cowley and co-workers[6] obtained an N inversion barrier for H_4N_2 at θ = 180° 11.5 kcal/mol higher than at the energy minimum of 91.5° in their 4–31G calculations at STO-3G geometry. They noted the problems with accurate calculation of this effect.

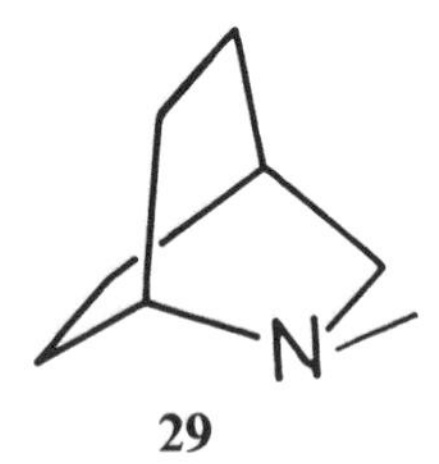

29

The "passing" and non-"passing" ring reversal barriers for **14** differ by 1.3 kcal/mol (see Figure 1-3). The non-"passing" barrier is indistinguishable from that for cyclohexane, indicating that the half-chair transition state involving the $(CH_2)_4$ unit, which can be reached without much changing the size of θ, is the lowest barrier. The "passing" ring reversal has a higher barrier, indicating that the eclipsed boat form in which the lone pairs and methyls pass each other has replaced the half-chair transition state as the highest barrier between conformations.

The near equality in energy of **14ee** and **14ae** deserves special comment. There is obviously steric destabilization for the axial *N*-methyl group in **14ae,** and in contrast to what was believed in 1974,[1] it is larger than in axial methylcyclohexane. Eliel and co-workers[37a] argued that axial *N*-methylpiperidine lies greater than 2.1 kcal/mol above the equatorial form, and Anet and co-workers estimated this number at 2.7 kcal/mol.[37b] The steric effect might be a little larger in **14** because an NN bond is shorter than a CC bond. The fact that **14ae** is only about 0.3 kcal/mol higher in energy than **14ee** must reflect some 2–4 kcal/mol of "electronic strain" in **14ee.** As pointed out above, a consequence of this electronic destabilization is a significant hybridization difference between **14ae** and **14ee.** Because the θ values would be similar for **14ae** and the much discussed[1] but never observed **14aa,** we argue that **14aa** would lie at least 2 kcal/mol above **14ae** in energy, and thus would not be a significant contributor to the equilibria.

The NMR work on **30** (methylated **14**)[38] is of interest in this context. Only the equatorial methyl compound was observed at low temperature, with an 11.1-kcal/mol barrier [-80°C] for an enantiomerization, which requires ring reversal and nitrogen inversion. This work illustrates the power of the technique introduced by Anet[37b,39] for estimates of $\Delta G°$ and $\Delta G^‡$ of unstable conformations. Line broadening due to the effects of the axial conformation **30a** were observed at -80°C, allowing $\Delta G° = 2.3$ and $\Delta G^‡$ for its conversion to **30e** of 7.8 kcal/mol to be calculated, allowing completion of the energy diagram in Figure 1-5. Alkylation of N_1 removes the vicinal lone pair, which effectively "turns off" the electronic interaction that destabilized **14ee,** because **30a** is 2.3 kcal/mol less stable than **30e.** This suggests that the electronic destabilization of **14ee** is about the 2-kcal/mol difference between $\Delta G°$ for **14** and **30.** The electronic part of the barrier difference for passing and nonpassing nitrogen inversion was estimated at 4 kcal/mol. The passing N inversion barrier for **14** involves p–sp^3, lone pair–lone pair interaction, but the electronic destabilization of **14ee** involves sp^3–sp^3 interaction. The latter should be somewhat smaller, making a 2-kcal/mol effect believable in size. There is apparently an anomeric stabilization when the remaining lone pair of **30** is aligned with a CN bond, because the ring reversal barrier at 11.2 kcal/mol is significantly higher than the nonpassing ring reversal barrier in **14,** although it is lower than the passing ring reversal barrier of **14.**

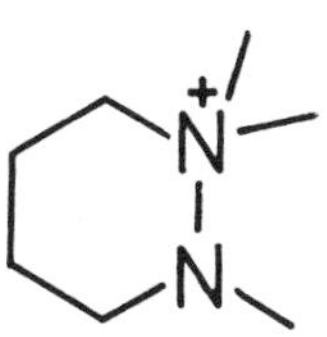

30 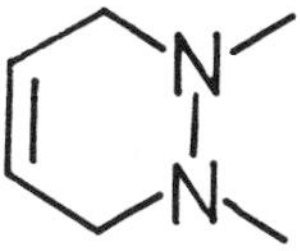 **31** 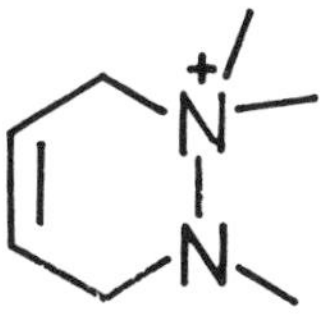 **32**

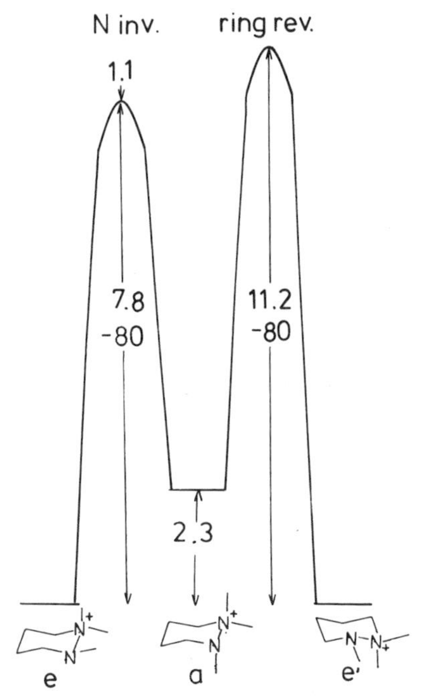

Figure 1-5. Comparison of NMR-derived barriers for conformational change in 1,1,2-trimethylhexahydropyridazine cation **(30^+)**.

The **ae**/**ee** equilibrium constant for hexahydropyridazines is extremely sensitive to substitution pattern. Introduction of a 4,5 double bond in **31** throws the equilibrium entirely to the **ae** form, as Anderson already knew from proton NMR work in 1969.[40] The amount **31ee** lies above **31ae** has not been measured directly. The methylated compound **32** has its axial form 1.2 kcal/mol above **32e** in energy and a ring reversal barrier of about 9.1 kcal/mol,[38] both estimated by the Anet maximum line-broadening technique.

Returning to hexahydropyridazines, it was pointed out above that PE measurements showed that both α-ring methylation (in **24**) and substitution of ethyl for methyl (in **25**) result in **ae** conformations predominating over **ee.** We expect that the predominance of **ae** in solution is even greater than the approximately 80–90% **ae** indicated by PES. Although the ratio of peak sizes observed in a PES experiment does not necessarily reflect the ratio of the species present (the cross sections for electron emission are not necessarily the same), the energy difference between **14ae** and **14ee** was obtained by variable temperature PES by Schweig and co-workers.[41] Although **14ee** is only 0.3 kcal/mol stabler than **14ae** in solution ($\Delta G^\circ \simeq 0.3$ kcal/mol, ΔH° 0 $\simeq 0.4$ kcal/mol), ΔE° in the gas phase is 1.2 kcal/mol. Both the size of the change on going from solution to the vapor phase and the dipole moments of the gauche forms are similar for **14ae** and for $ClCH_2CH_2Cl$, and the pref-

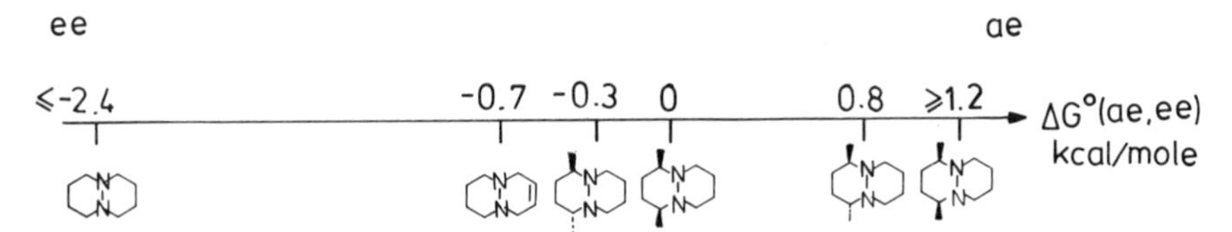

Figure 1-6. Comparison of $\Delta G°$ values for the **e–ae** equilibrium of some 1,6-diazabicyclo[4.4.0]decane derivatives. A positive $\Delta G°$ means that the **ae** form is stabler.

erential stabilization of the gauche form of these compounds in solution relative to the gas phase presumably has the same origin. In the gas phase, dipole effects are magnified compared to solution, since there is no solvent around to preferentially stabilize the more polar gauche form. We therefore expect a similar destabilization of the **ee** form relative to the gauche form in solution compared to the gas phase for **24** and **25.**

Both cis- and trans-1,2,3,6-tetramethylhexahydropyridazine have, as expected from the discussion above, only **ae** conformations detectable. Closing a second ring, however, strongly favors the **ee** conformation, and long acquisition time NMR on diazadecalin **(15)** indicates that $\Delta G°$ for the **ae–ee** equilibrium is at least -2.4 kcal/mol. Equilibrium experiments in which the **ee**-favoring decalin ring system is pitted against the **ae**-favoring unsaturation and methyl substitution in one ring[42] are summarized in Figure 1-6. trans-Dimethyl substitution favors the **ae** form less then cis substitution, because a cis-substituted compound can adopt conformation **33ae,** which lacks a gauche interaction present if the ring methyl groups are trans.

More interestingly, changing methyl to ethyl also greatly favors the **ae** form. NMR experiments show that replacing one methyl of **14** by ethyl **(25)** causes a change in $\Delta G°$ of 0.9 kcal/mol, and a lower limit for $\Delta\Delta G°$ of d. ethylhexahydropyridazine **(34)** is 1.8 kcal/mol. A large ethyl-for-methyl replacement effect appears to require two heteroatoms, because it is not seen for α-alkyl piperidines or for diethylcyclohexane,[43] but has been observed for other systems.[44] We note that the $\mathbf{T_{aa}}$-favoring effect of replacement of methyl by larger alkyl groups in pyrazolidines is somewhat similar. No explanation for this rather strong effect is currently available. In unpublished work, we have found a small preference for methyl being axial for **35.**[45] The second ring was necessary to find this out because we were unable

33ae **34** **35**

to freeze out the nonpassing N inversion in **31,** showing that its barrier is at least somewhat lower than in **14.** Interconversion of the ethyl and methyl axial **ae** conformations of **35** is a passing N inversion, which is easily frozen out.

A tetraalkylhydrazine that probably is forced to undergo nearly simultaneous, instead of sequential, N inversion processes has recently been studied.[46] Compound **36** must have rather flattened nitrogens because of its 3,3-dimethyl substitution, but its barrier for methyl group equilibration is 18.8 kcal/mol at 63°C, at least 13.3 kcal/mol higher than the maximum barrier estimated for **37** from conformational broadening observed at −150°C. The Newmann projection down the NN bond **(36A)** shows that NN rotation would have a huge barrier and that inversion at one nitrogen would also lead to impossible strain. The second nitrogen must at least start to invert to interconvert the methyls of **36.** Extremely large steric effects are clearly required to force both nitrogens to invert at once. The barrier for double-N inversion in **38** is only 10.6 kcal/mol, leading us to believe that the trans-fused compound is probably an undetected intermediate.[47]

Takeuchi and co-workers,[48] using a combination of ^{13}C NMR barrier measurements and detailed molecular mechanics calculations on the related hydrocarbons,[49] have convincingly differentiated passing nitrogen inversion and NN rotation for the first time in their studies of **39–41.** The barrier for

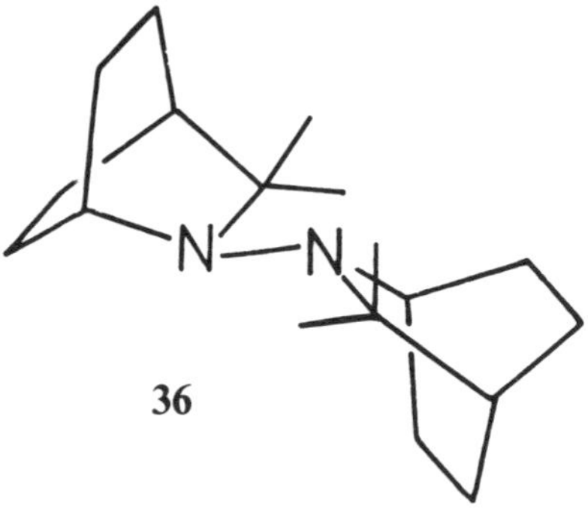

36 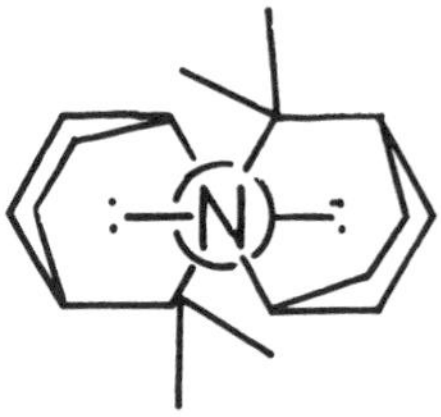 **36A**

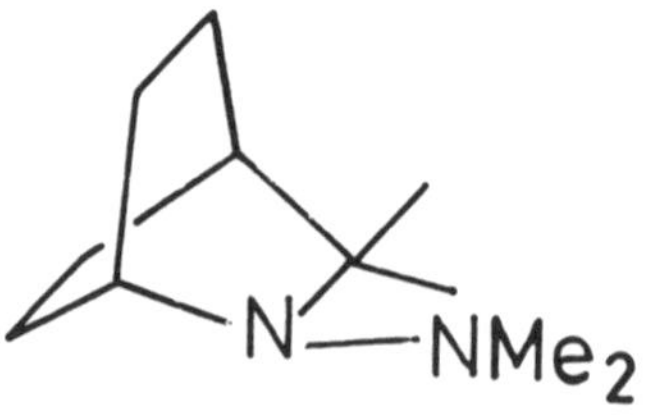

37 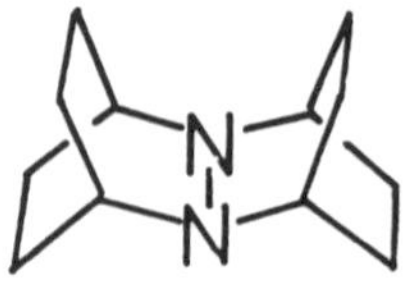 **38**

39 is 12.5 kcal/mol at −25°C, as expected for a passing N inversion, but these barriers go up dramatically in **40** (17.1 kcal/mol at +72°C) and **41** (19.0 kcal/mol at +97°C), because NN rotation becomes the lowest barrier when a 1,3-diaxial alkyl–alkyl interaction destabilizes the ring-reversed piperidine rings. The conformational picture is very complex, but completely worked out.[48a]

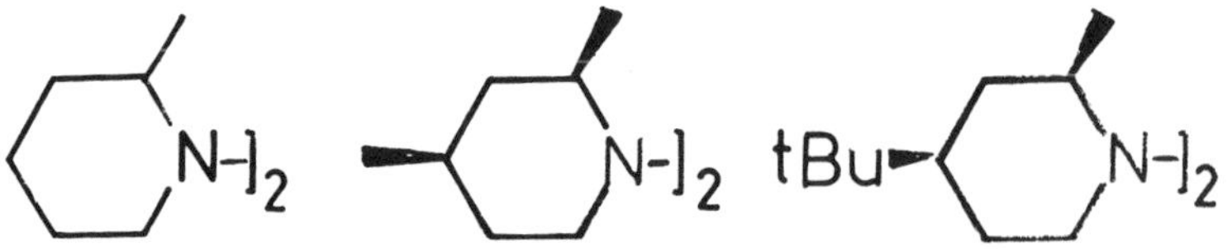

39 **40** **41**

3. HYDRAZINE RADICAL CATION CONFORMATIONS

A. X-Ray Studies and Calculations

Although there is no experimental support for an unsymmetrical structure, it is not required that the nitrogens of $R_4N_2^+$ necessarily be equivalent.[50] Localization of charge to give the amino-substituted amine cation radical implied by the single Lewis structure **III** could in principle occur, especially since counterion/solvation effects might favor higher charge density. All experimental and computational evidence, however, show that $R_4N_2^+$ exists with equivalent nitrogens bearing half a positive charge each, as given by the Linnett structure **IV**. The cation $R_4N_2^+$ may be described as having a "three-electron π bond," with two bonding π electrons, and one electron in the antibonding π^* orbital. Since amine radical cations are known to prefer planar structures (lone pair hybridization, pure p) and neutral amines pyramidal structures (lone pair hybridization, approximately sp^3) it is not obvious whether $H_4N_2^+$ should be bent or planar.

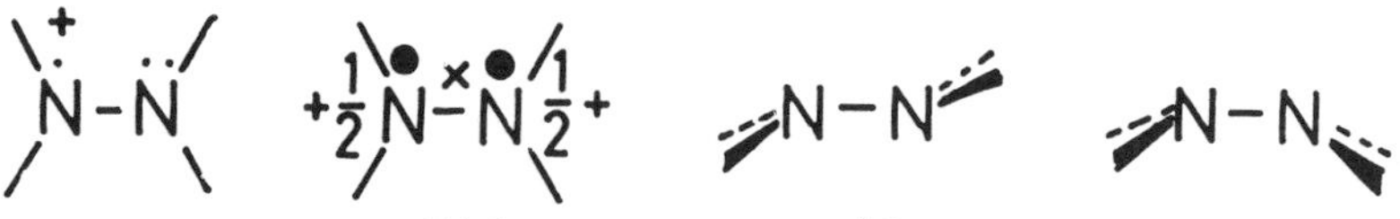

III IV anti bent syn bent

Semiempirical INDO calculations give slightly anti-bent structures as the minimum energy for a long NN bond length of 1.4 Å, but the true INDO open-shell energy minimum is a planar structure with *d*(NN) of 1.28 Å (for a constrained NH distance of 1.07 Å).[51] Although completely energy-minimized unrestricted Hartree–Fock (UHF) MNDO calculations get $H_4N_2^+$ somewhat bent [the minimum energy anti-bent form has β, the NN–HNH plane angle about 25°, and α(av), the average of the three bond angles, 117.7°; syn-bent β 8°, α(av) 118.2°; planar ($\beta \equiv 0°$, α(av) $\equiv$ 120°) with relative energies 0, 0.8, 0.7 kcal/mol], the structure is extremely easy to deform. Higher level 4–31G-UHF calculations get the planar form as stablest,[52] also with a very shallow energy minimum for bending, but HF/6–31G* level calculations give a nonplanar anti-bent minimum.[52b]

Tetraalkylhydrazine cations which have their α hydrogens constrained to lie near the nodal plane of the p orbital at nitrogen are isolably stable, and we published an X-ray structure of $\mathbf{10}^+PF_6^-$ claiming to prove a flat structure (β = 0, α(av) = 120°), with an anomalously short *d*(NN) of 1.27 Å.[53] This X-ray structure colored our thinking about $R_4N_2^+$ for several years, but we now know it to have undetected disorder. We discovered this when we tried to obtain the X-ray structure of $\mathbf{42}^+PF_6^-$, and found it to have the $(CH_2)_3$ and $(CH_2)_2$ bridges disordered. Going to the less symmetrical nitrate counterion, we obtained a structure with significant anti bending (β = 32.5°, α (av) = 115.6°), *d*(NN) = 1.323(4) Å.[54] The anomalously short NN bond length of $\mathbf{10}^+PF_6^-$ disappeared in the structure of $\mathbf{42}^+NO_3^-$, and we think it is obvious that it was an artifact of undetected disorder in the $\mathbf{10}^+PF_6^-$ structure; if β were 20° and *d*(NN) 1.32 Å, the measured bond length would have been that seen. The geometry change between neutral **42** and $\mathbf{42}^+$ is still substantial; *d*(NN) shortens 0.14_6 Å (9.9%), and α(av) goes from 107.9 in neutral **42** to 115.6° in $\mathbf{42}^+$, so α(av) changes from 1.6° less than the tetrahedral value of 109.5° to 6.1° greater, corresponding to 73% of the way from tetrahedral to planar. Compound $\mathbf{42}^+$ is doubtless significantly flatter in solution (see subsection 3B-b, "Ultraviolet Spectra"); crystal packing forces favor bending at N in this species. Energy-minimized MNDO calculations on anti $\mathbf{42}^+$ correctly give the diaxial species as the minimum, with a bond length of 1.30 Å and β of 17.3° (α(av) = 118.7°), encouragingly close to both the X-ray *d*(NN) and the β value estimated from electron nuclear double resonance (ENDOR) studies.[54]

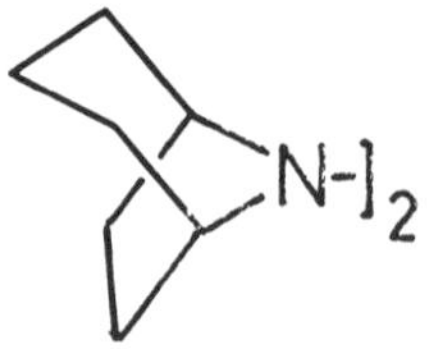

N-]$_2$

42

B. Bending and Twisting in Hydrazine Cation: Spectroscopic Studies

a. Electron-Spin Resonance

The nitrogen splitting, a(N), is the most instructive electron spin resonance (ESR) parameter for considering bending at nitrogen. Bending increases the s character of the spin-bearing orbital, so a(N) will increase monotonically as pyramidality increases. It is important to note that a(N) is a time average over a series of vibrational states for $R_4N_2^{+}$. The nitrogens of $H_4N_2^{+}$ have been argued to be flat at equilibrium since the first observation of this species,[55] and this conclusion has received further support from solid state ESR in both randomly oriented[56] and single-crystal-oriented[57] experiments. One might expect that because there is obviously hyperconjugative delocalization of spin from the nitrogen-centered lone pairs to attached alkyl groups (methyl proton splittings are about 12–13.5G in various methylated $R_4N_2^{+}$), a(N) ought to go down as H is replaced by methyl. Instead, it increases 16% from $H_4N_2^{+}$ to $Me_4N_2^{+}$. This requires that alkylation increase the ease of bend in $R_4N_2^{+}$, but it can be consistent with a planar equilibrium geometry for both species if the time-averaged amount of bend is larger for $Me_4N_2^{+}$. The temperature behavior for a(N) is entirely analogous to that for $a(^{13}C)$, which has been discussed in detail.[58] Raising the temperature will increase a(N) if the equilibrium geometry is planar, and positive slopes $d[a(N)]/dT$ are seen for **1**$^+$ (+1.4 mg/deg), **4**$^+$(+3.0), and $(PhCH_2)_4N_2^{+}$ (+1.8).[59] If these species are nonplanar, their inversion barriers must be small compared to vibrational spacing, which is very small indeed. Cyclic structures clearly can force nonplanar equilibrium geometries, as reflected in the size of a(N).

Compound **16**$^+$ is bent syn, with a 3.4-kcal/mol barrier (−110°C) for double nitrogen inversion determined by dynamic ESR.[50] Its temperature slope is −7.8 mG/deg, as expected for a bent equilibrium geometry. Compound **42** is unquestionably nonplanar at equilibrium and anti bent (β of 18° is estimated in solution); its temperature slope is very small, (<0.5 mG/deg). Some information on internal rotations of the NC bonds is available from β-hydrogen splitting constants. The cation $(PhCH_2)_4N_2^{+}$ freezes out at low temperature to two sets of $4H_\beta$, $a(4H) = 9.5_8$ and 6.3_3 G, which is consistent with conformations favoring C—Ph bonds nearly parallel to the lone pair axes as expected on steric grounds. A barrier to N—Bz rotation of about 4.9 kcal/mol (−70°C) was estimated.[59] Recent work[60] on **33**$^+$ has given a low-temperature spectrum with one axial CH and one equatorial CH bond, $a(H_a) = 27.5$ and (H_e) unobserved respectively, which is entirely consistent with the half-chair conformation **33A**$^+$. Interestingly, the *trans*-3,6-dimethyl cation **43**$^+$ shows $a(2H_e) = 3.60$ at low temperature, which seems to require that the CMe groups be axial as in **43A**$^+$. We had earlier argued non-half-chair conformations from the relative sizes of the proton splittings, but boat conformations for **33**$^+$ and **43**$^+$ would have equivalent methine splittings for the cis compound and nonequivalent ones for trans, which is not what is seen. Apparently buttressing effects in the all-equatorial conformation **$43B^+$** make it less stable than **$43A^+$**. The a(2H) splitting of **43^+** is remarkably temperature sensitive, increasing from 3.6 G at = 70°C to 8.1 G at 85°C. If **$43B^+$** were about 1 kcal/mol less stable than **$43A^+$**, such a large temperature variation would be expected.

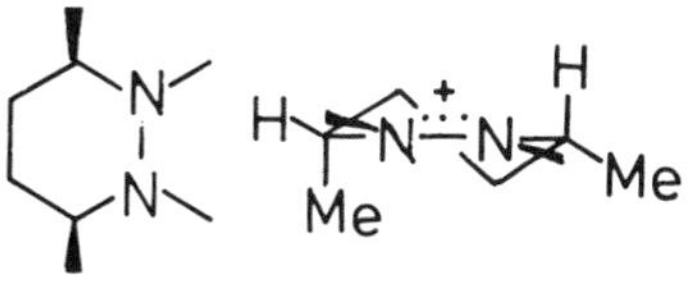

33 **$33A^+$**

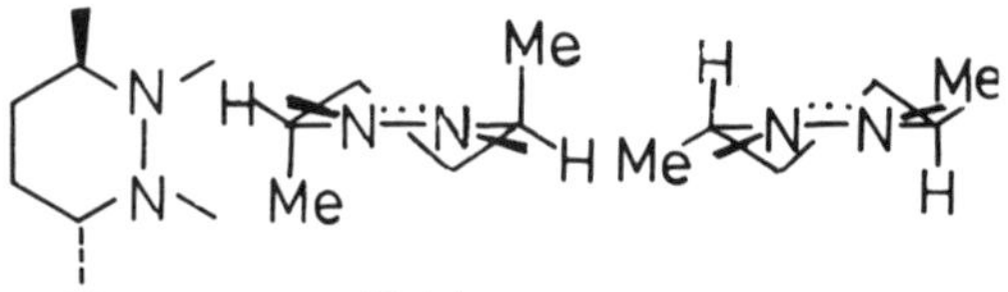

43 **$43A^+$** **$43B^+$**

Calculations show that bending $H_4N_2^+$ a given amount syn causes a significantly larger a(N) value than bending the same amount anti, and it is easily seen from Table 1-4 that higher a(N) values are found experimentally as well. The reason for this trend is the importance of σ–π mixing effects when the nitrogens of $R_4N_2^+$ are pyramidalized. The odd electron is in the π^* orbital as indicated at the center of Figure 1-7. No σ(CN)–σ(NN) mixing with π or π^* can occur for the planar species, but mixing will occur as the nitrogens pyramidalize. The important factor is that the σ orbital dominated by σ(NN)–σ(CN) antibonding combinations lies high in energy and has the right symmetry to mix with π upon syn pyramidalization, but with π^* upon anti pyramidalization. Because the size of this interaction depends on $1/\Delta E$, significantly more σ–π mixing occurs for syn bending than for anti bending, so more s character appears in the syn odd-electron MO than the anti one. The calculated size of the effect is large (see Figure 1-8). These calculations on $Me_4N_2^+$ had ρ_s(N) converted to a(N) values using the proportionality constant of 379 obtained by Pople and co-workers for INDO calculations.[61] It appears to be about the right size to use with MNDO as well.

b. Ultraviolet Spectra

The π–π^* transition of $R_4N_2^+$ appears in the near-UV region and proves to be also quite sensitive to bending effects. As indicated in Figure 1-9, bending syn does not decrease overlap of the p-rich lone pair orbitals as rapidly as bending anti, so from overlap considerations bending anti should decrease

TABLE 1-4. ESR Nitrogen Splitting and UV Data for the Hydrazine Cation

	Neutral compound	Cation bending type[a]	a(N) (G)[b]	λ(m) (nm)[c]
1	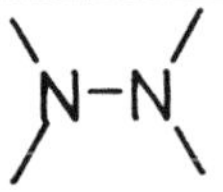	A	13.4	303
4	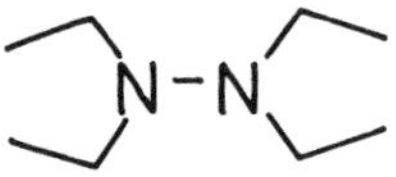	A	13.0	289
5	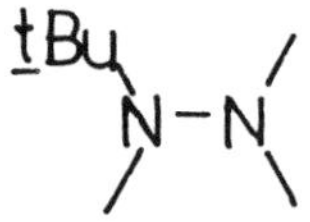 tBu N–N	A	[13.1?]	331
44	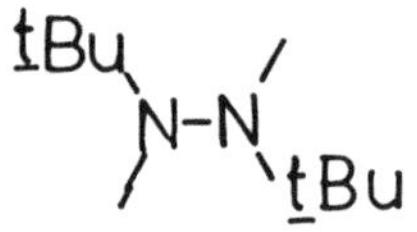 tBu N–N tBu	A	11.9	414
7	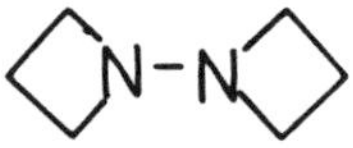	A	14.8	
8	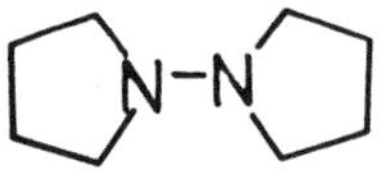	A	12.9[d]	
10	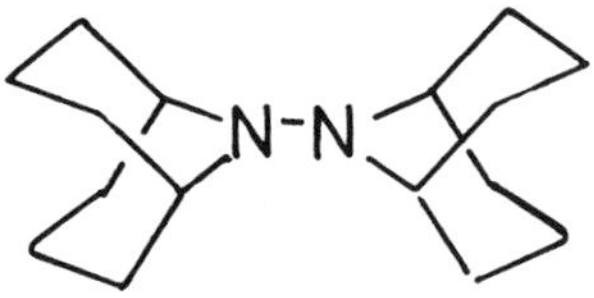	A	13.1_5	340
42	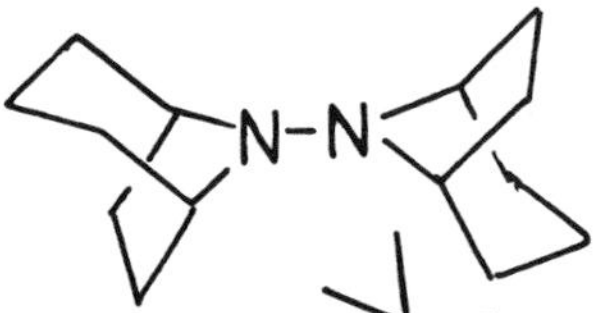	A	13.7[e]	334[e]
36	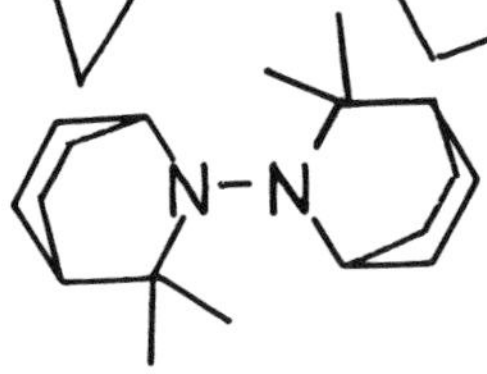	A	13.5[f]	303

TABLE 1-4. (continued)

	Neutral compound	Cation bending type[a]	a(N) (G)[b]	$\lambda(m)$ (nm)[c]
14	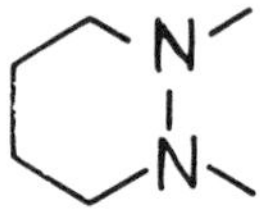	A?	13.1	260
15	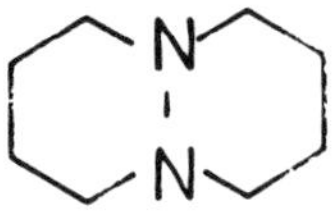	A?	13.9	255
12		S?	15.0[d]	
13	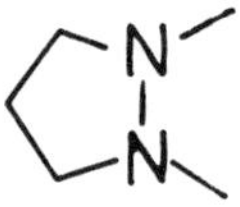	S	15.0	270
26	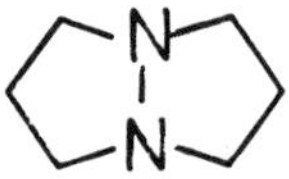	S	17.6	281
16	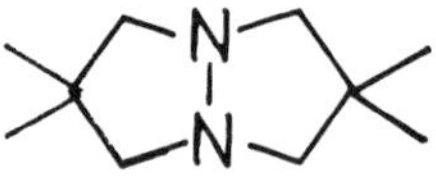	S	16.8	
17	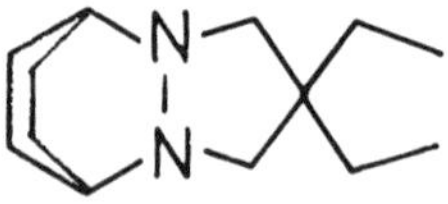	S	18.5	
18	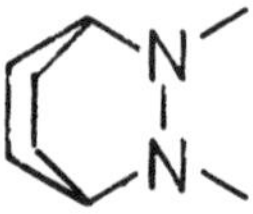	S	13.9	266

Table 1-4. (continued)

	Neutral compound	Cation bending type[a]	a(N) (G)[b]	$\lambda(m)$ (nm)[c]
19	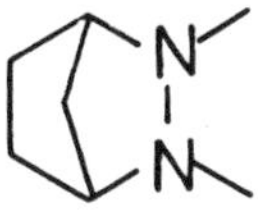	S	16.0	260
38	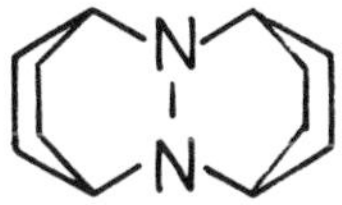	S	15.1[g]	266[g]
45	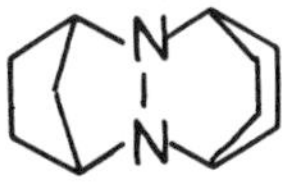	S	16.1[h]	283[h]
46		S	20.9[h][f]	321[h]
47	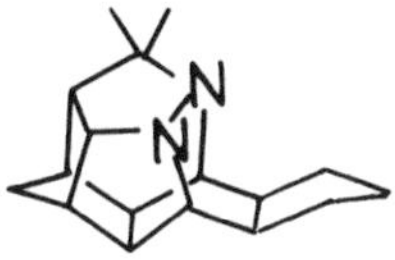	S	28[i]	454[i]

[a]A refers to anti bending; S to syn bending at nitrogen.
[b]From Reference 50 unless otherwise noted.
[c]From Reference 60 unless otherwise noted.
[d]Reference 19a.
[e]Reference 54.
[f]Reference 46.
[g]Reference 73a.
[h]Reference 73b.
[i]Reference 64b.

$\Delta E(\pi-\pi^*)$ more rapidly than bending syn. The opposite behavior is in fact calculated to occur; bending syn is calculated to decrease $\Delta E(\pi-\pi^*)$ more rapidly than bending anti (Figure 1-10). Figure 1-7 shows the reason for this, $\sigma-\pi$ mixing (again). The $\sigma-\pi$ mixing of syn-bent $H_4N_2^+$ causes a decrease in $\Delta E(\pi-\pi^*)$ over what it would be without $\sigma-\pi$ mixing, but that of anti increases $\Delta E(\pi-\pi^*)$ and the σ, π mixing effect overcomes the overlap effect, which is in the opposite direction. MNDO calculations of $\lambda(m)$ versus α(av) for $Me_4N_2^+$ are shown in Figure 1-10. The initial decrease in $\lambda(m)$ with syn bending is doubtless an artifact of the calculation, because it disappears in ab initio calculations. Experimentally,[62] $H_4N_2^+$ has $\lambda(m)$ less than 225 nm(probably not much less) in H_2O. There ought to be a red shift upon going from gas phase (which the calculation is for) to solution, but the shift

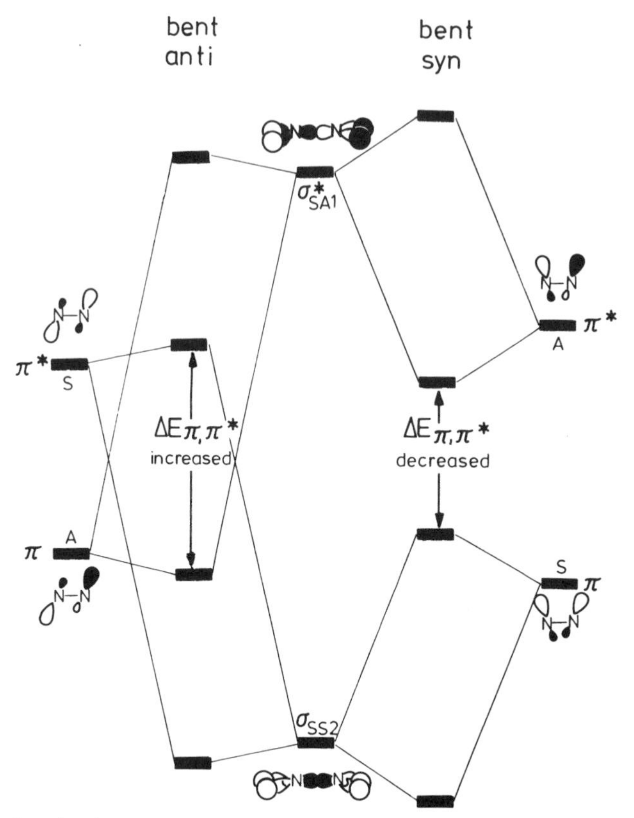

Figure 1-7. Qualitative energy diagram showing the difference in σ–π orbital mixing for anti- and syn-bent hydrazine cations.

is not very large.[63] For 1,4-diazabicyclooctane, the red shift is about 0.5 eV, corresponding to a $\lambda(m)$ red shift of 15 nm, and part of this red shift may be caused by a geometry change. There is expected to be a red shift associated with changing the substituents from H to alkyl, and probably also for enlarging the alkyl groups. The MNDO calculations give $\lambda(m)$231 nm for $H_4N_2^+$, much closer to the experimental value than ab initio calculations. The experimentally measurable quantities a(N) and $\lambda(m)$ are predicted to be different for syn- and anti-bent hydrazines, and they are plotted against each other in Figure 1-11. Figure 1-12 shows an experimental plot of the data of Table 1-4, which has encouraging similarities to Figure 1-11. It will be noted that **13, 18,** and **19,** which have their methyl groups trans in the neutral form, fall on the syn distribution curve in Figure 1-12 (**26, 38, 45, 46,** and **47** are all syn bent in the neutral form and will certainly also be syn bent as cations). This is reasonable, because if the internal ring CNNC dihedral angle is held near 0°, bending at nitrogen syn allows the lone pairs to remain coplanar, but bending anti twists them away from coplanarity.

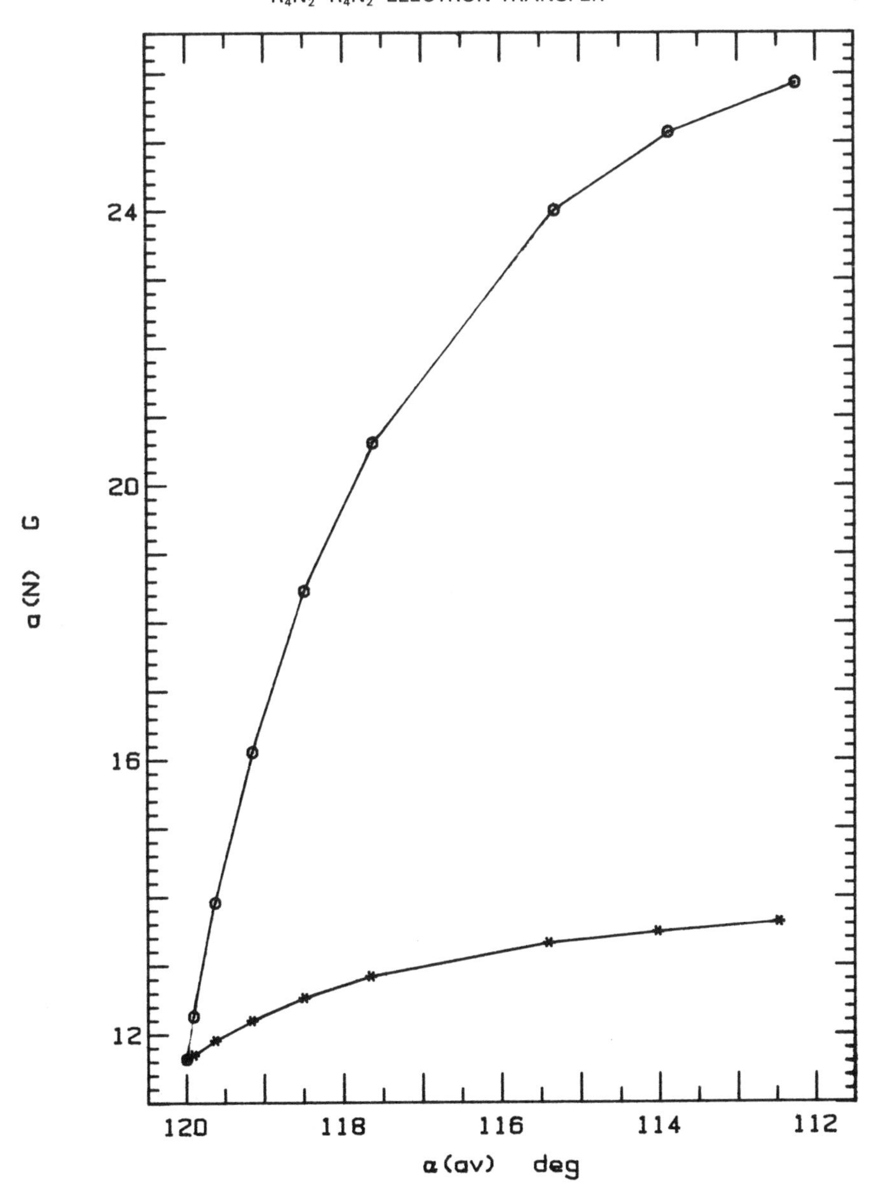

Figure 1-8. MNDO calculations of a(N) for $Me_4N_2^+$ as the nitrogens are bent; increasing bend at N lowers α(av). Circles, syn bent; asterisks, anti bent.

N ← bend anti — N — bend syn → N

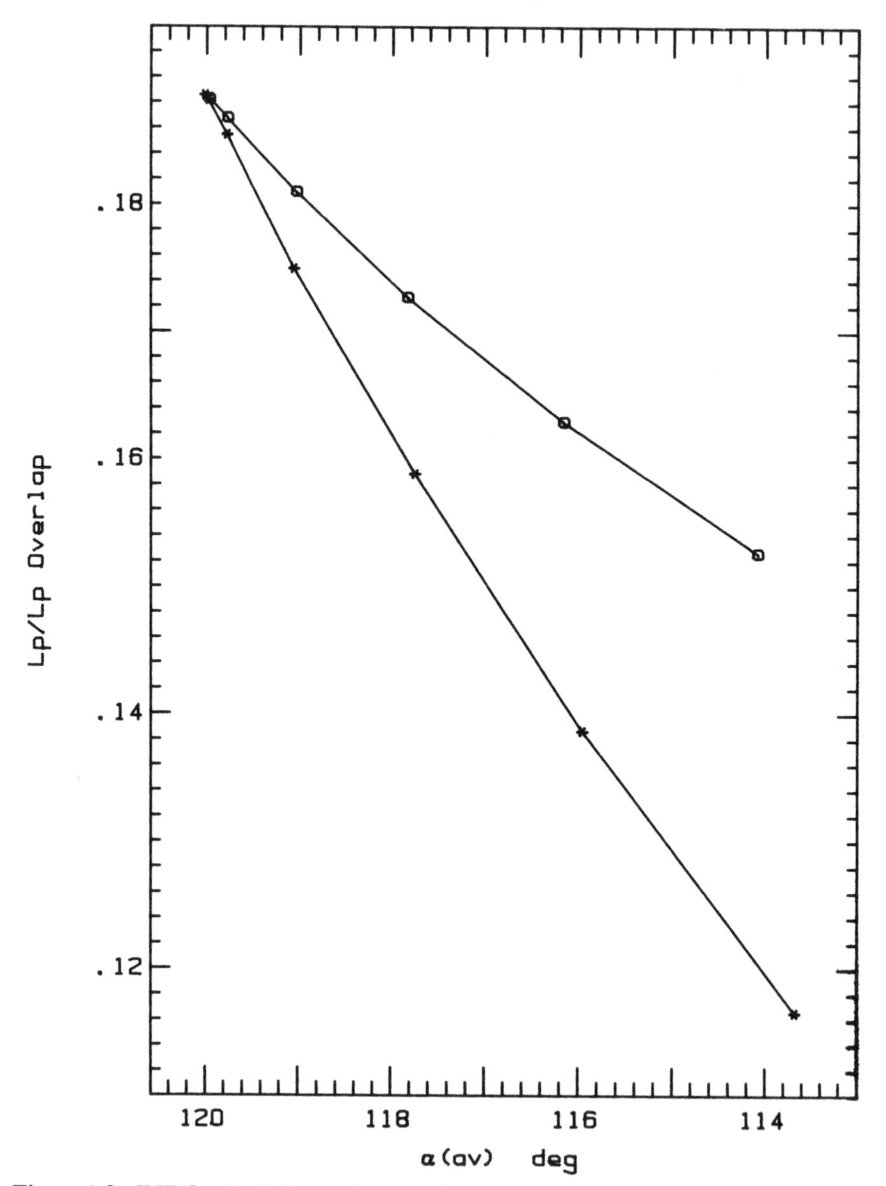

Figure 1-9. INDO calculations of lone pair–lone pair overlap for $H_4N_2^+$ as the nitrogens are bent; increasing bend at N lowers α(av). Circles, syn bent; asterisks, anti bent.

The ESR work discussed above was consistent with a half-chair conformation, and the short-wavelength $\lambda(m)$ for **14**$^+$ seems entirely consistent with a near-0° CNNC internal angle of a half-chair conformation with nearly planar nitrogens. The complexity of these ESR spectra would make it difficult to observe a minor conformation.

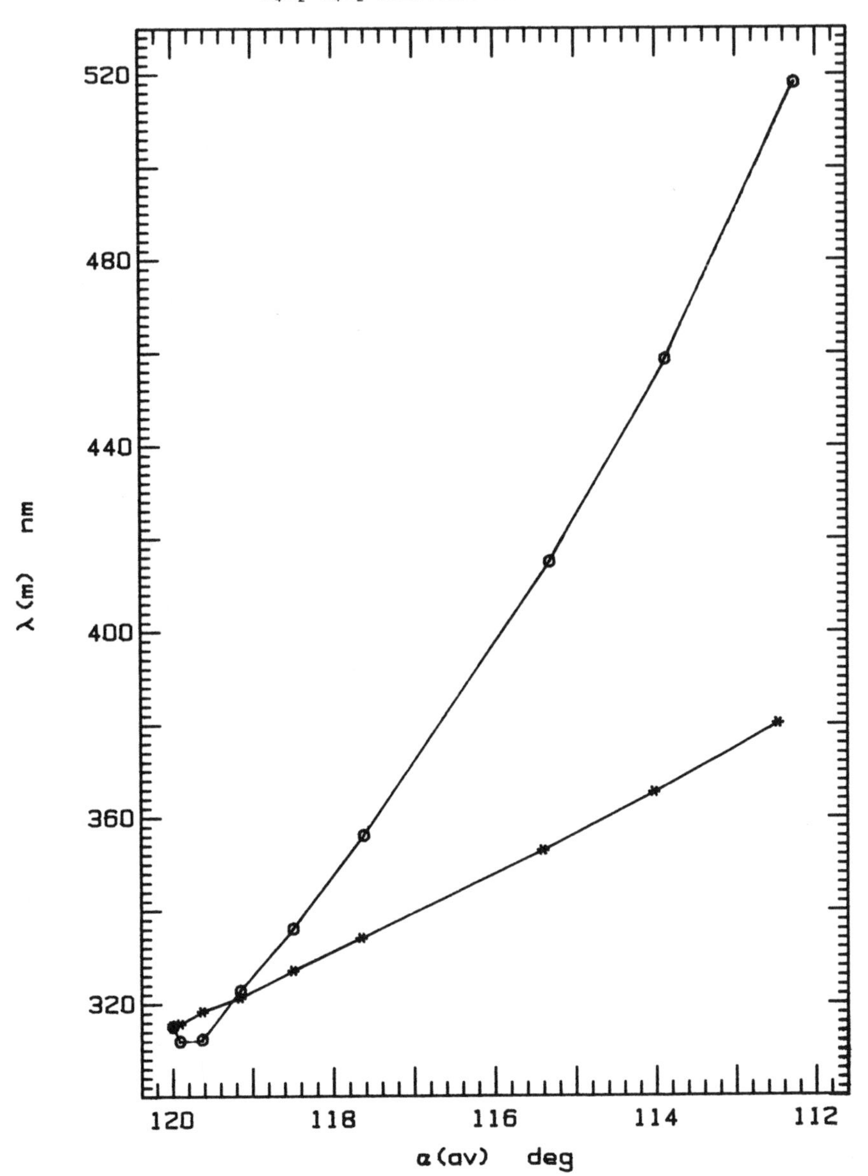

Figure 1-10. MNDO calculations of λ(*m*) for $Me_4N_2^+$ as the nitrogens are bent; increased bend at N lowers *α*(av). Circles, syn bent; asterisks, anti bent.

The UV spectra of **42**$^+$ are particularly revealing. Compound **42**$^+$ exists in two rotamers, the anti form, **42a**$^+$**,** which is 1.5 kcal/mol stabler than the syn form, **42s**$^+$. The λ(*m*) value for **42a**$^+$ is 337 nm, and that for **42s**$^+$ is 318 nm.[54] The ENDOR spectrum of **42a**$^+$ revealed that $a(H_b)$ is 4.49 G, corresponding to an averaged *β*(NN) of about 18°, significantly smaller than the

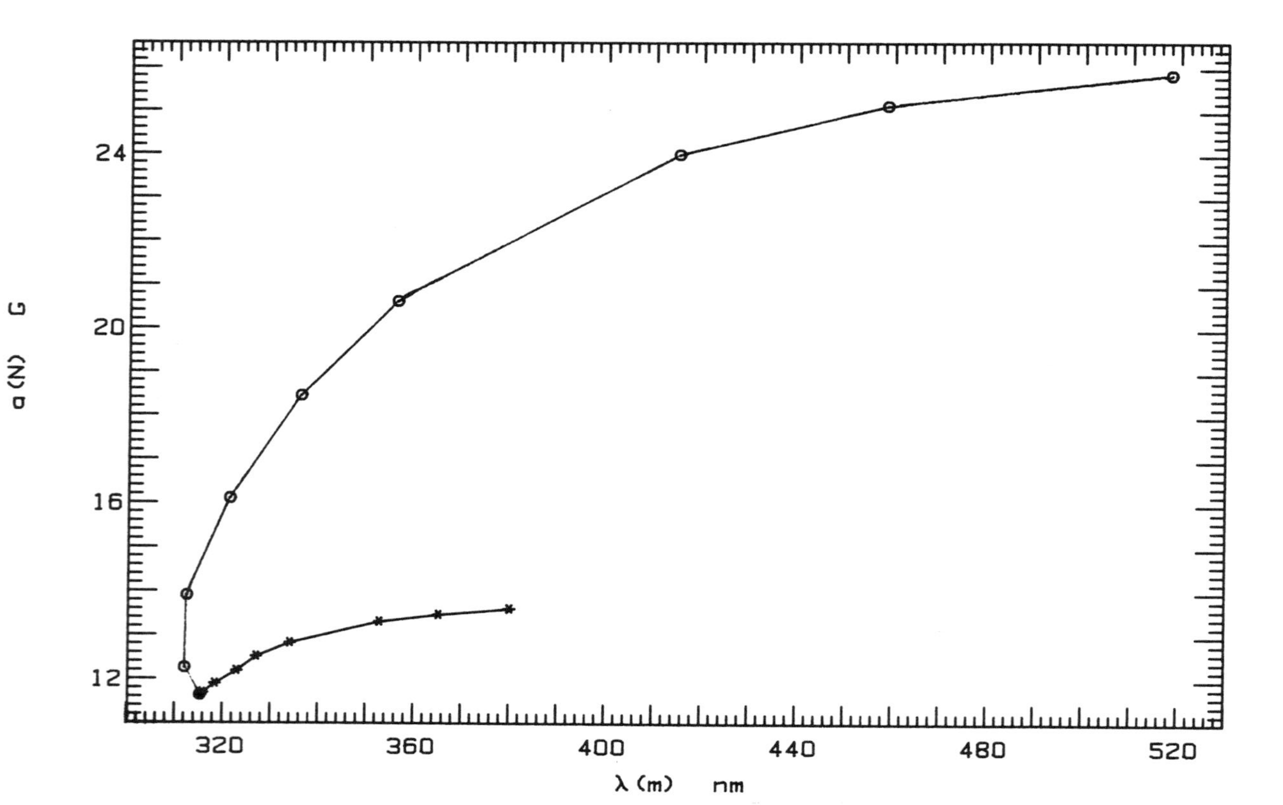

Figure 1-11. MNDO-calculated a(N) versus λ(m) for syn-bent (circles) and anti-bent (asterisks) $Me_4N_2^+$.

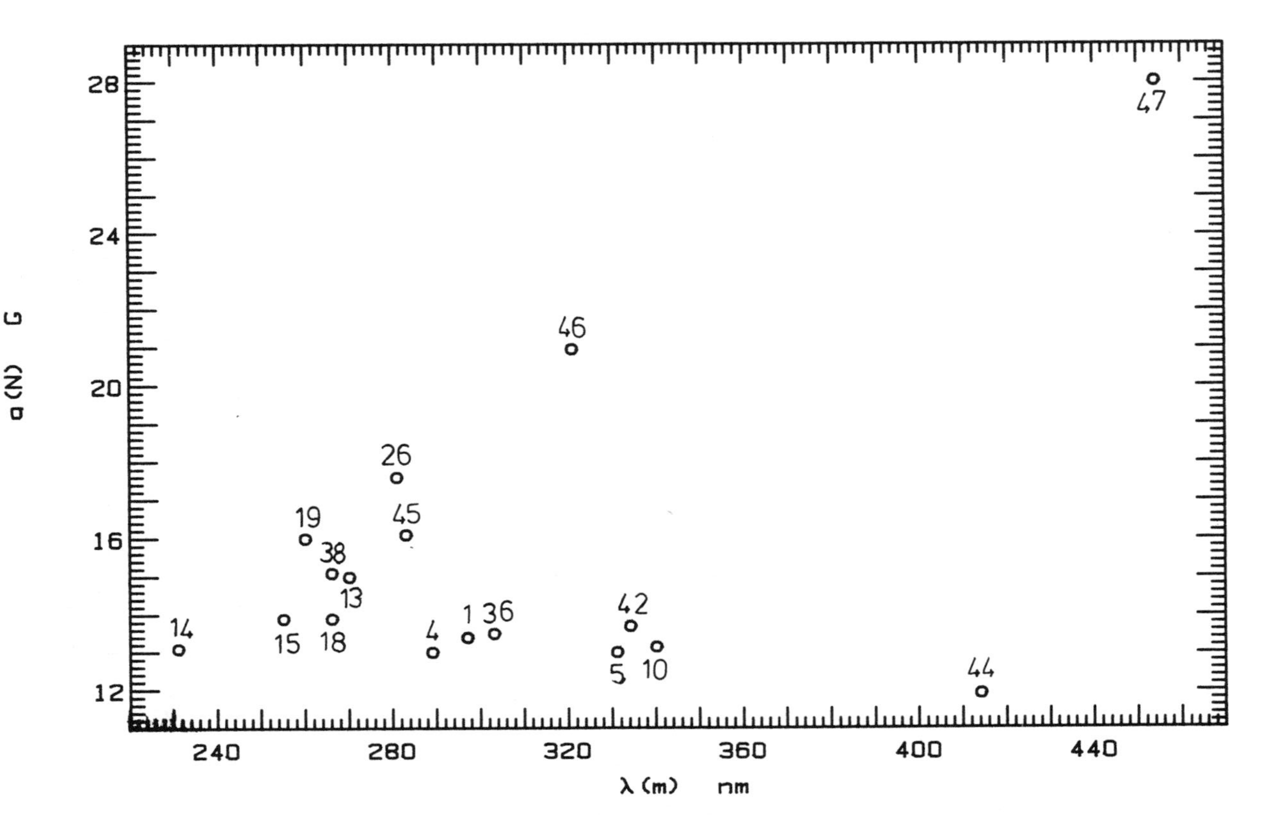

Figure 1-12. Experimental data for $R_4N_2^+$, a(N) versus $\lambda(m)$ plot.

32.5° for the solid **42a**$^+NO_3^-$.[54] For the solid the $\lambda(m)$ is 345 nm, consistent with greater bend. MNDO calculations on **42s**$^+$ indicate much less bending at the nitrogen bearing the NN bond equatorial to the piperidine ring and predict $\lambda(m)$ values of 334 and 321 nm for **42a**$^+$ and **42s**$^+$ respectively, surely fortuitously close to the observed values. The flattening at nitrogen when the NR_2 group is equatorial is caused by a larger interaction with the bridgehead hydrogen, which is tilted about 11–15° out of the CNC plane.

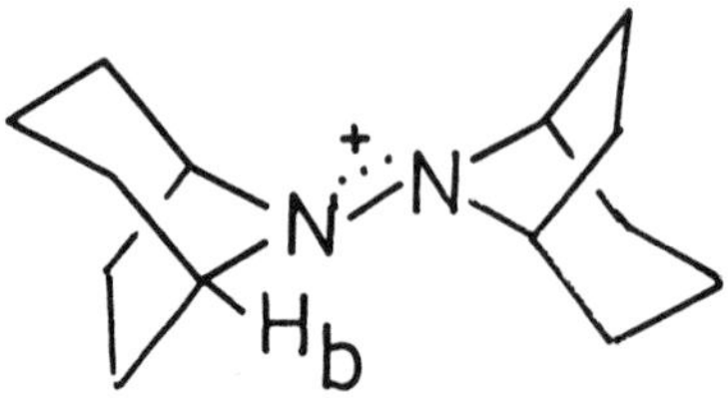

42a 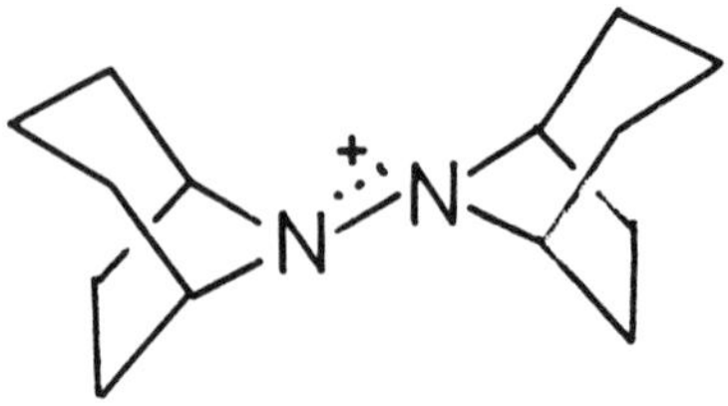 **42s**$^+$

The longest wavelength $\lambda(m)$ $R_4N_2^+$ example known is the cation from Hünig's dimethylpyrazolidine **(46)**,[64] which is both strongly bent and twisted about the NN bond some 25°, as estimated from MM2 calculations on the related hydrocarbon.[64b] A short-wavelength shoulder on the π–π^* absorption of **10**$^+$ has been attributed to charge transfer from the σ framework to the half-filled π^* MO, by analogy with olefin radical cations.[65]

C. Rotational Barrier

The best measure of the strength of the three-electron π bond of $R_4N_2^+$ is its rotational barrier. Calculations give a strong restoring force for coplanar nitrogen lone pairs. The highest level calculation of the transition state for rotation of $H_4N_2^+$ is an MP2 6–31G* calculation, which gives a barrier of 30.2 kcal/mol, at $\theta = 90°$.[66] The stablest $\theta = 90°$ conformation at all levels of calculation from MNDO to MP2 6–31G* has charge localization, with one nitrogen flattened, and the other strongly pyramidal. The MNDO calculation gives a rotation barrier of 26.6 kcal/mol (88% of the highest level calculation), and the MNDO energy versus θ curve is shown as Figure 1-13.[67] The solid circles are for calculations with the nitrogens allowed to have different amounts of bend. The asterisks show energies calculated with the nitrogens required to be the same.

The only experimentally measured rotational barrier for a hydrazine cation radical is for **42**$^+$, where the rotation interconverts syn and anti conformations. This reaction has been followed at 6°C by CV,[66] and in unpub-

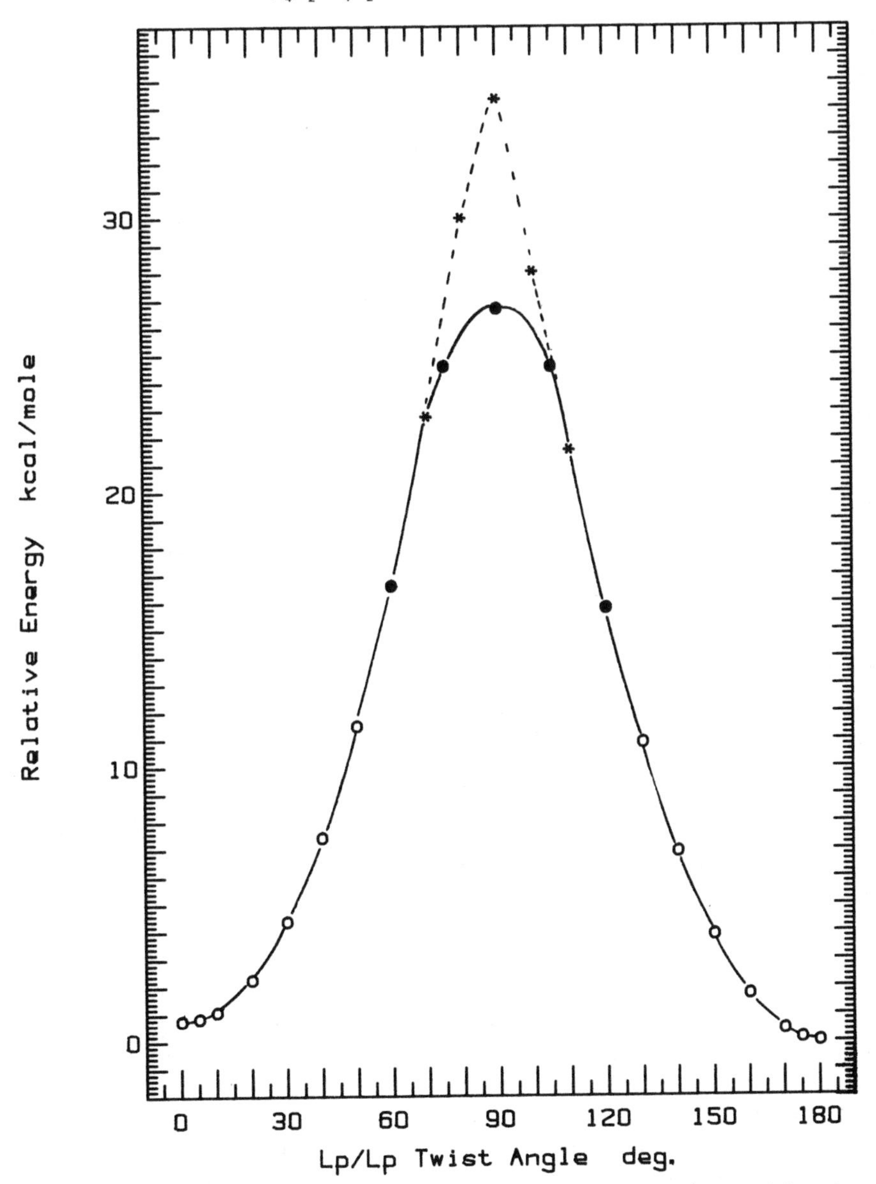

Figure 1-13. MNDO calculations of the energy for twisting $H_4N_2^+$. Asterisks, nitrogens required to be bent the same amount; solid circles, nitrogens allowed to have different amount of bend.

lished work[54] the temperature range has been extended to 41°C by UV analysis. The $\Delta G^{\ddagger}$ (25°C) is 22.0 kcal/mol for **42s** → **42a**$^+$, and since **42a**$^+$ is 1.5 kcal/mol stabler than **42s**$^+$,[66] the barrier is 23.5 kcal/mol in the reverse direction.

4. THE HYDRAZINE–HYDRAZINE RADICAL CATION ELECTRON TRANSFER

A. Thermodynamics of Electron Transfer in Solution

The effect of changing alkyl groups in R_4N_2 on the electron loss equilibrium has been studied in detail[17,68] and reviewed.[69,70] Although hydrazines show extensive CV wave broadening because of slow heterogeneous electron transfer at platinum and carbon electrodes, the rate is faster at gold,[71] and room temperature oxidation and reduction peak potential differences usually fall in the range of 65–90 mV. Under these circumstances, $(Ep^{ox} + Ep^{red})/2$ is within 10 mV of the formal potential $E^{\circ\prime}$, so comparison of $E^{\circ\prime}$ values measured by CV allows determination of the differences in ease of oxidation, $\Delta(\Delta G_e^\circ)$, by use of Equation 1-2. Table 1-5 shows $E^{\circ\prime}$ values for representative examples of R_4N_2, and also gives $\Delta(\Delta G_e^\circ)$ values relative to Me_4N_2, as well as Δ(vIP) values. The behavior of vIP with substituent change was discussed in detail in the subsection of Section 2 entitled "Conformational Studies by Photoelectron Spectroscopy"; it is complex.

We previously correlated $E^{\circ\prime}$ with vIP to attempt to allow for the expected decrease in ease of oxidation as the alkyl groups are enlarged, but such correlations introduce many other effects. Here we examine the $E^{\circ\prime}$ data for alkyl group effects by correlating with *n*(eff) (see above). A plot of all the R_4N_2 $E^{\circ\prime}$ data versus *n*(eff) appears in Figure 1-14 and shows tremendous scatter. In fact, simple alkyl group homologation (enlargement of methyl substituents by adding CH_2 groups) hardly charges $E^{\circ\prime}$ at all; the asterisks in Figure 1-14 are *n*-alkyl R_4N_2 examples. These data are fit well by a line $E^{\circ\prime} = 0.37–0.01[n(\text{eff})]$, although the data flatten out even more than this at high *n*(eff) values. We show the $E^{\circ\prime}$ data as deviations from the *n*-alkyl line in Figures 1-15 and 1-16; a positive deviation means that the hydrazine is harder to oxidize than an *n*-alkyl hydrazine of the same *n*(eff) value. The principal geometry changes upon electron removal are shortening of the NN bond, rotation of the lone pairs to $\theta = 0/180°$, and flattening at nitrogen. This eclipses the alkyl groups, increasing the $R_1N_1N_2R_2$ interaction (A) and decreasing the $R_1N_1R_1'$ interaction (B) (see **V**).

The only clear-cut trend for acyclics is an increase in $\Delta(\Delta G_e^\circ)$ when tert-butyls are present (Figure 1-15*a*); curiously, isobutyl groups cause a positive increment, and neopentyl groups a negative one, but the differences are quite small. The torsional effects causing a negative increment in $\Delta(\Delta G_e^\circ)$ for five- and seven-membered *N,N′*-cycloalkyl groups, as well as a decrease in the A strain in bipyrrolidine (**8**) are indicated by the data of Figure 1-15*b* and 1-15*c*. The increase in $E^{\circ\prime}$ for having azetidine rings is quite modest. 7,7′-bis (7-Azabicyclo[2.2.1]heptane) definitely shows an increase in $\Delta(\Delta G_e^\circ)$ larger than expected from its obvious CNC angle restriction.[72] Lehn[32] has pointed out that there is an increase in the activation energy for nitrogen inversion, $\Delta G_N^\ddagger$, over what one would expect on angle restriction grounds,

TABLE 1-5. Thermodynamic Data for Hydrazine–Hydrazine Cation Electron Transfer

	Compound	$E^{\circ\prime}$ (V)[a]	$\Delta(\Delta G_e^\circ)$ (kcal/mol)[b]	ΔvIP (kcal/mol)[b]
1	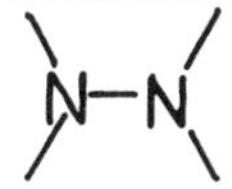 N–N	0.33	[0]	[0]
2	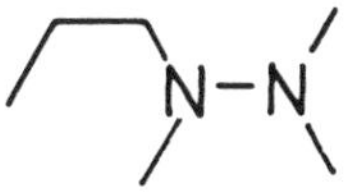 N–N	0.32	−0.2	−3.0
3	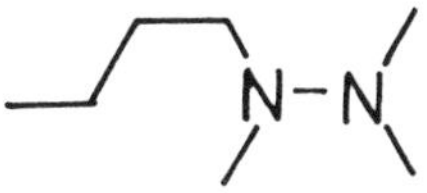 N–N	0.32	−0.2	−4.6
4	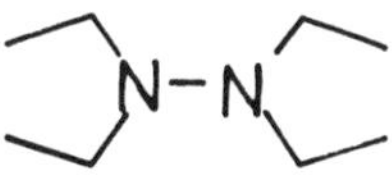 N–N	0.29	−0.9	−7.6
5	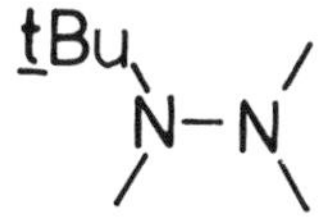 tBu N–N	0.49	+3.7	8.8
7	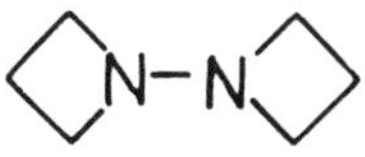 N–N	0.38[a]	+1.2	−0.5
8	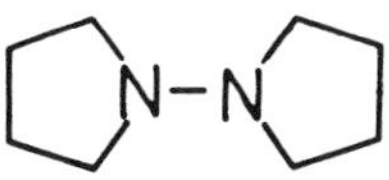 N–N	0.02	−7.1	−8.3
9	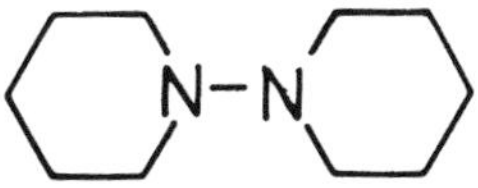 N–N	0.42	+1.6	−8.8
10	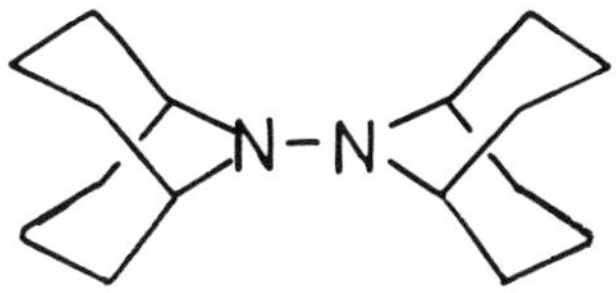 N–N	−0.01	−7.8[c]	−30.7

Table 1-5. (Continued)

	Compound	$E^{\circ\prime}$ (V)[a]	$\Delta(\Delta G^{\circ}_{e})$ (kcal/mol)[b]	ΔvIP (kcal/mol)[b]
12	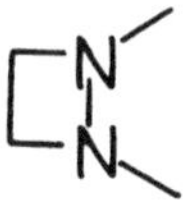	0.42[d]	+2.1	−3.4
13	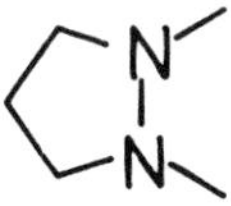	0.11	−5.1	−8.5
14	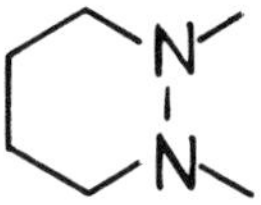	0.23	−2.3	−12.0
15	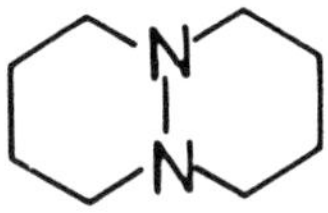	0.28	−1.2	+5.2
16	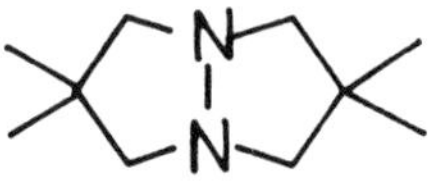	−0.01	−7.8	−17.1
17	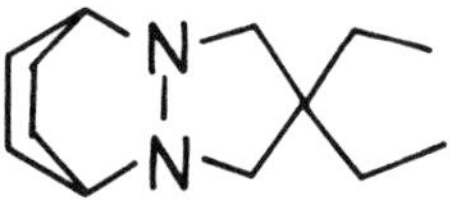	−0.20	−12.1	−31.1[e]
18	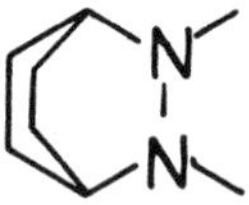	0.07	−6.0	−8.7
19	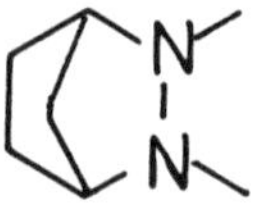	0.20	−4.1	−14.1
20	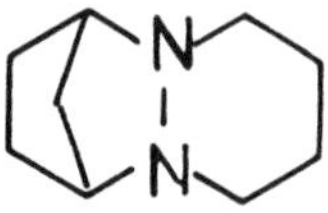	−0.08[f]	−9.5	−24.9[e]

TABLE 1-5. (Continued)

	Compound	$E^{\circ\prime}$ (V)[a]	$\Delta(\Delta G_e^\circ)$ (kcal/mol)[b]	ΔvIP (kcal/mol)[b]
21	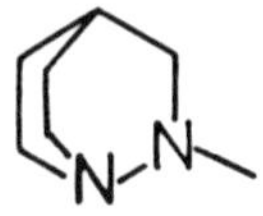	Irre v[f]	(> +12.9)	−5.8
22	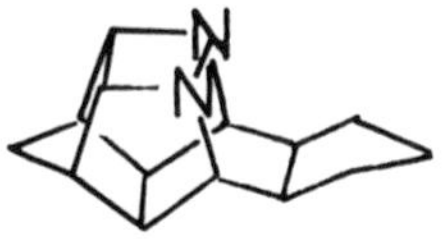	+0.81[f]	+11.1	−14.8
38	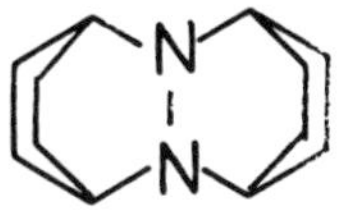	−0.53[g]	−19.8	—
47	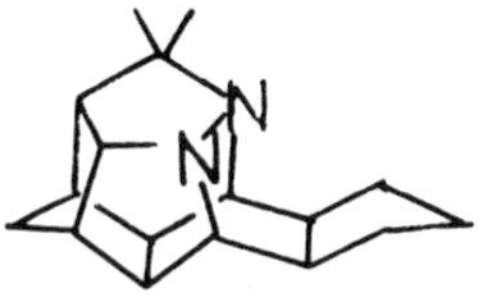	+0.59[h]	+6.0	—
45	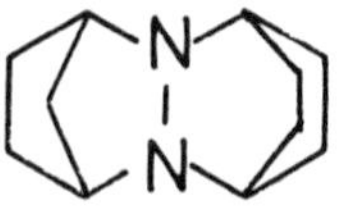	−0.26[i]	−13.6	—
46		+0.01[i]	−7.4	—

[a]CH_3CN versus standard calomel electrode. Reported for 0.1 M n-Bu_4NClO_4 supporting electrolyte. From Reference 17 unless otherwise noted.
[b]Relative to Me_4N_2 (**1**).
[c]Reference 19a.
[d]Reference 20.
[e]Reference 24.
[f]Reference 25b.
[g]Reference 73a.
[h]Reference 64b.
[i]Reference 37b.

but the origin of this effect is not known. Because the radical cation is flattened at nitrogen, effects that raise $\Delta G_N^\ddagger$ will raise $E^{\circ\prime}$, but it is not clear what portion of the increase in $\Delta E^{\circ\prime}$ is caused by neutral form, and what portion by cation radical effects. Its bicyclo[3.2.1]octyl (**42**) and bicyclo[3.3.1]nonyl (**10**) analogues are increasingly easy to oxidize. All three exist in the elec-

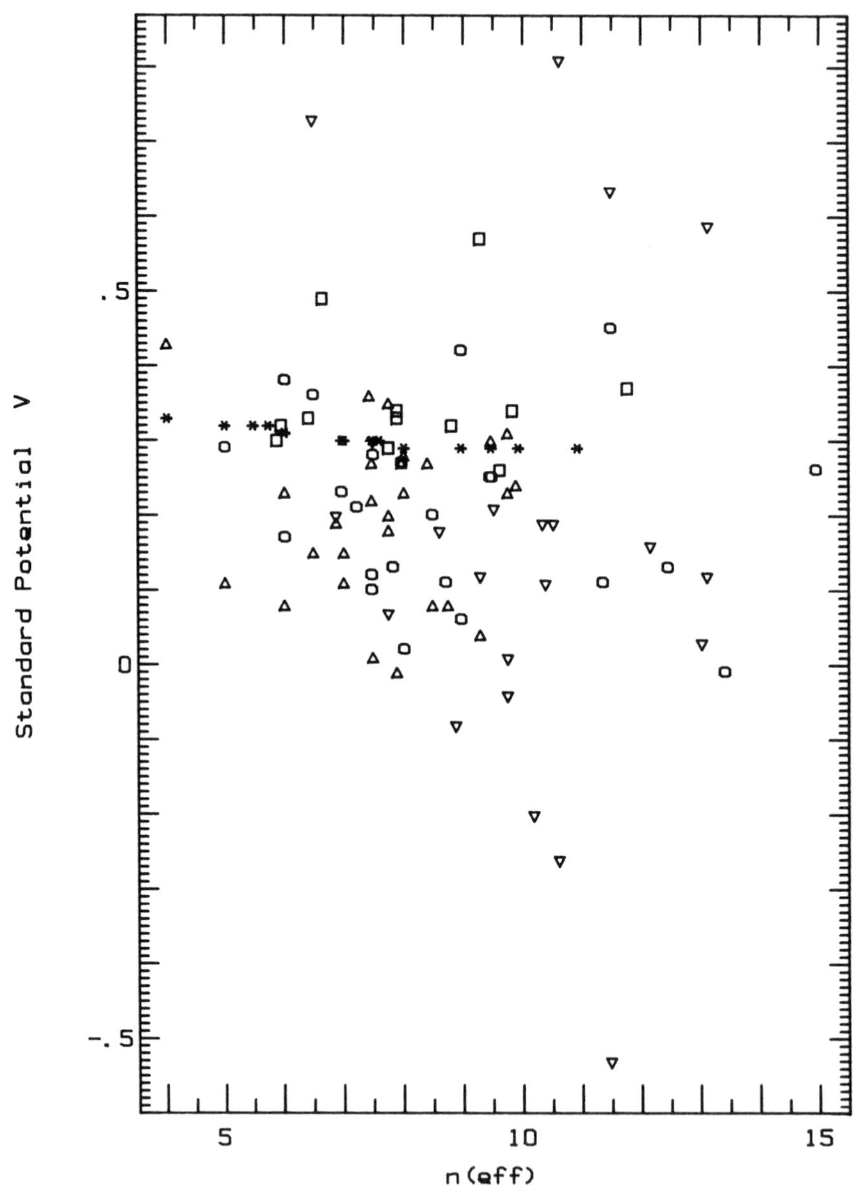

Figure 1-14. Plot of $E^{\circ\prime}$ versus *n*(eff) for tetraalkylhydrazines. Asterisks, tetra-*n*-alkyl compounds; squares, acyclic branched alkyl; circles, *N*,*N*-cycloalkyl; triangles, *N*,*N*′-cyclic; inverted triangles, *N*,*N*′,*N*,*N*′-biscyclic.

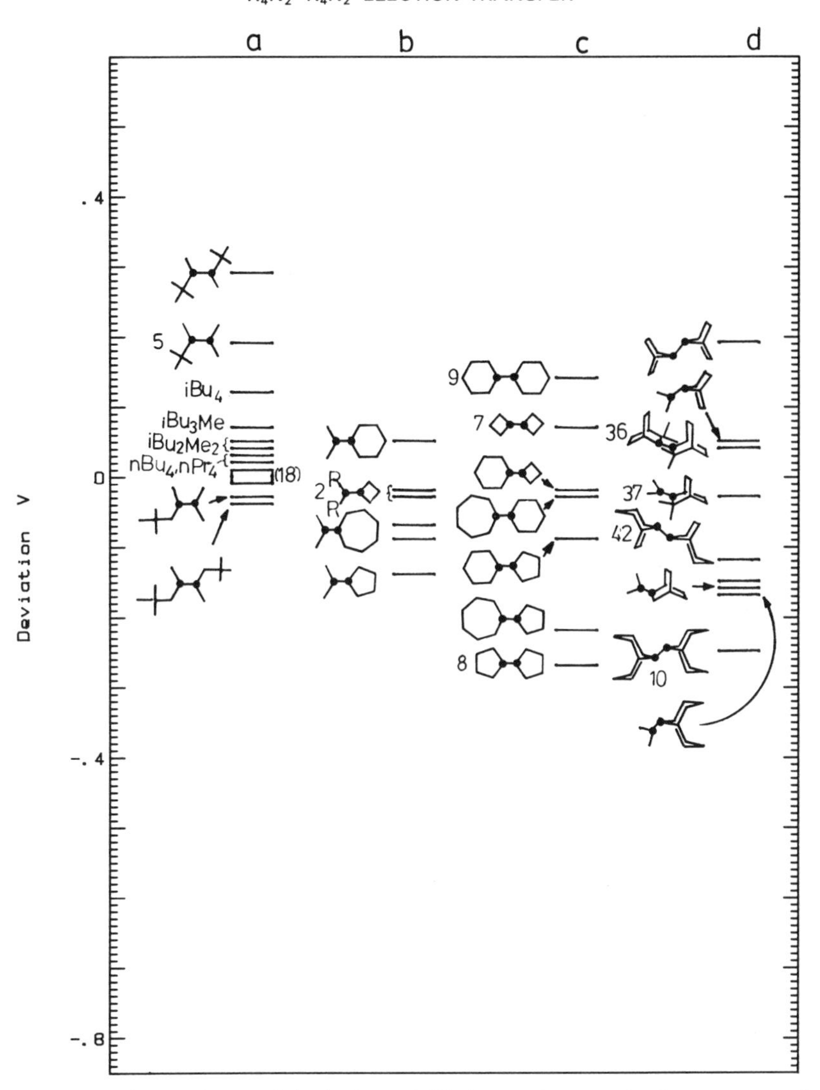

Figure 1-15. Comparison of deviation of $E^{\circ\prime}$ for various R_4N_2 from tetra-*n*-alkylhydrazine of the same *n*(eff) values. (*a*) acyclics. (*b*) *N*-Dialkylamines cyclics. (*c*) Bis-*N*,*N*-cycloalkyls (*d*) Bicyclic *N*,*N*-cycloalkyls.

tronically destabilized $\theta = 180°$ conformations. The torsional effects noted above also appear in the *N*,*N*-cycloalkyl compounds of Figure 1-16*a*, but there is a general decrease in $E^{\circ\prime}$ (negative increments in $\Delta(\Delta G_e^\circ)$) caused by decrease of the A eclipsing effect by having a ring. The 2,3-diazabicyclo[2.2.n] systems of Figure 1-16*b* also show the general decrease in $E^{\circ\prime}$ observed for α-branched systems, and replacing *N*-methyl by *N*-tert-butyl groups makes little change in $E^{\circ\prime}$. Presumably neutral form destabilization just about cancels the increase in steric interaction between the 2,3-dialkyl groups as the nitrogens try to flatten upon electron removal.

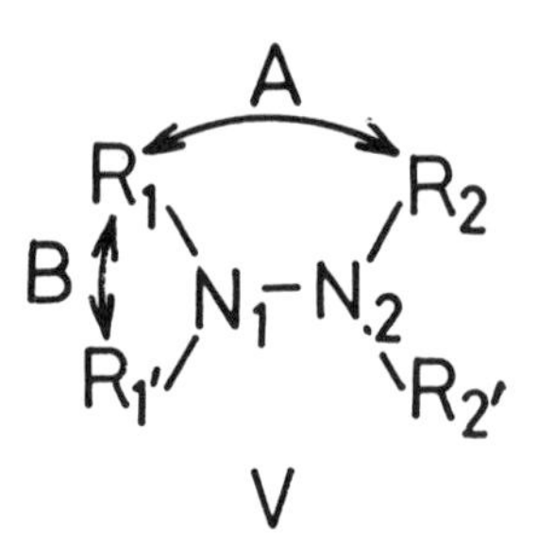

V

Bi-*N*,*N*′-cycloalkyl hydrazines (Figure 1-16*c*) are particularly easy to oxidize when the lone pairs are forced syn ($\theta \simeq 0°$) by two bicyclic rings that force both CNNC angles to be near 0°, and **38** is the most easily oxidized hydrazine yet reported.[73] Its nitrogens are probably unusually flattened in the ground state, because of the interaction of the two $(CH_2)_2$ bridges. Successively replacing one and two $(CH_2)_2$ bridges of **38** with CH_2 bridges to give **44** and **45** makes $E^{\circ\prime}$ substantially more positive, and electron removal from **45** is 12.4 kcal/mol more difficult than from **38.**[73b] Figure 1-16*d* contains compounds with special constraints at the nitrogens. Hünig's diazetidine **(22)**[25] has the highest $E^{\circ\prime}$ yet reported for a tetraalkyl hydrazine, 1.34 V (30.9 kcal/mol) higher than for **38.** Its polycyclic structure prevents much geometry charge upon electron removal. Hünig's dimethylpyrazolidine **47**[64] is 5.1 kcal/mol easier to oxidize and lacks a diazetidine ring. Compound **47** is estimated to have a θ value near 25° from MM2 calculations on the related hydrocarbon. 7-Methyl-1,7-diazabicyclo[3.2.1]octane **48** is quite hard to oxidize, $\Delta(\Delta G_e^\circ)$ being 9.7 kcal/mol. The bicyclic system imposes substantial twist on the NN bond of the cation. The even more twisted 2-methyl-1,2-diazabicyclo[2.2.2]octane **(21)** (θ = 90° if N_2 is flat in the cation) was unfortunately too short-lived to measure $E^{\circ\prime}$, but $\Delta(\Delta G_e^\circ)$ was estimated to be above 12.9 kcal/mol from its irreversible oxidation potential.[25] 2-Methyl-1,2-diazabicyclo[3.3.1]nonane **(49),** in contrast, is no harder to oxidize than an *N*-alkyl acyclic hydrazine of similar *n*(eff) value, so its cation is clearly not very destabilized. It is not necessary for $\mathbf{49}^+$ to have θ far from 0°, but the nitrogens must be bent. Bending at nitrogen in $R_4N_2^+$ is clearly not very costly in energy. For comparison, bridgehead olefin **50** has 12 kcal/mol of strain,[74] and the olefin analogue of **49** has proved to be unisolable.

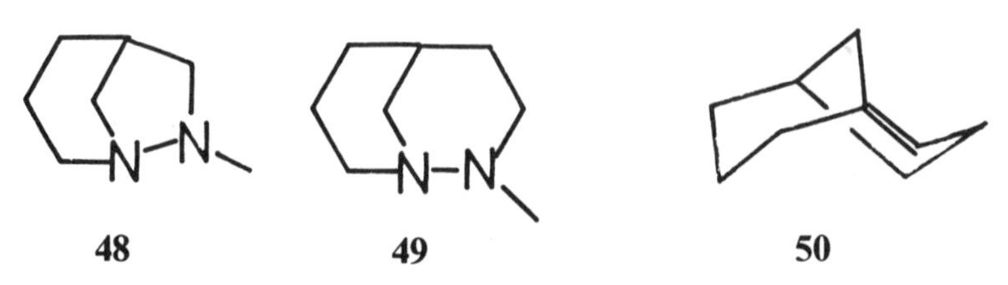

48 **49** **50**

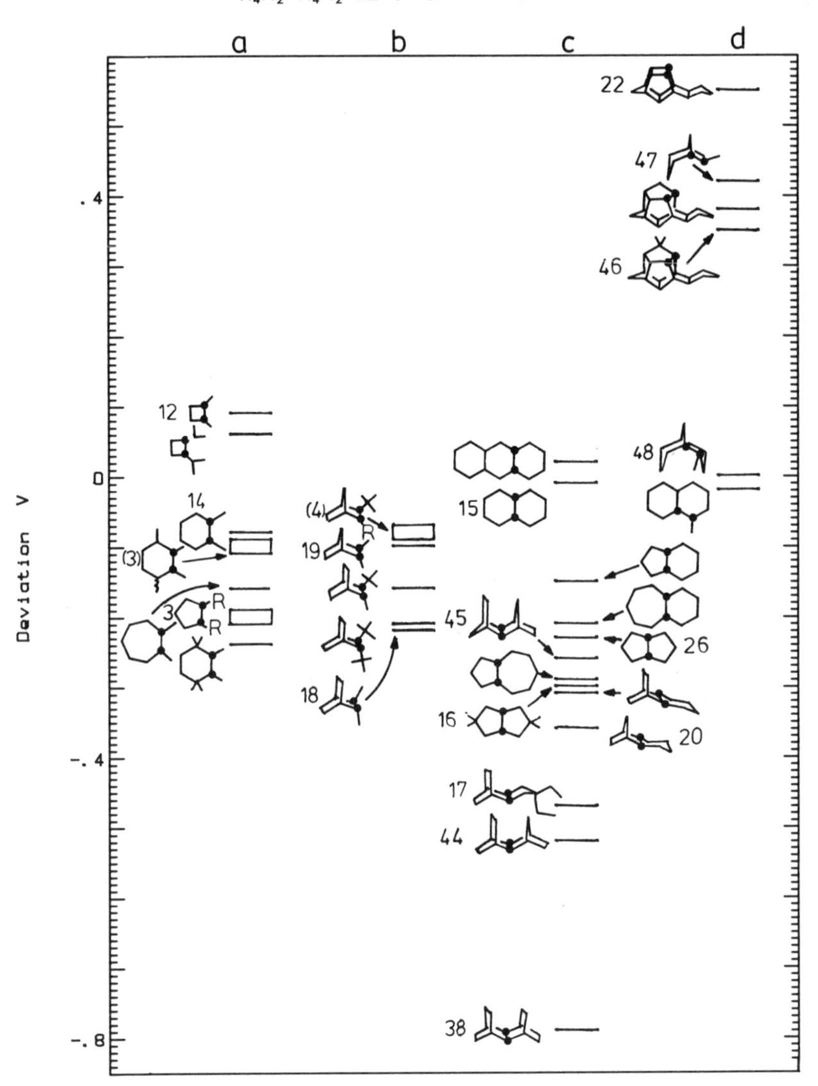

Figure 1-16. Comparison of deviation of $E^{\circ\prime}$ for various R_4N_2 from a tetra-*n*-alkylhydrazine of the same *n*(eff) value. (*a*) *N,N′*-Cycloalkyls. (*b*) 2,3-Dialkyl, 2,3-diazabicyclo[2.2.n] compounds. (*c*) Bridgehead diazabicyclo[m.n.0] compounds. (*d*) Compounds with restricted relaxation.

B. Thermodynamics of Electron Transfer in the Gas Phase

Because of the large geometry change upon electron removal from R_4N_2, there is an extremely bad mismatch of neutral and cation wave functions, and there is essentially no intensity for adiabatic ionization in the PE spectrum. It has recently been found that adiabatic ionization potentials may be determined for hydrazines by direct equilibration using high pressure mass

TABLE 1-6. HPMS Data for Tetraalkylhydrazine and Trimethylamine[a]

Compound	aIP (eV)[b]	ΔG_r° (eV)[c]
1 Me_2NNMe_2	6.78	1.49
2 *n*-$PrMeNMe_2$	6.64	1.50
3 *n*-$BuMeNNMe_2$	6.63	1.49
4 Et_2NNEt_2	6.50	1.44
5 *t*-$BuMeNNMe_2$	6.79	1.10
14	6.54	1.27
Me3N	7.76	0.77

[a]From Reference 75.
[b]Estimated relative error 0.04 eV.
[c]vIP − aIP.

spectroscopy (HPMS).[75] Although most saturated 0- and N-containing radical cations are deprotonated by neutral compound too rapidly for equilibration of two different radical cations to be achieved, this usually does not happen for R_4N_2, presumably because the resonance stabilization of the cations decreases their acidity. Absolute values of adiabatic ionization potential (aIP) were determined by equilibration with model compounds (substituted anilines). Data for only six tetraalkylhydrazines are currently available, and are summarized in Table 1-6.

The gas phase relaxation energies ΔG_r°[76] (see Figure 1-17) are directly available from this work, and it will be seen that ΔG_r° is about twice as large for R_4N_2 as for Me_3N. The Me_3N data are taken from the PE measurements by Aue and co-workers.[77] To understand the relationship between ΔG_r° of Me_3N and Me_4N_2, we refer to Figure 1-17. Figure 1-17*a* shows the situation for Me_3N, which is bent in the neutral form (B°); vIP is the energy gap to the bent cation B^+, and ΔG_r° is the relaxation in going to the relaxed planar radical $\mathbf{P}^+$, which is aIP above **B°**. To facilitate comparing with hydrazines, we will consider an alternate pathway for **B°** to $\mathbf{P}^+$, first planarization to P° (the transition state for N inversion), followed by a nearly vertical ionization to $\mathbf{P}^+$. We label the **P°**–$\mathbf{P}^+$ energy gap tIP, to signify transition state for nitrogen inversion. The relaxation process for $Me_4N_2^{+}$–$\mathbf{1}^+$, is much more complex. The neutral form is bent at both nitrogens and twisted 90° about the NN bond (designated BTB°). The energy gap to the vertical cation $\mathbf{BTB}^+$ is vIP, and aIP is the gap to the relaxed cation, which is planar at both nitrogens and has the alkyl groups eclipsed; it is labeled $\mathbf{PEP}^+$. Both nitrogens flatten, and there is a 90° twist about the NN bond, forming the three-electron π ond of $\mathbf{PEP}^+$. Nitrogen inversion in hydrazine occurs one nitrogen at a time, and the lone pairs are perpendicular at the transition state (see Subsection C of Section 2). The inversion barrier estimated is a little higher

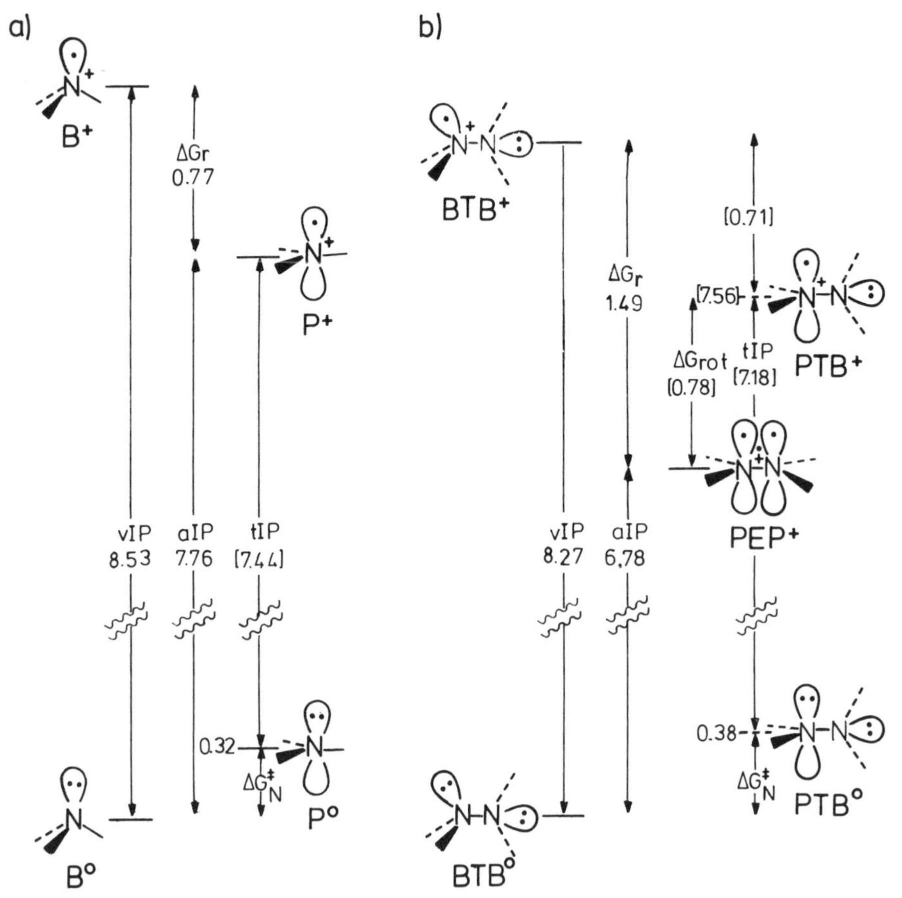

Figure 1-17. Comparison of vertical and adiabatic gas phase ionization for (*a*) trimethylamine and (*b*) tetramethylhydrazine: B and P refer to bent and planar nitrogens, and E and T to eclipsed ($\theta = 0°$) and twisted ($\theta = 90°$) lone pair orbitals.

than for Me_3N,[75] as shown in Figure 1-17*a*. A vertical ionization from the transition state for nitrogen inversion, **PTB°**, gives **PTB⁺**, which is the transition state for NN rotation (see Subsection C of Section 3). The vertical ionization from **PTB°** is estimated[75] to be slightly smaller than that for planar (P°) Me_3N, because vIP is lower for Me_4N_2 than for Me_3N, which gives an estimated rotational barrier for $Me_4N_2^+$ of 0.78 eV, labeled ΔG_{rot} in Figure 1-17. As pointed out in Subsection C of Section 3, $\Delta G^{\ddagger}$ for rotation in **42⁺** is 22 kcal/mol (0.95 eV), so the magnitude arrived at this way is correct, although the difference in substituents and phase makes quantitative comparison impossible. This analysis suggests that flattening at one nitrogen, and the NN bond rotation plus flattening at the other nitrogen contribute comparably to ΔG_r° for $R_4N_2^+$.

Because ΔaIP is the exact gas phase analogue of $\Delta E^{\circ\prime}$, if solvation energy

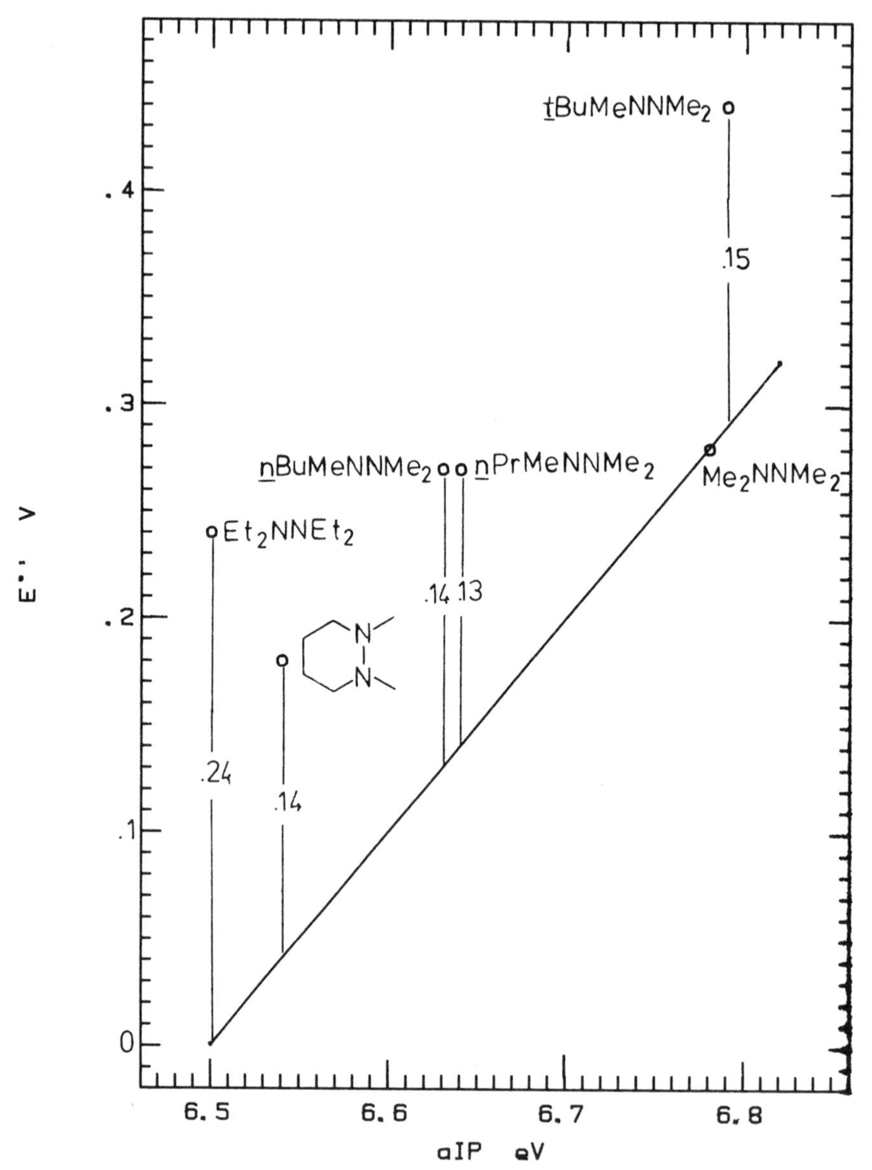

Figure 1-18. Plot of $E^{\circ\prime}$ versus aIP for six examples of R_4N_2. Vertical deviations from the line of slope 1 through the tetramethylhydrazine point represent solvation energy differences from that for the Me_4N_2–$Me_4N_2^+$ couple.

of all six R_4N_2 examples were the same, a plot of $E^{\circ\prime}$ versus aIP would be a straight line with a slope of 1. Figure 1-18 is such a plot with a line of slope 1 drawn through the Me_4N_2 point. It is seen that all the compounds with alkyl groups larger than methyl show positive $E^{\circ\prime}$ deviations. This suggests that enlarging the alkyl groups does indeed decrease solvation energy. A great deal more data will be required for meaningful discussion, but comparison of gas and solution phase electron transfer reactions of R_4N_2 ought to give significant thermodynamic information on solvation effects, eventually to be compared with the extensive material available on proton transfers.[78]

C. Kinetics for Electron Transfer

One of the basic tenets of the Marcus theory for electron transfer[79] is that because electrons move so much faster than nuclei, the intrinsic barrier for electron transfer represents the energy necessary to make the reactants in an electron transfer reaction isostructural with the products. The large difference in geometry between R_4N_2 and $R_4N_2^+$ documented above suggests that electron transfers involving these species would be unusually slow, and a variety of experiments show this to be the case.

The heterogeneous rate constant for electron transfer, k_s, proves to be unusually slow for various R_4N_2 examples, and rather dependent on the electrode material. A particularly useful fact is that k_s also shows a strong dependence on θ, as is demonstrated by work on hexahydropyridazine derivatives.[80,81] Although most hexahydropyridazines show typical CV oxidation waves at room temperature, low-temperature studies at fast scan rates often show an irreversible oxidation at higher potential in addition to a nearly reversible wave. The observed data have been shown to quantitatively fit Scheme 1, illustrated for **14ee.** Electron loss from **ee** conformations where θ is about 180° is chemically reversible and much more rapid than electron loss from **ae** conformations where θ is about 60°. When **ee–ae** thermal interconversion is rapid compared to the time scale of the CV experi-

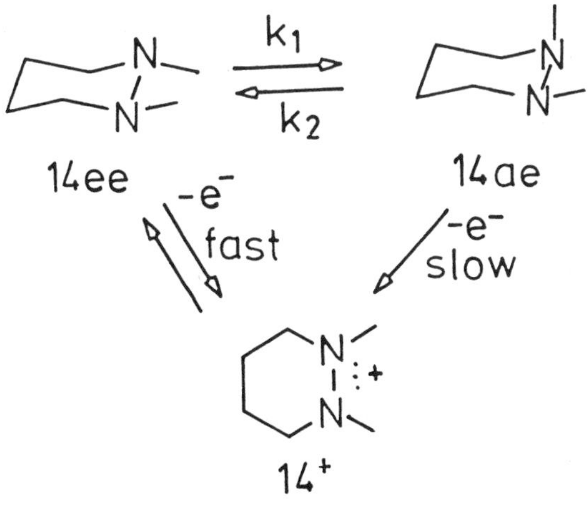

Scheme I

ment, only a single wave is observed, but at low temperature and/or fast scan rates, when **ee** and **ae** do not interconvert rapidly, there is a kinetic resolution of the nearly reversible **ee** oxidation wave from the low k_s and irreversible **ae** oxidation wave, even when **ee** and **ae** have nearly the same energy, so that $E^{\circ\prime}$ is essentially the same for both conformations. The relative sizes of the **ee** and **ae** waves has been shown to give the relative amounts of **ee** and **ae** conformations in the bulk solution under these conditions. When **ee** and **ae** are interconverting slowly during the time scale of the CV experiment, the size ratio of the two oxidation waves depends on scan rate, and digital simulation gives the quantity $K_{eq} \sqrt{k_1 + k_2}$, where $K_{eq} = k_1/k_2$, allowing measurement of rate constants for conformational interconversion. Thus CV provides a nonspectroscopic method for investigation of conformational interconversions of six-membered ring hydrazines, the only examples found to date for which high enough barriers between conformations with significantly different θ values to show the effect have been observed.[80]

Because the methods are so different, CV and DNMR frequently measure different rate constants, and can give complementary information. The rationale for seeing such different k_s values is that the transition state for electron transfer lies at an intermediate geometry between the neutral and radical cation forms, so it must be twisted about the NN bond for **ae** oxidation, but not twisted for **ee** oxidation, because the lone pairs are already coplanar in neutral **ee.** Twisting of hydrazine cation radicals is much more difficult than bending them, so the **ae** transition state lies higher in energy, leading to the small k_s values observed. The large dependence of electron transfer rate constant on θ suggests that the conformation oxidized under most conditions for most examples of R_4N_2 will be that with parallel lone pairs ($\theta = 0, 180°$), which is usually not the equilibrium conformation.

The monocation of bishydrazine **51** has been shown by ESR to have its positive charge localized in one hydrazine unit, requiring that the rate constant for electron transfer between the two hydrazine units be less than $4 \times 10^7\ s^{-1}$,[82] even though they are held close together. The transition state for intramolecular electron transfer needs equivalent geometry at both hydrazine units, which would require substantial flattening and twisting of the unoxidized hydrazine. This clearly is too costly in energy for a very fast intramolecular electron transfer.

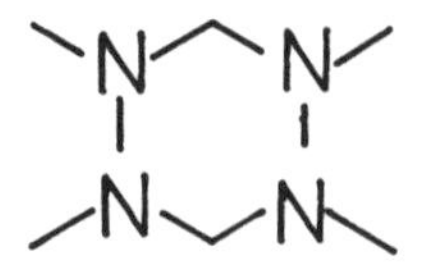

51

Measurement of intermolecular electron transfer rates between R_4N_2 and $R_4N_2^{+}$ has been difficult because k_2 (see Equation 1-4) is so low. No ESR line broadening is seen for most R_4N_2–$R_4N_2^{+}$ mixtures, although this is the standard technique for measuring k_2 values for most organic neutral, radical ion pairs, where k_2 values above 10^8 are observed.[83] For mixtures of **52** and **52**$^+$, no NMR line broadening was observed, showing that k_2 for this case is under $6 \times 10^2\ M^{-1}s^{-1}$.[70] The k_2 values for **42–42**$^+$ electron transfer have recently been measured[84] by two different methods based on the fact that syn and anti conformations of the cations are interconverted by the electron transfer shown in Equation 1-5. Neutral **42** exists in rapidly interconverting **s** and **a** conformations, and catalyzes isomerization of **42s**$^+$ to the stabler **42a**$^+$. Simulation of the CV of **42** at 22°C and following the conversion of **42s**$^+$ to **42a**$^+$ at 11°C by UV both gave barriers of $\Delta G^{\ddagger}_{2} = 12.3$, $\Delta G^{\ddagger}_{-2} = 14.0$ kcal/mol. The intrinsic barrier for the electron transfer at $\Delta G° = 0$ is therefore their average, 13.1_5 kcal/mol. This corresponds to a k_2 value of about 10^3 at 22°C, a factor of 10^7 below diffusion control, and 10^6 below k_2 for delocalized cases like tetramethyl-*p*-phenylenediamine. In other unpublished work,[67] the **38–38**$^+$ electron transfer has been shown to give easily detectable NMR line broadening, so this pair must show much faster k_2 values that either **51–51**$^+$ or **42–42**$^+$. We presume that the flattening at nitrogen introduced by interaction of the $(CH_2)_2$ bridges is responsible for this more rapid electron transfer.

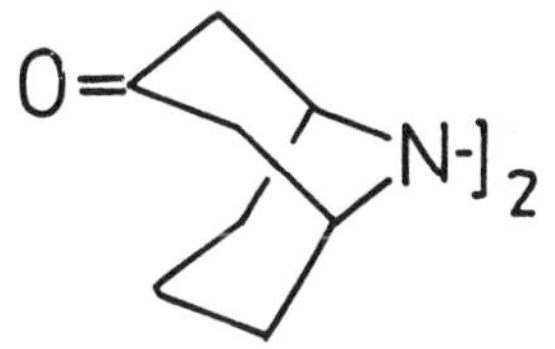

52

$$R_4N_2 + \mathbf{R_4N_2^{+}} \underset{k_{-2}}{\overset{k_2}{\rightleftarrows}} R_4N_2^{+} + \mathbf{R_4N_2} \tag{1-4}$$

$$\mathbf{42a} + \mathbf{42s}^+ \underset{k_{-2}}{\overset{k_2}{\rightleftharpoons}} \mathbf{42a}^+ + \mathbf{42s} \quad \Delta G° = -1.6_7 \text{ kcal/mol} \tag{1-5}$$

Vapor phase electron transfer between simple hydrazines and their cations has also been noted to be unusually slow, although quantitative studies have not yet been done.[75]

Despite the poor treatment of neutral H_4N_2 by the MNDO method (which gets $\theta = 180°$ as the lowest energy conformation), we think it is useful to

TABLE 1-7. Comparison of MNDO-Calculated H_4N_2, $H_4N_2^{+}$, and Electron Transfer Transition State Geometries

	anti H_4N_2	$H_4N_2^{+}$	Transition state	Location[a]
d(NN)	1.397 Å	1.300 Å	1.35 Å	44
d(NH)	1.021 Å	1.021 Å	1.02 Å	[b]
α(HNH)	103.08°	116.16°	108.4°	41
α(HNN)	107.18°	118.55°	116.7°	83
β	61.65°	25.34°	39.9°	60
α(av)	105.81°	117.75°	113.9°	67

[a]The percent of the change in geometry from neutral H_4N_2 (0%) to the radical cation (100%).
[b]The NH distance is insensitive to electron removal.

consider what such a calculation gives for the transition state for electron transfer between H_4N_2 and $H_4N_2^{+}$. We suggest that the obviously lower k_s values for **ae** than **ee** hexahydropyridazine demonstrates that θ must be near 180° anyway. The calculations were carried out by minimizing the energy of two molecules of H_4N_2 held far apart (20Å) with a symmetry plane between them, having an overall charge of +1 [84b]. As expected calculated is 239.8 kcal/mol, 11.7 kcal/mol above the energy of neutral and cation radical in their optimum geometries. The transition state obtained is compared with the equilibrium geometries in Table 1-7.

The transition state geometry obtained is rather different from what we would have predicted. Because bending at N in $H_4N_2^{+}$ is much easier than in H_4N_2, we would have expected bending at the transition state to be closer to that in H_4N_2; the calculation gives an amount of bend two-thirds of the way to the cation, using α(av) as the criterion for bending at N. Because compressing the NN bond of H_4N_2 ought to be easier than stretching that of $H_4N_2^{+}$, we would have expected the transition state to have *d*(NN) closer to that of $H_4N_2^{+}$; the calculation gives a *d*(NN) value only 44% of the way from H_4N_2 to $H_4N_2^{+}$. It would be interesting to see whether higher level calculations would give a similar answer, but transition state location is exceedingly time-consuming, and such calculations have not been carried out.

D. Tetraalkylhydrazine Preparation

Examples of R_4N_2 with a variety of alkyl substituents R were required for the work discussed above. Because alkylated hydrazine nitrogens alkylate much more rapidly than unalkylated ones, direct alkylation of hydrazines with alkyl halides is less useful than might have been hoped, although it was used to prepare **16** and **26**[17b] and (rather unsuccessfully) **12.**[19] The introduction of lithium aluminum hydride (LAH) as a synthetic reagent in the 1950s provided a reliable way of reducing the NN=C bonds of hydrazones and azines[85] without getting NN cleavage, as well as reducing NNC(=O)R to $NNCH_2R$.[85,86] Condon has worked out procedures for making unsymmetrically alkylated examples of R_4N_2 using acyl protecting groups and selectively alkylating either nitrogen.[87] The use of sodium cyanoborohydride[88] allows convenient reductive alkylation with aldehydes and ketones (used for **2–5** and **21**) and was used to prepare **44** from the oxadiazolidine.[50] The use of dicarbonyl compounds with *N,N'*-dialkylhydrazines provides annulation to pyrolidine and piperidine rings (**8, 9**) and with *N,N'*-dialkylhydrazines gives pyrazolidine, hexahydropyridazine, and diazacycloheptane rings (used for the compounds in Figure 1-6[42]).

Compounds with tetra- and hexahydropyridazine structures are usually most conveniently prepared by Diels-Alder addition of dicarbonyl-substituted azo compounds to dienes, the method employed for making **14, 15, 20, 24, 25, 27, 28, 30–35,** and **43.** The tetra-α-branched *N,N'*-bisbicyclic compounds **38, 45,** and **46** were prepared using Diels-Alder addition of protonated bicyclic azo compounds to cyclic dienes.[73] Attempted coupling of R_2NX compounds often leads to extensive formation of elimination products, but was used for biaziridine (**6**)[18] and bipiperidines (**39–41**).[48] When bicyclic structures prevent elimination of HX, chloroamines couple well upon treatment with *tert*-butyllithium, which was employed for **10,**[21] **36,**[46] and **42.**[66] The preparation of **21,**[89] **48,** and **49**[25b] employed catalytic hydrogenation of carbonyl-substituted pyridines to piperidines, followed by amination at nitrogen, closure to bicyclic *N*-aminohydrazides and LAH reduction. The cage compounds **22**[25a] and **47**[64a] were prepared by routes developed by Hünig's group.

ACKNOWLEDGMENTS

We thank the National Science Foundation and the National Institutes of Health for financial support for much of the work discussed in this chapter. The hydrazine work done at Wisconsin has been carried out over a period of 15 years by a large number of skilled and dedicated co-workers, listed here in rough chronological order: P.J. Hintz, J.M. Buschek, R.T. Landis II, G.R. Weisman, L. Echegoyen, E.L. Clennan, W.C. Hollinsed, C.R. Kessel, L.A. Grezzo, W.P. Parmelee, P.M. Gannett, D.J. Steffek, G.T. Cunkle, N.P. Yumibe, M.R. Willi, S.C. Blackstock, D.T. Rumack, and T.B. Frigo. We have also enjoyed the collaboration of several other groups, including those of Prof. D.H. Evans (Wisconsin), Prof. F. Gerson (Basel), Prof. H. Kurreck and Dr. B. Kirste (Free University of Berlin), Dr. T. Clark (Erlangen), Prof. F. Weinhold (Wisconsin), Dr. M. Meot-Ner (Mautner) (National Bureau of Standards), Prof. A. Schweig (Marburg), Prof. K.-D. Asmus (Hahn-Meitner-Institut, Berlin), and Prof. M. Kaftory (Technion, Israel) on various aspects of this work.

REFERENCES

1. Shvo, Y. In "Chemistry of Hydrazo, Azo, and Azoxy Groups", Patai, S., Ed.; Wiley: New York, 1975; Part 2, Chapter 21, pp. 1017–1095.
2. Kohata, K.; Fukuyama, T.; Kuchitsu, K. *J. Phys. Chem.* **1982,** *86,* 602.
3. Chiu, N.S.; Sellers, H.L.; Sclidfer, L.; Kohata, K. *J. Am. Chem. Soc.* **1979,** *101,* 5883.
4. Kohata, K.; Fukuyama, T.; Kuchitsuk, K. *Chem. Lett.* **1979,** 257.
5. Brunck, T.K.; Weinhold, F. *J. Am. Chem. Soc.* **1979,** *101,* 1700.
6. Cowley, A.H.; Mitchell, D.J.; Whangbo, M.-H.; Wolfe, S. *J. Am. Chem. Soc.* **1979,** *101,* 5224.
7. Vitkouskaya, N.M.; Dolgnicheva, O.Yu.; Frolov, Yu.L.; Kieko, U.V.; Voronkov, M.G. *Dok. Acad. Nauk SSSR* **1977,** *235,* 843; Engl transl., 749.
8. Rademacher, P. *Angew. Chem.* **1973,** *85,* 410; *Int. Ed. Engl. 12,* 408.
9. Rademacher, P. *Tetrahedron Lett.* **1974,** 83.
10. Rademacher, P. *Chem. Ber.* **1975,** *108,* 1548.
11. Rademacher, P.; Koopman, H. *Chem. Ber.* **1975,** *108,* 1557.
12. Nelsen, S.F.; Buschek, J.M. *J. Am. Chem. Soc.* **1973,** *95,* 2011.
13. Nelsen, S.F.; Buschek, J.M.; Hintz, P.J. *J. Am. Chem. Soc.* **1973,** *95,* 2013.
14. Nelsen, S.F.; Buschek, J.M. *J. Am. Chem. Soc.* **1974,** *96,* 2392.
15. Nelsen, S.F.; Buschek, J.M. *J. Am. Chem. Soc.* **1974,** *96,* 6982.
16. Nelsen, S.F.; Buschek, J.M. *J. Am. Chem. Soc.* **1974,** *96,* 6987.
17. (a) Nelsen, S.F.; Peacock, V.E. Weisman, G.R. *J. Am. Chem. Soc.* **1976,** *98,* 5269. (b) Nelsen, S.F.; Hintz, P.J. *J. Am. Chem. Soc.* **1972,** *94,* 7108.
18. (a) Rademacher, P.; Lüttke, W. *Angew. Chem.* **1970,** *82,* 258; *Int. Ed. Engl. 9,* 245. (b) Rademacher, P. *Acta Chem. Scand.* **1972,** *26,* 1981.
19. (a) Nelsen, S.F.; Peacock, V.E.; Weisman, R.G.; Landis, M.E.; Spencer, J.A. *J. Am. Chem. Soc.* **1978,** *100,* 2806. (b) Kirste, K.; Lüttke, W.; Rademacher, P. *Angw. Chem.* **1978,** *90,* 726. (c) Hall, J.H.; Brigard, W.S. *J. Org. Chem.* **1978,** *43,* 2784.
20. Nelsen, S.F.; Kessel, C.R.; Brien, D.J. *J. Am. Chem. Soc.* **1980,** *102,* 702.
21. Nelsen, S.F.; Hollinsed, W.C.; Kessel, C.R.; Calabrese, J.C. *J. Am. Chem. Soc.* **1978,** *100,* 7876.
22. Förterer, M.; Rademacher, P. *Chem. Ber.* **1980,** *113,* 221.
23. Studied along with **26** by IR Raman spectroscopy comparison: Koopman, H.-P.; Rademacher, P. *Spectrochim. Acta* **1976,** *32A,* 157.
24. Nelsen, S.F.; Hollinsed, W.C.; Grezzo, L.A.; Parmelee, W.P. *J. Am. Chem. Soc.* **1979,** *101,* 7347.
25. (a) Bernig, W.; Hünig, S. *Angew. Chem. Int. Ed. Engl.* **1977,** *16,* 277. (b) Nelsen, S.F.; Kessel, C.R.; Brace, H.L. *J. Am. Chem. Soc.* **1979,** *101,* 1875.
26. (a) Cocksey, B.J.; Eland, D.H.; Danby, W. *J. Chem. Soc. B* **1971,** 790. (b) For a somewhat similar treatment of photoionization data, see Levitt, L.S.; Alding, H.F. *Prog. Phys. Org. Chem.* **1976,** *12,* 119.
27. Nelsen, S.F. *J. Org. Chem.* **1984,** *49,* 1891.
28. Taft, R.W., Jr. In "Steric Effects in Organic Chemistry", Newman, M.S., Ed.; Wiley: New York, 1956, p. 594.
29. Aue, D.H.; Webb, H.M.; Bowers, M.T. *J. Am. Chem. Soc.* **1975,** *97,* 4136.
30. Nelsen, S.F.; Hollinsed, W.C.; Calabrese, C.J. *J. Am. Chem. Soc.* **1977,** *99,* 4461. For similar results on **ae** and **ee** tetraazacyclohexane derivatives, see Katritzky, A.R.; Baker V.J.; Camalli, M.; Spagna, R.; Vaciago, A. *J. Chem. Soc. Perkin 2,* **1980,** 1753.
31. Nelsen, S.F.; Hollinsed, W.C. *J. Am. Chem. Soc.* **1980,** *45,* 3609.
32. For a review, see Lehn, J.M. *Fortschr. Chem. Forsch.* **1971,** *15,* 311.
33. Anderson, J.E.; Lehn, J.M. *J. Am. Chem. Soc.* **1967,** *89,* 81.
34. Dewar, M.J.S.; Jennings, W.B. *J. Am. Chem. Soc.* **1973,** *95,* 1562.
35. Jones, R.K.Y.; Katritzky, A.R.; Scattergood, R. *J. Chem. Soc. Chem. Commun.* **1971,** *644.*
36. Nelsen, S.F. *Acc. Chem. Res.* **1978,** *95,* 1562.
37. (a) Eliel, E.L.; Kardasamy, D.; Yen, C.-Y.; Hargrave, K.D. *J. Am. Chem. Soc.* **1980,** *102,* 3698. (b) Anet, F.A.L.; Vavari, I.; Ferguson, H.; Katritsky, A.R.; Morenes-Manas, H.; Robinson, M.T. *J. Chem. Soc. Chem. Commun.* **1976,** *399.*
38. Nelsen, S.F.; Gannett, P.M. *J. Am. Chem. Soc.* **1981,** *103,* 3300.
39. Anet, F.A.L.; Basus, V.J. *J. Magm. Resonance* **1978,** *32,* 339.

40. Anderson, J.E. *J. Am. Chem. Soc.* **1969,** *91,* 6374.
41. Schweig, A.; Thon, N.; Nelsen, S.F.; Grezzo, L.A. *J. Am. Chem. Soc.* **1981,** *102,* 7438.
42. Nelsen, S.F.; Clennan, E.L. *J. Am. Chem. Soc.* **1978,** *100,* 4004.
43. Weisman, G.R.; Nelsen, S.F. *J. Am. Chem. Soc.* **1976,** *98,* 7007.
44. (a) Katritzky, A.R.; Baker, V.J.; Brito-Palma, F.M.S. *J. Chem. Soc. Perkin Trans. 2* **1980,** 1739. (b) Katritzky, A.R.; Baker, V.J.; Brito-Palma, F.M.S.; Ferguson, I.J.; Angiolini, L. ibid, **1980,** 1746.
45. Cunkle, G.T. Unpublished work.
46. Nelsen, S.F.; Gannett, P.M. *J. Am. Chem. Soc.* **1982,** *104,* 5292.
47. Blackstock, S.C. Unpublished work.
48. (a) Ogawa, K.; Takeuchi, Y.; Suzuki, H.; Nomura, Y. *J. Am. Chem. Soc.* **1984,** *106,* 831. (b) Ogawa, K.; Takeuchi, Y.; Suzuki, H. Nomura, Y. *Chem. Lett.* **1981,** 690. (c) Ogawa, K.; Takeuchi, Y.; Suzuki, H. Nomura, Y. *J. Chem. Soc. Chem. Commun.* **1981,** 1015.
49. For molecular mechanics calculations on cyclohexylpiperidines that make the same point, see Jaime, C.; Osawa, E. *J. Chem. Soc. Chem. Commun.* **1983,** 708; *J. Chem. Soc. Perkin Trans. 2.* **1984,** 995.
50. Nelsen, S.F.; Weisman, G. R.; Hintz, P.J.; Olp, D.; Fahey, M.R. *J. Am. Chem. Soc.* **1974,** *96,* 2918.
51. Bock, H.; Kaim, W.; Nöth, H.; Semkow, A. *J. Am. Chem. Soc.* **1980,** *102,* 442.
52. (a) Clark, T. Private communication, see also Reference 66. (b) Frisch, M.J.; Raghavachari, K.; Pople, J.A.; Bouna, W.J.; Radom, L. *Chem. Phys.* **1983,** *75,* 323.
53. Nelsen, S.F.; Hollinsed, W.C.; Kessel, C.R.; Calabrese, J.C. *J. Am. Chem. Soc.* **1978,** *100,* 7876.
54. Nelsen, S.F.; Cunkle, G.T.; Evans, D.H.; Haller, K.J.; Kaftory, M.; Kirste, B.; Kurreck, H.; Clark, T. *J. Am. Chem. Soc.* **1985,** *107,* 3829.
55. Adams, J.O.; Thomas, J.R. *J. Can. Phys.* **1963,** *39,* 1904.
56. Brivati, J.A.; Gross, J.M.; Symons, M.C.R.; Timling, D.J.A. *J. Chem. Soc* **1965,** 6504.
57. (a) Marquardt, C.L. *J. Can. Phys.* **1970,** *53,* 3248. (b) Reilley, M.H.; Marquardt, C.L. *J. Chem. Phys.* **1970,** *53,* 3257.
58. (a) Lisle, J.B.; Williams, L.F.; Wood, D.E. *J. Am. Chem. Soc.* **1976,** *98,* 227. (b) Krusic, P.J.; Meaken, P. ibid, **1976,** *98,* 226. (c) Krusic, P.J.; Burgham, R.C. ibid, **1976,** *98,* 230.
59. Nelsen, S.F.; Echegoyen, C. *J. Am. Chem. Soc.* **1975,** *97,* 4930.
60. (a) Yumibe, N.P. Unpublished work. (b) Nelsen, S.F.; Blackstock, S.C.; Yumibe, N.P.; Frigo, T.B.; Carpenter, J.E.; Weinhold, F. *J. Am. Chem. Soc..* **1985,** *107,* 143.
61. Pople, J.A.; Beveridge, D. dL.; Dobosh, P.A. *J. Am. Chem. Soc.* **1966,** *90,* 4201.
62. Hayon, E.; Sunic, M. *J. Am. Chem. Soc.* **1972,** *94,* 42.
63. (a) Shida, T.; Nosaka, Y.; Kato, T.J. *J. Chem. Phys.* **1978,** *82,* 695. (b) Haselbach, E.; Bally, T.; Gschwind, R.; Klemm, U.; Layinova, Z. *Chimia* **1979,** *33,* 405. (c) Haselbach, E.; Klemm, U.; Buser, V.; Gschwind, R.; Jurgen, H.; Kloster-Jensen, E.; Maier, J.P.; Marothaler, O.; Christen, H.; Baertschi, P. *Helv. Chim Acta.* **1981,** *65,* 823.
64. (a) Hünig, S.; Prokschy, R. *Chem. Ber.* **1984,** *117,* 2099. (We thank Prof. Hünig for a preprint.) (b) Willi, M.R. Unpublished results.
65. Nelsen, S.F.; Teasley, M.F.; Kapp, D.L.; Kessel, C.R.; Grezzo, L.A. *J. Am. Chem. Soc.* **1983,** *105,* 5928.
66. Nelsen, S.F.; Cunkle, G.T.; Evans, D.H.; Clark, T. *J. Am. Chem. Soc.* **1983,** *105,* 5928.
67. Blackstock, S.C. Unpublished work.
68. Nelsen, S.F.; Peacock, V.E.; Kessel, C.R. *J. Am. Chem. Soc.* **1978,** *100,* 7017.
69. Nelsen, S.F. *Isr. J. Chem.* **1979,** *18,* 45.
70. Nelsen, S.F. *Acc. Chem. Res.* **1981,** *14,* 131.
71. Evans, D.H.; Kinlen, P.J.; Nelsen, S.F. *J. Electroanal. Chem.* **1979,** *97,* 265.
72. Steffek, D.J.; Blackstock, S.C. Unpublished work.
73. (a) Nelsen, S.F.; Blackstock, S.C.; Frigo, T.B. *J. Am. Chem. Soc.* **1984,** *106,* 3366. (b) Blackstock, S.C.; Frigo, T.B. Unpublished work.
74. Lesko, P.M.; Turner, R.B. *J. Am. Chem. Soc.* **1968,** *90,* 6888.
75. Meot-Ner (Mautner), M.; Nelsen, S.F.; Willi, M.R.; Frigo, T.B. *J. Am. Chem. Soc.* **1984,** *106,* 7384.
76. Although vIP and aIP are supposed to represent ΔH values, the aIP values are in fact $\Delta G°$ measurements at 550 K, and entropy effects have been more or less ignored in arriving at aIP.[75]
77. Aue, D.H.; Webb, H.M.; Bowers, M.T. *J. Am. Chem. Soc.* **1976,** *98,* 311.

78. (a) Aue, D.H.; Webb, H.M.; Bowers, M.T. *J. Am. Chem. Soc.* **1976,** *98,* 318. (b) Taft, R.W. In "Proton Transfer Reactions", Calden, E.F.; and Gold, V., Ed.; Chapman and Hall: London, 1975, Chapter 2, pp. 31–77. (c) Aue, D.H.; Bowers, M.T. In "Gas Phase Ion Chemistry", Vol. 2; Bowers, M.T., Ed.; Academic Press: New York, 1979, Chapter 9, pp. 1–15.
79. For a recent discussion, see Eberson, L. In "Advances in Physical Organic Chemistry", Vol. 18; Gold, V.; and Bethell, D., Eds.; Academic Press: New York, 1982.
80. (a) Nelsen, S.F.; Echegoyen, L.; Evans, D.H. *J. Am. Chem. Soc.* **1975,** *97,* 3530. (b) Nelsen, S.F.; Echegoyen, L.; Clennan, E.L.; Evans, D.H.; Corrigan, D.A. ibid, **1977,** *99,* 1130. (c) Nelsen, S.F.; Clennan, E.L.; Evans, D.H. ibid, **1978,** *100,* 4012.
81. For a review see Ref. 79, and Evans, D.H.; Nelsen, S.F. In "Characterization of Solutes in Nonaqueous Solvents", Mamantov, G., Ed.; Plenum Press: New York, 1978, p. 131.
82. Nelsen, S.F.; Hintz, P.J.; Buschek, J.M.; Weisman, G.R. *J. Am. Chem. Soc.* **1975,** *97,* 4933.
83. For a review, see: Szwarc, M.; Jagur-Grodzinski, J. In "Ions and Ion-Pairs in Organic Reactions", Szwarc, M., Ed.; Wiley: New York, 1975, pp. 1–150.
84. (a) Recent work with G.T. Cunkle and D.H. Evans. (ref. 54). (b) These calculations were carried out by S.C. Blackstock, following a suggestion of J.A. Thompson-Colon of minimizing two molecules with a symmetry plane.
85. (a) Class, J.B.; Aston, J.G.; Oakswood, T.S. *J. Am. Chem. Soc.* **1953,** *75,* 2937. (b) Renaud, R.; Leitch, L.C. *Can. J. Chem.* **1954,** *32,* 545.
86. Beltrami, R.T.; Bissel, E.R. *J. Am. Chem. Soc.* **1956,** *78,* 2467.
87. (a) Condon, F.E. *J. Org. Chem.* **1972,** *37,* 3615. (b) Condon, F.E.; Thakkan, D.C.; Goldstein, T.B. *Org. Prep. Proc. Int.* **1973,** *5,* 233.
88. Nelsen, S.F.; Weisman, G.R. *Tetrahedron Lett.* **1973,** 2321.
89. Nelsen, S.F.; Gannett, P.M.; Steffek, D.J. *J. Org. Chem.* **1980,** *45,* 3857.

CHAPTER 2

Stabilization and Destabilization of Aromatic and Antiaromatic Compounds

Gerald R. Stevenson

Illinois State University, Normal, Illinois

CONTENTS

1. INTRODUCTION

Since Faraday's discovery of benzene in 1825,[1] many chemists have searched for a satisfactory structure for this compound. Kékulé[2] was first to

MOLECULAR STRUCTURE AND ENERGETICS, Vol. 3

suggest a cyclic hexagon with alternating double bonds. Subsequent experimental work has shown that all the positions in the ring are equivalent, which led Kékulé[3] to propose "resonating" structures (2-1). Many other structures were proposed, such as the "para" bonded system (**2**) by Dewar and Wislicenus,[4] and the "diagonal" formula (**3**) and (**4**) by Claus.[5] Their structural theories, however, did not explain the lack of reactivity of benzene.

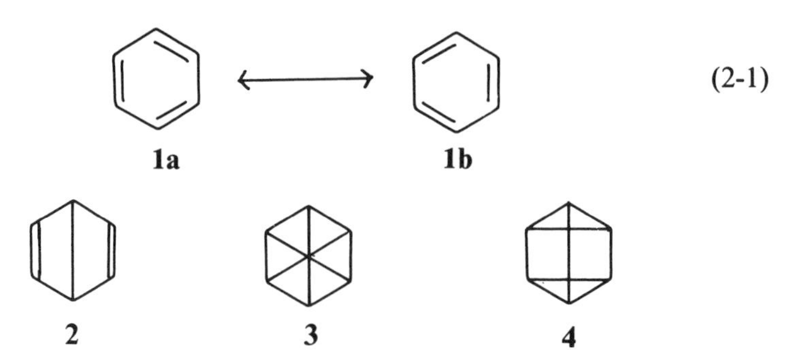

One of the first attempts to account for the observed lack of reactivity of benzene, and thus provide a structural basis for aromaticity, was advanced by Thiele in 1899.[6] Thiele's theory of partial valency (that a partial double bond exists between each pair of adjacent carbon atoms) predicted that all cyclic compounds with alternating double and single bonds would exhibit the enhanced stability indicative of an aromatic compound. Therefore, cyclobutadiene and cyclooctatetraene should have properties similar to those of benzene. However, the synthesis of cyclooctatetraene by Willstätter and co-workers[7] in 1911 led to evidence suggesting that it is olefinic. This coupled with the fact that cyclobutadiene remained a fugitive molecule (until very recently) led to the demise of Thiele's theory.

The advent of quantum mechanics in the 1920s dramatically increased our understanding of the benzene system. The first and still most successful application of quantum mechanics to aromatic systems was developed by Erich Hückel in the mid-1930s.[8] Hückel molecular orbital (HMO) theory (for a review of the HMO method, see Reference 9) provides a theoretical basis for the stability of benzene. The HMO approach leads us to the conclusion that every planar annulene with $4n$ (n is an interger) π electrons has a degenerate pair of nonbonding MOs that are partially filled. The remaining monocyclic annulenes have $(4n + 2)$ π electrons and are closed-shelled compounds. The HMO solutions for [4], [6], and [8]annulenes are compared in Figure 2-1. Thus, Hückel's rule simply states that planar monocyclic hydrocarbons with alternating double and single bonds containing $(4n + 2)$ π electrons are aromatic and those with $4n$ π electrons are antiaromatic. Calculations of the resonance energies of the planar annulenes show that they generally increase as the number of π electrons increases, but this increase is much smaller as we proceed from a $(4n + 2)$ system to an adja-

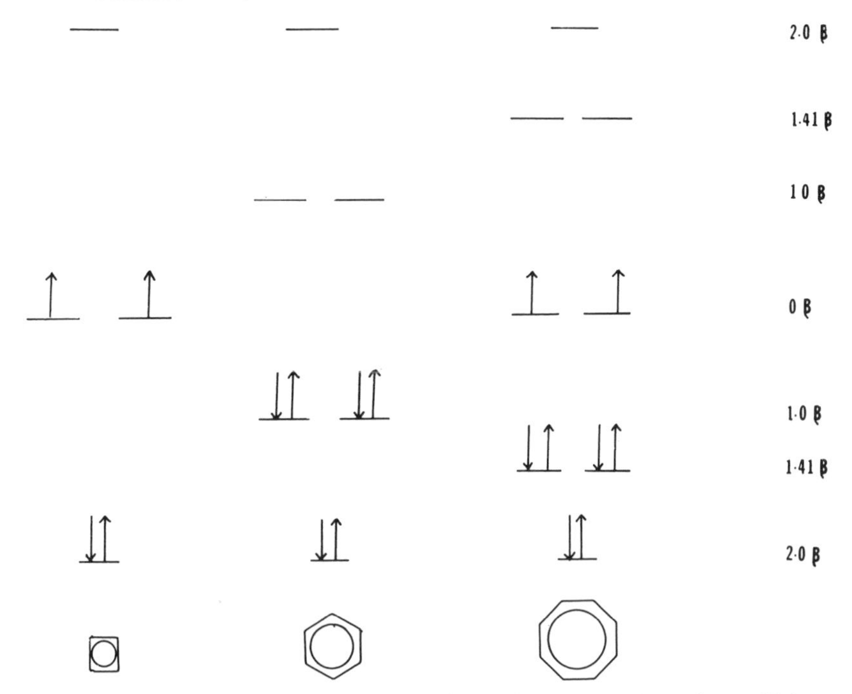

Figure 2-1. Hückel molecular orbital diagrams for (*a*) cyclobutadiene, (*b*) benzene, and (*c*) cyclooctatetraene. The relative energies for the molecular orbitals are given in units of β, and α is considered to be zero.

cent $4n$ system than from a $4n$ system to an adjacent $(4n + 2)$ π-electron system.

Since the entire concept of aromaticity exists simply because benzene is much more stable than one would expect for the hypothetical cyclohexatriene, it seems reasonable that the experimental determination of aromatic character would involve measurements of either thermodynamic or kinetic stability. This does not turn out to be a very practical empirical test for aromatic character, in view of the often fleeting and tenuous existence of some of the most interesting molecular systems.

The accepted empirical test for aromatic character involves an NMR study of the compound in question. Aromatic annulenes are considered to be those that have a diamagnetic ring current resulting in a downfield shift (relative to the resonance field for olefinic protons) of external protons and an upfield shift of internal protons. The chemical shift for the protons on benzene is 7.2 ppm, which is definitely downfield from the normal olefinic proton (5.6 ppm). The paratropic nature of antiaromatic compounds results in an upfield shift for the external protons and a downfield shift for the internal protons. Using the benzene model, in which the π electrons precess in orbitals over and below the ring, a magnetic field H' is created that opposes the applied field $H°$. The lines of force resulting from the induced field are shown below, and the effect of this induced field results in a decrease in the apparent field inside the ring and an increase in the apparent field outside the ring. Thus, external protons resonate at lower fields than protons uninfluenced by the induced field, whereas internal protons will resonate at higher fields. This model was originally proposed by Pople.[10]

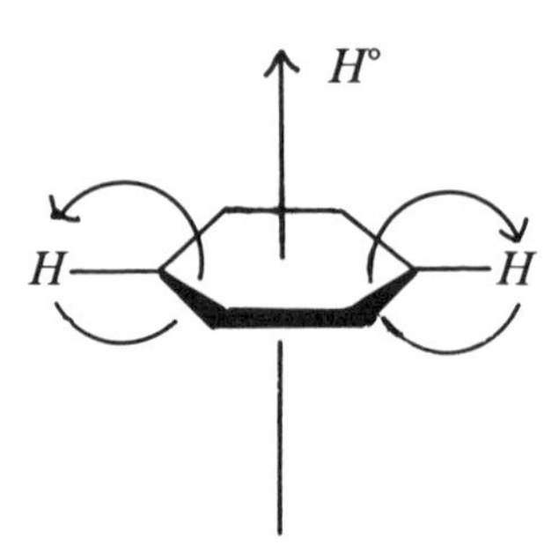

In reality it isn't quite consistent to have two independent definitions of aromaticity: one based on the number of π electrons (Hückel) and the other based on NMR data. To be fair, we should call compounds with paramagnetic ring currents paratropic and those with diamagnetic ring currents diatropic. However, "paratropic" and "diatropic" are often used interchangeably with "aromatic" and "antiaromatic" in the literature. For a comprehensive review of the NMR characteristics of the annulenes and the annulene dianions, see Reference 11. The remainder of this chapter is devoted to exploring the relationship between the structure and aromatic character in the annulenes, the homoannulenes, and their anion radicals and dianions. The considerable literature concerning dications and cation radicals is not covered here.

2. THERMODYNAMIC STABILITY OF THE ANNULENES

The main consequence of HMO theory is the concept of electron delocalization. Because of electron delocalization, aromatic compounds become stabilized relative to their nondelocalized analogues, while antiaromatic compounds become destabilized relative to their nondelocalized analogues. In other words, benzene (**5**) is more stable than the hypothetical cyclohexatriene (**6**), whereas the planar delocalized form of cyclooctatetraene (COT: **7**) is less stable than the planar, nondelocalized form (**8**).

The energy difference between **6** and **5** is called the resonance energy, and it represents the amount of stabilization imparted to **5** upon aromatization. Of course, **6** cannot be generated, so the resonance energy of benzene can be obtained only by indirect means. Heats of hydrogenation,[12] heats of combustion,[13] and heats of atomization[14] have all been used to obtain values for the resonance energy of benzene. Each of these methods yields a value for the resonance energy that is between 25 and 40 kcal/mol.

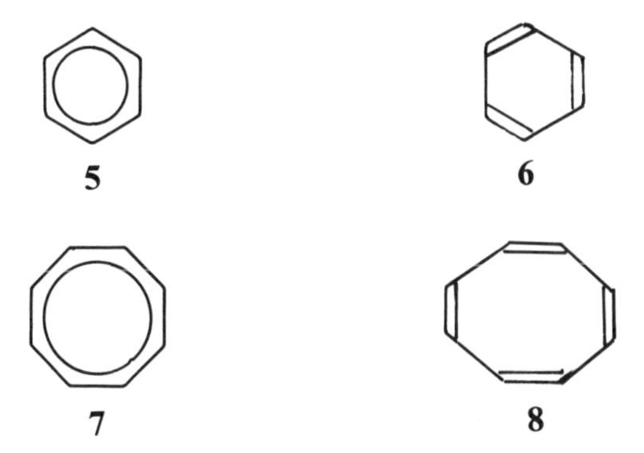

The heat of hydrogenation of benzene to form 1,3-cyclohexadiene (2-2) is +5.6 kcal/mol,[15] while that of cyclohexene to form cyclohexane (2-3) is −28.6 kcal/mol.

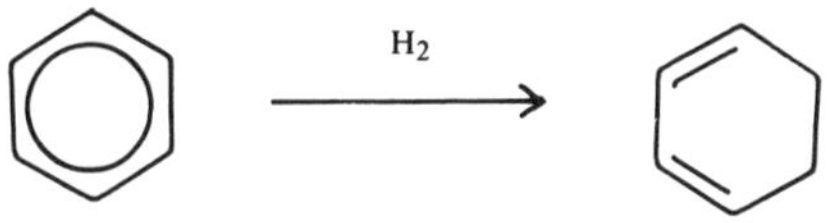

H_2

$\Delta H^\circ = +5.6$ kcal/mol (2-2)

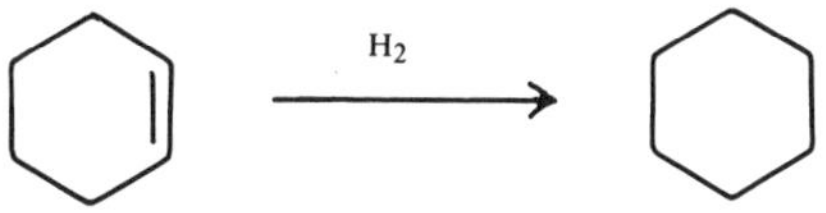

H_2

$\Delta H^\circ = -28.6$ kcal/mol (2-3)

From this, one calculates a resonance energy of −34.2 kcal/mol for benzene. We must remember that strain energies have been neglected in this calculation.

Similar, but more involved, procedures can be utilized to obtain the destabilizing or "negative" resonance energy for the symmetrical, planar, fully conjugated cyclooctatetraene. Electron diffraction and X-ray crystallographic[16,17] studies have unequivocally established that COT and its simple monosubstituted derivatives exist in the D_{2d} tub conformation (**9**) having alternate single and double bonds. Thus COT is neither aromatic nor antiaromatic but is nonaromatic.

9

This tub structure of COT allows it to undergo two dynamic processes, namely, bond shift (2-4) and ring inversion (2-5).

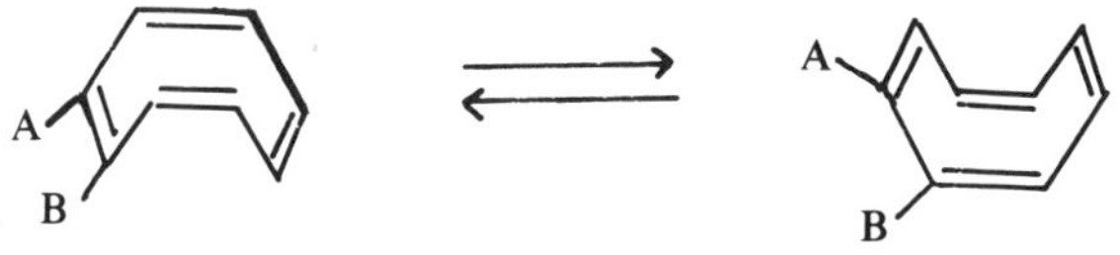

A B (2-4)

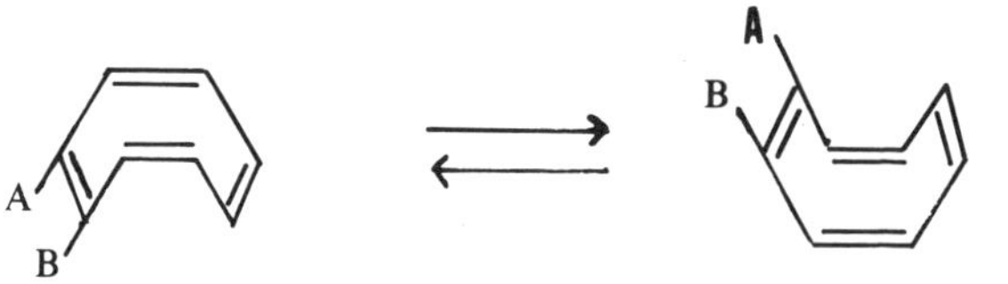

A B (2-5)

Utilizing a substituent with prochirality, Anet[18] has been able to show that both ring inversion (2-5) and bond shift (2-4) can be detected via NMR methods and that monosubstitution of the ring has little effect on the barriers to these processes. Based on the NMR studies of Anet and others,[19,20] it is clear that the free energy of activation for ring inversion (2-5) is less than that for bond shift (2-4) by about 2.4 kcal/mol, $\Delta G\ddagger = 12.5\text{–}14.8$ kcal/mol for ring inversion and $\Delta G\ddagger = 14.9\text{–}17.4$ kcal/mol for bond shift.[21] Utilizing a variety of substituted COT systems, Paquette and co-workers[22] have shown that the best value for the difference in the enthalpies of activation for bond shift and ring inversion in the unsubstituted [8]annulene is 3.5–4 kcal/mol. This is the best value for the difference in the energies of **7** and **8** and of the "negative" resonance energy of planar symmetrical [8]annulene.

Since [8]annulene (**9**) consists simply of a benzene ring expanded by one ethylene group with elimination of the aromaticity, the resonance energy of benzene can be estimated by comparing the heat of combustion of [8]annulene minus one ethylene with that of benzene.[23] This method avoids some of the problems of strain energies and hydrogenation of hypothetical molecules, since both COT in its tub conformation and benzene are almost free of bond angle strain. The heats of combustion of [6] and [8]annulenes (Table 2-1), are given by Equations 2-6 and 2-7, respectively, where all materials are in their standard states.

$$C_6H_6(l) + {}^{15}\!/_2O_2(g) \rightarrow 6CO_2(g) + 3H_2O(l) \quad \Delta H^\circ_6 = -780.9 \text{ kcal/mol} \qquad (2\text{-}6)$$

$$C_8H_8(l) + 10\,O_2(g) \rightarrow 8CO_2(g) + 4H_2O(l) \quad \Delta H^\circ_7 = -1086 \text{ kcal/mol} \qquad (2\text{-}7)$$

Since [8]annulene is effectively unconjugated and contains four ethylene groups, the enthalpy of combustion of three ethylene functions would be $\Delta H^\circ_8 = 0.75\,\Delta H^\circ_7$.

$$C_6H_6(\text{unconjugated})(l) + {}^{15}\!/_2O_2(g) \rightarrow 6CO_2(g) + 3H_2O(l) \quad \Delta H^\circ_8 = -814.5 \text{ kcal/mol} \qquad (2\text{-}8)$$

The difference between ΔH°_6 and ΔH°_8 represents the resonance energy of the benzene ring, assuming the resonance energy of the cyclooctatetraene ring system is zero and that ΔH°_v is constant per carbon. Actually the COT ring system has about 4 kcal/mol of resonance energy.[24] Thus, our value is

TABLE 2-1. Heats of Combustion, ΔH_c°, and Heats of Hydrogenation, $\Delta H_h^{\circ a}$

Compound	Heats (kcal/mol)	
	ΔH_c°	ΔH_h°
[6]Annulene (l)	−780.97	+5.6
[8]Annulene (l)	−1086.5	−25.6
[16]Annulene (s)	−2182	−28
[18]Annulene (s)	−2347	
Cyclooctane (l)	−1258.4	
Cyclohexadecane (l)	−2501.4	
Cyclooctene (l)		−23.5

[a]Data from Reference 23.

too low by 0.75 × 4 kcal/mol. This comparison of benzene to [8]annulene yields a resonance energy for benzene of 814.5 − 780.9 + 3 = 36.6 kcal/mol. For an energy diagram of the C_6H_6–COT system, see Reference 11.

Since the Reppe synthesis of COT in 1948,[25] the first annulene containing $4n$ π electrons to be synthesized is [16]annulene.[26] Unlike [8]annulene, which exists in a tub conformation, [16]annulene has been shown to exist as a dynamic equilibrium between two interconverting configurations (2-9), each having double and single bond alternation.[27]

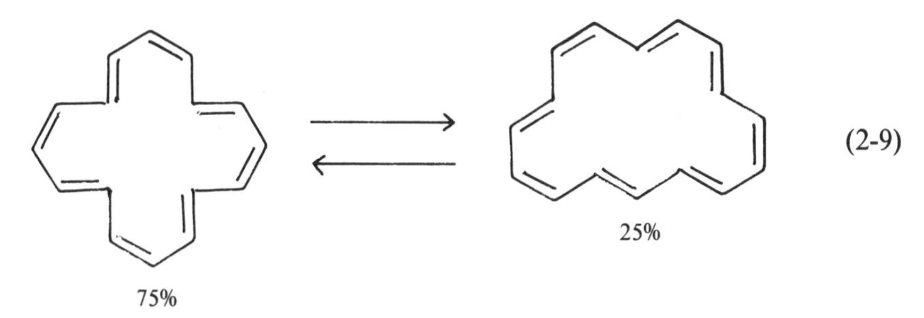

(2-9)

It is apparent from the fact that the heat of combustion of [16]annulene is very close to twice that of [8]annulene (Table 2-1) that [16]annulene does not possess any more extra stabilization due to π-electron delocalization than does COT. In fact, the heat of combustion of [16]annulene is about 9 kcal/mol more exothermic than that for two [8]annulenes (2-10).

2 (l) ⟶ (s) $\Delta H^\circ = +9$ kcal/mol (2-10)

We do not know the heat of vaporization of [16]annulene, but it is a hydrocarbon, and there is no reason to suspect that it is very different from that for two [8]annulenes. Thus, [16]annulene is close to being thermoneutral with two [8]annulenes even when an estimate of 4 kcal/mol for the heat of sublimation (eg, ΔH_s, diphenylmethane) of [16]annulene is included.

A similar comparison can be made between [6]annulene and [18]annulene. Both are aromatic in the Hückel sense [they are $(4n + 2)$ π-electron systems], and both exhibit a diatropic ring current. The NMR spectrum of [18]annulene at −60°C exhibits a multiplet at 10.72 ppm due to the 12 external protons and a multiplet at −3.99 ppm due to the six internal protons.[28] The internal and external protons are rapidly (on the NMR time scale) interchanged at higher temperatures. Only a single average line can be observed at +110°C. The heat of combustion of [18]annulene is just 4 kcal/mol more exothermic than three times that for [6]annulene,[29] indicating that the diatropic ring current observed in [18]annulene does indeed lead to stabilization of the system (2-11). If one also assumes about 4 kcal/mol for the sublimation enthalpy of [18]annulene and assumes equal ΔH_v per carbon, then Equation 2-11 is endothermic by 8 kcal/mol for all-gaseous reactants and products.

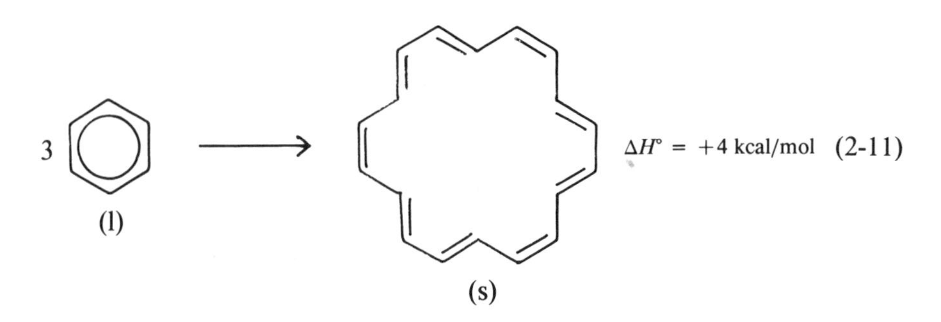

A comparison of the heat of combustion data reported in Table 2-1 shows that for the four annulenes discussed above, the diatropic ring current does indeed lead to thermodynamic stabilization due to aromatization. The enthalpy of reaction 2-11 coupled with the accepted value for the resonance energy of benzene suggests that the stabilization due to aromaticity of [18]annulene is about 100 kcal/mol.[30] The "negative" resonance energy for [8]annulene, evaluated from kinetic data for ring inversion and bond shift, appears to result in a destabilization by about 4 kcal/mol. For philosophical reasons, it has been thought that the "negative" resonance energy of **7** should be comparable to the "positive" resonance energy of benzene.[31] Indeed, calculations by modified intermediate neglect of differential overlap (MINDO)[31] suggest that the difference in energy between **7** and **9** is much larger (~14 kcal/mol) than that measured by Anet[18] and Paquette.[22] Similar quantum mechanical opposition has surfaced concerning the resonance energy of [18]annulene. Semiempirical ASMO-CI calculations indicate that [18]annulene is distorted from planarity and has much less resonance energy than three times that of benzene, or about 100 kcal/mol.[32] In fact Van-Catledge and Allinger[32] suggest that dynamic bond alternation exists in the [18]annulene system. This bond alternation hypothesis is in contrast with the X-ray analysis,[33] which shows 12 inner bonds of mean length 1.382 Å and six outer bonds of mean length 1.419 Å, with deviations from planarity of 0.085 Å. Thus, it was suggested that the isolated molecule is planar.[33]

The actual resonance energies of both [18]annulene and planar symmetrical [8]annulene remain controversial. However, for both cases the workers carrying out the experiments generally accept the analysis detailed above; the controversy arises from those carrying out computer chemistry only.

3. ANNULENE DIANIONS

Chemists have known for some time that it is easier to add extra electrons to $4n$ π-electron annulenes than to aromatic systems. The dianions of both [8]annulene[34] and [16]annulene[35] are readily generated via alkali metal reduction of the neutral molecular systems. However, the dianion of benzene [a $(4n + 2)$ π-electron annulene] is still unknown. The dianion of fused polyaromatic systems can be generated, but with more difficulty than those for the $4n$ π-electron annulenes.[36] The reduction potentials of the $4n$ π-electron annulenes are less negative than those of the aromatic systems because the addition of the two electrons leads to an aromatic annulene in the $4n$ π-electron systems and to a divergence from aromatic character in the aromatic systems.

Treatment of COT with alkali metal in tetrahydrofuran (THF),[34] hexamethylphosphoramide,[37] or liquid ammonia[37] leads to the formation of the dianion, which exhibits a single ^{1}H NMR signal at 5.6 ppm. The effect of the diamagnetic ring current in the dianion is canceled by the increased shielding due to the added electrons; thus the dianion's NMR resonance is in the same place as that for the neutral molecule. However, the chemical shift for the dianion does vary slightly because of ion association with the alkali metal cation.[36,37] Solvent-free solid salts of the alkali metal[38] and alkaline earth metal[39] COT dianions can be isolated and stored, without decomposition, for many months at room temperature. However, these salts do explode upon contact with moist air.

The dianion of [16]annulene can be generated in an identical manner to that of COT. In THF this aromatic dianion yields NMR signals at 9.83, 7.45, and −8.17 ppm.[35] The signal that is on the highfield side of the TMS signal is due to the internal protons. Both the solvated dianions of [16]annulene and [8]annulene will comproportionate with a small amount of the respective neutral molecule to generate the planar anion radicals (2-12 and 2-13).

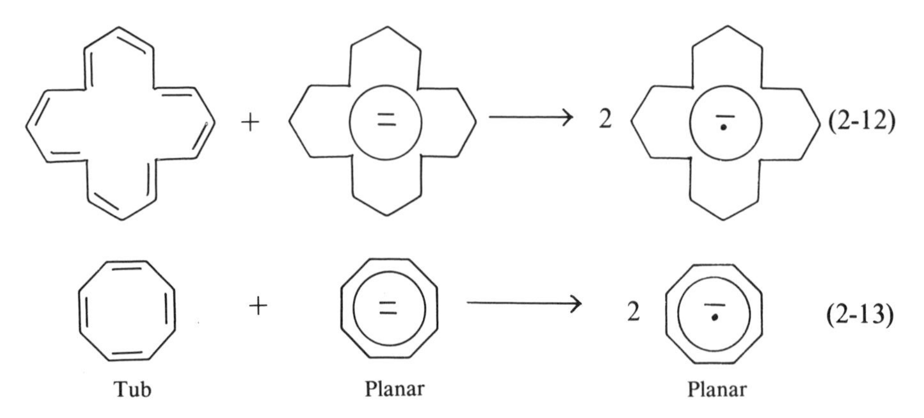

Neutral [12]annulene has been observed only at low temperature, as it rearranges irreversibly to its bicyclic tautomer at −40°C (2-14).[40] The unstable annulene can be reduced either polarographically or via lithium metal in THF to give the dianion (2-14).[41] This dianion is kinetically stable, and its ^{1}H NMR signal remains unaltered at temperatures between −90 and +30°C (NMR signals for the external protons at 6.98 and 6.23, and −4.6 ppm for the internal protons). The anion radical of [12]annulene is also known, but unlike the dianions generated from the other two $4n$ π-electron annulenes, it is not planar.[42] This nonplanarity results from the close proximity of the three internal protons.

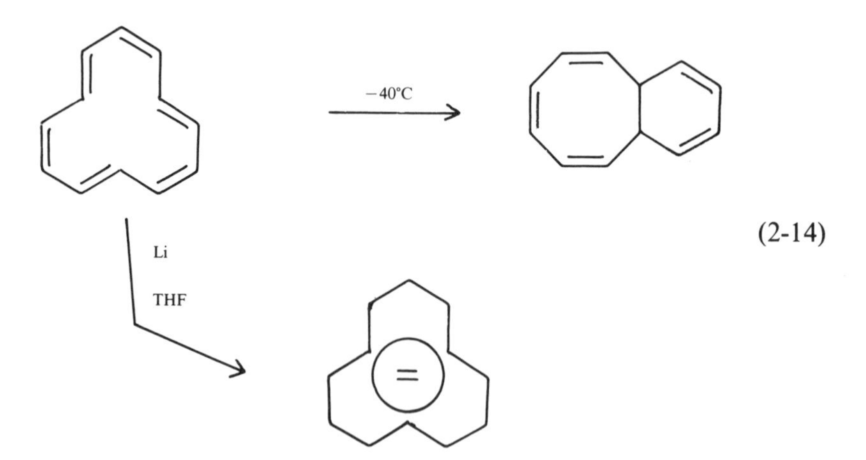

Only one dianion generated from a $(4n + 2)$ π-electron dianion is known. [18]Annulene will accept two electrons, over a period of days, to form its dianion in THF. This dianion exhibits a rather dramatic paramagnetic ring current effect upon NMR analysis. Signals for the internal protons are found at 29.5 and 28.1 ppm.[43] The external protons are at resonance at −1.13 ppm. The NMR spectra for both the [18]annulene neutral molecule (10.43 ppm for the external protons and 5.4 ppm for the internal protons)[44] and dianion are interpreted in terms of two structures in dynamic equilibrium.

It might seem surprising that these annulene dianions can be generated at all, especially for the smaller systems. In the COT dianion the electron–electron repulsion energy exhibits a destabilizing effect of 106 kcal/mol.[36] Clearly the stabilizing effect due to the aromatic character of the dianion is not sufficient to overcome this electron–electron repulsion. That this is the case is evidenced by the fact that no annulene dianion has ever been generated in the gas phase. In the condensed phases, solvation or crystal lattice effects couple with the stabilization due to aromatic character to render these annulene dianions thermodynamically stable.

4. THERMODYNAMICS OF THE [8]ANNULENE DIANION

As mentioned earlier, benzene will not accept two more electrons under any experimental conditions yet realized. This is unfortunate, because the comparison of the heat of formation of the benzene dianion to that of the benzene neutral molecule would afford the "perfect" measurement of the difference in stability between aromatic and antiaromatic systems. The only difference between the benzene dianion and benzene is the number of π electrons, and it is this factor that the Hückel definition of aromaticity is all about.

The experimental comparison of the heats of formation of the COT neutral molecule and dianion is possible, since the heat of formation of the neutral molecule is known, and the alkali metal salts of the dianion can be isolated and purified. The most common method for obtaining the heat of formation of any compound involves the measurement of the heat of combustion. This method requires very high precision to give useful results, and this high precision can be obtained only under very special conditions for organometallic compounds.[45] It is much more feasible to measure a heat of reaction involving the alkali metal salts of the COT dianion.

The heat of reaction of the disodium and dipotassium salts with water to yield aqueous alkali metal hydroxide and liquid cyclooctatriene are -56.1 ± 0.7 and -40.5 ± 1.4 kcal/mol respectively (2-15).[38]

$$M_2^+ [C_8H_8^{2-}](s) + 2H_2O(l) \rightarrow 2MOH(aq) + C_8H_{10}(l) \qquad (2\text{-}15)$$

$$\Delta H^\circ = -40.5 \text{ kcal/mol for M} = \text{K}$$
$$\Delta H^\circ = -56.1 \text{ kcal/mol for M} = \text{Na}$$

These enthalpies can be placed into a simple thermochemical cycle[38] to yield the heat of formation (from the metal and COT in their standard states) of the solid [8]annulene dianion salt (2-16).

$$2M(s) + [\text{COT}](l) \rightarrow M_2^+ [\text{COT}^{=}](s) \quad (2\text{-}16)$$

$\Delta H° = -79.5 \pm 2.0$ kcal/mol for M = K
$\Delta H° = -57.7 \pm 1.5$ kcal/mol for M = Na

Similar studies were carried out on the THF-solvated dianion. The enthalpy of reaction of the solvated ion pair with water of -29.4 ± 1.1 kcal/mol yields a heat of reaction of potassium with COT in THF (heat of formation of the solvated salt from the COT and potassium in THF) of -90.2 ± 2.0 kcal/mol.[45]

$$2K(s) + [\text{COT}](l) \rightarrow [\text{COT}^{=}], 2K^+(\text{THF}) \quad (2\text{-}17)$$

$\Delta H° = -90.2 \pm 2$ kcal/mol

Note that COT accepts an electron from potassium metal to form either the solvated or solvent-free dianion salt exothermically. In fact the reaction between [8]annulene and potassium metal in THF is about as exothermic as the reaction between potassium metal and water, which has an enthalpy of -94.4 kcal/mol. Clearly part of the driving force behind this reaction is the formation of a planar, $(4n + 2)$ π-electron, aromatic dianion from the tub-shaped, $4n$ π-electron, nonaromatic, neutral COT. The difference in energy between the planar, symmetrical, antiaromatic COT and tub COT is about 16 kcal/mol (see Section 1); thus the heat of reaction of antiaromatic COT with potassium to form the solvated dianion salt is exothermic by 106 kcal/mol.

$$[\text{planar COT}](l) + 2K(s) \rightarrow [\text{COT}^{=}], 2K^+(\text{THF}) \quad (2\text{-}18)$$

$\Delta H° = -106$ kcal/mol

For the formation of the solid salts, these enthalpies are as follows.

$$[\text{planar COT}](l) + 2M(s) \rightarrow M_2^+ [\text{COT}^{=}](s) \quad (2\text{-}19)$$

$\Delta H° = -96$ kcal/mol for M = K
$\Delta H° = -74$ kcal/mol for M = Na

Aromaticity certainly accounts for part of the very exothermic nature of these reactions, given that the benzene dianion cannot even be formed. However, several other factors (heats of solvation or lattice energies, metal ionization potentials, etc) contribute significantly to these enthalpies. To gain some understanding of these "other factors," we need to compare the enthalpies of reactions 2-18 and 2-19 with similar values for analogous gas phase reactions. Fortunately these gas phase enthalpies can be evaluated. This is the topic of the next section.

5. DIANIONS IN THE GAS PHASE

As mentioned earlier, gas phase dianions are unknown. However, the energetics of putting two extra electrons into the [8]annulene moiety is relatively simple, and the energy difference between the gas phase neutral and dianion systems can be accurately evaluated from a simple semiempirical calculation.

Since the energy difference between the planar, symmetrical, antiaromatic COT and the tub conformation of COT is 16 kcal/mol, the electron affinity (EA) of the antiaromatic COT is equal to 16 plus the EA of the ground state COT (13 kcal/mol).[65] The antiaromatic form of COT accepts an electron in the gas phase exothermically by 29 kcal/mol.[36]

$$\text{COT (g)} + e^-\text{(g)} \rightarrow \text{COT}^{\overset{-}{\cdot}}\text{ (g)} \qquad (2\text{-}20)$$

$$\Delta H^\circ = -29 \text{ kcal/mol}$$

In the Hückel approximation a second added electron must go into the same molecular orbital as did the first extra electron. Thus, the second EA of COT is simply 29 kcal/mol minus the electron–electron repulsion energy (E_{rep}). The electron–electron repulsion energy simply represents the work done to overcome the Coulombic repulsive force as a second electron is forced into the anion radical. To carry out a calculation of the second E_{rep}, all one needs to know is the electron distribution in the anion radical, which we have from the electron paramagnetic resonance (EPR) coupling constants. Dewar[31] has established that the value for E_{rep} in the COT dianion is 106 kcal/mol. From this it is clear that it would take $106 - 2 \times 29$ kcal/mol to add two electrons to COT to form the aromatic gas phase dianion (2-21).

$$\text{COT (g)} + 2e^-\text{(g)} \rightarrow \text{COT}^{=}\text{ (g)} \qquad (2\text{-}21)$$

$$\Delta H^\circ = +48 \text{ kcal/mol}$$

The addition of two electrons to benzene can be visualized in a similar manner. The electron affinity of benzene is −26.6 kcal/mol[46,66] and E_{rep} is 162 kcal/mol.[36] This means that the addition of two gas phase electrons to benzene is endothermic by 162 − 2(−26.6) kcal/mol (2-22).

$$\text{C}_6\text{H}_6\,(g) + 2e^-(g) \rightarrow \text{C}_6\text{H}_6^{=}\,(g) \qquad (2\text{-}22)$$

$$\Delta H^\circ = +215.2 \text{ kcal/mol}$$

The ΔH° of 215.2 kcal/mol simply represents the sum of the first two EA's ($EA_1 + EA_2$) of benzene. The enthalpy for the transfer of two electrons from the antiaromatic benzene dianion to the antiaromatic COT is simply the difference between the enthalpies for reactions 2-21 and 2-22 (2-23).

$$\text{C}_6\text{H}_6^{=}\,(g) + \text{COT}\,(g) \rightarrow \text{C}_6\text{H}_6\,(g) + \text{COT}^{=}\,(g) \qquad (2\text{-}23)$$

$$\Delta H^\circ = 48 - 215.2 = -167.2 \text{ kcal/mol}$$

This is a very exothermic gas phase electron transfer, but aromaticity and antiaromaticity are not the only driving forces, since E_{rep} is 56 kcal/mol larger in the benzene dianion than it is in the COT dianion. This means that the concepts of aromaticity and antiaromaticity account for 167.2 − 56 = 111.2 kcal/mol of the enthalpy of generation of two aromatic systems from two antiaromatic systems. The enthalpy of generation of one aromatic system from one antiaromatic must be half this value, or about 56 kcal/mol. If one assumes that the destabilization due to antiaromaticity is the same in magnitude but opposite in sign as the stabilization due to aromatic character, then half of 56 (28 kcal/mol) should be close to the value for a change from a nonaromatic system to an aromatic one. This is fairly close to the values obtained from heats of hydrogenation (34.2 kcal/mol) and heats of combustion (36.6 kcal/mol); see Equations 2-2, 2-3, and 2-6–2-8. Finally, this argument is identical with the following statement: the average stabilization due to aromatic character in benzene and the COT dianion is simply the average of the absolute values of their respective EA's, or (29 + 26.6)/2 = 28.

6. DIANIONS IN THE SOLID PHASE

The dianion of benzene is unknown. However, dianions of most of the polyacenes have been generated, and many of them are stable in the solvent-free solid state.[36] This is true even though the addition of two electrons to any of the polyacene systems results in a divergence from aromatic character.

The polyacene dianion sodium salts can be reacted with water to yield the dihydropolyacene and aqueous sodium hydroxide in a calorimeter just as described for the COT dianion systems (2-24).[36]

$$Na_2^+Ar^{2-}(s) + 2H_2O(l) \rightarrow ArH_2(s) + 2NaOH(aq) \qquad (2\text{-}24)$$

The heats of reaction 2-24, where Ar is anthracene (AN), tetracene (TE), pentacene (PT), pyrene (PY), perylene (PE), and benz(a)pyrene (BPY) can be treated in the same manner as was the heat of reaction of the COT dianion with water (reaction 2-15) to yield the heats of formation of the solid polyacene dianion salts from sodium metal and the polyacene (Table 2-2).

It is important to note that this enthalpy of formation from the hydrocarbon and sodium metal in their standard states is more negative for the COT system than for any of the polyacenes, even for systems containing five fused rings. This is partly because COT^{2-} is aromatic and the addition of electrons to the polyacenes results in a divergence from aromatic character. The electron–electron repulsion energy in the tetracene dianion is almost identical to that in the COT dianion (see Table 2-3), yet the transfer of two electrons from TE^{2-} to COT is quite exothermic even though the COT ring must be flattened upon the addition of the two electrons (2-25).

$$Na_2^+TE^{2-}(s) + COT(l) \rightarrow Na_2^+COT^{2-}(s) + TE(s) \qquad (2\text{-}25)$$

$$\Delta H^\circ = -32.6 \text{ kcal/mol}$$

Crystal lattice energies are an important contribution to the energetics of this reaction, but aromaticity considerations account for the negative character of ΔH°. Clearly the most important contribution to the thermodynamic stability of all these dianion salts is the crystal lattice energy. Crystal lattice energies can be obtained by simply combining these heats of formation with the heats of vaporization of Ar and Na, the ionization potential of Na (118.4 kcal/mol), and the energy necessary to add two electrons to Ar in the gas phase.[36] This latter enthalpy represents the heat of reaction (2-26).

$$Ar(g) + 2e^-(g) \rightarrow Ar^{2-}(g) \qquad (2\text{-}26)$$

The ΔH° for reaction 2-26 is the sum of the first and second electron affinities ($EA + EA_2$), which is simply $E_{rep} - 2EA$. The resulting crystal lattice energies U° are shown in Table 2-3. From Table 2-3 it is clear that all the dianion salts have about the same crystal lattice energy. The average and standard deviation of the crystal lattice energies are 420 ± 18 kcal/mol. It is remarkable that the standard deviation in the crystal lattice energies of all the systems is less than 3%. The consistency of these lattice energies means that U° for $Na^+Bz^{2-}(s)$ must be very close to 420 kcal/mol. Using this value, we can now back-calculate the heat of formation of the benzene dianion salt from benzene and sodium metal (2-27).

$$Bz(l) + 2Na(s) \rightarrow Na_2^+Bz^{2-}(s) \qquad (2\text{-}27)$$

$$\Delta H^\circ = +92 \pm 18 \text{ kcal/mol}$$

TABLE 2-2. Enthalpies of Reaction

Reaction[a]	ΔH° (kcal/mol) for Ar =						
	AN	TE	PT	PY	PE	BPY	COT
$ArH_2 + 2NaOH(aq) \rightarrow Na_2^+Ar^{2-} + 2H_2O$	79.1	75.6	68.7	43.2	66.8	70.9	56.1
$2Na + 2H_2O \rightarrow 2NaOH(aq) + H_2$	−88.2	−88.2	−88.2	−88.2	−88.2	−88.2	−88.2
$Ar + H_2 \rightarrow ArH_2$	−17.0	−12.5	−15.0	−2.7	−15.0	−15.0	−25.6
$2Na + Ar \rightarrow Na_2^+Ar^{2-}$	−26.1	−25.1	−34.5	−47.7	−36.4	−32.3	−57.7

[a]All materials are in their standard states unless otherwise indicated.

Table 2-3. E_{rep}, EA,[a] and Crystal Lattice Energies U° for Sodium Dianion Salts

Hydrocarbon	Enthalpies (kcal/mol) E_{rep}	EA	EA + EA_2	U° (for Na^+Ar^{2-})
AN	117	12.7	91.6	430
TE	107	16.0	75.0	419
PT	99	19.0	61.0	419
PY	108	13.3	81.4	440
PE	100	21.1	57.8	413
BPY	101	19.1	62.8	412
COT (antiaromatic)	106	29	48.0[b]	405
Benzene	162	−26.6	215.2[c]	—

[a] Enthalpy for the addition of two electrons (reaction 2-26) to gas phase hydrocarbons.
[b] See reaction 2-21.
[c] See reaction 2-22.

Equations 2-28 and 2-29 show the transfer of two electrons and two cations from the solid benzene dianion salt both to COT and to the antiaromatic form of COT.

$$Na_2^+Bz^{2-} + COT(l) \rightarrow Na_2^+COT^{2-} + Bz(l) \qquad (2\text{-}28)$$

$$\Delta H^{\circ} = -150 \text{ kcal/mol}$$

$$Na_2^+ \text{(benzene dianion)}(s) + \text{(COT)}(l) \rightarrow Na_2^+ \text{(COT dianion)}(s) + \text{(benzene)}(l) \qquad (2\text{-}29)$$

$$\Delta H^{\circ} = -150 - 16 = -166 \text{ kcal/mol}$$

The heat for reaction 2-29 differs from that for reaction 2-23 only because of the uncertainty in the crystal lattice energy of $Na_2^+Bz^{2-}$.

7. STABILITIES OF HOMOAROMATICS AND THEIR CATIONS

To account for the high stability of molecules in which conjugation is interrupted in one or more places by aliphatic groups, the concept of aromatic character was extended to include homoaromaticity.[47] The essential feature of homoaromatic character is the union of an unsaturated linear segment with a cyclopropane moiety as shown in structure **10**.

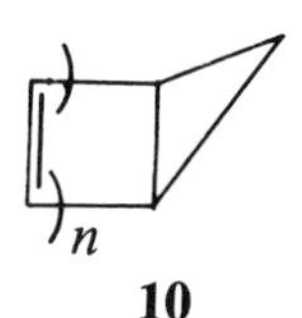

10

The first suggestion of homoaromaticity was advanced by Thiele to explain the decreased acidity of the methylene protons in cycloheptatriene (2-30) relative to those of cyclopentadiene.[11]

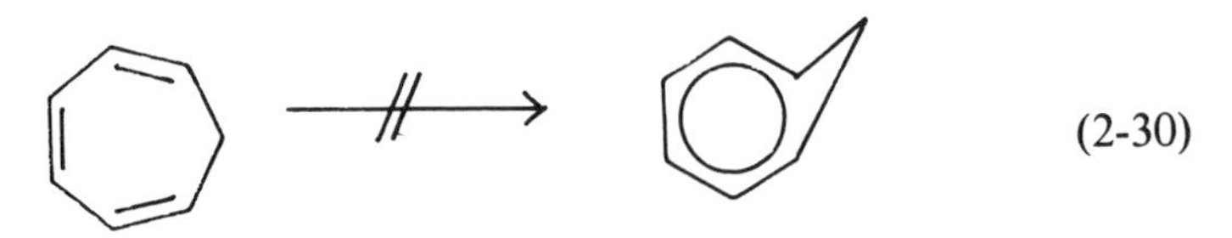

(2-30)

This explanation is, of course, incorrect, but homoaromaticity has turned out to be very important in controlling the chemistry of a number of charged organic systems.[49] The large difference in the chemical shift for the exomethylene versus endomethylene protons ($\delta H_{\text{ext}} - \delta H_{\text{int}} = \Delta\delta = 5.8$ ppm) in the homotropylium cation **(11)** is interpreted in terms of the delocalization of the two electrons in the fusion bond into a 6π-electron, homoaromatic molecule. Thus, **11a** is a better interpretation than **11b.**[50]

11a **11b**

The bicyclo[3.1.0]hexenyl cation **(12)** has been generated and shown by Berson[51a] to have a slight negative difference in the chemical shifts of the exo and endo protons. The value of -0.3 ppm suggests a small paramagnetic ring current.

12

The lowest homologue of the series (the homocyclopropenyl cation) was generated by Olah and co-workers.[51b] The 270-MHz ^{1}H NMR spectra[51] show temperature dependence, and at -110°C, the puckered homoaromatic ion was "frozen out" ($\delta\Delta = 0.82$ ppm). NMR line shape analysis yielded a free energy of activation ($\Delta G^{\ddagger}$) of 8.4 kcal/mol for the ring-flipping process (2-31).

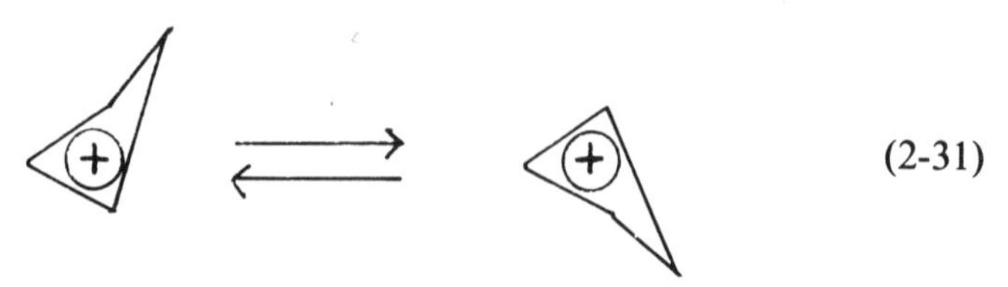

(2-31)

This suggests that the difference in the energies of the planar (with 1,3 π overlap) and the nonplanar species is 8.4 kcal/mol. This can be compared to the analogous ring flipping for the homotropylium cation, which has a $\Delta G^{\ddagger}$ of 22.3 kcal/mol (2-32).

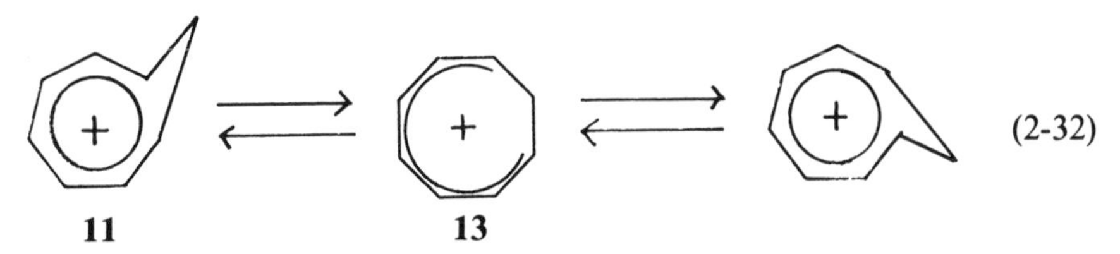

The difference in energy between **13** and **11** is the reason that structure **13** has never been observed, regardless of the substitution pattern.

Vogel and co-workers[52] have found a $\Delta\delta$ value of 1.7 ppm in the norcaradiene system **(14).** However, the degree of homoaromaticity is diminished in the neutral systems relative to the cationic systems.

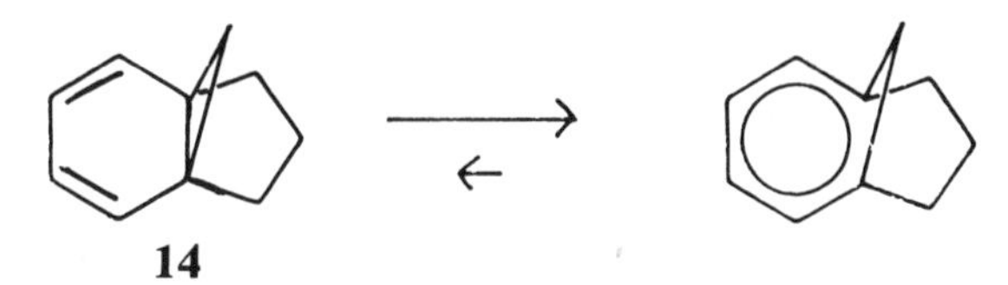

Masamune successfully isolated isomers **15** and **16** of [10]annulene at $-80°C$, thus ending the long search for these fugitive compounds.[53] Both these isomers undergo ring closure upon warming, so that the planarity necessary to allow π-electron delocalization can be achieved. Isomer **15** suffers from severe bond angle (144°) strain, whereas **16** and **17** suffer from nonbonded steric repulsion of the internal protons. After the work of Masamune, it remained clear that both **15** and **16** are highly reactive, nonplanar, olefinic systems with no aromatic stabilization.

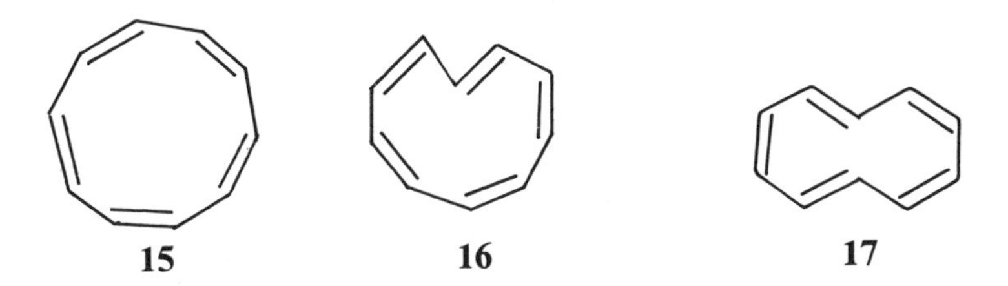

The supposition that nonbonded interactions must negate any gain in energy attained by π-electron delocalization has been supported by the synthesis of 1,6-methano[10]annulene **(18)** by Vogel and Roth in 1964.[54] The problem of steric nonbonded interactions was eliminated by the introduction of a methylene bridge across the 1,6 positions. X-Ray diffraction work has revealed the near-planar geometry of **18.** The C—C bond lengths in the

TABLE 2-4. Heats of Reaction

Reaction	ΔH° (kcal/mol)	Ref.
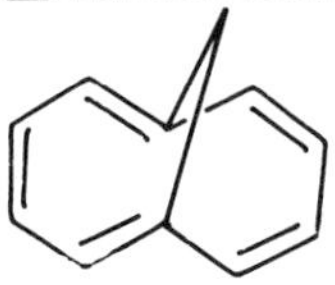 [structure] $+ \frac{27}{2} O_2 \rightarrow 11CO_2 + 5H_2O$	−1436.9	58
$11CO_2 + 7H_2O \rightarrow \frac{29}{2} O_2 +$ 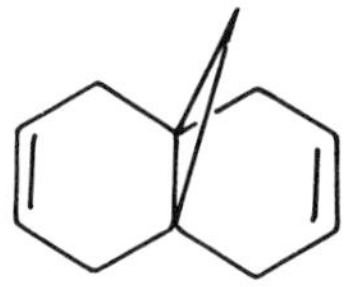 [structure]	+1525.9	55
$2H_2 + O_2 \rightarrow 2H_2O$	−136.6	
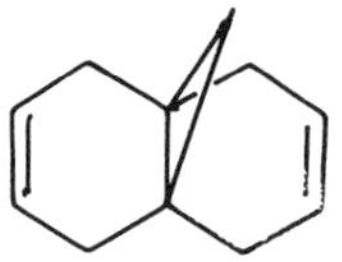 [structure] $\rightarrow 2H_2 +$ [structure] 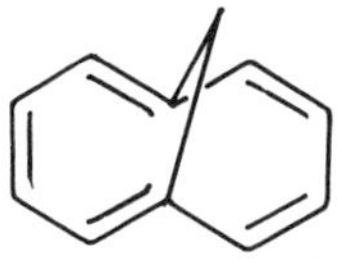	+53.3	55
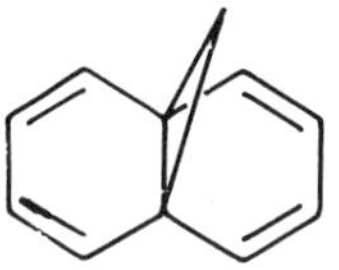 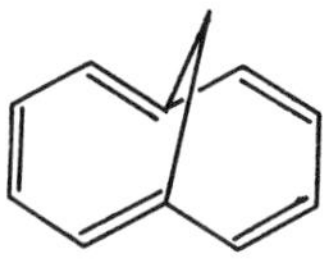 [structure] $\rightarrow$ [structure]	+5.7	

periphery of the ring exhibit no significant alternation and lie in the area of 1.40 Å.[55] The ^1H NMR resonances at 7.2 ppm for the ring protons and −0.5 ppm for the methylene protons are very indicative of a diatropic ring current.[54] Certainly the X-ray and NMR evidence indicates that the 1,6 σ bond is not intact, and the equilibrium constant for the tautomerization to dinorcaradiene (2-33) is very small.

18 $\rightleftharpoons$ (2-33)

The heat of combustion of both **18** and dihydro-**18** have been measured,[56,57] and these can be combined in a thermochemical cycle to determine the enthalpy for reaction 2-33 (Table 2-4). However, the necessary thermochemical cycle requires the heat of hydrogenation of dinorcaradiene. The heat of hydrogenation of cyclohexadiene needed to yield cyclohexene is −26.65 kcal/mol, and the additions of two hydrogens to dinorcaradiene must have an enthalpy that is very close to twice this value. From Table 2-4, the enthalpy for reaction 2-33 is about 5.7 kcal/mol, which leads to an equilibrium constant of about 10^{-4} for reaction 2-33. It should be pointed out that the full [10]annulene ring structure in this 10-membered ring system is not a necessary requirement to keep the three-membered ring open, as evidenced by the fact that dihydromethano[10]annulene also exists in the open form **(19)**.[55]

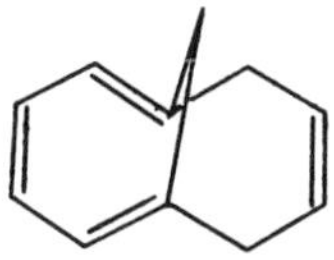

19

Before changing the topic to homoaromatic anions, it should be mentioned that decamethyl[5]pericyclyne **(20)** has been synthesized recently.[56] The authors do not use the term "aromatic" in their paper. However, the spectrum of this novel compound obtained by photoelectron spectroscopy clearly indicates that it exists in a cyclic "homoconjugated" form.

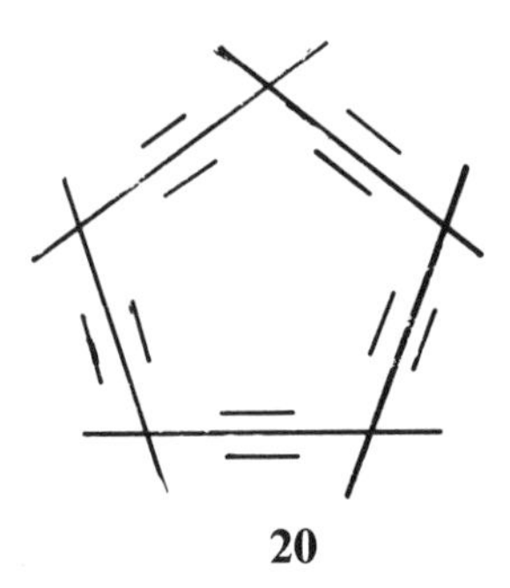

20

8. STABILITIES OF A HOMOAROMATIC DIANION AND A HOMOANTIAROMATIC NEUTRAL MOLECULE

One of the most notable anion homoaromatic systems is the homo[8]annulene dianion ($HCOT^{2-}$).[57] The chemical reduction of *cis*-bicyclo[6.1.0]nona-2,4,6-triene (CBN) proceeds through the 9π-electron homoconjugated anion radical ($HCOT^{-\cdot}$) to the delocalized (homoaromatic) monohomocyclooctatetraene dianion (2-34).[57]

e^- e^- (2-34)

In contrast to the dianion of COT, which can be stored at room temperature in THF for years, $HCOT^{2-}$ will not endure in this solvent except at low temperatures. Indeed, it has been shown that $HCOT^{2-}$ is a much stronger base than COT^{2-}, due to the former's inability to delocalize charge as efficiently as does the planar π analogue.[57b] The kinetic instability of $HCOT^{2-}$ in THF is due to its ability to deprotonate the solvent.[57c]

The $E_{1/2}$'s of COT and CBN in THF with 0.2 M tetra-*n*-butylammonium perchlorate are -1.96 and -2.55 V, respectively.[59] The difference between these two corresponds to 13.5 kcal/mol, but we have to keep in mind that this value is affected by ion association with the tetra-*n*-butylammonium ion and by entropy effects.

Recently mixtures of CBN and the anion radical of anthracene ($AN^{-\cdot}$) have been equilibrated in dimethoxyethane (DME) with respect to electron transfer (2-35).[60]

$$AN^{-\cdot}, K^+ + CBN \rightarrow CBN^{-\cdot}, K^+ + AN \qquad (2\text{-}35)$$

$$\Delta H^\circ = +3.9 \text{ kcal/mol}$$

The enthalpy of reaction 2-35 can be placed into a thermochemical cycle to obtain the electron affinity of CBN.[60] The only assumption utilized in the calculation of the EA from the enthalpy of reaction 2-35 is that the heat of solvation of $HCOT^{-\cdot} + K^+$ is -168 kcal/mol. However, it has been shown that the enthalpy of solvation of gas phase anion radicals plus the gas phase alkali metal cation is almost invariant with the choice of organic anion radical.[61,62] This solvation enthalpy turns out to be -168 kcal/mol when DME and K^+ serve as the solvent and the cation.[62] Reactions 2-36 and 2-37 depict the EA's of CBN and COT.

$$COT + e^- \rightarrow COT^{-\cdot} \qquad (2\text{-}36)$$

$$\Delta H^\circ = -13 \text{ kcal/mol}$$

$$CBN + e^- \rightarrow CBN^{-\cdot} \qquad (2\text{-}37)$$

$$\Delta H^\circ = +3.7 \text{ kcal/mol}$$

Notice that the difference in the two electron affinities (16.7 kcal/mol) is in excellent agreement with the difference in the the $E_{1/2}$ values (13.5 kcal/mol). This is the case despite entropy and ion association effects.

The fully conjugated (homoantiaromatic) homo[8]annulene (HCOT) should have an electron affinity that is very close to that for the planar symmetrical COT (28 kcal/mol), since the electron will be delocalized in eight p orbitals occupying the same relative positions in the molecule. By subtracting this vertical electron affinity of HCOT (EA_{open} = 29 kcal/mol) from the adiabatic EA of CBN (−3.7 kcal/mol), we find the difference in the heats of formation of CBN and HCOT to be about 33 kcal/mol. This 33 kcal/mol represents the energy difference between CBN and the homoantiaromatic system (2-38).

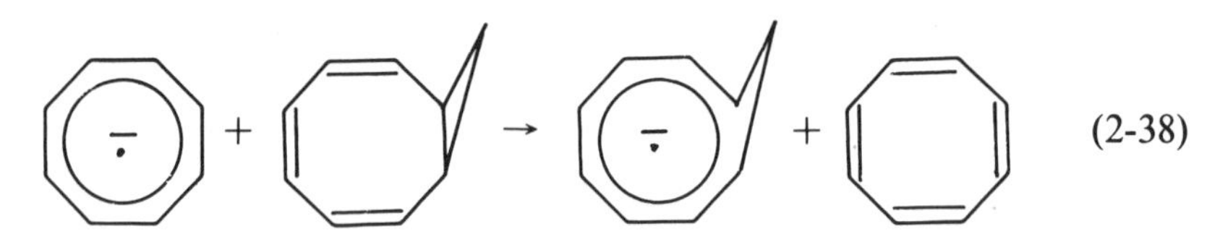

(2-38)

ΔH° = +33 kcal/mol

The very endothermic nature of reaction 2-38 is accounted for by the antihomoaromatic character of HCOT and by steric effects. The 33 kcal/mol means that the equilibrium constant is about 10^{-24}.

The second EA of CBN is simply the EA of HCOT minus the electron–electron repulsion energy (139 kcal/mol).[60] Thus, the energy required to add two electrons to gas phase CBN to generate the homoaromatic dianion (Scheme I) is +112 kcal/mol. From Scheme I and the analogous scheme for COT (Scheme II), it is clear that the enthalpy for reaction 2-39 in the gas phase must be 50 kcal/mol.

+ e^- ⟶ +3.7 kcal

+ e^- ⟶ +139 − EA_{OPEN}

+ $2e^-$ ⟶ 143 − EA_{OPEN}

Scheme I.

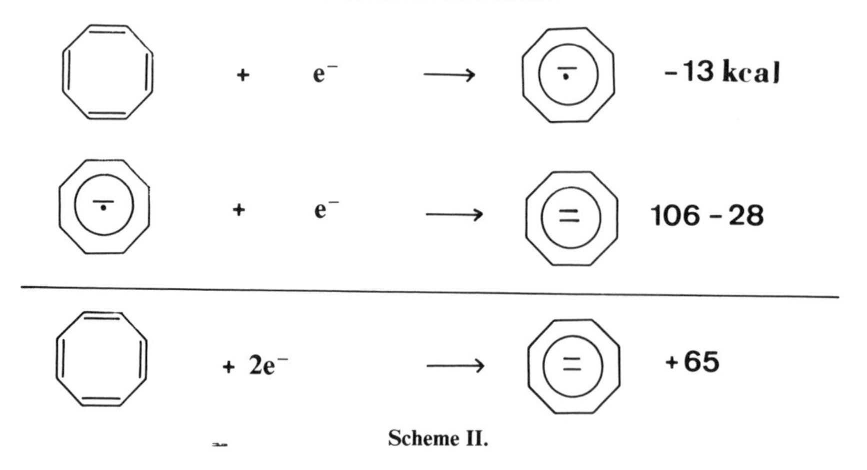

Scheme II.

$$\text{(2-39)}$$

$\Delta H^\circ = +50$ kcal/mol (gas phase)
$\Delta H^\circ = +79$ kcal/mol (solid with cation $= K^+$)

The heat of reaction 2-39 means that $HCOT^{2-}$ is much less thermodynamically stable than is COT^{2-} relative to their respective neutral molecules. This has been verified in the condensed phases. In fact, the solid potassium salt of the HCOT dianion ($K_2^+HCOT^{2-}$) reacts with water (2-40) almost three times as exothermically as does $K_2^+COT^{2-}$ (2-15).

K_2^+ (s) + $2H_2O \rightarrow$ (l) + 2KOH(aq) (2-40)

$\Delta H^\circ = -119.1$ kcal/mol

The heats of reactions 2-15 (-40.5 kcal/mol) and 2-40 can be interpreted to mean that the enthalpy for the two-electron transfer from COT to CBN in the solid state, when potassium serves as the cation, is endothermic by 79 kcal/mol. The electron transfer is more endothermic in the solid state than in the gas phase due to the smaller crystal lattice energy of $K_2^+HCOT^{2-}$ relative to $K_2^+COT^{2-}$ caused by the protruding methylene group.[60]

9. THE AROMATIC PRIORITY

"One of the most striking phenomena related to aromaticity is the unusual tendency of aromatic systems to remain so."[63] Pericondensed polycycles can, in principle, accommodate more than one mode of electron delocalization over their π frameworks. Two intriguing examples of pericondensed polycyclics are the aceheptylene **(21)** and acenaphthalene **(22)** systems.

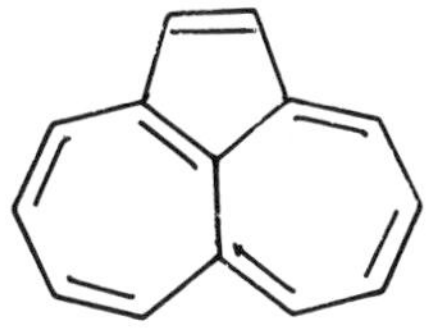

21 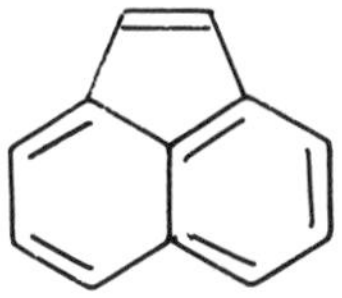 **22**

In the case of **21,** if the path of conjugation is established over the entire periphery, the molecule would be considered to maintain a 13-π-electron periphery [a $(4n + 1)$ system]. However, if only part of the periphery (the heptylene moiety) is involved in the π conjugation, the molecule would have a 12-π-electron periphery and would be antiaromatic in the Hückel sense. In the case of **22,** the situation is reversed. Conjugation over the napthalene periphery only results in Hückel-type aromaticity. If the conjugated path extends over the entire periphery, however, the result is an 11 or $(4n + 3)\pi$-electron system. According to the criterion suggested by Platt,[64] both $(4n + 3)$- and $(4n + 1)\pi$-electron systems should be considered to be nonaromatic.

Considering only aromaticity arguments, we might expect delocalization around the entire periphery of **21** with a resulting nonaromatic molecule as opposed to an antiaromatic molecule. One the other hand, we would expect delocalization over the naphthalene moiety only in **22,** with resulting aromatic character, as opposed to delocalization around the entire molecule with resulting nonaromatic character. NMR evidence suggests that these predictions have been borne out.

The ^{1}H NMR spectrum of **21** shows vinyl protons between 5.17 and 6.98 ppm.[63] This is the expected atropic behavior. In contrast to this, the ^{1}H NMR bands of **22** appear between 7.13 and 7.89 ppm, clearly indicating a diatropic system.[63] Thus, **22** is said to be an aromatic system that suffers a small perturbation because of an outer double bond.[63]

Alkali metal reduction of **21** yields a dianion. The center of gravity of the ^{1}H NMR spectrum of this dianion is diatropically shifted with respect to that of **21** by 0.75 ppm because of diamagnetic ring current.[63] Thus, the neutral molecule, which has conjugation over the entire periphery to avoid antiaromaticity, shows conjugation only over the heptylene moiety in the dianion. This gives the dianion aromatic character.

Similar reductions of **22** yield a dianion that is clearly atropic.[63] Thus the neutral molecule, which has conjugation over the naphthylene moiety to obtain aromatic character, yields a dianion that is conjugated over the entire periphery. In this way the dianion avoids antiaromatic character and is nonaromatic.

This interesting work by Rabinovitz and co-workers[63] clearly demonstrates how the path of conjugation is "forced" to vary so that the systems involved can avoid antiaromaticity and approach aromaticity.

REFERENCES

1. Faraday, M. *Phil Trans.* **1825,** *CXV,* 440.
2. Kékulé, A.M. *Annalen* **1872,** *CXLII,* 77.
3. Kékulé, A.M. *J. Chem. Soc.* **1872,** *X,* 612.
4. Dewar, J.; Wislicenus, J. *Proc. R. Soc. Edinburgh* **1867,** *VI,* 82.
5. Claus, M. "Theoretische Betrachtungen und deren Anwendung zur Systematik der organischen Chemie". Freiburg, **1867**, p. 207.
6. Thiele, J. *Annalen* **1899,** *87,* 306.
7. (a) Willstätter, R.; Waser, E. *Berichte* **1911,** *44,* 3423. (b) Willstätter, R.; Heidelberger, H. ibid, **1913,** *46,* 517.
8. (a) Hückel, E., *Z. Physik. 1931, 70,* 204. (b) Hückel, E. *Z. Electrokhim.* **1937,** *43,* 752.
9. Roberts, J.D. "Notes on Molecular Orbital Calculations." Benjamin: New York, 1962.
10. Pople, J.A. *J. Chem. Phys.* **1956,** *24,* 1111.
11. Garrat, P.J. "Aromaticity". McGraw-Hill: London, 1971.
12. Kistiakowsky, G.B.; Ruhoff, J.R.; Smith, H.A.; Vaughn, W.E. *J. Am. Chem. Soc.* **1936,** *58,* 237.
13. Stevenson, G.R. *J. Chem. Educ.* **1972,** *49,* 781.
14. Prosen, E.J., Gilmont, R., Rossini, F.D. *J. Res. Natl. Bur. Stand.* **1945,** *34,* 65.
15. Kistiakowsky, G.B.; Ruhoff, J.R.; Smith, H.A.; Vaughn, W.E. *J. Am. Chem. Soc.* **1935,** *57,* 876.
16. (a) Karle, I.L. *J. Chem. Phys.* **1952,** *20,* 65. (b) Tratteberg, M. *Acta Chem. Scand.* **1966,** *20,* 1724.
17. Bordner, J.; Parker, R.G.; Stanford, R.H. *Acta Crystallogr. Sect. B* **1972,** *28,* 1069.
18. (a) Anet, F.A.L.; Bourn, A.J.R.; Lin, Y.S. *J. Am. Chem. Soc.* **1964,** *86,* 3576. (b) Anet, F.A.L.; Bock, L.A. ibid, **1968,** *90,* 7130.
19. (a) Oth, J.F.M.; Merenyi, R.; Martini, T.; Schroder, G. *Tetrahedron Lett.* **1966,** 3087. (b) Luz, Z.; Meiboom, S. *J. Chem. Phys.* **1973,** *59,* 1077.
20. Gwynn, D.E.; Witesides, G.M.; Roberts, J.D. *J. Am. Chem. Soc.* **1965,** *87,* 2862.
21. Paquette, L.A.; Gardlik, J.M. *J. Am. Chem. Soc.* **1980,** *102,* 5016.
22. Paquette, L.A.; Gardlik, J.M. *J. Am. Chem. Soc.* **1980,** *102,* 5033.
23. Stevenson, G.R.; Forch, B.E. *J. Am. Chem. Soc.* **1980,** *102,* 5985.
24. Turner, R.B.; Meador, W.R.; Doering, W.v.E.; Knox, L.H.; Meyer, J.R.; Willey, D.W. *J. Am. Chem. Soc.* **1957,** *79,* 4127.
25. Reppe, W. Schlichting, O.; Klager, K; Toepel, T. *Annalen* **1948,** 560.
26. Sondheimer, F.; Gaoni, Y. *J. Am. Chem. Soc.* **1961,** *83,* 4863.
27. Oth, J.F.M.; Gille, J.M. *Tetrahedron Lett.* **1968,** 6259.
28. (a) Jackman, L.M.; Sondheimer, Y.; Amiel, Y.; Ben-Efraim, A.; Gaoni, Y.; Wolovsky, R; Bothner-By, A.A. *J. Am. Chem. Soc.* **1962,** *84,* 4307. (b) Gaoni, Y.; Melera, A.; Sondheimer, F.; Wolovsky, R. *Proc. Chem. Soc.* **1964,** 397.
29. Beezer, A.E.; Mortimer, C.T.; Springall, H.D.; Sondheimer, F.; Wolovsky, R. *J. Chem. Soc.* **1965,** 6970.
30. Beezer, A.E.; Mortimer, C.T.; Springell, H.D.; Sondheimer, F.; Wolovsky, R. *J. Chem. Soc.* **1965,** 216.
31. Dewar, M.J.S.; Harget, A.; Haselbach, E. *J. Am. Chem. Soc.* **1969,** *91,* 7521.
32. Van-Catledge, F.A.; Allinger, N.L. *J. Am. Chem. Soc.* **1969,** *91,* 2582.

33. Hirshfeld, F.L.; Rabinovich, D. *Acta Crystallogr.* **1965,** *19,* 235.
34. Strauss, H.L.; Katz, T.J.; Fraenkel, G.K. *J. Am. Chem. Soc.* **1963,** *85,* 2360.
35. Oth, J.F.M.; Bauman, H.; Gilles, J.M.; Schröder, G. *J. Am. Chem. Soc.* **1972,** *94,* 3498.
36. Stevenson, G.R.; Zigler, S.S.; Reiter, R.C. *J. Am. Chem. Soc.* **1981,** *103,* 6057.
37. (a) Wiedrich, C.R.; Catlett, D.L.; Sedgwick, J.B.; Stevenson, G.R. *J. Phys. Chem.* **1983,** *87,* 578. (b) Smentowski, F.J.; Stevenson, G.R. *J. Phys. Chem.* **1969**, *73,* 340.
38. Stevenson, G.R.; Schock, L.E.; Reiter, R.C.; Hansen, J.F. *J. Am. Chem. Soc.* **1983,** *105,* 6078.
39. Stevenson, G.R.; Schock, L.E. Unpublished results.
40. Oth, J.F.M.; Rottele, H.; Schröder, G. *Tetrahedron Lett.* **1970,** *37,* 3251.
41. Oth, J.F.M.; Schröder, G. *J. Chem. Soc. B,* **1971,** 904.
42. Stevenson, G.R.; Concepcion, R.; Reiter, R.C. *J. Org. Chem.* **1983,** *48,* 2777.
43. Oth, J.F.M.; Woo, E.P.; Sondheimer, F. *J. Am. Chem. Soc.,* **1973,** *95, 7337.*
44. Sondheimer, F.; Wolovsky, R.; Amiel, Y. *J. Am. Chem. Soc.* **1962,** *84,* 274.
45. Stevenson, G.R., Schock; L.E.; Reiter, R.C. *J. Phys. Chem.* **1984,** *88,* 5417.
46. Lawler, R.G.; Tabit, C.T. *J. Am. Chem. Soc.* **1969,** *91,* 5671.
47. Winstein, S.; Adam, R. *J. Am. Chem. Soc.* **1948,** *70,* 838.
48. Jorgensen, W.L. *J. Am. Chem. Soc.* **1976,** *98,* 6784.
49. (a) Grutzner, J.B.; Jorgensen, W.L. *J. Am. Chem. Soc.* **1981,** *103,* 1373. (b) Haddon, R.C. *J. Org. Chem.* **1979,** *44,* 3608. (c) Paquette, L.A.; Kekula, M.J.; Ley, S.V.; Traynor, S.G. *J. Am. Chem. Soc.* **1977,** *99,* 4756.
50. (a) Warner, P.; Harris, D.L.; Bradely, C.H.; Winstein, S. *Tetrahedron Lett.,* **1970,** 4013. (b) Keller, C.E.; Petit, R. *J. Am. Chem. Soc.* **1966,** *88,* 604, 606.
51. (a) Vogel, P.; Saunders, M.; Hasty, N.M., Jr.; Berson, J.A. *J. Am. Chem. Soc.* **1971,** *93,* 1551. (b) Olah, G.A.; Staral, J.S.; Spear, R.J.; Liang, G. *J. Am. Chem. Soc.* **1975,** *97,* 5489.
52. Bleck, W.E.; Grimme, W.; Gunther, H.; Vogel, E. *Angew. Chem. Int. Ed. Engl.* **1970,** *9,* 303.
53. Masamune, S.; Hojo, K.; Bigman, G.; Rabinstein, D.L. *J. Am. Chem. Soc.* **1971,** *93,* 4966.
54. Vogel, E.; Roth, H.D. *Angew. Chem. Int. Ed. Engl.* **1964,** *3,* 228.
55. Stevenson, G.R.; Zigler, S. *J. Phys. Chem.* **1983,** *97,* 895.
56. Scott, T.L.; DeCicco, G.J.; Hyun, J.L.; Reinhart, G. *J. Am. Chem. Soc.* **1983,** *105,* 7761.
57. (a) Smentowski, F.J.; Owens, R.M.; Faubion, B.D. *J. Am. Chem. Soc.* **1968,** *90,* 1537. (b) Katz, T.J.; Talcott, C. ibid, **1966,** *88,* 4732. (c) Ley, S.V.; Paquette, L.A. ibid, **1974,** *96,* 6670.
58. Bremeser, W.; Hagen, R.; Heilbronner, E.; Vogel, E. *Helv. Chim. Acta* **1969,** *52,* 418.
59. Anderson, L.B.; Broadhurst, M.J.; Paquette, L.A. *J. Am. Chem. Soc.* **1973,** *95,* 2198.
60. Concepcion, R.; Reiter, R.C.; Stevenson, G.R. *J. Am. Chem. Soc.* **1983,** *105,* 1778.
61. Stevenson, G.R.; Schock, L.E.; Reiter, R.C. *J. Phys. Chem.* **1983,** *87,* 4004.
62. Stevenson, G.R.; Chang, Y. *J. Phys. Chem.* **1980,** *84,* 2265.
63. Minsky, A.; Meyer, A.Y.; Hafner, K.; Rabinovitz, M. *J. Am. Chem. Soc.* **1983,** *105,* 3975.
64. Platt, J.R. *J. Chem. Phys.* **1954,** *22,* 1448.
65. Wentworth, W.E.; Ristau, W. *J. Phys. Chem.* **1969,** *73,* 2126.
66. Jordan, K.D.; Barrow, P.D. *Acc. Chem. Res.* **1978,** *11,* 341.

CHAPTER 3

Structural Limitations in Cyclic Alkenes, Alkynes, and Cumulenes

Richard P. Johnson*

Iowa State University, Ames, Iowa

CONTENTS

1. INTRODUCTION

A. The Problem

One of the most fundamental, yet persistent, questions in organic chemistry involves the structural limitations nature has imposed on carbon compounds. These limitations long have fascinated and challenged organic

*Current address: Department of Chemistry, University of New Hampshire, Durham, New Hampshire

MOLECULAR STRUCTURE AND ENERGETICS, Vol. 3

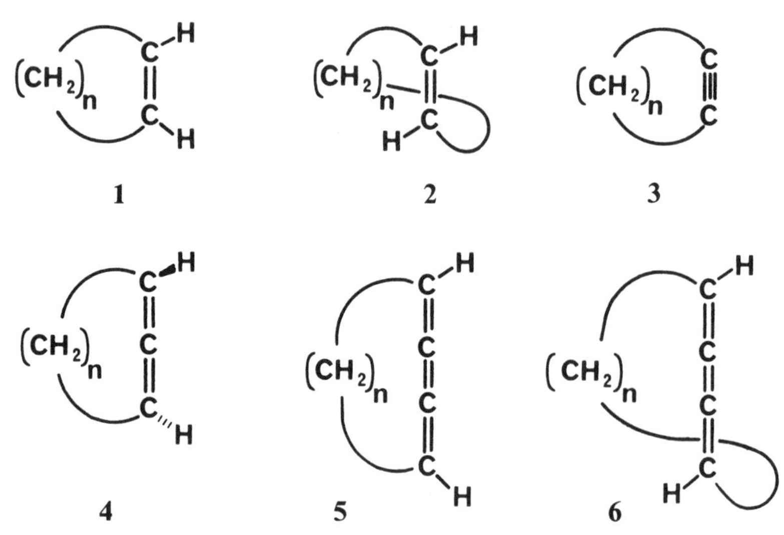

Figure 3-1. Cyclic alkenes, alkynes, and cumulenes.

chemists; they continue to be explored through a sophisticated array of experimental and theoretical techniques.[1–10]

This chapter treats one fundamental aspect of organic structure limitations, the incorporation of double and triple bonds in carbocyclic rings. These ring constraints often engender significant torsional strain or angle strain, and result in high kinetic reactivity. The range of structures encompassed here is shown in Figure 3-1. Bicyclic and more complex structures are dealt with in a less comprehensive fashion. The reader is forewarned that the author has an unabashed bias toward hydrocarbons.

Nonconjugated *cis* double bonds **(1)** afford stable compounds in carbocyclic rings of any size, although cyclopropene and cyclobutene are quite strained and reactive.[2,7] Small-ring *trans*-alkenes **(2)**, however, are far more kinetically reactive, due to the badly twisted π bond.[2] In the alkyne series **(3)**, ring constraints bend the normally linear C—C≡C—C moiety, ultimately resulting in significant diradical character in the in-plane π bond.[8,11] The allenic moiety **(4)** prefers a linear, orthogonal (ie, perpendicular) geometry; constraints of small rings cause both bending and twisting toward planarity.[2] Deformations in *cis*- and *trans*-butatrienes (**5** and **6**) produce strain contributions that resemble those of both alkenes and alkynes. These are considered in more detail in later sections.

It is essential to exercise two criteria in describing structural limitations: existence and isolability. With regard to the first criterion, does the structure proposed correspond to a stationary point on a potential energy surface, or will it rearrange to a more stable structure through a simple vibration? An excellent example is cyclopropyne, which has been predicted to rearrange spontaneously to vinylidene carbene.[11] On a more practical note, can the compound be isolated and characterized under "ordinary" laboratory conditions? For example, 1,2-cyclononadiene is the smallest cyclic allene that can be distilled and worked with under ordinary conditions.[12] Between these two extremes lie many substances that are stable only in low-temperature matrices, due to high kinetic reactivity.

As noted by two recent reviewers, "... organic chemists usually recognize a strained molecule when they see one."[2b] The six-membered ring compounds **7–11** illustrate the range of unusual molecular structure types considered here. Except for *cis*-cyclohexene (**7**; available from any chemical stockroom) all are instantly recognizable as highly strained, and all are believed to be incapable of isolation under ordinary conditions. Nevertheless, structures **8–11** all have been the subject of both experimental and theoretical investigations, and various types of evidence suggest *(see below)* that all represent energy minima. Thus, these structures fulfill the criterion of existence, but not isolability.

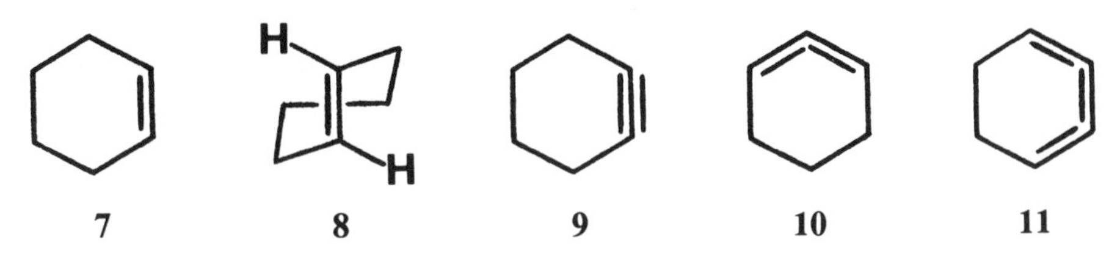

B. Theoretical Background

Theoretical methods provide information that usually is complementary to experiment.[13] This information may include geometries and energetics for reactive intermediates or other structures, molecular properties such as vibrational frequencies, dipole moments, or heats of formation for direct comparison to experiment, estimates of strain, or detailed studies of transition states and reaction paths. The sophistication and accuracy (relative to experiment) of theoretical methods has made a number of quantum leaps (pun intended) over the past decade. Nevertheless, the approach to "reality" (ie, chemical accuracy) is a difficult one, as is the recognition that one has arrived there. Fortunately, many of the fundamental problems posed by organic chemists can be adequately resolved at relatively low levels of theory.

Theoretical methods may be roughly categorized in order of increasing sophistication as follows: (a) "back of the envelope," (b) Hückel and extended Hückel, (c) semiempirical valence electron calculations such as complete or modified neglect of differential overlap (CNDO or MNDO), and (d) ab initio (or, as some prefer, nonempirical) molecular orbital methods. Distinct from these are: (e) the nonquantum mechanical force-field ("molecular mechanics") methods.[14] Each method is elaborated briefly; the reader is referred to the references for recent state-of-the-art applications. It is essential to recognize that each method has its strengths and weaknesses.

One must be careful not to place too much faith in computations that are not in some way "calibrated" versus experimental results. The chemical literature is replete with computational papers of questionable accuracy. (Of course, there also are many published experiments of questionable validity.)

"Back of the envelope" approaches usually are based on arithmetic, simple resonance or VSEPR (Valence shell–electron pair repulsion) theory, symmetry, intuition, or all the above. One can often estimate structure and energetics from bond dissociation energies, and through comparison to other structures. These methods probably are underutilized by organic chemists. The group increment approach of Benson is a powerful tool,[15] as is the idea of isodesmic reactions.[16] "Macroincrementation" is discussed in Chapter 6 of this volume (see also Reference 17).

Simple Hückel molecular orbital theory, most generally applied to π-electron systems, had a profound effect on the way chemists viewed electronic structure and reactivity.[18] This was later elaborated by Hoffmann in the extended Hückel approach, which uses a three-dimensional basis of Slater-type functions for valence atomic orbitals, and empirically derived values for Coulombic integrals.[19] Neither of these methods is of sufficient quantitative accuracy for the problems presented here, because neither include electron repulsion and neither can provide accurate geometries and energies.

Semiempirical valence shell MO methods are characterized by neglect of certain types of electron repulsion integrals (which greatly simplifies the computation), and by some degree of empirical parameterization.[20,21] The first of these was the CNDO method[21] of Pople and Segal. This was succeeded by (among others) CNDO/2, INDO (intermediate NDO),[22] various levels of MINDO (modified INDO), and more recently MNDO (modified neglect of diatomic overlap).[23] One key point of the philosophy behind the widely used MINDO and MNDO approaches of Dewar and co-workers has been to adjust computational parameters to reproduce experimental heats of formation and geometries.[24a] Not surprisingly, the MNDO method usually performs quite well on prediction of geometries, heats of formation, and ionization potentials for ground state molecules,[24b] although limitations are noted for certain classes of compounds, such as strained hydrocarbons. Because of its computational efficiency, this is an excellent method for large molecules. As an example, complete geometry optimization for phenylallene (C_9H_8) with MNDO consumed about 400 s of CPU time on a mainframe IBM 370 computer system. A single point ab initio calculation with a minimal (STO-3G) basis set consumed about the same CPU time,[25] much of which was required for the computation of more than 5.6 million two-electron integrals.

On the other hand, the author believes that these methods must be used with considerable caution in the calculation of reaction coordinates, transition states, and so on. A good example is the Cope rearrangement.[26] MINDO calculations predict a metastable biradical intermediate,[26a,b] cyclohexane-1,4-diyl, and Dewar has argued that such multibond reactions cannot normally be synchronous.[26c] In sharp contrast, recent split-valence multiconfiguration SCF (MCSCF) calculations by Borden and co-workers[26d] predict a concerted process, which seems in better agreement with a large body of experiments.[26e]

Ab initio computational methods are potentially the most accurate, but have the disadvantage of requiring large expenditures of computer time. This also is the most flexible approach.[27] One can choose among a sophisticated array of basis sets, SCF (self-consistent-field) methods, and means of including correlation energy.[27] Linear contractions of Gaussian-type orbitals (GTOs) of the general form $\exp(-\alpha r^2)$ (where α is the orbital exponent) are the most commonly used and best-tested basis functions.[27] These have been extensively developed by the Pople group,[28] and most organic chemists today are familiar with such acronyms for Slater-type orbitals as "4–31G" and "STO–3G." STO–3G is a minimal basis set; that is, it is composed of one function (composed of three GTOs) per atomic orbital. Beyond this are commonly used "double-zeta" (DZ) basis sets (two functions per atomic orbital)[29] or "split-valence" bases such as 4–31G or 3–21G.[28] Further improvement usually comes from the addition of d orbitals (so-called polarization functions) on carbon or other heavy atoms, and p orbitals on hydrogen.[29] It is commonly observed that minimal basis sets yield good geometries[27c,30] (although improvement is observed with better basis sets[28]) but unreliable relative energies. The commonly used STO-3G basis tends to underestimate the stability of π bonds; this is especially significant when the species to be compared differ in the number of π bonds. The energetics of three-membered rings also is badly described by the STO-3G basis. An excellent compilation of geometries calculated with different basis sets is given in a recent paper by Wiberg and Wendoloski.[31]

The proper choice of SCF type depends on the problem at hand. For example, the Cope rearrangement transition state, mentioned above,[26] necessitates a multiconfiguration wave function. Molecular ground states of ordinary molecules usually can be described with a single configuration RHF (restricted Hartree–Fock) wave function. As an example, the geometries of ethylene, allene, and butatriene, which are dervied from various levels of RHF calculation,[23,32] are summarized in Table 3-1. In general, HF calculations yield bond lengths that are too short.

Transient species such as radicals and carbenes, or transition states, usually require a more complex wave function for a believable description. Excellent examples are diradicals such as tetramethylene **(12)**[33a] or trimethylenemethane **(13)**,[33b,34], which need to be treated with at least a TCSCF (two-configuration self-consistent-field) approach to prevent localization of the wave function. MCSCF methods are becoming increasingly popular and are a very powerful tool for the study of reactive intermediates and reaction pathways.[26d,33,35] These avoid many of the problems inherent in a single-configuration wave function. As Ruedenberg and co-workers have noted,

TABLE 3-1. Geometric Parameters for Cumulenes

Compound and symmetry	Parameter	Bond length (Å) or angle (deg) by: STO−3G[a]	4-31G[b]	MNDO[c]	Experiment
$H_2C{=}CH_2$ (D_{2h})	C−C	1.306	1.316	1.335	1.339
	C−H	1.082	1.073	1.089	1.085
	HCH	115.6	116.0	113.6	117.8
$H_2C{=}C{=}CH_2$ (D_{2d})	C−C	1.288	1.294	1.306	1.308[d]
	C−H	1.083	1.073	1.090	1.087
	HCH	116.2	117.0	114.2	118.2
$H_2C{=}C{=}C{=}CH_2$ (D_{2h})	C_1-C_2	1.296	1.301	1.310	1.318[e]
	C_2-C_3	1.257	1.262	1.270	1.283
	CH	1.085	1.073	1.090	1.089

[a,b]Reference 32a.
[c]Reference 23.
[d]Reference 32b.
[e]Reference 32c.

MCSCF results lend themselves to relatively straightforward interpretation.[35a] A good example is the detailed analysis of dihydrogen transfer from ethane to ethylene.[35a]

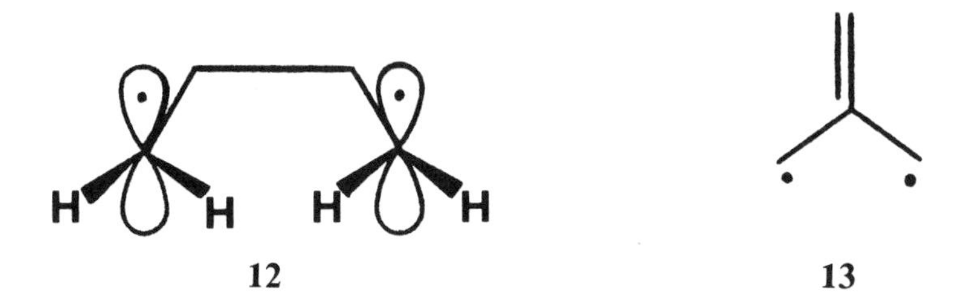

12 **13**

Davidson has recently given an excellent discussion of the computational effort necessary to arrive at a believable result with carbenes, that is, one likely to correlate with experiment.[36] The triplet of (eg) methylene ($^3B_1$) can be adequately described with a single configuration ROHF (restricted open-shell Hartree–Fock) calculation. However, the singlet (1^1A_1) requires correlation of the in-plane electron pair through a TCSCF calculation. This corresponds to mixing 1^1A_1 and 2^1A_1. Of course, both these calculations will recover only a small fraction of the *total* electron correlation energy; however, this singlet TCSCF/triplet ROHF approach should account for the major correlation energy *difference* between the singlet and the triplet. Empirically, this gives good agreement with experiment.[36]

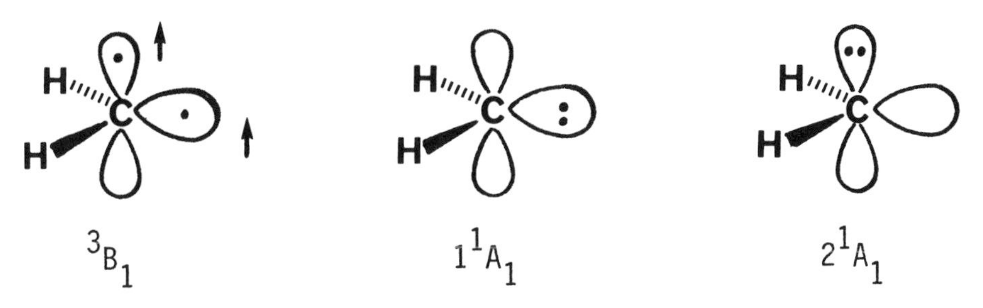

Cyclopropylidene **(14)** provides a recent application of this simple approach.[37,38] The ground state multiplicity is not known from experiment, primarily because of facile ring opening to allene. However, in keeping with the general belief that singlet and triplet states of carbenes will cross as the angle grows smaller,[36] cyclopropylidene *should* have a singlet ground state. Earlier theoretical work provides only confusion of counsel; studies using MINDO/2,[37a] STO-3G first-order configuration interaction (CI),[37b] and 4-31G RHF/UHF[37c] predict a triplet ground state, two others (INDO[37d] and STO-3G RHF plus 2×2 CI/UHF)[37e] a singlet, with S–T energy gaps varying from 2.3 to 22 kcal/mol. Of these five studies, only that of Baird and Taylor[37e] employed a properly correlated singlet wave function, but with a minimal basis set. Stierman and Johnson used a split-valence plus polarization basis set and the TCSCF/ROHF approach, with gradient optimization, to predict a gap of 14.8 kcal/mol, favoring the singlet.[38] More extensive basis sets and highly correlated wave functions lower this to 12.3 kcal/mol.[39] This result is at least a believable one, and the ball is now in the experimentalists' court.

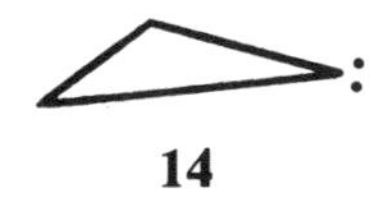

14

2. CYCLIC ALKENES

A. Introduction: Theoretical Studies

The π bond in simple ethylene derivatives normally is ascribed a strength of about 60–65 kcal/mol; this corresponds to the experimental ground state rotational barrier, which has been measured for 1,2-dideuteroethylenes.[40] Numerous structures contain torsionally distorted π bonds.[2] Most notable are the large collection of bridgehead alkenes; this area has recently been reviewed by Szeimies.[4a] Small-ring *trans*-alkenes also will have badly twisted and somewhat pyramidalized[41–42] π bonds.

Twisted π bonds have been the subject of many theoretical investigations, usually with the goal of predicting rotational barriers. Theoretical studies indicate that the excited states of twisted π bonds, which usually correspond to excited potential surface minima, are distinctly zwitterionic, while the ground states are biradicaloid.[43,44] Salem has referred to this excited state phenomenon as "sudden polarization."[43a]

Molecular mechanics has been extensively applied to prediction of geometry and strain in bridgehead alkenes and, less extensively, *trans*-cyclic alkenes.[45–47] Schleyer has recently defined the term "olefin strain" (OS) as the difference in strain between an alkene and its corresponding alkane derivative.[46]

Other π-bond deformations include in-plane bending, such as occurs in small rings, and pyramidalization (ie, rehybridization). This is believed to occur even in relatively simple systems, such as norbornene, and there has been considerable interest in the effects of pyramidalization on reactivity of π-bond faces.[5a] Twisted π bonds undoubtedly will pyramidalize somewhat, a phenomenon recognized in two early papers by Mock, and by Radom, Pople, and Mock.[41,42] There have been a number of theoretical studies of pyramidalization.[5a,7,41,42,46,48–50] Excellent reviews may be found in recent articles by Houk[5a] and by Wiberg.[7] It is now apparent that this is a very general phenomenon, which may be predominantly "torsional" in origin.[5a] Among strained systems, Hehre and Pople first predicted that bicyclobutene **(15)** would prefer a distinctly nonplanar geometry.[48] Schleyer and co-workers extended this to homologous alkenes **16** and **17.**[49] The authors suggested from model calculations on ethylene that at H—C—C angles less than 100°, such as those enforced by ring constraints in **15–17,** a nonplanar (pyramidalized) structure is preferred because of favorable interactions between σ* and π orbitals. In contrast to **15–17, 18** is believed to prefer a planar structure.[7]

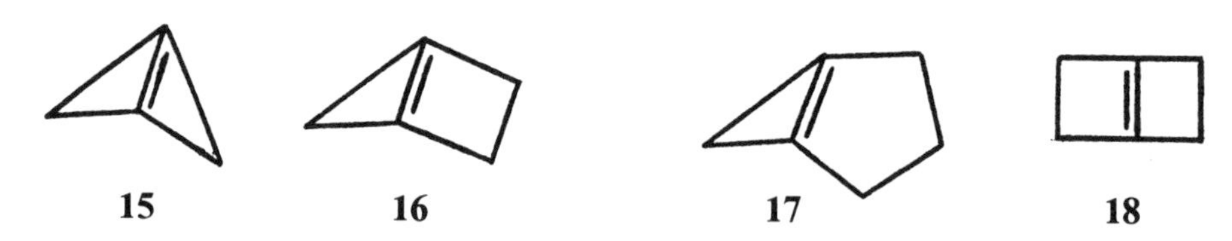

Volland, Davidson, and Borden have performed model calculations on anti- and syn-pyramidalized ethylene.[50] A prediction that interactions between π* and filled σ MOs would favor anti-pyramidalization was confirmed by these calculations. An earlier ab initio study by Radom, Pople, and Mock showed that torsional motion and pyramidalization should be intimately linked.[41]

Wiberg and co-workers performed a series of ab initio and molecular mechanics calculations on the remarkable structure tricyclo[4.2.2.2^{2,5}]dodeca-1,5-diene **(19).**[51] Resultant geometries, which showed significantly nonplanar double bonds, compared favorably with the structure determined by X-ray crystallography. The intramolecular distance between the two π bonds was observed to be 2.395 Å and predicted (STO-3G basis) to be 2.399 Å.[51] An experimental pyramidalization angle of 27.3° was reported.[51]

19

Recent ab initio theoretical studies on **20–22,** referred to as "columnenes," have led to the prediction that these should have significantly pyramidalized π bonds.[52] The pyramidalization angles were estimated to be 47.3°, 29.3°, and 18.2° in **20–22,** respectively, based on STO-2G geometry optimizations.[52] The pyramidalization in **20–22** is believed to arise because of a combination of electronic and torsional effects. These compounds are predicted to have strong through-space π–π interactions.

20

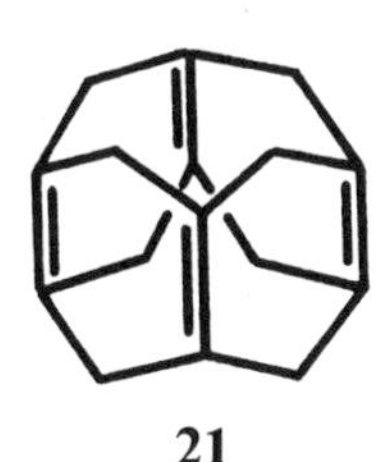

21

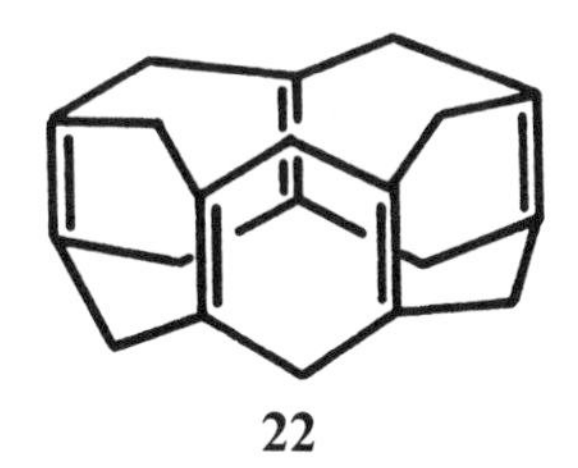

22

B. (Z)-Alkenes

Planar cis (or Z) double bonds may be incorporated in carbocyclic rings of any size. The strain energy in cyclopropene (**23:** $\simeq$ 55 kcal/mol) undoubtedly is the origin of its high thermal and photochemical reactivity.[53] This is substantially greater than the strain in cyclopropane ($\simeq$ 30 kcal/mol). Cyclobutene **(24)** also is quite strained, but still requires an additional 33 kcal/mol of activation for its symmetry allowed conrotatory ring opening to 1,3-butadiene.[54]

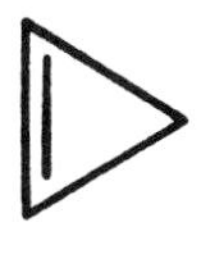

23

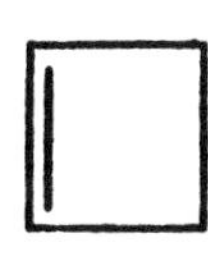

24

25 X = O

26 X = S

27 X = NH

Heterocyclic analogues of cyclopropene **25–27** are prototypal antiaromatic (4 π electron) systems and are generally quite unstable.[55–58] Among the unsubstituted compounds, only **26** has been spectroscopically observed.[58] Oxirene is an important intermediate in isomerizations related to α-ketocarbenes and ketenes,[56] however, until recently, oxirene derivatives had eluded detection. Strausz and co-workers have recently reported evidence for bis(trifluoromethyl)oxirene in an argon matrix.[57] High-level ab initio calculations by several groups predict the parent oxirene **(25)** to correspond to a shallow ($<$ 5 kcal/mol) potential surface minimum.[56] Krantz and Laureni have reported that matrix photolysis of 1,2,3-thiadiazole yields thiirene **(26)**.[58]

C. (E)-Alkenes

Small-ring *trans*- (or E) alkenes have been the subject of numerous investigations.[2a,4a] An excellent discussion of early work is given by Greenberg and Liebman.[2a] These compounds are of interest both because of their strain and because of their inherent chirality. It is well known that the smallest isolable homologue is *trans*-cyclooctene, although there is convincing evidence (see below) for the intermediacy of *trans*-cycloheptenes and *trans*-cyclohexenes.[2] *trans*-Cyclopentene **(29e)** does not appear to have received serious consideration.

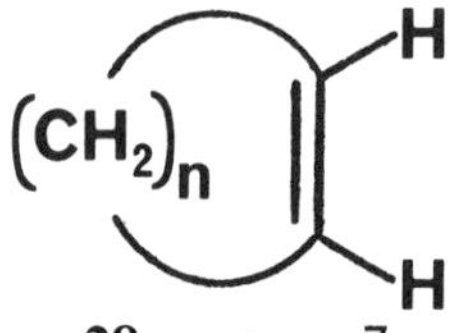

28a n = 7 **29a** n = 7

b n = 6 **b** n = 6

c n = 5 **c** n = 5

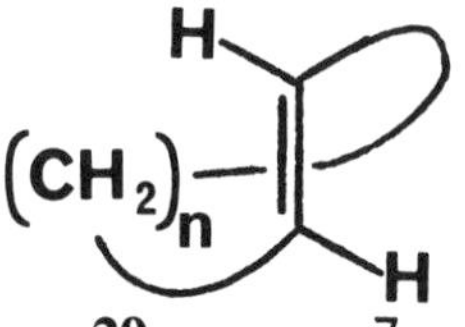

d n = 4 **d** n = 4

e n = 3 **e** n = 3

One of the more reliable predictors of the isolability of bridgehead alkenes was the assertion by Wiseman and Pletcher[59] that this should parallel behavior of monocyclic alkenes; that is, bridgehead alkenes[4] with a trans double bond in a ring smaller than eight carbons would be too kinetically reactive to be isolable. As one recent example, Wiseman has reported that (Z)-alkene **30** photoisomerizes to its (E) isomer, which is trapped by methanol. **(E)-31** contains a *trans*-cyclohexene moiety.[60]

30 —hν, CH_3OH→ **31** ——→ **32** (CH_3O)

The chirality of *trans*-cyclic alkenes is one of the most fascinating aspects of their structure. Stereochemical aspects of twisted double bond systems have recently been reviewed by Nakazaki and co-workers.[4c] This inherent chirality was first predicted by Blomquist in 1952,[61] and then experimentally confirmed by Cope, through the resolution of *trans*-cyclooctene.[62] Optically active *trans*-cyclononene could be prepared at low temperature, but this compound rapidly racemizes at ambient temperature.[63] This racemization does not occur by π-bond rotation but by the —$(CH_2)_n$— chain looping around different sides of the π bond, a motion that interconverts enantiomers. In small *trans*-cycloalkenes, this confirmational change is impossible. *trans*-Cyclodecene could not be resolved,[64] although Marshall and co-workers later prepared optically active 1,2-dimethylcyclodecene and 1,2-dimethylcycloundecene.[65] The 12-carbon homologue racemized easily. More recently, Marshall has reported a clever Sharpless epoxidation approach to optically active 1,2-disubstituted *trans*-cycloalkenes and betweenanenes.[66–68] *trans*-Cyclooctene **(29b)** was first reported by Cope, and has been the subject of numerous studies.[62] The groups of Allinger[45] and of Ermer[69] have reported the results of force-field calculations on **29b** and related structures. From the relative heats of hydrogenation, Allinger found the cis–trans energy difference to be 11.4 kcal/mol.[70] The total strain in *trans*-cyclooctene was estimated to be 22.0 kcal/mol, as opposed to 10.4 kcal/mol for the cis isomer. Electron diffraction studies yielded a double-bond torsional angle of 136°.[71]

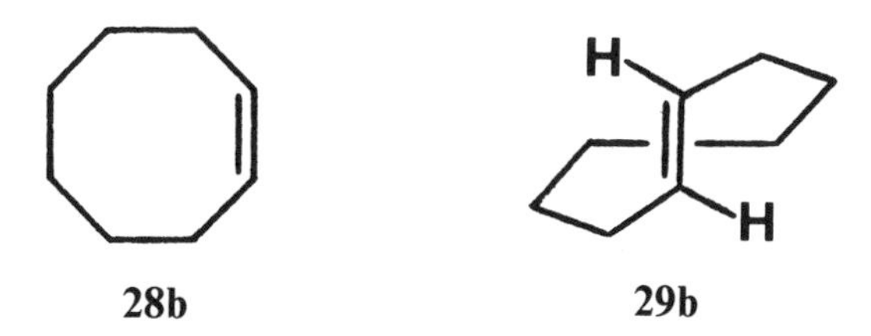

28b **29b**

There is convincing evidence for the existence of smaller *trans*-alkenes as transients, or as metal complexes. *trans*-Cycloheptene has been generated both thermally and photochemically.[72,73] Through the use of a dark reaction of photogenerated *trans*-cycloheptene with acidified methanol, Inoue and co-workers recently estimated an ambient temperature lifetime of 47 s and an activation energy of 18.7 kcal/mol for cis–trans isomerization.[73] As is typically observed for such highly strained species, reaction with acidified methanol yields methoxycycloheptane. Bonneau and co-workers have generated a number of *trans*-cycloalkenes by the use of laser flash photolysis

TABLE 3-2. UV Spectral Data for cis- and trans-Cycloalkenes[a]

Cycloalkene	λ_{max} (nm)	
	cis	trans
1-Phenylcyclohexene	248	385
1-Phenylcycloheptene	248	300
2-Cycloheptenone	222	265
Acetylcycloheptene	236	282
Cyclooctene	~185	196

[a]From Reference 74.

and have compiled the UV spectral data given in Table 3-2.[74] As expected, there is substantial red shift for the trans relative to the cis isomers. The same group estimated a thermal activation energy exceeding 17 kcal/mol for trans–cis isomerization in cycloheptenone,[74] surprisingly about the same as that for the parent structure.[73] The infrared spectrum of *trans*-cyclohepten-one had previously been reported by the groups of Eaton and of Corey.[75,76]

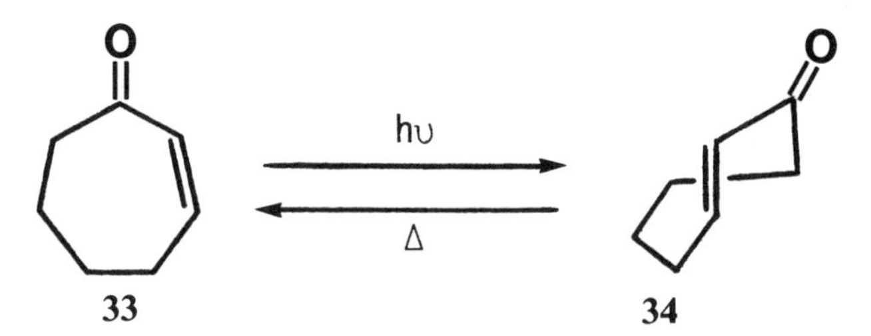

Evers and Mackor have reported that the complex of copper(I)triflate with *cis*-cycloheptene undergoes photoisomerization to a complex of the trans isomer.[77] Michl and co-workers have recently reported dynamic NMR stud-ies of these complexes, as well as MNDO and force-field calculations.[78]

Other *trans*-cycloheptene derivatives have been postulated as reaction intermediates. Schemes I and II show representative examples. Christl and

150°
36 37 39
or
38

Scheme I.

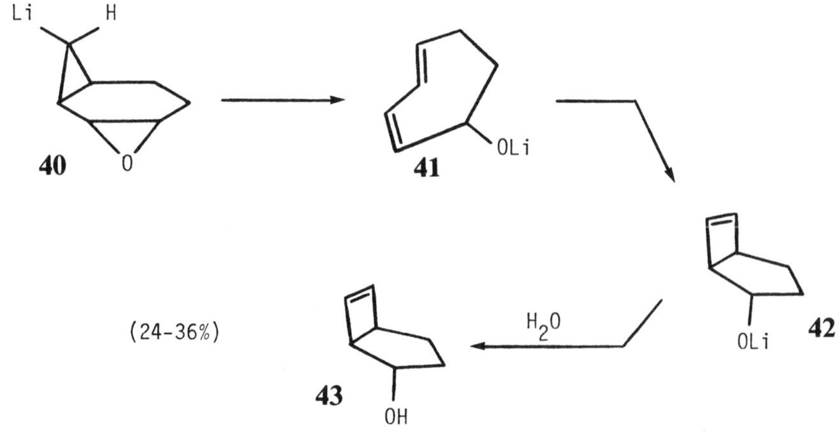

Li H

40 O

41 OLi

42 OLi

(24-36%)

43 OH

H_2O

Scheme II.

Brüntrup postulated *trans*-cycloheptatriene as an intermediate in the **36** → **39** thermal rearrangement (Scheme I).[79] Coates and Last suggested a *trans*-cycloheptadiene intermediate in the reaction shown in Scheme II.[80] Conrotatory closure in both cases is symmetry allowed.

Relatively little work has been done on *trans*-cyclohexenes, although there seems little doubt that these exist as transient species. Simple derivatives are trapped as ethers, when photogenerated in methanol.[81] An excellent discussion of numerous experiments is found in the review by Kropp on alkene photoreactions.[81] *trans*-1-Phenylcyclohexene (**32b,** where R = Ph) undoubtedly is the best characterized derivative.[82,83] This compound dimerizes through addition of the trans bond to the styryl moiety of another cis molecule.[82] Joussot-Dubien observed **32b** (R = Ph) as a transient upon flash photolysis of the cis isomer in methanol.[83] An ambient temperature lifetime of 9 μs increased to 500 μs at -70°C, and a thermal barrier to trans–cis isomerization of *ca.* 7 kcal/mol was estimated.[83] Dauben and co-workers have obtained a variety of kinetic evidence for this species.[82]

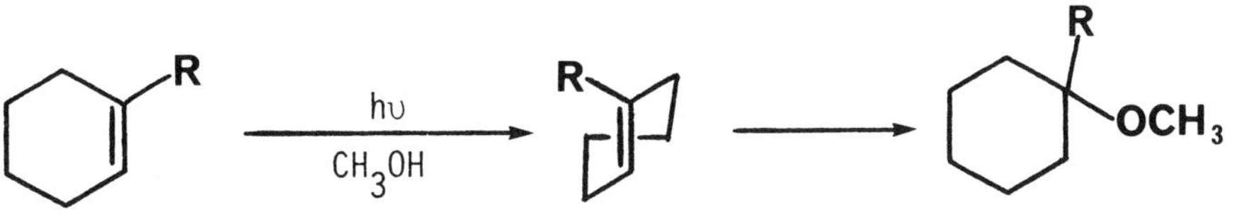

R hυ CH_3OH R R OCH_3

32a (R=Ph) **32b** **35**

Photodimerization of cyclohexene catalyzed by copper triflate has been suggested to arise from cis–trans isomerization of a complex with Cu(I).[84]

In earlier work on cyclic enones, Eaton and co-workers did not obtain evidence for *trans*-cyclopentenone or *trans*- cyclohexenone, however, 1-acetylcyclohexene did yield a cycloadduct with cyclopentadiene, which was believed to arise from a trans intermediate.[76]

D. Pyramidalized π Bonds

Another common deformation of π bonds is pyramidalization; theoretical studies of this are described in Subsection A, above. Even simple alkenes such as norbornene are believed to undergo some degree of pyramidalization.[5a]

There are quite a number of known strained pyramidalized alkenes, all of which are syn pyramidalized. In this deformation, attached ligands are not coplanar with the π bonds; the result is significant rehybridization toward sp^3. Greene and co-workers originally reported synthesis of alkene **44,** and a dihydro analogue.[85] Although the double bonds are quite pyramidalized, compound **44** is stable, and its crystal structure determination was reported. By contrast, **45,** reported by Borden and co-workers, did not prove to be isolable.[86] This compound readily [2 + 2] dimerizes, but can be trapped by diphenylisobenzofuran.[86] Szeimies has reported generation of the truly remarkable hydrocarbons **47–49** as transient species. Bicyclobutene **(47)** rearranges to 1,2,3-cycloheptatriene,[87] while **49** apparently opens to benzene. By contrast, **48** did not rearrange and was itself trapped by diphenylisobenzofuran. Alkene **46,** reported by Borden and co-workers, underwent facile dimerization via ene reaction.[89] Maier and Schleyer estimated [46] an out-of-plane deformation for **46** of 50° and olefin strain (OS) of 21.1 kcal/mol.

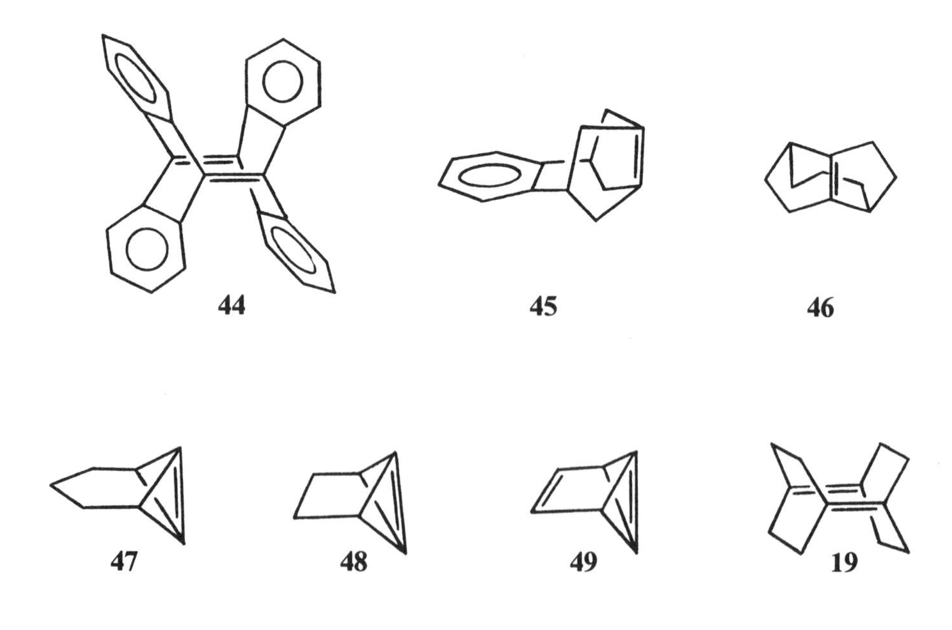

Wiberg and co-workers have reported the first synthesis of diene **19**, and a dihydro analogue.[90] As with **44**, this proved to be readily isolable. The X-ray structure showed the olefinic carbons to be deformed 0.40 Å out of the plane defined by the four attached allylic carbons; the resultant pyramidalization angle is 27.3°.

What are the limitations of syn pyramidalization in alkenes? In effect, there may be none within the range of possible carbon structures. This is perhaps best indicated by the successful trapping of bicyclobutene **(48)**, another truly remarkable hydrocarbon.[87,88] Because some degree of π–π overlap always will be maintained, the π bond will be severely weakened, but intact. In terms of isolability, the limitations of pyramidalization have not yet been defined; however, structures **44** and **19** demonstrate that buttressing by large groups is crucial, as might be expected.

3. CYCLIC ALKYNES

A. Introduction

Incorporation of a triple bond in a small-ring carbocyclic skeleton will lead to enormous deformation **(50** → **51)** from the preferred linear geometry.[2,3,8,91,92] Small-ring cycloalkynes should have considerable diradical character. It is well known that cyclooctyne is the smallest isolable unsubstituted cycloalkyne.[2] A smaller homologue, 3,3,7,7-tetramethylcycloheptyne, presumably owes its stability to steric effects of the methyl groups.[92] Since Krebs and Wilke recently have reviewed the abundant literature in this field,[8] only some very recent developments are described here. *o*-Benzyne is considered in Section 6.

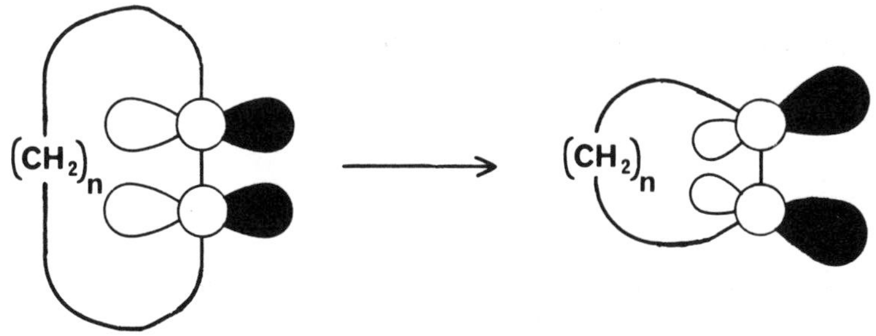

$(CH_2)_n$ ⟶ $(CH_2)_n$

50 **51**

B. Three-, Four-, and Five-Membered Ring Cycloalkynes

The limiting ring size in cycloalkynes is a problem of long-standing interest. For cyclopropyne[11] and cyclobutyne,[93,94] the best available data thus far have resulted from theoretical studies. Experimental attempts to generate cyclobutyne **(53)** have been unsuccessful.[95,96] In the case of cyclopentyne, experiments have provided convincing results.[3,97–101]

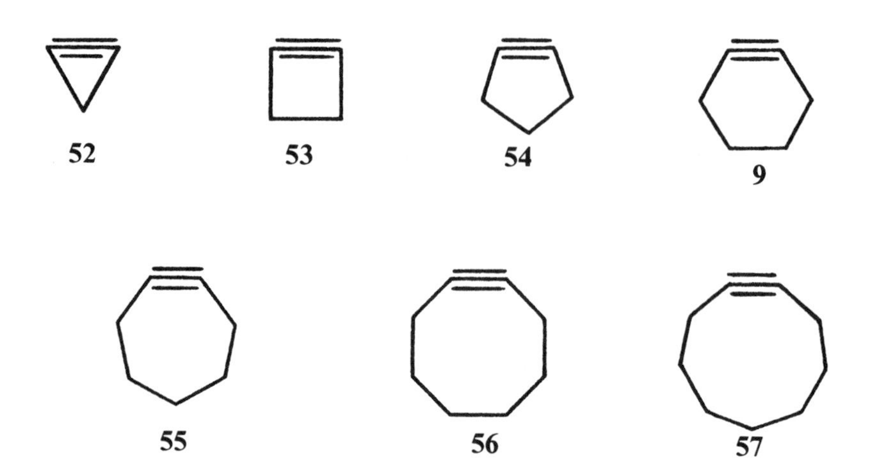

Schaefer and co-workers have recently applied high-level ab initio methods to the problems of cyclopropyne[11a,b] and cyclobutyne.[93] A previous MINDO/3 study by Dewar had led to conclusions that cyclobutyne is not a local minimum.[94] Geometries for singlets of **52** and **53** were optimized with a DZ basis set and TCSCF wave function; the latter approach is necessary to account for the high degree of diradical character expected for the weakened in-plane π bond. In the case of cyclopropyne, it was concluded that the singlet is a *transition state* for degenerate rearrangement of propadienylidene **(58)**, lying about 45 kcal/mol higher than **58.** By contrast, triplet cyclopropyne is predicted to be an energy minimum, but still is 22.5 kcal/mol above the triplet propadienylidene. These calculations are of sufficiently high level (DZ + polarization basis; well-correlated wave function) that there seems little reason to doubt Schaefer's conclusions.[11a,b]

$:C{=}C{=}CH_2 \rightleftharpoons$ (52: H_2C over C≡C ring) $\rightleftharpoons H_2C{=}C{=}C:$; 59: $\ddot{Si}$ over C≡C ring

58 **52** **58** **59**

Remarkably, recent theoretical[11c] and experimental[11d] studies indicate that silacyclopropyne **(59)** may exist. If so, this qualifies as the smallest stable cyclic structure that contains a triple bond.

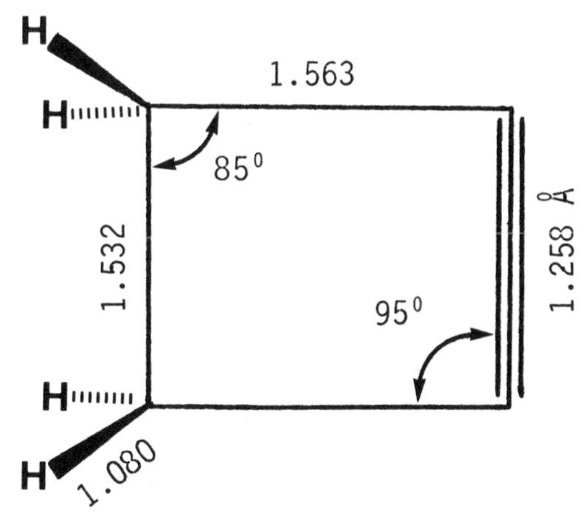

Figure 3-2. Ab initio TCSCF-optimized geometry for cyclobutyne.

Experimental studies on cyclobutyne and its derivatives have proved to be inconclusive.[95,96] However, theoretical studies strongly suggest that this may be the limiting ring size for simple cycloalkynes. Schaefer has used ab initio calculations to predict that both singlet and triplet cyclobutyne will correspond to energy minima.[93] The optimized geometry is shown in Figure 3-2. The triple bond in singlet cyclobutyne is expected to be quite weak (vibrational frequency $\sim$ 1700 cm^{-1}) and this species is predicted to be about 78 kcal/mol above vinylacetylene, the lowest energy C_4H_4 structure. The depth of the energy well containing singlet cyclobutyne is not known, and one might still expect facile ring opening to butatriene (ie, **53** → **60**).

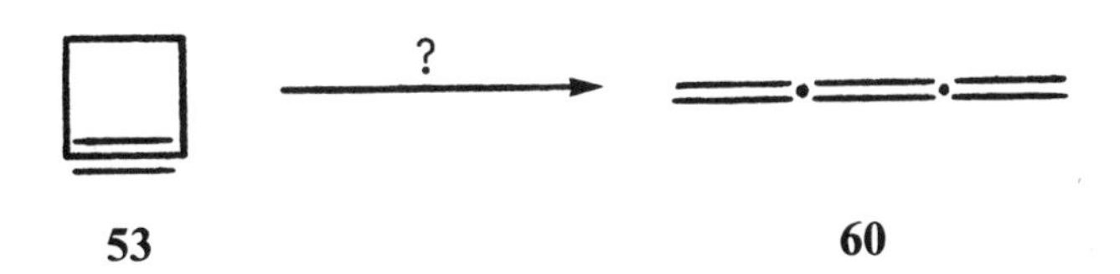

One remarkable aspect of both cyclopropyne and cyclobutyne is the extremely small gap between the closed-shell singlet ground state (which is not an energy minimum for cyclopropyne) and the lowest triplet state.[11,93] This is predicted to be 9.6 kcal/mol in cyclopropyne and about 13 kcal/mol in cyclobutyne, undoubtedly a consequence of the extremely weak in-plane π bond.

Cyclopentyne and various derivatives, such as norbornyne and acenaphthyne, have been generated by diverse routes.[3,8,97–104] A number of early solution phase studies convincingly demonstrate the intermediacy of this

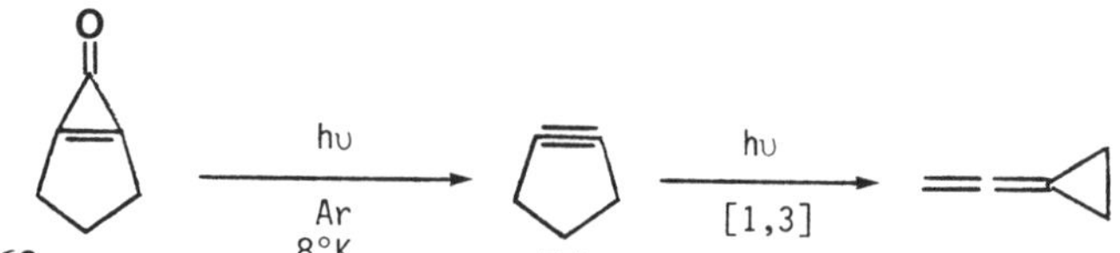

Scheme III.

parent species, **54.**[8] Chapman has generated cyclopentyne through photochemical extrusion of CO from cyclopropenone **(63)** in an argon matrix at 8 K (Scheme III). In early experiments, rapid secondary photoisomerization to allene **64** prevented observation of the infrared spectrum of cyclopentyne,[99a] however, this difficulty has been overcome.[99b]

Two groups have recently reported on the remarkable stereospecificity of [2 + 2] cycloadditions of cyclopentyne.[100,101] Fitjer and Modaressi generated cyclopentyne (Scheme IV) from dibromide **65.** Trapping with *cis*- or *trans*-2-butene afforded modest yields (36–38%) of cyclobutenes **66** or **68,** respectively.[101] Trapping with 1,3-butadiene yielded a [2 + 2] adduct. The authors suggested involvement of an "antisymmetric singlet ground state," that is one with a highest occupied molecular orbital (HOMO) as in **69.**[101] Such antisymmetric structures have commonly been proposed for benzynes. However, ab initio TCSCF calculations on cyclopentyne (at an MNDO-

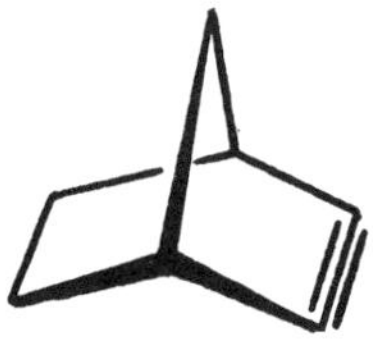

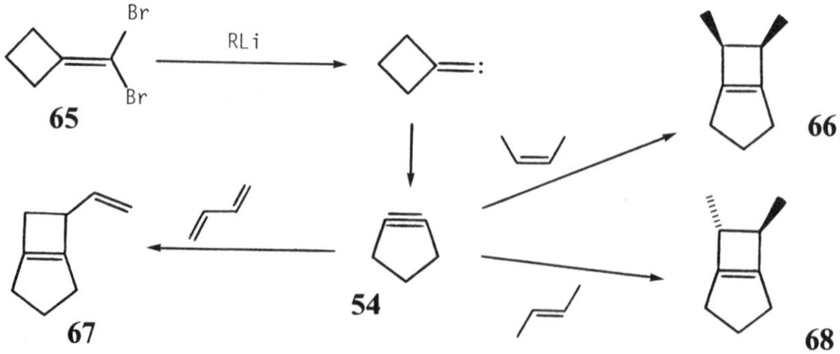

Scheme IV.

optimized geometry) show a symmetrical HOMO.[25] The same was observed for cyclobutyne.[93]

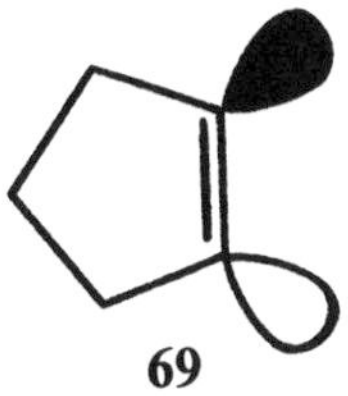

69

Gilbert and Baze have reported generation of cyclopentyne through base-promoted reaction of cyclobutanone (Scheme V) with dialkyl(diazomethyl)phosphonates.[100] Evidence for stereospecific reaction with *cis*- and *trans*-1-methoxy-1-propenes was observed.[100] To explain this result, the authors considered "antarafacial participation of cyclopentyne in a concerted cycloaddition." Gilbert also noted that MINDO/3 calculations suggested cyclopentyne to be *transition state* linking two π complexes, rather than an energy minimum. This probably is incorrect, an artifact of the method of calculation.[100c]

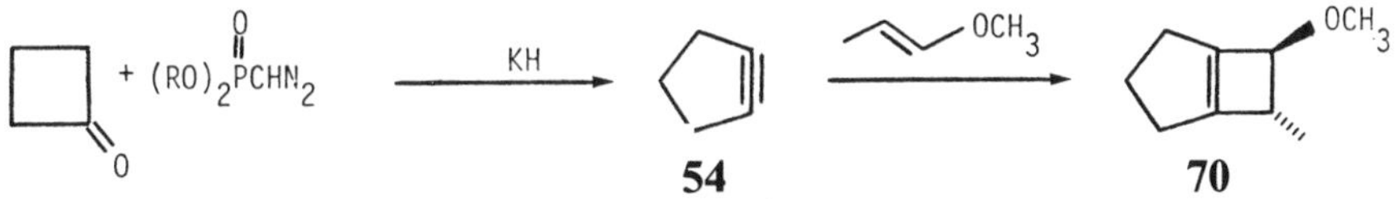

+ $(RO)_2\overset{O}{\overset{\|}{P}}CHN_2$ —KH→ 54 —(OCH_3)→ 70

Scheme V.

Bolster and Kellog have reported generation of thiacyclopentyne **(71)** from a bisdiazo precursor.[102] This proved to be readily trapped by a variety of reagents.

71

Acenaphthyne **(61)** has been the target of a number of groups.[99a,103] Nakayama generated this elusive alkyne (Scheme VI) from bisdiazo compound **72.** More recently, the same group used a Ramberg-Bäcklund approach for generation of **61.**[103] Trimerization to decacyclene **(76)** was observed in both studies.

Chapman and co-workers have generated acenaphthyne from irradiation of cyclopropenone **74** in an argon matrix.[99a] Among the reactions observed upon warming the matrix were trimerization to **76** and reaction with water to form ketone **75.**

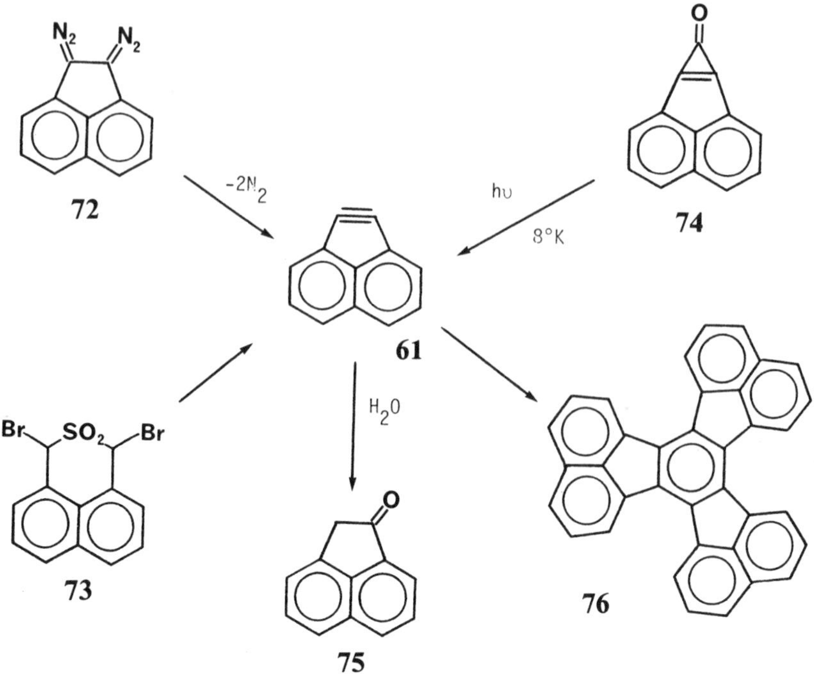

72 -2N2 61 hν 8°K 74
73 H2O 75 76

Scheme VI.

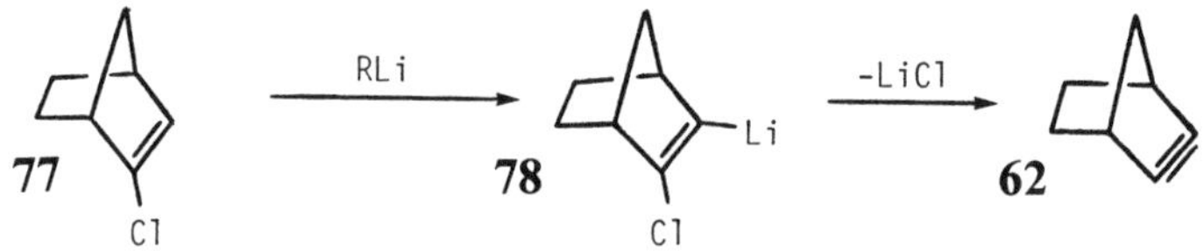

77 Cl RLi 78 Cl Li -LiCl 62

Scheme VII.

Gassman and Gennick have generated norbornyne **(62)** efficiently from treatment (Scheme VII) of 1-chloronorbornene with alkyllithiums.[104] In the absence of various trapping agents, this substance yields two trimers.[22]

C. Higher Cycloalkynes

In rings of greater than five carbons, there seems little difficulty in accepting a cycloalkyne structure.[3,8] There are numerous studies on the chemistry of cyclohexyne and cycloheptyne.[3,8] Cyclooctyne is a stable, isolable substance. Comprehensive discussions are given in the recent review by Krebs and Wilke[8] and in a monograph by Wentrup.[3] There is an abundant and growing literature on the synthesis and properties of very large cyclic polyalkynes.[91,92]

For example, Scott and co-workers recently reported the preparation of **79,** the first example of a "pericyclyne."[105a] Sakurai and co-workers have prepared a series of analogous compounds with dimethylsilyl and acetylene units in cyclic arrays.[105b] Perhaps some future review will deal with the question of the *largest* cyclic polyalkyne that might be prepared!

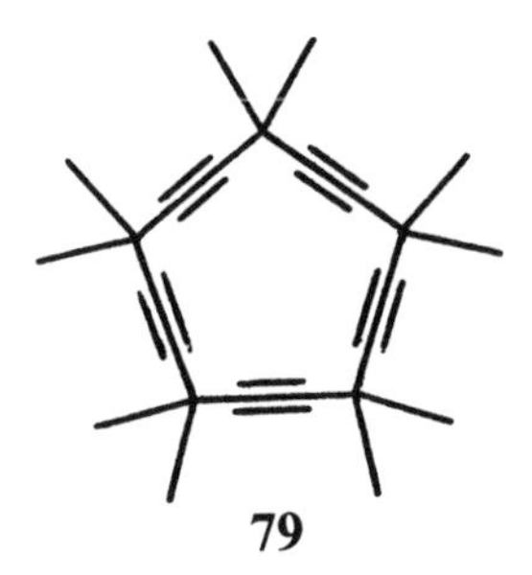

79

4. CYCLIC ALLENES

A. Introduction

As every first-year organic chemistry student learns, the equilibrium geometry for allene is linear, with the two π bonds and CH_2 groups orthogonal. Incorporation of an allene in a ring smaller than 10 carbons will bend and twist the allene (Scheme VIII), thus introducing strain and resultant kinetic reactivity.[106] These ring constraints should decrease the barrier to π-bond rotation which, in cyclic allenes, interconverts enantiomers.[44,107,108] Eventually, the ring constraints may enforce a planar allene geometry. Thus, one can pose two questions: What is the limiting ring size for chiral equilibrium structure, and what is the electronic nature of ground state planar allene? Neither of these questions is easily answered; indeed early attempts to settle both issues[109,110] led to what now are regarded as incorrect solutions. Exper-

80 **81** **82**

(* = ·, +, -)

Scheme VIII.

imentally, it is well known that the limiting size for an *isolable* cyclic allene is 1,2-cyclononadiene **(87)**.[2,106]

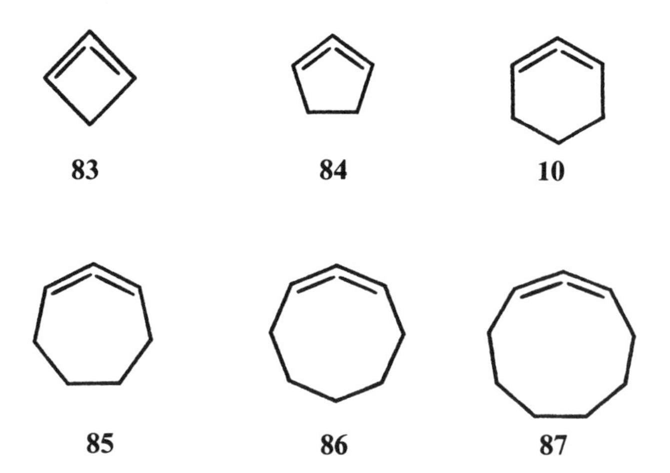

Renewed interest in these questions resulted from efforts fo extend Salem's "sudden polarization" phenomenon to allenes.[43,44] Briefly stated, Salem recognized[43a] that twisted singlet excited states of alkenes, which usually correspond to energy minima, will be distinctly zwitterionic, rather than diradical. For example, excited singlet ethylene is predicted to twist to a C_s geometry (**88**; one end pyramidalized), which is strongly polarized.[43b] Extension of this concept to allene led to the conclusion that two polarized minima, **89** and **90**,[44] should exist. The C_{2v} bent minimum corresponds to the geometry expected for a cyclic system.

$$H_2C{=}CH_2 \xrightarrow{h\nu} \mathbf{88}$$

$$H_2C{=}C{=}CH_2 \xrightarrow{h\nu} \mathbf{89} \quad \mathbf{90}$$

Figure 3-3 shows the molecular orbitals and configurations for C_{2v} planar allene.[44,107a,111] One might initially expect the ground state to be the zwitterion 1^1A_1, and the open-shell diradical 1^1A_2 to be an excited state. However,

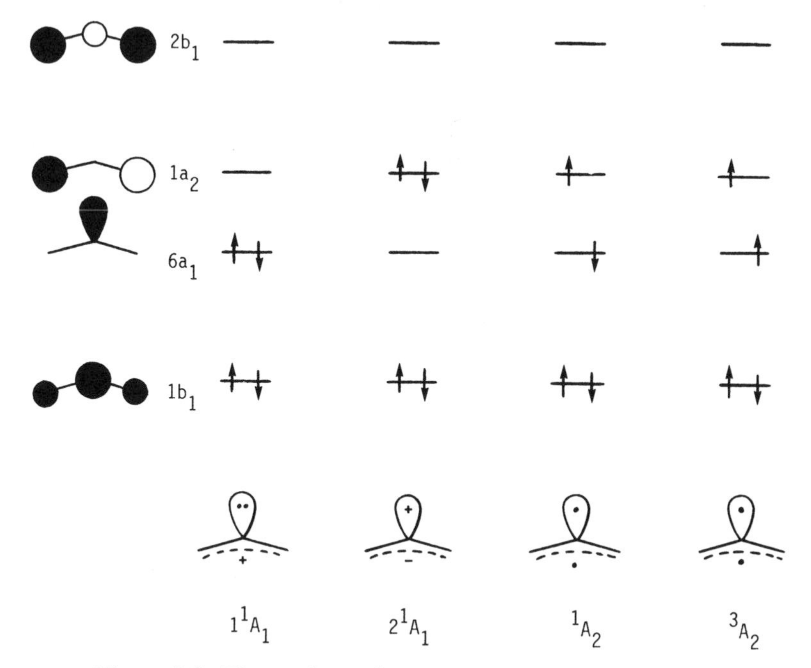

Figure 3-3. Electronic configurations for planar allene (C_{2v}).

the two nonbonding molecular orbitals $6a_1$ and $1a_2$ are quite close in energy and, at the one-electron level of theory, these two states would be nearly degenerate. Ab initio SCF and MCSCF calculations very clearly give the order $^1A_2 < {}^1A_1$ with a gap of about 1 eV.[44,107a] The simplest explanation for this result is that the closed-shell state is of higher energy due to an additional intrapair electron repulsion.

To complicate matters further, the open-shell singlet and triplet ($^1A_2$ and $^3A_2$) are nearly degenerate. This occurs because the exchange integral between the two odd electrons is extremely small, as a result of the different spatial localization of the two singly occupied molecular orbitals.[107a] however recent MCSCF calculations predict the singlet to be lower by about 5 kcal/mol.[111]

In sum, the order of the four states shown in Figure 3-3 is predicted to be $1^1A_2 < 1^3A_2 < 1^1A_1 < 2^1A_1$. Figure 3-4 shows the correlation of the three lowest states with those of a chiral C_2 allene, such as would exist in a strained cyclic system.[111] It is worth noting that the thermal racemization barrier in allene is predicted[107a] to be a bent (134.3° bending from MCSCF calculations) $^1A_2$ species,[111] and the in-plane bending potential for this state is quite soft.[44] Thus, while a chiral (C_2) allene may be highly strained in a small ring because of weakening of the π bond, the geometry that corresponds to the π-bond rotational barrier will be essentially unstrained. As a consequence, the decrease in the rotational barrier in a cyclic allene, relative

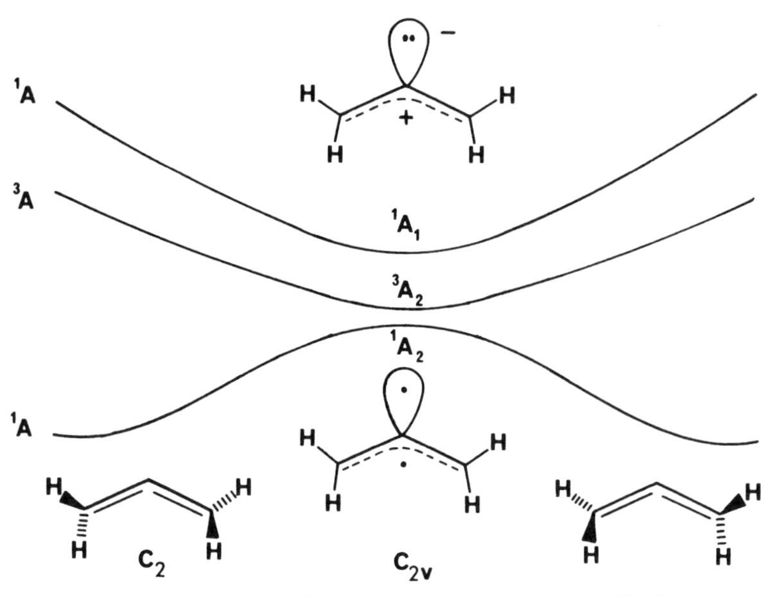

Figure 3-4. State correlations for racemization in a C_2 allene.

to allene (~ 50 kcal/mol[107,108]), should be a very good approximation to allene strain. Of course, the ground state rotational barrier in parent allene is not known experimentally; however, Roth and co-workers determined a value of 46.3 kcal/mol for the (chiral) dimethyl derivative.[108] This led to an estimate of about 50 kcal/mol for unsubstituted allene. Early theoretical estimates for this barrier were quite high (> 80 kcal/mol) because of the incorrect use of an RHF wave function (which yields a zwitterionic state) and D_{2h} geometry. More recent calculations give numbers that agree well with experiment.[107,111]

B. 1,2-Cyclohexadiene: The Belated Convergence of Theory and Experiment

The structure of 1,2-cyclohexadiene **(10)** has generated considerable interest—and controversy. This archetypal strained allene may be prepared by a surprising number of routes, summarized in Scheme IX. In the absence of other trapping agents, a dimer and two tetramers are formed.[109] Early workers do not apper to have seriously considered a chiral allenic structure **(10),** and the planar allene structures **91–94** have been proposed by various

10 **91** **92** **93** **94**

Scheme IX.

groups. These correspond to the elctronic configurations for planar allene shown in Figure 3-3. Moore and Moser first proposed the zwitterionic structure **91,** but noted that the triplet **(94)** may be the ground state.[109] They concluded that either representation implied high reactivity. The idea of a singlet zwitterionic structure was supported by model INDO calculations by Dillon and Underwood, which showed this zwitterion to be an energy minimum.[110] This *is* a minimum—but on the excited-state potential surface (Figure 3-4)![44] These authors also noted that the triplet might be of lower energy. Bottini and co-workers have performed extensive studies of the chemistry of 1,2-cyclohexadiene.[112] One conclusion was that a singlet diradicallike structure **(93)** was involved in dimerization, and trapping with dienes. Nucleophilic trapping was suggested to occur through zwitterion **92,** or a "badly twisted allene carbene," drawn as **95.** Balci and Jones recognized that, among the structural possibilities, only **10** is chiral, and they successfully trapped (Scheme X) optically active **10** with diphenylisobenzofuran.[113] This clever experiment was based on the use of an isotope effect with optically active vinyl bromide **(96);** that is, HBr should be eliminated in preference to DBr. At higher temperatures, competitive racemization of the allene was inferred from a decrease in optical activity of the cycloadducts.[113] Balci and Jones thus concluded that 1,2-cyclohexadiene exists as a bent, chiral structural, which is easily racemized through an achiral geometry.[113] More recently, Wentrup has trapped pyrolytically generated **10** in a cryogenic matrix and has observed its infrared spectrum.[114]

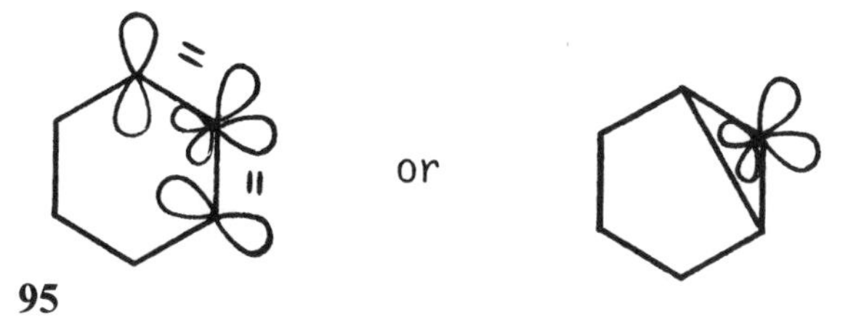

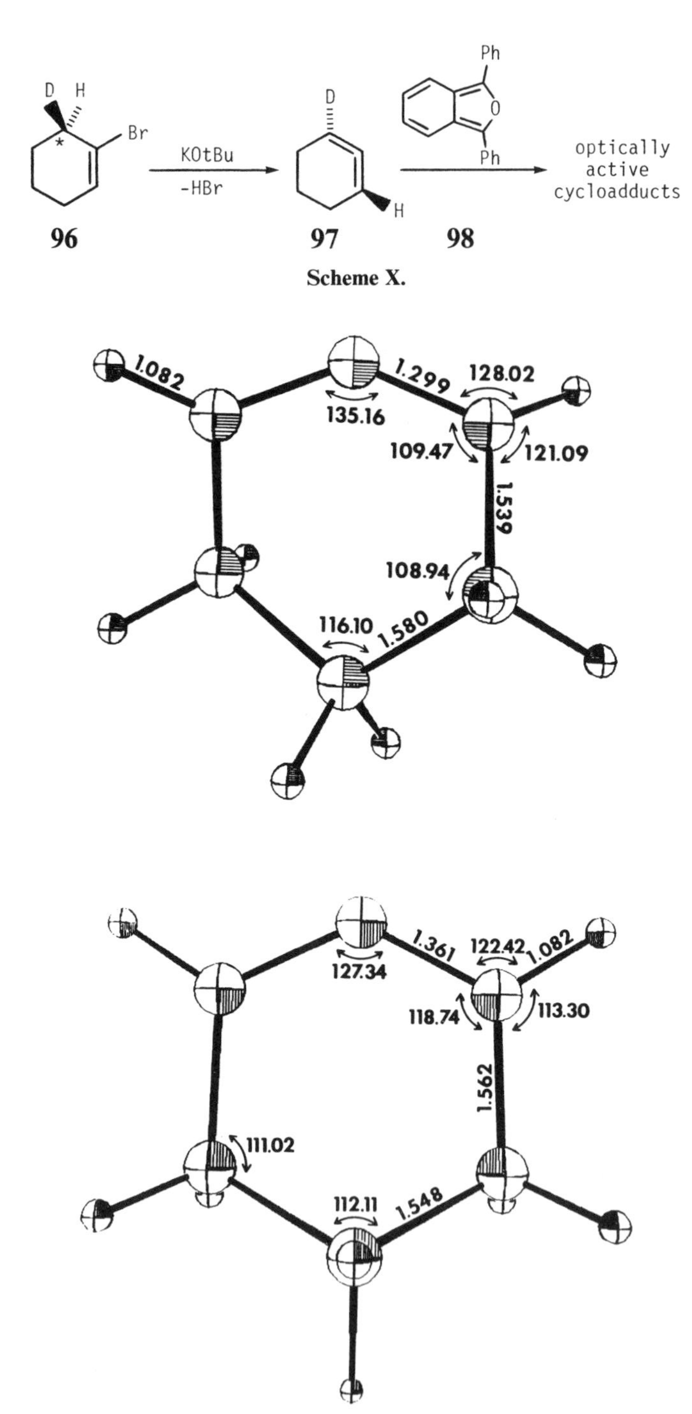

Scheme X.

Figure 3-5. STO−3G optimized geometries for chiral 1,2-cyclohexadiene (**10**) and the geometry (**93**) corresponding to its barrier to thermal racemization.

Theoretical interest in this problem was rekindled by Jones's elegant experiments, and the realization that the zwitterionic structure **91,** initially proposed by Moore and Moser,[109] should be an excited state of **93.** This was shown by Pople's calculations on the allene rotational barrier,[107a] as well as MCSCF and CI studies of planar allene.[44,111] Thus, the two extremes on the ground state potential surface should include chiral structure **10** and planar biradical **93.** Initial MNDO calculations, with complete geometry optimization, afforded a prediction that **93** should be 1.5 kcal/mol above chiral structure **10.**[115] Ab initio SCF calculations at the 3–21G + polarization // STO-3G level increased this barrier to 3.9 kcal/mol.[111,115] The optimized geometries for **10** and **93** are shown in Figure 3-5. Full optimized reaction space (FORS-MCSCF) or Möller-Plesset calculations increased this estimated inversion barrier to 15.0 or 23.7 kcal/mol, respectively.[115]

At this point, we may summarize that first experiment[113] and now theory[111,115] have converged on the remarkable conclusion that 1,2-cyclohexadiene should maintain a chiral equilibrium geometry. This is a very unusual and highly strained structure, but certainly no more so than many bridgehead alkenes, propellanes, or other strained molecules.[2]

Several heterocyclic analogues of 1,2-cyclohexadiene have been described. 1,2-Dioxacyclohexa-3,4-dienes **(99)** have recently been postulated as intermediates (Scheme XI) in the addition of singlet oxygen to but-1-ene–3-ynes.[116] This strained peroxide fragments to afford the observed products. The authors briefly reported results of ab initio geometry optimization on **99** (R = H), which was consistent with a chiral equalibrium geometry. An inversion barrier of 28 kcal/mol was predicted, although no description of the computational methodology was given.[116]

Cuthbertson and co-workers have reported the results of MNDO calculations on allene **100a** and higher homologues.[117a] In all cases, a chiral allenic structure was preferred, even though a planar zwitterionic structure would

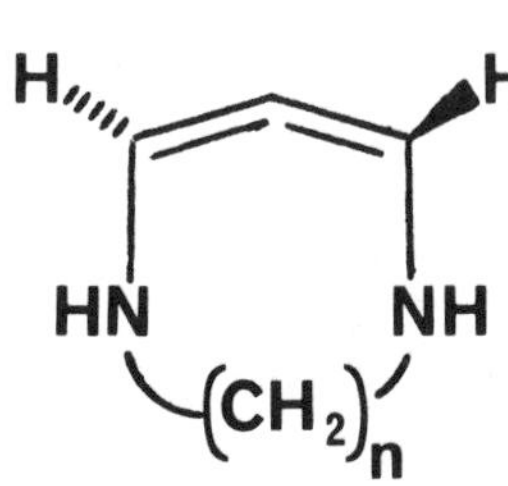

100a n = 1

b n = 2

c n = 3

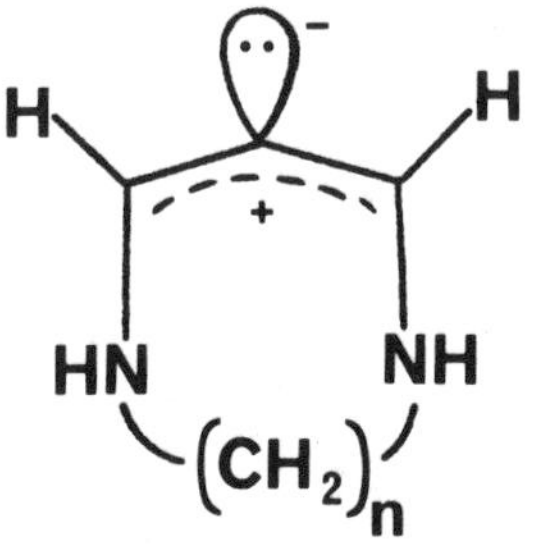

101 n = 1,2,3

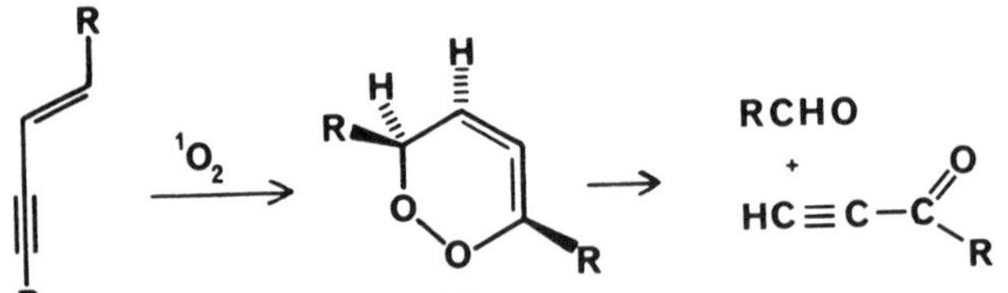

99

Scheme XI.

enjoy significant resonance stabilization. Experimental evidence for these structures has been described, although these data do not mandate chirality.[117b] Protonation at the central allenic carbon is generally observed.

C. Four- and Five-Membered Rings

Can an allene moiety exist as anything but a planar structure in rings with fewer than six carbons? At present, there are insufficient experimental data to resolve this question. Solution phase experiments that might have yielded 1,2-cyclopentadiene **(84)** instead gave evidence for its isomer cyclopentyne.[118a–c] Attempts at matrix isolation of **84** are as yet inconclusive.[118d]

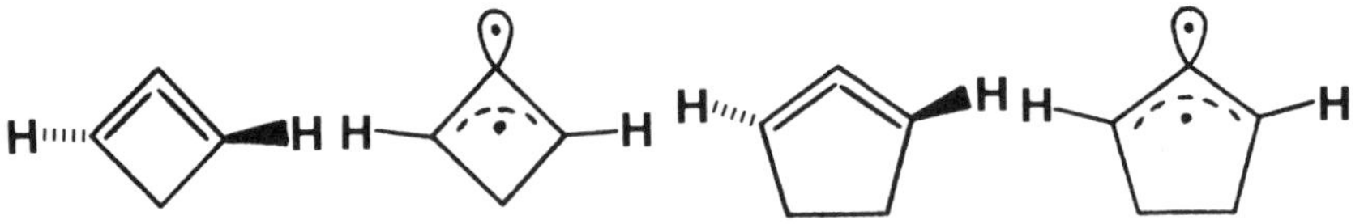

83 **102** **84** **103**

Recent theoretical studies on 1,2-cyclopentadiene (Table 3-3) suggest that it may have a chiral ground state, with a barrier to inversion (via **103**) of 2–5 kcal/mol.[111] This number is derived from MCSCF calculations on **84** and **103.**[111] The optimized structure for **84** (STO-3G RHF) has the allenic hydrogens bent 21° out of the plane defined by the three allenic carbons. The authors noted that the inversion barrier is well within the confidence limits, that might be expected for this level of calculation.[111] One hopes that definitive experimental results will be forthcoming. Unfortunately, biradical **103** probably would be a ground state singlet, which seems to preclude the use of electron spin resonance (ESR) spectroscopy. Additionally, the inversion barrier probably is too small to have a chance of intercepting chiral 1,2-cyclopentadiene. If this compound can be generated in a matrix, it may be partially resolvable through irradiation with circularly polarized light.

One novel possibility is that cyclopentadienylidene and its benzannelated derivative **107** could exist as allenic structures **106** and **108.** In the case of cyclopentadienylidene, this cannot be the ground state, since ESR studies have shown a ground state triplet,[119] however, many reactions of cyclopen-

Table 3-3. Theoretical Studies of Strained Cyclic Allenes

Structure	Allenic bending angle		Force field[b]	Out-of-plane hydrogen bending		MNDO, ΔH_f (kcal/mol)	Predicted inversion barrier (kcal/mol)	Estimated allene strain (kcal/mol)
	STO−3G[a]	MNDO[a]		STO−3G	MNDO			
84	110.6°	121.0°		20.9°	13.0°	104.3	2–5	41.3
10	135.2°	138.5°		30.4°	22.9°	67.7	15	31.2
85	—	153.4°		—	27.6°	44.2	—	
86	—	161.5°	158–160°	—	31.0°	32.7	—	~15
87	—	170.4°	171°	—	33.7°	29.2	—	

[a]References 111, 115.
[b]Reference 126.

tadienylidene are attributed to a singlet.[120a–c] Previous theoretical studies have considered only planar singlets.[120d] Intriguingly, MNDO calculations show an energy minimum (ΔH_f = 144.7) for chiral allene **106,** but this is predicted to be 21.2 kcal/mol above the triplet **104** (ΔH_f = 123.5 kcal/mol).[25] These energies are not quantitatively reliable, but they do suggest that allene **106** might be involved in singlet reactions of cyclopentadienylidene.[120] Further theoretical studies seem warranted, although these would not definitely resolve this question.

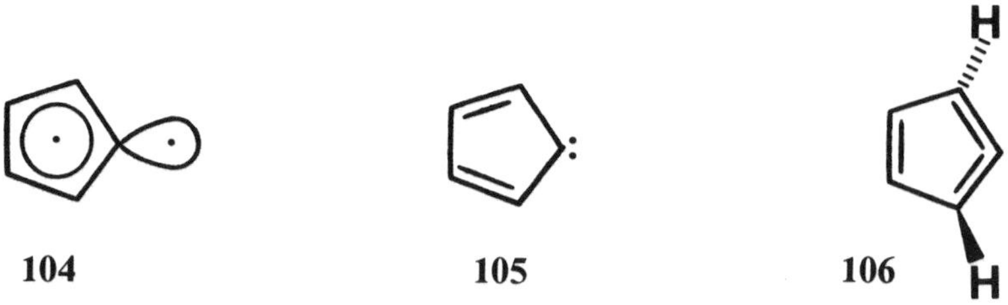

104 **105** **106**

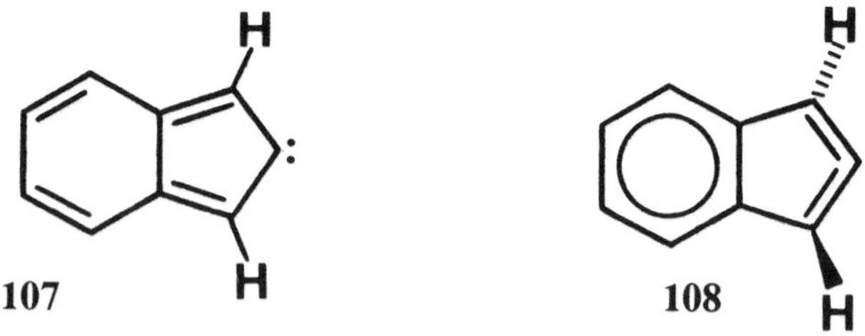

107 **108**

Tolbert and Siddiqui have reported the photogeneration of a diphenyl derivative of **107.**[121] An allene structure such as **108** was briefly considered, but did not seem to be consistent with electrophilicity of the intermediate.

In sum, there are no experiments that demonstrate the intermediacy of a chiral five-membered ring allene, but additional studies to resolve this question seem mandated. Of course, "1,2-cyclopropadiene" is simply cyclopropenylidene; Maier and co-workers have recently trapped this compound in a low-temperature matrix.[123]

We are aware of no reported attempts to generate 1,2-cyclobutadiene **(83).** There are a few theoretical studies on **83** or its related cyclopropylidene,[94,111,122] but these are not sufficiently high level to provide convincing results.

D. Larger Monocyclic Rings

In rings larger than six carbons, there seems to be little difficulty in accepting the possibility of a chiral allenic structure.[2,106] As confirmation, Balci and Jones have generated optically active 1,2-cycloheptadiene by the same method that was used for the six-carbon allene and have isolated active cycloadducts.[113] No evidence for competitive racemization was observed. Structural data from MNDO calculations on larger allenes are summarized in Table 3-3.

Visser and Ramakers have isolated stable platinum complexes of 1,2-cycloheptadiene and 1,2-cyclooctadiene.[124a] When the seven-carbon allene was liberated at −25°C by ligand displacement with CS_2, only the allene dimer was observed. By contrast, 1,2-cyclooctadiene proved sufficiently stable that its ^{1}H NMR spectrum and dimerization kinetics could be studied.[124] Jones and co-workers have recently isolated a fluxional Fp^+ complex [Fp = $C_5H_5Fe(CO)_2$] of 1,2-cycloheptadiene.[125] The barrier between chiral allenic and planar structures was estimated to be greater than 13.9 kcal/mol. Two distinct allenic vinyl protons were observed at −40°C; these coalesced at 29°C. Spin saturation transfer was employed to demonstrate rapid interconversion of the two hydrogens. 1,2-Cyclononadiene is the smallest unsubstituted cyclic allene that is kinetically stable at ambient temperature. This compound is readily prepared in large quantity and better than 99% purity by treatment of 9,9-dibromobicyclo[6.1.0]nonane with methyllithium.[12] Molecular mechanics calculations have been reported for 1,2-cycloocta- and cyclononadienes.[126] Results agree very well with more recent MNDO studies.[111]

E. Doubly Bridged Bicyclic Allenes

Bicyclic allenes of type **109** (Scheme XII) have been the subject of a number of synthetic attempts but only one successful synthesis. Nakazaki and co-workers prepared bridged allene **109a** from a carbenoid precursor.[127] The authors referred to this molecule as [8][10]screw[2]ene, a descriptive departure from IUPAC nomenclature. All experiments that might have yielded lower homologues of **109** by similar carbenoid routes have been diverted by

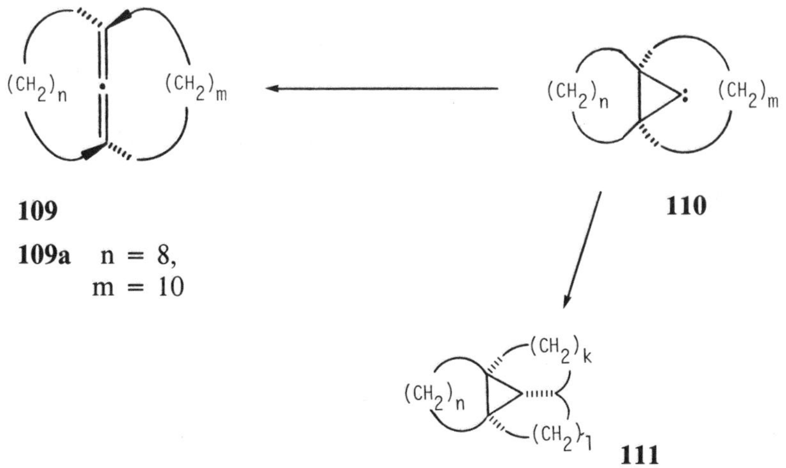

$(CH_2)_n$ $(CH_2)_m$ ← $(CH_2)_n$ $(CH_2)_m$

109 **110**

109a n = 8, m = 10

$(CH_2)_k$ $(CH_2)_n$ $(CH_2)_l$ **111**

Scheme XII.

Cl Cl

RLi

or

112 113 114

Scheme XIII.

C—H insertion, which yields tricyclics **(111).**[128–130] This is an excellent route to bicyclobutanes.[129,130] In more highly unsaturated systems, complex rearrangements occur.[131,132] Levin has suggested that chloride **112** yields (Scheme XIII) a bicyclic allene **113,** a closed structure of which **(114)** is trapped with diphenylisobenzofuran.[132] Warner and Chang have provided experimental evidence in closely related systems that argues against the intermediacy of allenes such as **113.**[133]

To summarize, structural limitations in bicyclic allenes remain predominantly unknown. The one example prepared to date, allene **109a,** is essentially unstrained.[127] Further advances here probably will require an allene synthesis that does not involve cyclopropylidene carbenoids. Molecular models suggest that three-membered ring opening in such systems will be severely restricted.

F. Cycloheptatetraene: Carbene or Allene?

Cycloheptatetraene **(116)** and its related carbenic structure cycloheptatrienylidene **(115)** have been the subject of more experimental study than any other small-ring cyclic allene. This species, which is a homologue of cyclopentadienylidene **(105** or **106),** may be generated through diverse routes (Scheme XIV),[134–136] the most novel of which is argon matrix irradiation of phenylcarbene.[135] Possible structures include the chiral allene **(116)** (C_2 symmetry) or planar carbene. Planar structure **115** contains an aromatic 6π-electron system and might initially be expected to be quite stable.

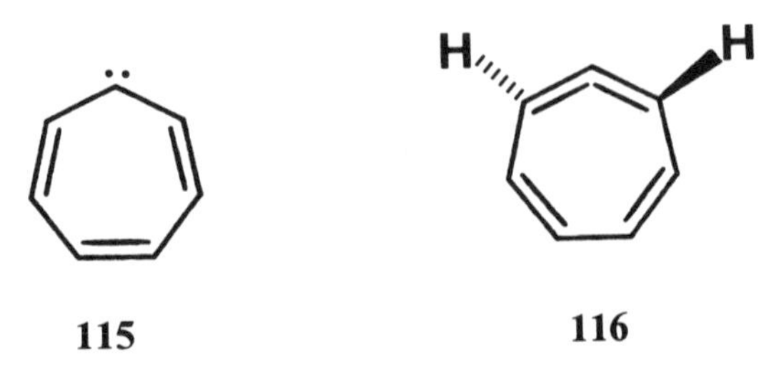

There have been four theoretical studies on the structure and energetics of structures **115** and **116.**[137–140] Results are collected in Table 3-4. Jones and

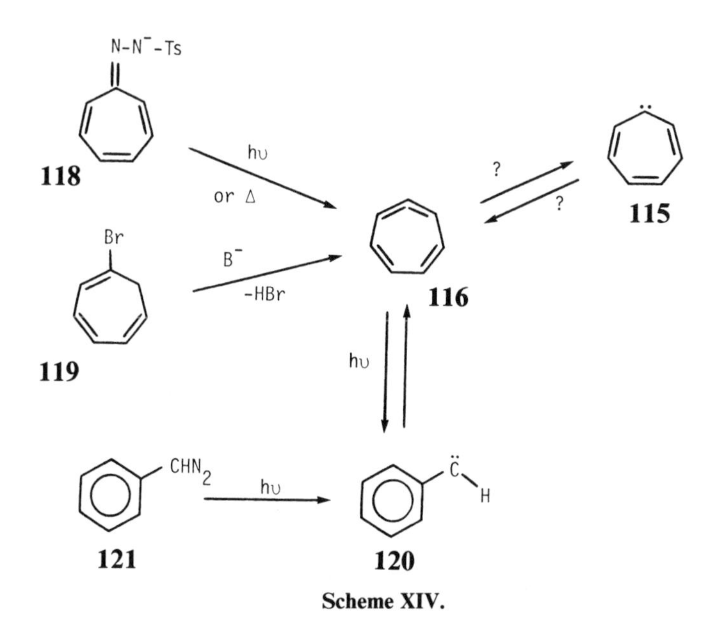

Scheme XIV.

TABLE 3-4. Theoretical Studies on Cycloheptatetraene

Structure	Relative energy (kcal/mol)[a]			
	INDO[b]	MINDO/3[c]	MNDO[d]	4–31G//STO–3G[e]
116	0.0	0.0 (87.0)	0.0 (91.5)	0.0 [0.0][f]
115	13.8	—	22.9 (114.4)	15.8
117	—	−3.5 (83.5)	5.9 (97.4)	0.0, 1,1[g] [30][f]

[a]Predicted heats of formation (kcal/mol) are given in parentheses.
[b]From Reference 137.
[c]From Reference 138.
[d]From reference 139.
[e]From Reference 140.
[f]Qualitative empirical correction, based on comparison of similar computational methods to methylene.
[g]Two different triplet states.

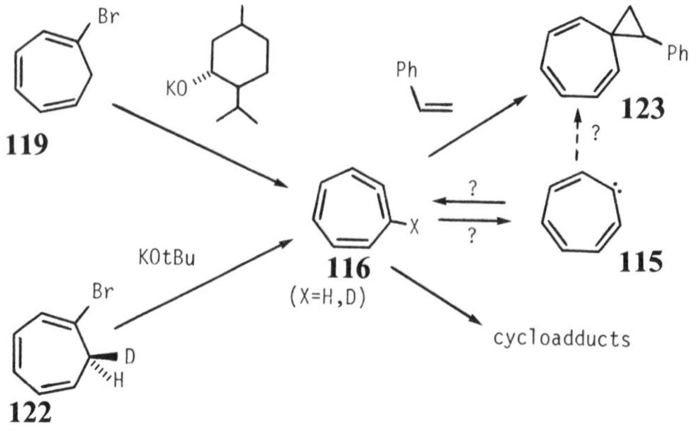

Scheme XV.

co-workers initially used the semiempirical INDO method to predict that the chiral allene will be 14 kcal/mol below cycloheptatrienylidene.[137] This prediction has withstood both higher level theory and experimental studies. Dewar and Landman later optimized this chiral structure with MINDO/3 and predicted that the planar carbene triplet might be slightly lower in energy.[138] Waali used the MNDO method to predict that cycloheptatrienylidene **(115)** should not correspond to an energy minimum, but will undergo deformation to the chiral structure.[139] At about the same time, Radom, Schaefer, and Vincent performed the first ab initio calculations on these systems.[140] Again, a chiral equilibrium geometry was predicted.

From a purely theoretical viewpoint, none of these studies is very convincing with respect to the relative energetics of cycloheptatetraene and cycloheptatrienylidene. This is in part because in no case were the correlation energy differences taken into account. Although the chiral structure **116** seems most likely to be of lower energy, we speculate that calculations that include correlation energy may show a shallow energy minimum for the carbene **115.** This problem is not one that can be confidently resolved with single-determinant wave functions.

Experimental studies on these intriguing molecules have provided more definitive results to some questions. Harris and Jones have observed that two approaches (Scheme XV) to optically active allene, one using a chiral base, the second employing an optically active monodeuterated substrate, both yielded chiral [2 + 4] cycloadducts with diphenylisobenzofuran.[134] These results were taken as unequivocal evidence for a chiral allenic precursor. To complicate this result, the spirononatriene **123,** which is isolated in the presence of styrene, showed no measurable rotation.[134] This result suggests that an achiral carbene **(115?)** is being trapped, although the authors noted that it is not definitive, since stepwise addition may occur.

Further argument for a carbenic structure is due to recent experiments by Kirmse.[136] Irradiation (Scheme XVI) of the tropone tosylhydrazone sodium

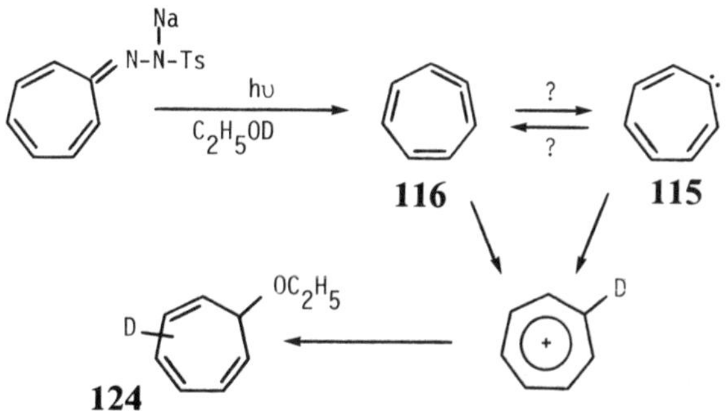

Scheme XVI.

salt in C_2H_5OD yielded 7-ethoxycycloheptatriene **(124)** in which statistical deuterium scrambling had occurred. A carbenic structure would be expected to protonate in this fashion, however, the authors noted that strained allenes generally produce vinyl ethers.[112]

There are several possible explanations for this result. First, the carbenic structure **115** is significantly more polar (MNDO dipole moment 4.03 debye)[139] than the allene **116** (0.83 debye) and may be more strongly stabilized in methanol. This could shift the **115–116** equilibrium (if this exists) such that protonation of the carbenic structure becomes favorable. As an alternative explanation, cycloheptatetraene **116** differs significantly from allenes such as 1,2-cyclohexadiene, because C-2 protonation yields an aromatic species, whereas nucleophilic attack at C-2 might yield a transient antiaromatic ion. Thus, Kirmse's results are far from unequivocal in supporting carbene **115.**[136]

Chapman and co-workers have generated cycloheptatetraene from matrix irradiation of phenylcarbene at 10 K.[135] The allenic structure was consistent with the infrared spectrum (absorptions at 1824 and 1816 cm^{-1}). A deuterium labeling study ruled out a bicyclo[4.1.0]hepta-2,4,6-triene intermediate.

G. Bicyclo[3.2.1.]octatrienes

In 1970 Bergman and Rajadhayaksha reported that base treatment of vinyl bromide **125,** or generation of a carbene from pyrolysis of a tosylhydrazone salt **126,** both yielded (Scheme XVII) **129,** *endo*-6-ethynylbicyclo[3.1.0]hex-2-ene.[141] The authors porposed the intermediacy of a "homoconjugated carbene," structure **127,** since the allene structure **128** seemed too strained. The zwitterionic structure **127** would have six electrons in a homoconjugated π system and was considered to be homoaromatic.

A year later, Klumpp and van Dijk reported isolation of the same product from photolysis (Scheme XVIII) of carbon suboxide with norbornadiene.[142] More recently, Freeman and Swenson reported[143] that carbene **131** yields

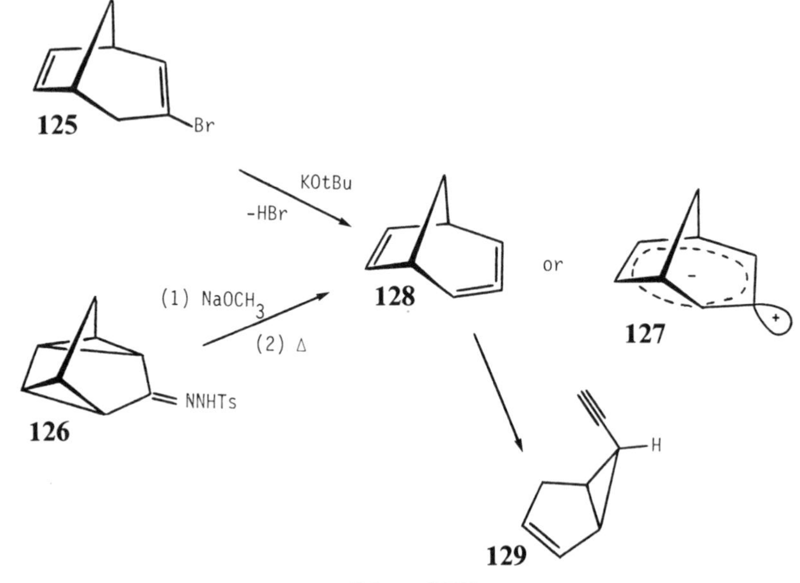

Scheme XVII.

alkyne **129** (3–17% of products), presumably as a result of 1,2-H migration to allene **128,** and subsequent rearrangement.

Balci and Harmandar recently reported generation (Scheme XIX) of a benzannelated derivative **(132),** which did not rearrange and was readily trapped.[144] In the absense of trapping agent, *t*-butyl ether **(133)** was isolated.

Balci and Jones have once again used the inherent chirality of allenes to approach this problem.[145] Treatment of vinyl bromide **(125)** with potassium menthoxide yielded optically active cycloadducts, with the activity decreasing as the reaction temperature was raised. Additionally, alkyne **129,** isolated in low yield (2–3%) in the absence of trap, proved to be optically active.[145] These experiments were consistent with a chiral allene, which undergoes facile racemization, competitive with rearrangement to **129.**

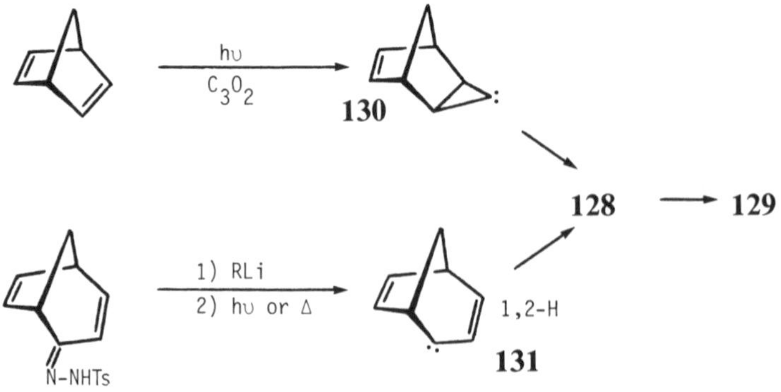

Scheme XVIII.

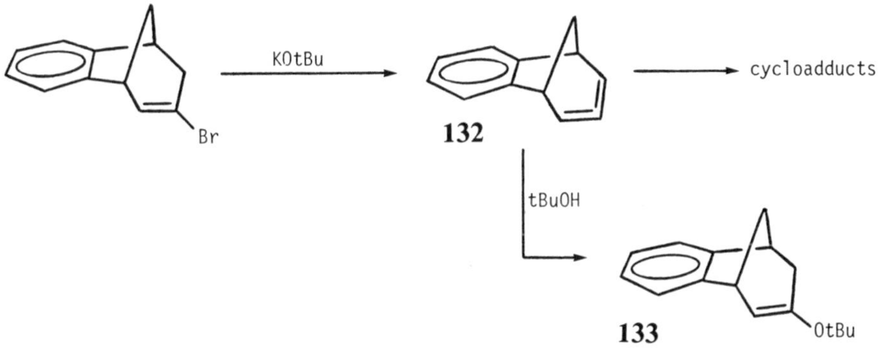

Br KOtBu **132** cycloadducts

tBuOH **133** OtBu

Scheme XIX.

Although allene **128** (C_8H_8) is currently too large for reliable ab initio calculations, the MNDO method should provide qualitatively useful results. However, by comparison to MNDO results for allene and 1,2-cyclohexadiene,[111] the predicted rotational barrier should be too small. Figure 3-6 shows an end-on view of the optimized structure for bicyclo[3.2.1.]octa-2,3,6-triene ($\triangle H_f$ = 122.8 kcal/mol).[111] As expected, the allene (formally a 1,2-cyclohexa- or cycloheptadiene) is strongly bent (111.7°) and twisted. This is a wholly remarkable structure.

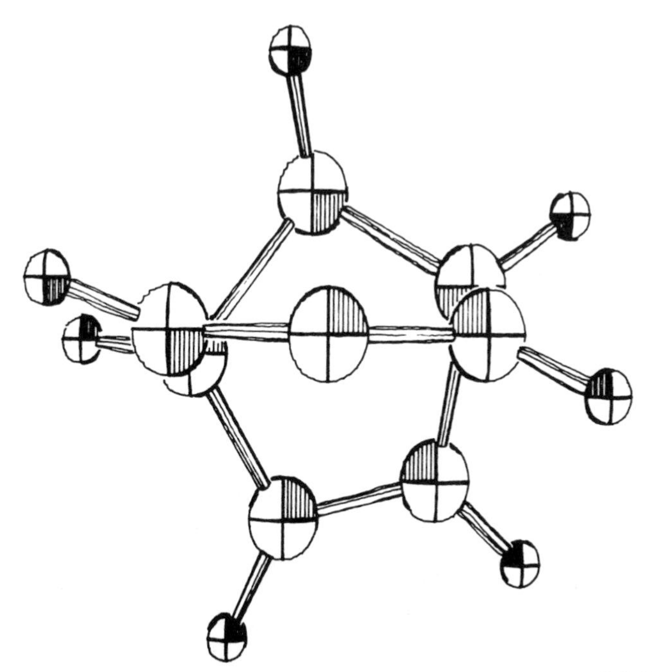

Figure 3-6. MNDO optimized geometry for bicyclo[3.2.1]octa-2,3,6-triene.

Racemization of **128** should proceed through the diradical **134.** MNDO geometry optimzation for **134** did not converge satisfactorily,[111] but the best structure was about 1 kcal/mol below that for **128.** By analogy to results for 1,2-cyclohexadiene, where MNDO significantly underestimated the inversion barrier,[111,115] we expect that high-level ab initio calculations would simply reverse this order.

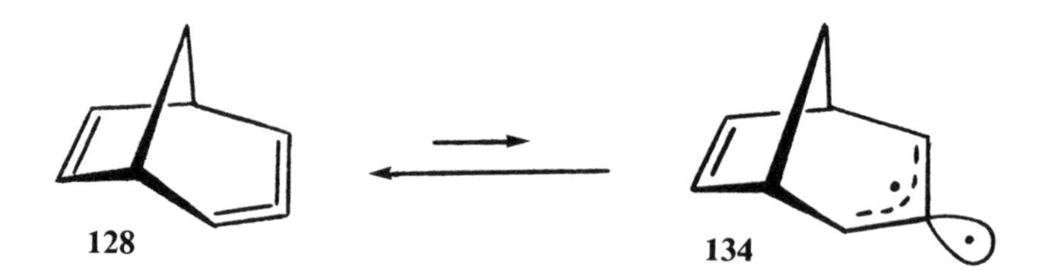

H. Related Heteroallene Structures

Studies of cyclic heterocumulenes provide some fascinating comparisons with the carbon analogues. Structural limitations in some heteroallenes have been clearly defined, at least with respect to isolability. Heterocumulenes are more readily bent than their hydrocarbon analogues, presumably because of their weaker π bonds.

Firl and co-workers have reported the synthesis of cyclic ketenimine 1-aza-1,2-cyclooctadiene.[146] In contrast to the hydrocarbon analogue **86,**[124] this proved to be isolable under ordinary conditions. This compound can exist as two diastereomers, **135a** and **135b,** which were observed to undergo ready thermal interconversion. The inversion barriers in ketenimines usually are less than 15 kcal/mol.

$PhCH_2$ N CH_2Ph ⇌ $PhCH_2$ N CH_2Ph

135a **135b**

Richter and co-workers very recently described an investigation of a homologous series of cyclic carbodiimides, synthesized as shown in Scheme XX.[147] The eight-membered ring carbodiimide **139** proved to be isolable and could be stored for several days, again in contrast to the hydrocarbon analogue **86.**[124] 1,3-Diazacyclohepta-1,2-diene **(138)** could not be isolated, but instead yielded a crystalline [2 + 2] dimer **140.** Attempts to prepare and

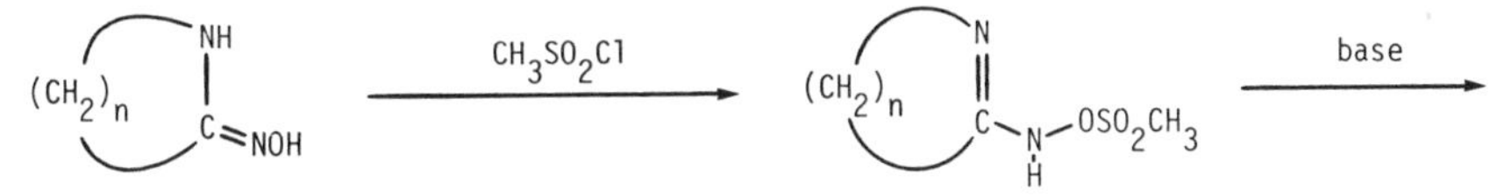

136 → (CH_3SO_2Cl) → **137** → (base)

138 → (n = 4) → **140** + trimer

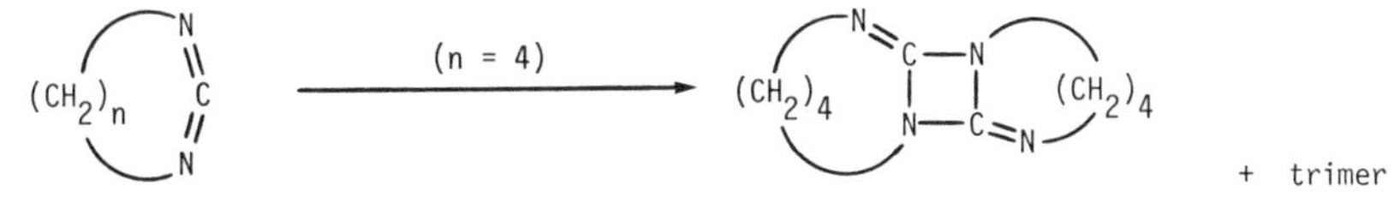

138 n = 4 (dimerizes)
139 n = 5 (isolable)
141 n = 3 (not generated)

Scheme XX.

trap 1,3-diazacyclohexa-1,2-diene **(141)** by a different route were unsuccessful.[147]

Carbodiphosphoranes ($R_3P{=}C{=}PR_3$) apparently have a very "soft" bending potential and can be incorporated into amazingly small rings. Schmidbaur and co-workers recently reported the synthesis and *isolation* of carbodiphosphoranes **143a–143c** (Scheme XXI).[148] As might be expected, stability of these compounds decreased with dimishing ring size. An X-ray structure of **143b** revealed a P=C=P angle of 117°; the authors suggested that this compound might best be described as **144**, as indicated from its poor solubility in organic solvents.[148]

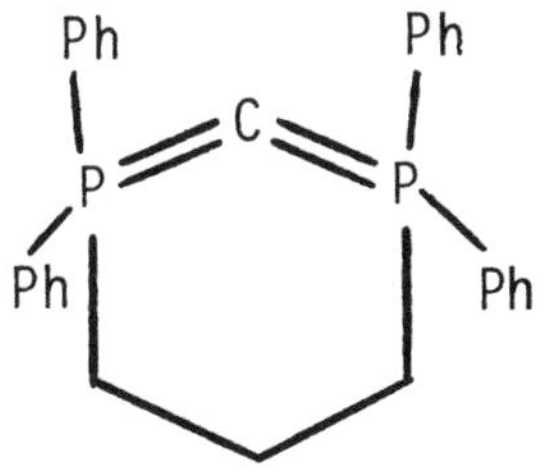

143b 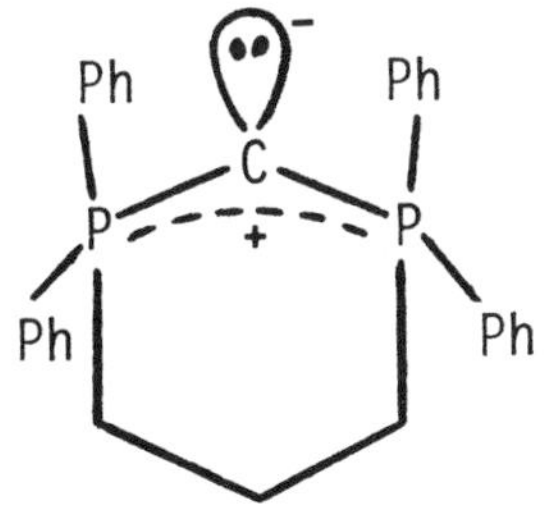 **144**

Cyclic heteroallene **145** is a well-known intermediate that is readily accessible from carbene or nitrene precursors.[3,149] Matrix isolation studies, in particular the infrared spectrum, were consistent with the ketenimine structure **145.**[149] This is quite analogous to cycloheptatetraene (Subsection F of Section 3). Wentrup and co-workers have reported evidence for intermediacy of cyclic carbodiimide **149** and a dibenzannelated derivative.[3,150] As noted

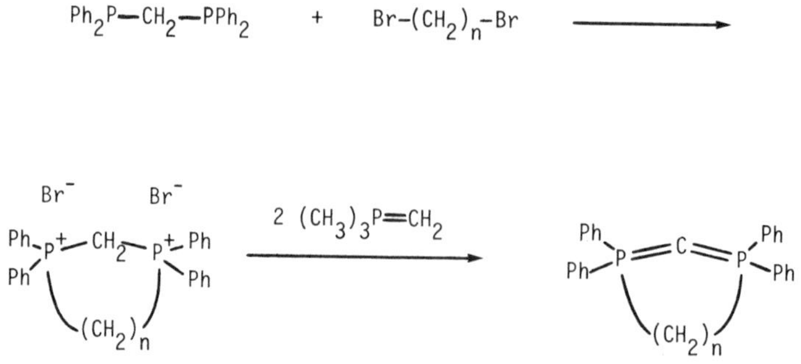

$Ph_2P{-}CH_2{-}PPh_2$ + $Br{-}(CH_2)_n{-}Br$ ⟶

142 ⟶ [2 $(CH_3)_3P{=}CH_2$] 143a n = 2, 143b n = 3, 143c n = 4

Scheme XXI.

above, the parent seven-membered ring carbodiimide **(138)** did not prove to be isolable, but readily dimerized.[147]

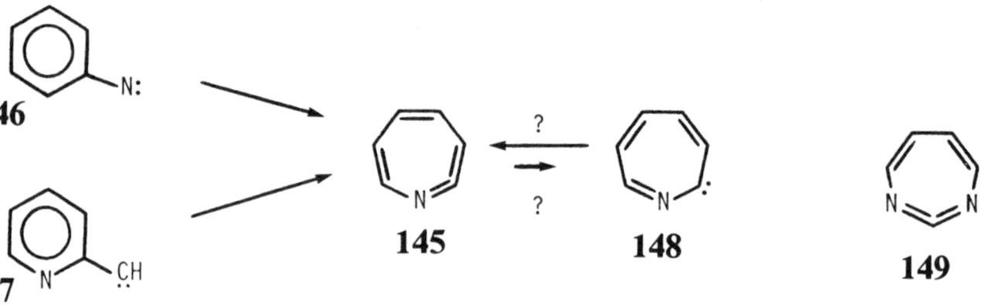

146, 147 ⟶ 145 ⇄ 148 (?); 149

5. CYCLIC BUTATRIENES

A. Introduction

1,2,3-Butatriene is geometrically similar to ethylene; that is, it is planar, and 1,4-disubstituted derivatives can exist as cis–trans isomers. Incorporation of a butatriene in a small ring will not alter the planarity, but will cause in-plane bending and resultant angle strain. This does not significantly affect the out-of-plane butadiene-like π system, but it will introduce strain due to decreased overlap in the in-plane "ethylene" moiety. The resultant situation, shown in Scheme XXII, is similar to that in orthobenzyne or other small-ring alkynes. In part because only one of the three π bonds is affected, one might expect the bending potential to be initially rather soft. To date, there are no reported examples of *trans*-cyclic butatrienes.

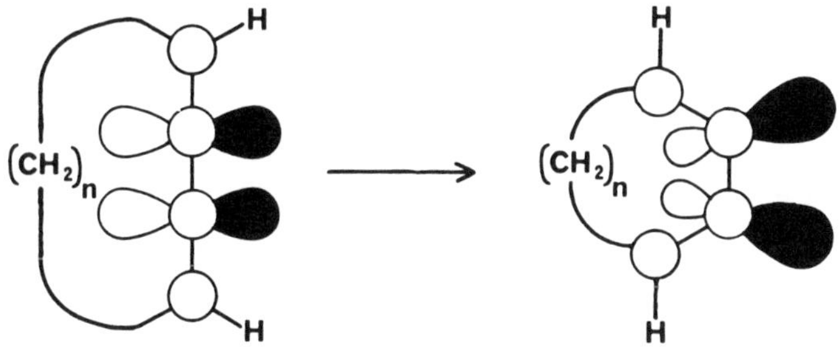

(CH2)n H H 150 → (CH2)n H H 151

Scheme XXII.

B. Theoretical Studies

There has been one report of theoretical studies on cyclic butatrienes.[151a] Figure 3-7 shows the estimated bending potential for 1,2,3-butatriene. This was calculated by optimizing the geometry with MNDO, using a restricted bending angle, followed by SCF calculation at various levels of theory. The authors noted surprisingly good agreement among the different levels.

Figure 3-8 shows the MNDO-optimized geometries for cyclic butatrienes from five-to-nine carbons.[151a] Heats of formation, strain estimates, and other salient geometric data are summarized in Table 3-5. Two conformations (C_s and C_2 symmetry) were considered for 1,2,3-cyclononatriene. Strain estimates are derived from plotting the optimized bending angle for a given molecule versus the bending potential in Figure 3-7. Of course this assumes only a single source of strain, which would underestimate the strain in smaller homologues. Nevertheless, these estimates provide argument that strain in the butatriene moiety roughly doubles with each decreasing carbon. 1,2,3-Cyclononatriene **(155)** has remarkably little predicted strain (5 kcal/mol for the more stable conformation).

C. Experimental Studies

Experimentally, little is known about the chemistry of cyclic butatrienes. Moore and Ozretich prepared 1,2,3-cyclodecatriene **(157)** in 1965, by carbenoid ring expansion (Scheme XXIII) of 1,2-cyclononadiene.[152] Nearly 10 years later, Bhagwat and Devaprabhakara reported preparation of 1,2,3-cyclotridecatriene **(159)** from dibromocyclopropane **158**, presumably through base-catalyzed isomerization of vinylallene **160**.[153] Although cis and trans isomers of **159** are possible, according to models, the detection of only one isomer was reported.[153]

More recently, Angus and Johnson have prepared 1,2,3-cyclononatriene **(155)**.[151a] This compound proved to be stable under ordinary laboratory con-

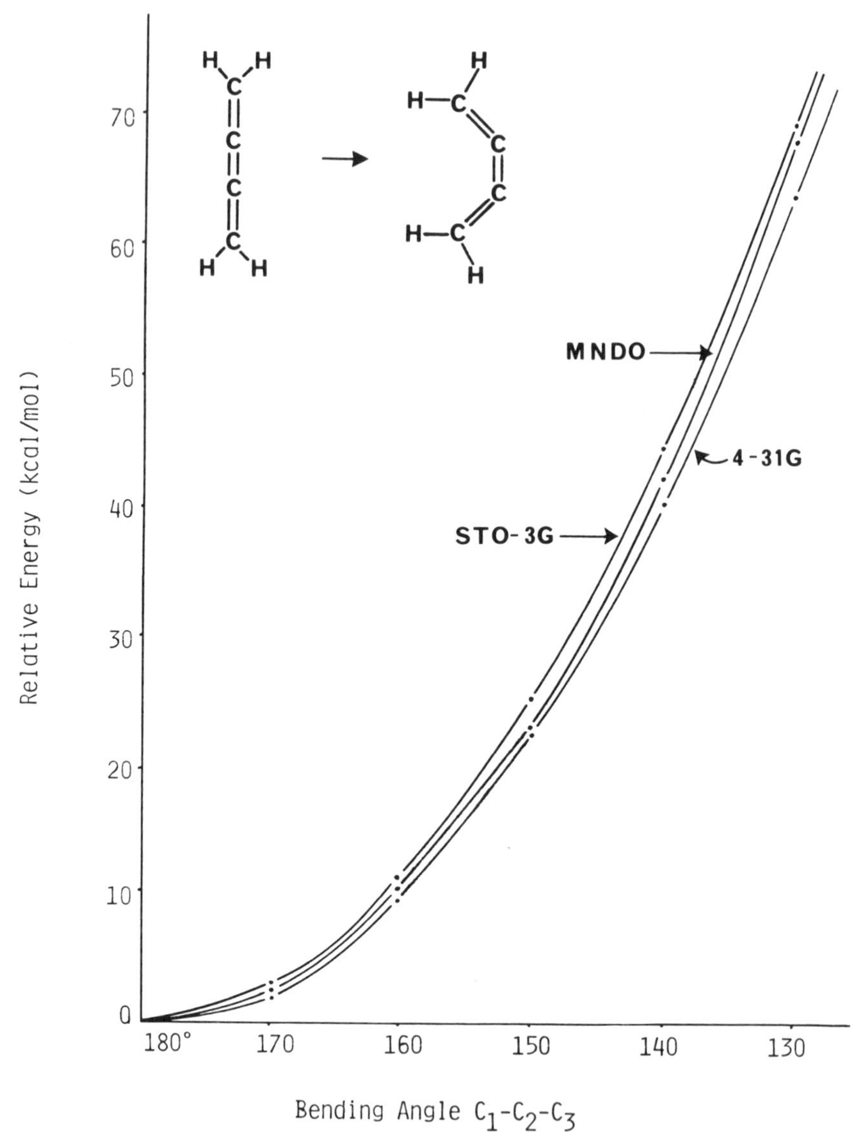

Figure 3-7. Model calculations for in-plane bending in a 1,2,3-butatriene.

ditions but, like many butatrienes, easily polymerizes when not in solution. The key to this synthesis (Scheme XXIV) is the addition of dibromocarbene to kinetically unstable 1,2-cyclooctadiene at −78°C. More recently, a crystalline rhodium complex of **155** has been prepared.[151b] The authors suggested that cyclononatriene might be the smallest isolable cyclic butatriene.[151]

Szeimies has successfully generated 1,2,3-cycloheptatriene **(153)** from the rearrangement (Scheme XXV) of bicyclobutene **47**.[87] Both **47** and **153** could

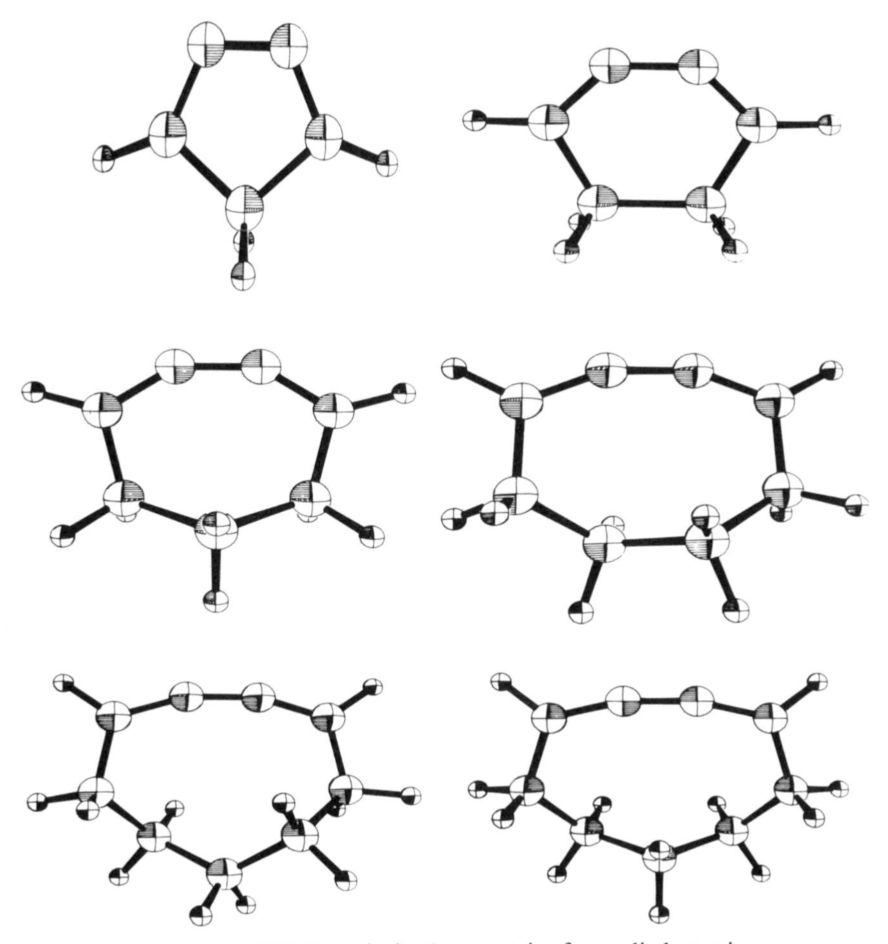

Figure 3-8. MNDO-optimized geometries for cyclic butatrienes.

be trapped as cycloadducts with diphenylisobenzofuran or 9-methoxyanthracene. Lowering the concentration of trapping agent led to an increase in the relative yield of cycloadduct from **47.** The authors noted that the **47** → **153** rearrangement is forbidden by orbital symmetry, but a stepwise route is quite feasible. A carbenic structure such as **165** seemed to be the most likely precursor to 1,2,3-cycloheptatriene.

47 ⟶ **164** ⟶ **165** ⟶ **153**

TABLE 3-5. Theoretical Studies on Cyclic Butatrienes[a]

Compound	$C_1-C_2-C_3$ Bending angle	MNDO, ΔH_f (kcal/mol)	Estimated butatriene strain (kcal/mol)
152	116.4°	186.3	~130.0
11	132.2°	129.5	60.5
153	145.0°	98.0	31.5
154	156.3°	74.9	14.0
155	165.1° 161.9°	60.5(C_s) 64.8(C_2)	5.0 8.0

[a]Compound **155** has two different conformations.

Subsequent efforts by the same group to generate 1,2,3-cyclohexatriene **(11)** through rearrangement of bicyclobutene **(48)** were unsuccessful.[88] The authors suggested that the lack of rearrangement of **48** is due to decreased exothermicity.[88] This is consistent with the observation that **11** has about double the predicted strain (61 vs 32 kcal/mol) of **153.**[151] We also speculate that if 1,2,3-cyclohexatriene were generated, it would undergo an allowed electrocyclic ring opening to dienyne **166.**

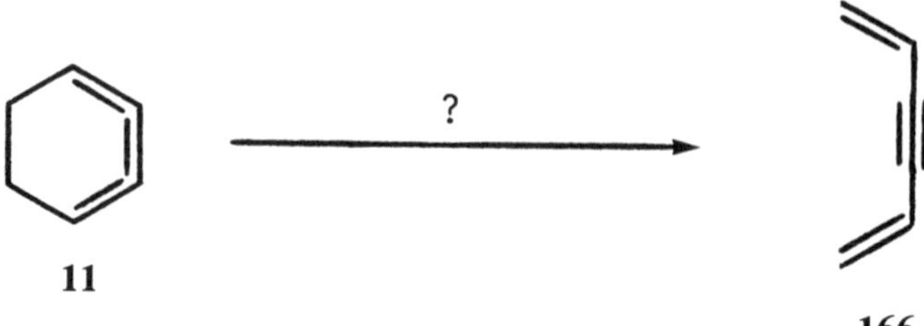

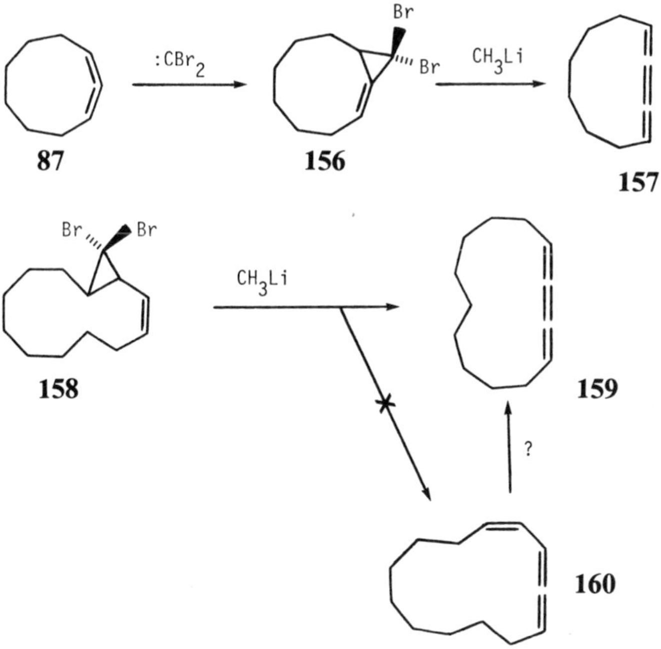

Scheme XXIII.

Interestingly, the Szeimies group showed that benzyne or naphthyne could be generated (Scheme XXVI) through the same route from benzvalene or naphthvalene precursors.[88] In these cases, one expects product aromaticity to be a powerful driving force.

We are aware of no efforts to generate either the parent 1,2,3-cyclopentatrienes or 1,2,3-cyclooctatrienes. The five-ring butatriene **152** is so strained

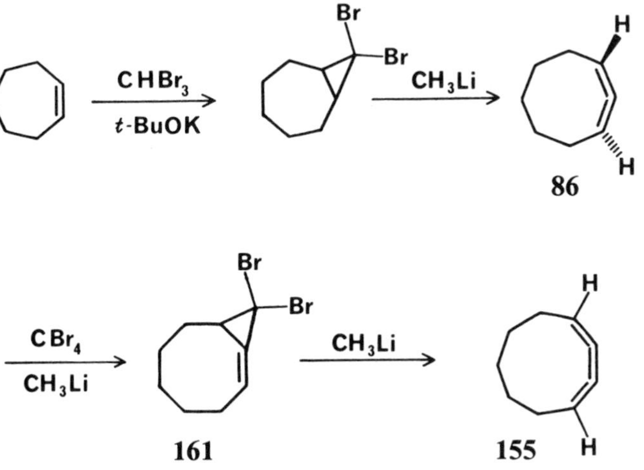

Scheme XXIV.

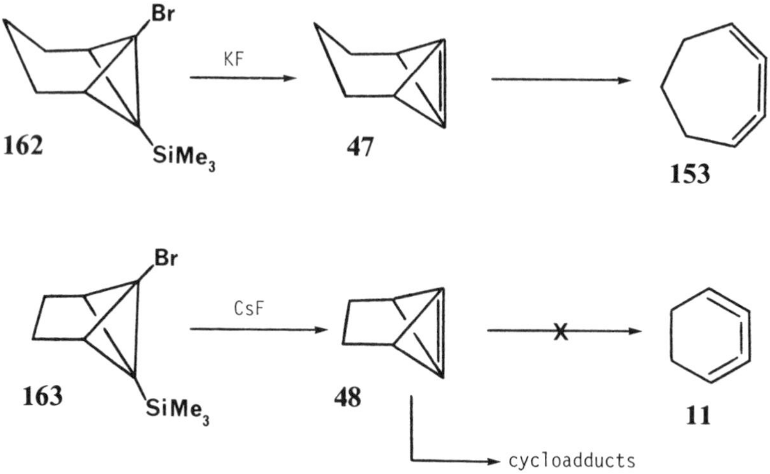

Scheme XXV.

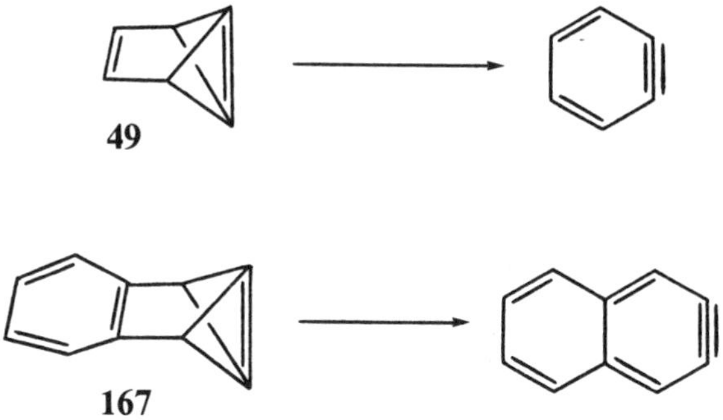

Scheme XXVI.

(> 100 kcal/mol) that it might be expected to rearrange (Scheme XXVII) via ring opening. Cyclooctatriene **(154)** would be an interesting target for synthesis. The butatriene moiety is predicted to be bent about 24° from linearity, although the resultant strain estimate of 14 kcal/mol is not high. Thus although this molecule should be thermodynamically stable, it may be too kinetically reactive to be isolated, except at low temperature. Mataga and co-workers have reported photochemical generation of the unusual

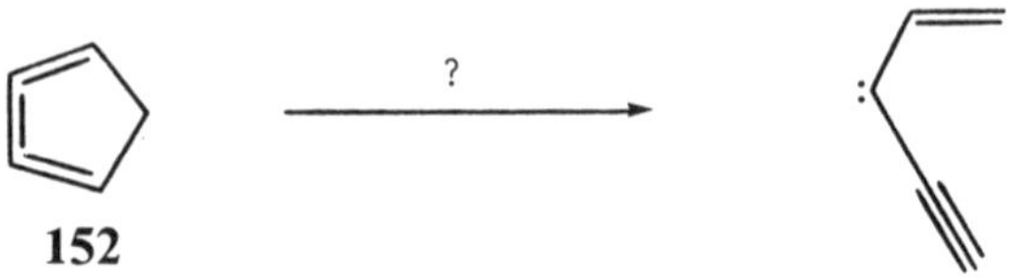

Scheme XXVII.

cyclooctatriene derivative **168.**[154] This compound yielded an equally unusual radialene dimer.

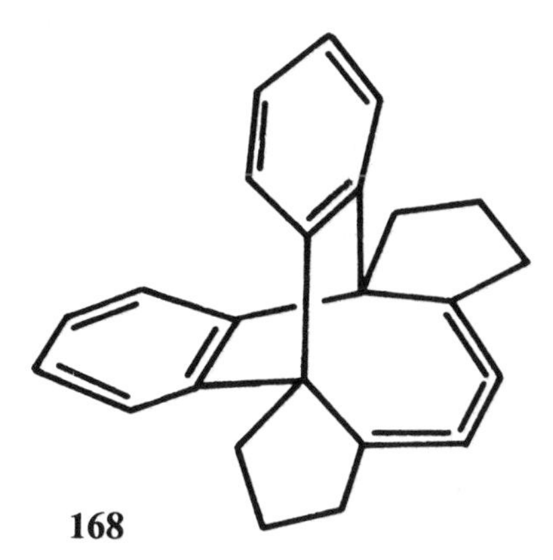

In the large-ring butatrienes, one expects cis–trans isomers to be isolable. These might be interconverted either thermally or photochemically. Roth and Exner have measured a ground state rotational barrier of 31 kcal/mol for 2,3,4-hexatriene.[155a] Dreiding molecular models show that the smallest ring that can accommodate an unstrained (ie, planar) *trans*-butatriene has 12 carbons. We are attempting to verify this experimentally.[155b]

6. BENZYNES AS CUMULENES?

The three isomeric dehydrobenzenes, *o*-, *m*-, and *p*-benzyne, comprise the subjects of a large literature.[10,156–167] Their precise structures, however, remain open to some speculation, although all appear to be ground state singlets.

Three distinct structures can be considered for *o*-benzyne; these might be characterized as aromatic **(169)**, aryne **(170)**, or cumulenelike **(171).** Each structure implies a specific geometry and, in principle, accurate bond lengths would resolve this question. Those who have studied this problem generally agree that the C_1-C_2 bond should be quite short, but the degree of alternation in the remaining bonds is still being debated.

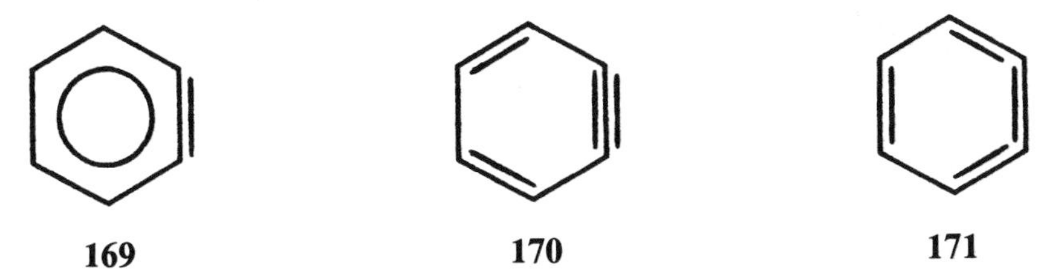

Arguments for each structure as a significant contributor have been advanced by different groups. In the experimental camp, Chapman and co-workers have measured the infrared spectrum for photogenerated *o*-benzyne in an argon matrix at 8 K.[156] Stretching frequencies were interpreted to support a structure **(170)** with "more cyclohexatriene character as far as the π system is concerned." Based on a normal coordinate analysis of Chapman's data, Laing and Berry concluded: "We can say with considerable confidence that the benzyne molecule is properly called a cycloalkyne."[157] Both studies ruled out cumulenelike structure **171.**

Early π-electron calculations on *o*-benzyne by Coulson also favored the aryne geometry **(170).**[158] Several subsequent calculations by Hoffmann (extended Hückel)[160] and by Haselbach (MINDO/2)[159] suggested a significant resonance contribution from cumulene **171.**[159,160] However, more recent theoretical studies favor an "aromatic" structure, with relatively small bond length alternation.[161–163] This is best depicted as **169.** As the most recent example, Figure 3-9 shows the result of a 6–31G* HF geometry optimiza-

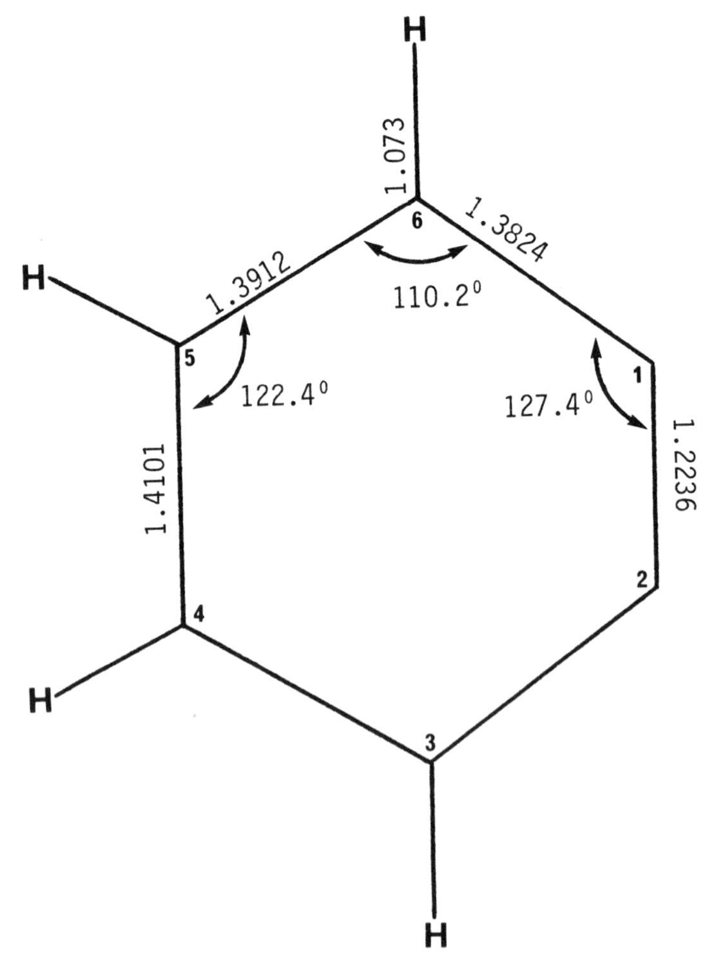

Figure 3-9. 6–31G*-optimized geometry for *o*-benzyne.

tion, reported by Bock and co-workers.[162] Based on these and other results, the authors concluded that *o*-benzyne possesses a "highly strained aromatic structure."

Which structure for *o*-benzyne is correct? Is this just a question of resonance semantics? Experimental groups favor aryne structure **170**,[156,157] while theoreticians argue for something that may be best described as a strained aromatic, **169**.[162] Of course, the "real" structure must be a hybrid of these. In the author's opinion, neither side has yet made a definitive case. *o*-Benzyne has now been generated by Lineberger and co-workers through photodetachment from its anion. It seems likely that interpretation of the well-resolved photoelectron spectrum will resolve structural questions for *o*-benzyne.[164]

Thermolysis of enediyne **(172)** gives rise to a degenerate rearrangement that is believed by Bergman and co-workers to proceed through *p*-benzyne.[165,166] A ground state singlet with a small singlet–triplet gap has been

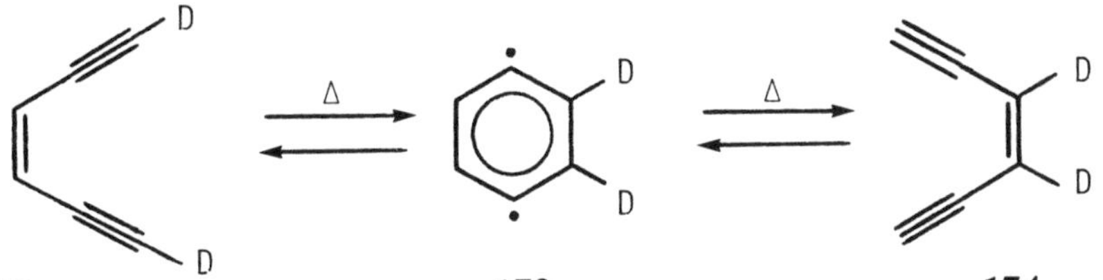

D D D D D D
Δ Δ
172 **173** **174**

supported by experimental[166] and theoretical[161,163] studies. Breslow has reported evidence for generation of butalene **(175)**, a high-energy antiaromatic isomer of *p*-benzyne.[167]

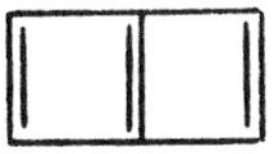

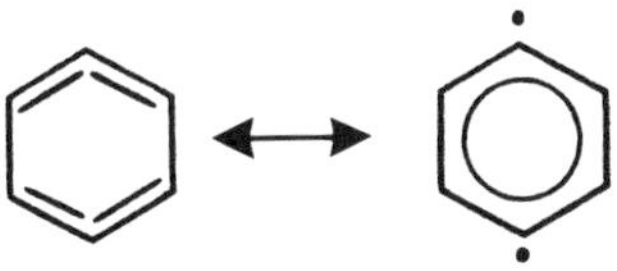

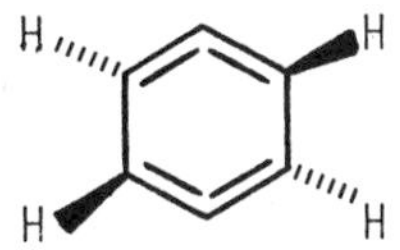

H H H H
175 **176** **177** **178**

What is the structure for *p*-benzyne? To date, the leading candidates would appear to be planar biradical **(177)** or a bisallene **(176)**. There are no reports of ab initio optimizations on *p*- benzyne that employed appropriate correlated wave functions, but MNDO/CI yields a geometry with C_1-C_2 bond lengths much shorter (1.37Å) than C_2—C_3 (1.43Å). This implies a significant contribution from allenic structure **176**. Extending this one step further, one wonders whether a planar geometry is mandated, as seems to have been assumed by all previous studies. Intriguingly, MNDO calculations do indicate an energy minimum (ΔH_f = 183 kcal/mol, D_2 symmetry) corresponding to **178**,[168] but this is significantly higher in energy than the value reported for planar **177** (131.6 kcal/mol). Higher level calculations will be necessary to accurately determine the energetics of **177** and **178**.

7. CONCLUSIONS

What are the structural limitations that nature has imposed on cyclic alkenes, alkynes and cumulenes?

Among simple *trans*-cycloalkenes, there is good evidence for the transient existence of *trans*-cyclohexene; this probably has a very low barrier (< 15 kcal/mol) for conversion to the cis isomer. An optimistic interpretation of (flexible) molecular models suggests that *trans*-cyclopentene might represent a shallow energy minimum. At present, there appear to be neither experimental[81] nor theoretical results to support this.

Cyclopentyne is the smallest cycloalkyne for which unambiguous experimental support is available. High-level theoretical studies suggest that cyclobutyne might exist as a stable structure, but the barrier to rearrangement undoubtedly is quite small. In the realm of simple allenes, 1,2-cyclononadiene long has been recognized as the smallest example that is isolable under ordinary conditions.[12,106] The smallest cyclic allene for which a *chiral* structure has unequivocal support is 1,2-cyclohexadiene.[113–115] Both the elegant trapping experiments of Balci and Jones[113] and ab initio calculations[115] clearly implicate a chiral equilibrium geometry, with a low barrier to π– bond rotation. If, as the author prefers to do, one defines an allene as being capable of chirality (ie, not planar), then the jury is still out on 1,2-cyclopentadiene. Ab initio calculations suggest a slight advantage for the chiral structure,[111] but the energy difference (< 5 kcal/mol) is within what one might expect for anything but a very high level of theory. It will take clever experiments to resolve this question. Numerous other strained cyclic allenes can be envisioned, or have been prepared. Two of the best studied have been cycloheptatetraene **(116)**[134] and bicyclo[3.2.1]octa-2,3,6-triene **(128)**.[141–145] There is convincing evidence that both have chiral equilibrium geometries.

Butatrienes are equally interesting but have received far less study. 1,2,3-Cyclononatriene **(155)** has now been prepared and may well be the smallest cyclic butatriene that will be stable under ordinary conditions.[151] The seven-ring butatriene has already been trapped by Szeimies as a result of a remarkable rearrangement,[87] and it probably won't be too many years before a synthesis of 1,2,3-cyclooctatriene is reported. The six-ring butatriene (strain estimate ~ 60 kcal/mol[151]) may be too unstable to exist beyond a cryogenic matrix, and one questions whether the five-ring structure will truly correspond to an energy minimum.

ACKNOWLEDGMENTS

I am very grateful to Richard Angus and Michael Schmidt, my collaborators in cyclic cumulene chemistry, for their excellent experimental and theoretical work. Our efforts have been supported by the National Science Foundation, the donors of the Petroleum Research fund, administered by the

American Chemical Society, and by Ames Laboratory, U.S. Department of Energy. I thank Dr. M. Yoshimine and Professor W.M. Jones for preprints of References 39 and 125, respectively, and I am grateful to D.G. Leopold, and W.C. Lineberger for discussions of their work.

REFERENCES

1. There is a vast literature on the synthesis of strained and otherwise unusual organic structures. For a random assortment of recent references, see: (a) McMurry, J.E.; Haley, G.J.; Matz, J.R.; Clardy, J.C.; Van Duyne, G. *J. Am. Chem. Soc.* **1984,** *106,* 5018. (b) Wiberg, K.B.; Walker, F.H. ibid, **1982,** *104,* 5239. (c) Eaton, P.E.; Or, Y.S.; Branca, S.J. ibid, **1981,** *103,* 2134. (d) Vogel, E.; Markowitz, G.; Schmalsteig, L.; Ito, S.; Breuckmann, R.; Roth, W.R. *Angew Chem. Int. Ed. Engl.* **1984,** *23,* 719. (e) Marshall, J.A.; Peterson, J.C.; Lebioda, L. *J. Am. Chem. Soc.* **1984,** *106,* 6006. (f) Gassman, P.G.; Bonser, S.M. ibid, **1983,** *105,* 667. (g) Luyten, M.; Keese, R. *Angew, Chem. Int. Ed. Engl.* **1984,** *23,* 390. Recent reviews on some specific types of strained organic structures are given in References 2–10.
2. Comprehensive review: (a) Greenberg, A.; Liebman, J.F. "Strained Organic Molecules". Academic Press: New York, 1978. (b) Liebman, J.F.; Greenberg A. *Chem. Rev.* **1976,** *76,* 311.
3. For an excellent review of strained rings and many other neutral reactive intermediates, see: Wentrup, C. "Reactive Molecules". Wiley-Interscience: New York, 1984.
4. Bridgehead alkenes: (a) Szeimies, G. In "Reactive Intermediates", Vol. 3, Abramovitch, R.A. Ed.; Plenum Press: New York, 1983, Chapter 5, p. 299. (b) Shea, K.J. *Tetrahedron* **1980,** *36,* 1683. (c) Chiral cyclic alkenes: Nakazaki, M.; Yamamoto, K.; Naemura, K. *Top. Curr. Chem.* **1984,** *125,* 1.
5. Pyramidalized alkenes: (a) Houk, K.N.; Rondan, N.G.; Brown, F.K. *Isr. J. Chem.* **1983,** *23,* 3. (b) Szeimies, G. *Chimia* **1980,** *35,* 243. See also Reference 7.
6. Polyquinanes (including dodecahedrane): Paquette, L.A. *Top. Curr. Chem.* **1984,** *119,* 1.
7. Strained small-ring alkenes: Wiberg, K.B.; Bonneville, G.; Dempsey, R. *Isr. J. Chem.* **1983,** *23,* 85.
8. Cycloalkynes: Krebs, A.; Wilke, J. *Top. Curr. Chem.* **1983,** *109,* 189.
9. Aromatic valence bond isomers: Kobayashi, Y.; Kumadaki, I. *Top. Curr. Chem.* **1984,** *123,* 103.
10. Benzyne and related structures: (a) Hoffmann, R.W. "Dehydrobenzene and Cycloalkynes", Academic Press: New York, 1967. (b) Levin, R.H. In "Reactive Intermediates", Vol. 3; Jones Jr., M. and Moss, R.A., Eds.; Wiley: New York, 1985, Chapter 1.
11. (a) Saxe, P.; Schaefer, H.F., III. *J. Am. Chem. Soc.* **1980,** *102,* 3239. (b) Fitzgerald, G.; Schaefer, H.F., III. *Isr. J. Chem.* **1983,** *23,* 93. (c) Grev, R.S.; Schaefer, H.F., III. *J. Chem. Phys.* **1984,** *80,* 3552. (d) Michalopoulos, D.L.; Geusic, M.E.; Langridge-Smith, P.R.R.; Smalley, R.E. *J. Chem. Phys.* **1984,** *80,* 3556.
12. Skattebøl, L.; Solomon, S. *Org. Synth.* **1969,** *49,* 35.
13. For a general discussion of theoretical studies on strained molecules, see: Newton, M.D. in "Applications of Electronic Structure Theory," Schaefer H.F., III, Ed.; Plenum Press: New York, 1977, Chapter 6, p. 223.
14. For an excellent review of molecular mechanics, see: Ōsawa, E.; Musso, H. *Angew. Chem. Int. Ed. Engl.* **1983,** *22,* 1.
15. Benson, S. W. "Thermochemical Kinetics: Methods for the Estimation of Thermochemical Data and Rate Parameters", 2nd ed.; Wiley: New York, 1976.
16. Hehre, W.J.; Ditchfield, R.; Radom, L.; Pople, J.A. *J. Am. Chem. Soc.* **1970,** *92,* 4796. For leading references to more recent developments, see: Schulman, J.M.; Disch, R.L. ibid, **1984,** *106,* 1202.
17. Rosenstock, H.M.; Dannacher, J.; Liebman, J.F. *Radiat. Phys. Chem.* **1982,** *20,* 7.
18. Heilbronner, E.; Bock, H. "The HMO Model and Its Application", Wiley-Interscience: New York, 1976.
19. Hoffmann, R. *J. Chem. Phys.* **1963,** *39,* 1397.

20. Borden, W.T. "Modern Molecular Orbital Theory for Organic Chemists". Prentice-Hall: Englewood Cliffs, N.J., 1975.
21. Pople, J.A.; Beveridge, D.L. "Approximate Molecular Orbital Theory". McGraw-Hill: New York, 1970.
22. Segal, G.A., Ed. "Semiempirical Methods of Electronic Structure Calculation", Parts A (Techniques) and B (Applications); Plenum Press: New York, 1977.
23. Dewar, M.J.S.; Thiel, W. *J. Am. Chem. Soc.* **1977,** *99,* 4899.
24. (a) Dewar, M.J.S. *J. Mol. Struct.* **1983,** *100,* 41. (b) Dewar, M.J.S.; Ford, G.P. *J. Am. Chem. Soc.* **1979,** *101,* 5558.
25. Johnson, R.P. Unpublished results.
26. (a) Dewar, M.J.S.; Ford, G.P.; McKee, M.L.; Rzepa, H.S.; Wade, L.E. *J. Am. Chem. Soc.* **1977,** *99,* 5069. (b) Komornicki, A.; McIver, J.W. ibid, **1976,** *98,* 4553. (c) Dewar, M.J.S. ibid, **1984,** *106,* 209. (d) Osamura, Y.; Kato, S.; Morokuma, K.; Feller, D.; Davidson, E.R.; Borden, W.T. ibid, **1984,** *106,* 3362. (e) Gajewski, J.J. "Hydrocarbon Thermal Isomerizations". Academic Press: New York, 1981.
27. For an introduction, see: (a) Szabo, A.; Ostlund, N. "Modern Quantum Chemistry". Macmillan: New York; 1982. (b) Richards, W.G.; Cooper, D.L. "Ab Initio Molecular Orbital Calculations for Chemists". Clarendon Press: Oxford, 1983. (c) Hehre, W.J.; Radom, L.; Schleyer, P.v.R.; Pople, J.A. "Ab Initio Molecular Orbital Theory". Wiley: New York, 1985. (d) Clark, T. "A Handbook of Computational Chemistry." Wiley: New York, 1985.
28. Binkley, J.S.; Pople, J.A.; Hehre, W.J. *J. Am. Chem. Soc.* **1980,** *102,* 939, and references therein.
29. Dunning, T.H., Hay, P.J. In "Methods of Electronic Structure Theory", Schaefer III, H.F., Ed.; Plenum Press: New York, 1977, p. 1.
30. Pople, J.A. In "Applications of Electronic Structure Theory", Schaefer, H.F., III, Ed.; Plenum Press: New York, 1977, p. 1.
31. Wiberg, K.B.; Wendoloski, J.J. *J. Am. Chem. Soc.* **1982,** *104,* 5679.
32. (a) Radom, L. *Aust. J. Chem.* **1978,** *31,* 1. (b) Maki, A.G.; Toth, R.A. *J. Mol. Spectrosc.* **1965,** *17,* 136. (c) Almenningen, A.; Bastiansen, O.; Traetteberg, M. *Acta Chem. Scand.* **1961,** *15,* 1557.
33. (a) Doubleday, C., Jr.; McIver, J.W., Jr.; Page, M. *J. Am. Chem. Soc.* **1982,** *104,* 3768. (b) Feller, D.F.; Tanaka, K.; Davidson, E.R.; Borden, W.T. *J. Am. Chem. Soc.* **1982,** *104,* 967.
34. Borden, W.T., Ed. "Diradicals". Wiley: New York, 1982, p. 1.
35. Recent examples of MCSCF studies on organic reactions and intermediates: (a) Feller, D.F.; Schmidt, M.W.; Ruedenberg, K. *J. Am. Chem. Soc.* **1982,** *104,* 960. (b) Osamura, Y.; Borden, W.T.; Morokuma, K. ibid, **1984,** *106,* 5112. (c) Karlstrom, G.; Roos, B.O.; Carlsen, L. ibid, **1984,** *106,* 1557. (d) Doubleday Jr., C.; Camp, R.N.; King, H.F.; McIver, J.W., Jr.; Mullally, D.; Page, J. ibid, **1984,** *106,* 447. (e) Bernardi, F.; Robb, M.A., Schlegel, H.B., Tonachini, G. ibid, **1983,** *106,* 1198.
36. Davidson, E.R. In "Diradicals", Borden, W.T., Ed.; Wiley: New York, 1982, Chapter 2, p. 73.
37. (a) Bodor, N.; Dewar, M.J.S.; Wasson, J.S. *J. Am. Chem. Soc.* **1972,** *94,* 9095. (b) Minato, T.; Osamura, Y.; Yamabe, S.; Fukui, K. ibid, **1980,** *102,* 581. (c) Pasto, D.J.; Haley, M.; Chipman, D.M. ibid, **1978,** *100,* 5272. (d) Dillon, P.W.; Underwood, G.R. ibid, **1977,** *99,* 2435. (e) Baird, N.C.; Taylor, K.F. ibid, **1978,** *100,* 1333.
38. Stierman, T.J.; Johnson, R.P. *J. Am. Chem. Soc.,* **1985,** *107,* 3971.
39. Honjou, N.; Pacansky, J.; Yoshimine, M. *J. Am. Chem. Soc.,* **1985,** *107,* 5332.
40. Douglas, J.E.; Rabinovitch, B.S.; Looney, F.S. *J. Chem. Phys.* **1955,** *23,* 315.
41. Radom, L.; Pople, J.A.; Mock, W.L. *Tetrahedron Lett.* **1972,** 479.
42. Mock, W.L. *Tetrahedron Lett.* **1972,** 475.
43. (a)Salem, L. *Acc. Chem. Res.* **1979,** *12,* 87. (b) Brooks, B.R.; Schaefer, H.F., III; *J. Am. Chem. Soc.* **1979,** *101,* 308. (c) Buenker, R.J.; Bonacic-Koutecky, V.; Pogliani, L. *J. Chem. Phys.* **1980,** *73,* 1836. (d) Malrieu, J.-P.; Trinquier, G. *Theor. Chim. Acta* **1979,** *54,* 59. (e) Johnson, R.P.; Schmidt, M.W. *J. Am. Chem. Soc.* **1981,** *103,* 3244.
44. Lam, B.; Johnson, R.P. *J. Am. Chem. Soc.* **1983,** *105,* 7479.
45. Allinger, N.L.; Sprague, J.T. *J. Am. Chem. Soc.* **1972,** *94,* 5734.
46. Maier, W.F.; Schleyer, P.v.R. *J. Am. Chem. Soc.* **1981,** *103,* 1891.
47. Warner, P.M.; Peacock, S. *J. Comput. Chem.* **1982,** *3,* 417; *Tetrahedron Lett.* **1983,** 4169.

48. Hehre, W.J.; Pople, J.A. *J. Am. Chem. Soc.* **1975,** *97,* 6941.
49. Wagner, H.-U.; Szeimies, G.; Chandrasekhar, J.; Schleyer, P.v.R.; Pople, J.A.; Binkley, J.S. *J. Am. Chem. Soc.* **1978,** *100,* 1210.
50. Volland, W.V.; Davidson, E.R.; Borden, W.T. *J. Am. Chem. Soc.* **1979,** *101,* 533.
51. Wiberg, K.B.; Matturro, M.G.; Okarma, P.J.; Jason, M.E. *J. Am. Chem. Soc.* **1984,** *106,* 2194.
52. Angus Jr., R.O.; Johnson, R.P. *J. Org. Chem.* Accepted for publication.
53. Steinmetz, M.G.; Srinivasan, R.; Leigh, W.J. *Rev. Chem. Intermed.* **1984,** *5,* 57, and references therein.
54. Carr, R.W., Jr.; Walters, W.D. *J. Phys. Chem.* **1965,** *69,* 1073.
55. Carsky, P.; Hess, B.A., Jr.; Schaad, L.J. *J. Am. Chem. Soc.* **1983,** *105,* 396, and references therein.
56. Lewars, E.G. *Chem. Rev.* **1983,** *83,* 519, and references therein.
57. Torres, M.; Bourdelande, J.L.; Clement, A.; Strausz, O.P. *J. Am. Chem. Soc.* **1983,** *105,* 1698.
58. Krantz, A.; Laureni, J. *J. Am. Chem. Soc.* **1981,** *103,* 486.
59. Wiseman, J.R.; Pletcher, W.A. *J. Am. Chem. Soc.* **1970,** *92,* 956.
60. Wiseman, J.R.; Kipp, J.E. *J. Am. Chem. Soc.* **1982,** *104,* 4688.
61. Blomquist, A.T.; Liu, L.H.; Bohrer, J.C. *J. Am. Chem. Soc.* **1952,** *74,* 3643.
62. Cope, A.C.; Howell, C.F.; Knowles, A. *J. Am. Chem. Soc.* **1962,** *84,* 3190; Cope, A.C., Ganellin, C.R., Johnson, H.W. ibid, **1962,** *84,* 3191; Cope, A.C., Ganellin, C.R., Johnson, H.W., Van Auken, T.V., Winkler, H.J.S. ibid, **1963,** *85,* 3276.
63. Cope, A.C.; Banholzer, K.; Keller, H.; Pawson, B.A.; Whang, J.J.; Winkler, H.J.S. *J. Am. Chem. Soc.* **1965,** *87,* 3644.
64. Cope, A.C.; Pawson, B.A. *J. Am. Chem. Soc.* **1965,** *87,* 3649; Binsch, G.; Roberts, J.D. *J. Am. Chem. Soc.* **1965,** *87,* 5157.
65. Marshall, J.A., Konicek, T.R., Flynn, K.E. *J. Am. Chem. Soc.* **1980,** *102,* 3287.
66. Marshall, J.A., Flynn, K.E. *J. Am. Chem. Soc.* **1982,** *104,* 7430.
67. Marshall, J.A. *Acc. Chem. Res.* **1980,** *13,* 213.
68. Marshall, J.A.; Flynn, K.E. *J. Am. Chem. Soc.* **1984,** *106,* 723.
69. Ermer, O. *Struct. Bonding* **1976,** *27,* 196–198.
70. Rogers, D.W.; v. Voithenberg, H.; Allinger, N.L. *J. Org. Chem.* **1978,** *43,* 360.
71. Tratteberg, M. *Acta Chem. Scand. Ser. B.* **1975,** *29,* 29.
72. (a) Corey, E.J.; Carey, F.A.; Winter, R.A.E. *J. Am. Chem. Soc.* **1965,** *87,* 934. (b) Kropp, P.J. *J. Am. Chem. Soc.* **1969,** *91,* 5783. (c) Inoue, Y.; Takamuka, S.; Sakurai, H. *J. Chem. Soc. Perkin Trans 2* **1977,** 1635.
73. Inoue, Y.; Ueoka, T.; Kuroda, T.; Hakushi, T. *J. Chem. Soc. Chem. Commun.* **1981,** 1031; *J. Chem. Soc. Perkin 2* **1983,** 983.
74. Bonneau, R.; Joussot-Dubien, J.; Yarwood, J.; Pereyre, J. *Nouv. J. Chim.* **1977,** *1,* 31; *Tetrahedron Lett.,* **1977,** 235, and references therein.
75. (a) Corey, E.J.; Tada, M.; LeMahieu, R.; Libit, L. *J. Am. Chem. Soc.* **1965,** *87,* 2051. (b) Eaton, P.E.; Lin, K. ibid, **1965,** *87,* 2052.
76. Eaton, P.E. *Acc. Chem. Res.* **1968,** *1,* 50.
77. Evers, J.T.M.; Mackor, A. *Rec. Trav. Chim. Pays-Bas* **1979,** *98,* 423.
78. Wallraff, G.M.; Boyd, R.H.; Michl, J. *J. Am. Chem. Soc.* **1983,** *105,* 4550.
79. Christl, M.; Brüntrup, G. *Angew. Chem. Int. Ed. Engl.* **1974,** *13,* 208.
80. Coates, R.M.; Last, L.A. *J. Am. Chem. Soc.* **1983,** *105,* 7322.
81. Review: Kropp, P.J. *Org. Photochem.* **1979,** *4,* 1.
82. Dauben, W.G.; van Riel, H.C.H.A.; Robbins, J.D.; Wagner, G.J. *J. Am. Chem. Soc.* **1979,** *101,* 6383, and references therein.
83. Bonneau, R.; Joussot-Dubien, J.; Salem, L.; Yarwood, A.Y. *J. Am. Chem. Soc.* **1976,** *98,* 4329.
84. Reviews: (a) Salomon, R.G. *Tetrahedron* **1983,** *39,* 485. (b) Kutal, C.; Grutch, P.A. *Adv. Chem. Serv.* **1979,** *173,* 325.
85. Viavattene, R.L.; Greene, F.D.; Cheung, L.D.; Majeste, R.; Trefonas, L.M. *J. Am. Chem. Soc.* **1974,** *96,* 4342.
86. Greenhouse, R.; Borden, W.T.; Hirotsu, K.; Clardy, J. *J. Am. Chem. Soc.* **1977,** *99,* 1664.
87. Zoch, H.-G.; Szeimies, G.; Römer, R.; Schmitt, R. *Angew. Chem. Int. Ed. Engl.* **1981,** *20,* 877.

88. Schlüter, A.-D.; Belzner, J.; Heywang, V.; Szeimies, G. *Tetrahedron Lett.* **1983,** *24,* 891.
89. Greenhouse, R.; Borden, W.T.; Ravindranathan, T.; Hirotsu, K.; Clardy, J. *J. Am. Chem. Soc.* **1977,** *99,* 1664.
90. (a) Wiberg, K.B.; Matturro, M.; Adams, R. *J. Am. Chem. Soc.* **1981,** *103,* 1600. (b) Wiberg, K.B.; Matturro, M.G.; Okarma, P.J.; Jason, M.E. ibid, **1984,** *106,* 2194.
91. Reviews: (a) Nakagawa, M. In "The Chemistry of the Carbon-Carbon Triple Bond," Patai, S., Ed.; Wiley: New York, 1978, p. 635. (b) Krebs, A. In "Chemistry of Acetylenes," Viehe, H.G., Ed.; Dekker, New York, 1969. See also References 3, 8, and 10.
92. Krebs, A.; Kimling, H. *Angew. Chem. Int. Ed. Engl.* **1971,** *10,* 509.
93. Fitzgerald, G.; Saxe, P.; Schaefer, H.F., III. *J. Am. Chem. Soc.* **1983,** *105,* 690.
94. Kollmar, H.; Carrion, F.; Dewar, M.J.S.; Bingham, R.C. *J. Am. Chem. Soc.* **1981,** *103,* 5292.
95. Montegomery, L.K.; Roberts, J.D. *J. Am. Chem. Soc.* **1960,** *82,* 4750.
96. Wittig, G.; Wilson, E.R. *Chem. Ber.* **1965,** *98,* 451.
97. Baxter, G.J.; Brown, R.F.C. *Aust. J. Chem.* **1978,** *31,* 327.
98. Gilchrist, T.L.; Rees, C.W. "Carbenes, Nitrenes and Arynes", Nelson: London, 1969.
99. (a) Chapman, O.L.; Gano, J.; West, P.R.; Regitz, M.; Maas, G. *J. Am. Chem. Soc.* **1981,** *103,* 7033. (b) Chapman, O.L. Private communication.
100. (a) Gilbert, J.C.; Baze, M.E. *J. Am. Chem. Soc.* **1984,** *106,* 1885. (b) Gilbert, J.C.; Baze, M.E. ibid, **1983,** *105,* 664. (c) Gilbert, J.C. Private communication.
101. Fitjer, L.; Modaressi, S. *Tetrahedron Lett.* **1983,** 5495.
102. Bolster, J.M.; Kellog, R.M. *J. Am. Chem. Soc.* **1981,** *103,* 2868.
103. Nakayama, J.; Ohshima, E.; Ishii, A.; Hoshino, M. *J. Org. Chem.* **1983,** *48,* 60, and references therein.
104. Gassman, P.G.; Gennick, I. *J. Am. Chem. Soc.* **1980,** *102,* 6863, and references therein.
105. (a) Scott, L.T.; DeCicco, G.J.; Hyun, J.L.; Reinhardt, G. *J. Am. Chem. Soc.* **1983,** *105,* 7760. (b) Sakurai, H.; Nakadaira, Y.; Hosomi, A.; Eriyama, Y.; Kabuto, C. ibid, **1983,** *105,* 3359.
106. For an excellent and comprehensive discussion of earlier work on cyclic allenes, see: Hopf, H. In "The Chemistry of Ketenes, Allenes, and Related Compounds", Part 2, Patai, S., Ed.; Wiley: New York, 1980, Chapter 20, p. 779.
107. For theoretical studies of the barrier to π-bond rotation in allene, see: (a) Seeger, R.; Krishnan, R.; Pople, J.A.; Schleyer, P.v.R. *J. Am. Chem. Soc.* **1977,** *99,* 7103; (b) Dykstra, C.E. ibid, **1977,** *99,* 2060; (c) Staemmler, V. *Theor. Chim. Acta.* **1977,** *45,* 89; (d) Krogh-Jespersen, K. *J. Comput. Chem.* **1982,** *3,* 571, and references therein.
108. Experimental rotational barrier in allene: Roth, W.R.; Ruf, G.; Ford, P.W. *Chem. Ber.* **1974,** *107,* 48.
109. Moore, W.R.; Moser, W.R. *J. Am. Chem. Soc.* **1970,** *92,* 5469.
110. Dillon, P.W.; Underwood, G.R. *J. Am. Chem. Soc.* **1974,** *92,* 779.
111. Angus Jr., R.O.; Schmidt, M.W.; Johnson, R.P. *J. Am. Chem. Soc.* **1985,** *107,* 532.
112. (a) Bottini, A.T.; Cabral, L.J.; Dev, V. *Tetrahedron Lett.* **1977,** 615. (b) Bottini, A.T.; Carson, F.P.; Fitzgerald, R.; Frost, K.A. *Tetrahedron* **1972,** *28,* 4338. (c) Bottini, A.T.; Hilton, L.L.; Plott, J. ibid, **1975,** *31,* 1997. (d) Bottini, A.T.; Corson, F.B.; Fitzgerald, R.; Frost, K.A. *Tetrahedron Lett.* **1970,** 4753, 4757.
113. Balci, M.; Jones, W.M. *J. Am. Chem. Soc.* **1980,** *102,* 7608.
114. Wentrup, C.; Gross, G.; Maquestiau, A.; Flammary, R. *Angew. Chem. Int. Ed. Engl.* **1983,** *27,* 542.
115. Schmidt, M.W., Angus, R.O., Jr.; Johnson, R.P. *J. Am. Chem. Soc.* **1982,** *104,* 6838.
116. Lee-Ruff, E.; Maleki, M.; Duperrovzel, P.; Lien, M.H.; Hopkinson, A.C. *J. Chem. Soc. Chem. Commun.* **1983,** 346.
117. (a) Cuthbertson, A.F.; Glidewell, C.; Lloyd, D. *J. Chem. Res. (S)* **1982,** 80; **1983,** 48. (b) Lloyd, D.; Mackie, R.K.; Richardson, G.; Marshall, D.R. *Angew. Chem. Int. Ed. Engl.* **1981,** *20,* 190, and references therein.
118. (a) Montgomery, L.K.; Scardiglia, F.; Roberts, J.D. *J. Am. Chem. Soc.* **1965,** *87,* 1917. (b) Montgomery, L.K.; Applegate, L.E. ibid, **1967,** *89,* 2952. (c) Wittig, G.; Heyn, J. *Liebigs Ann. Chem.* **1972,** *756,* 1. (d) Chapman, O.L. Private communication.
119. Wasserman, E.; Barash, L.; Trozzolo, A.M.; Murray, R.W.; Yager, W.A. *J. Am. Chem. Soc.* **1964,** *86, 2304.*
120. (a) Moss, R.A. *J. Org. Chem.* **1966,** *31,* 3296. (b) Moss, R.A.; Przybla, J.R. ibid, **1968,** *33,*

3816. (c) Durr, H.; Werndorff, F. *Angew. Chem. Int. Ed. Engl.* **1974,** *13,* 483. (d) Shepard, R.; Simmons, J. *Int. J. Quant. Chem. Symp.* **1980,** *14,* 345.
121. Tolbert, L.M.; Siddiqui, S. *J. Am. Chem. Soc.* **1984,** *106,* 5538.
122. Hehre, W.J.; Pople, J.A. *J. Am. Chem. Soc.* **1975,** *97,* 6941.
123. Reisenauer, H.P.; Maier, G.; Riemann, A.; Hoffmann, R.W. *Angew. Chem. Int. Ed. Engl.* **1984,** *23,* 641.
124. (a) Visser, J.P.; Ramakers, J.E. *J. Chem. Soc. Chem. Commun.* **1972,** 178. (b) Marquis, E.T.; Gardner, P. *Tetrahedron Lett.* **1966,** 2793.(c) Wittig, G.; Dorsch, H.L.; Menske-Schüler, J. *Liebigs Ann. Chem.* **1968,** *711,* 55. (d) Reese, C.B.; Shaw, A. *J. Chem. Soc. Chem. Commun.* **1972,** 787.
125. Manganiello, F.J.; Oon, S.M.; Radcliffe, M.D.; Jones, W.M. *J. Organomet. Chem.* Submitted for publication.
126. (a) Yavari, I. *J. Mol. Struct.* **1980,** *65,* 169. (b) Anet, F.A.C.; Yavari, I. *J. Am. Chem. Soc.* **1977,** *99,* 7640.
127. Nakazaki, M.; Yamamoto, K.; Maeda, M.; Sato, O.; Tsutsui, T. *J. Org. Chem.* **1982,** *47,* 1435. See also Reference 4c.
128. Chang, S.-C. Doctoral dissertation, Iowa State University: Ames, 1980.
129. Vaidyanathaswamy, R.; Devaprabhakara, D. *Chem. Ind. (London)* **1968,** 515.
130. Paquette, L.A.; Chamot, E.; Browne, A.R. *J. Am. Chem. Soc.* **1980,** *102,* 637.
131. Warner, P.M.; Chang, S.-C.; Powell, D.R.; Jacobson, R.A. *J. Am. Chem. Soc.* **1980,** *102,* 5125, and references therein.
132. (a) Carlton, J.B.; Levin, R.H. *Tetrahedron Lett.* **1976,** 3761. (b) Carlton, J.B.; Levin, R.H.; Clardy, J.C. *J. Am. Chem. Soc.* **1976,** *98,* 6068.
133. Warner, P.M.; Chang, S.-C. *Tetrahedron Lett.* **1979,** 4141.
134. Harris, J.W.; Jones, W.M. *J. Am. Chem. Soc.* **1982,** 104, *7329,* and references therein.
135. West, P.R.; Chapman, O.L.; LeRoux, J.-P. *J. Am. Chem. Soc.* **1982,** *104,* 1779.
136. Kirmse, W.; Loosen, K.; Sluma, H.D. *J. Am. Chem. Soc.* **1981,** *103,* 5935.
137. Tyner, R.L.; Jones, W.M.; Öhrn, Y.; Sabin, J.R. *J. Am. Chem. Soc.* **1974,** *96,* 3765.
138. Dewar, M.J.S.; Landman, D. *J. Am. Chem. Soc.* **1977,** *99,* 6179.
139. Waali, E.E. *J. Am. Chem. Soc.* **1981,** *103,* 3604.
140. Radom, L.; Schaefer, H.F., III; Vincent, M.A. *Nouv. J. Chim.* **1980,** *4,* 411.
141. Bergman, R.G.; Rajadhyaksha, V.J. *J. Am. Chem. Soc.* **1970,** *92,* 2163.
142. Klumpp, G.W.; van Dijk, P.M. *Rec. Trav. Chim. Pays-Bas* **1971,** *90,* 381.
143. Freeman, P.K.; Swenson, K.E. *J. Org. Chem.* **1982,** *47,* 2033.
144. Balci, M.; Harmandar, M. *Tetrahedron Lett.* **1984,** *25,* 237.
145. Balci, M.; Jones, W.M. *J. Am. Chem. Soc.* **1981,** *103,* 2874.
146. Firl, J.; Schink, K.; Kosbahn, W. *Chem. Lett.* **1981,** 527.
147. Richter, R.; Tucker, B.; Ulrich, H. *J. Org. Chem.* **1983,** *48,* 1694.
148. Schmidbaur, H.; Costa, T.; Milewski-Mahrla, B.; Schubert, U. *Angew. Chem. Int. Ed. Engl.* **1980,** *19,* 555.
149. Chapman, O. *Pure Appl. Chem.* **1979,** *51,* 331, and references therein. An excellent discussion of the chemistry of **145** is given in the monograph by Wentrup.[3]
150. Wentrup, C.; Winter, H.W. *J. Am. Chem. Soc.* **1980,** *102,* 6159.
151. (a) Angus, R.O., Jr.; Johnson, R.P. *J. Org. Chem.* **1984,** *49,* 2880. (b) Angus, R.O., Jr.; Johnson, R.P. To be submitted.
152. Moore, W.R.; Ozretich, T.M. *Tetrahedron Lett.* **1967,** 3205.
153. Bhagwat, M.M.; Devaprabhakara, D. *Indian J. Chem.* **1975,** *13,* 506.
154. Hayashi, T.; Mataga, N.; Inoue, T.; Kaneda, T.; Irie, T.; Misumi, S. *J. Am. Chem. Soc.* **1977,** *99,* 523.
155. (a) Roth, W.R.; Exner, H.-D. *Chem. Ber.* **1976,** *109,* 1158. (b) Price, J.D.; Johnson, R.P. Unpublished work.
156. (a) Chapman, O.L.; Mattes, K.; McIntosh, C.L.; Pacansky, J. *J. Am. Chem. Soc.* **1973,** *95,* 6134. (b) Chapman, O.L.; Chang, C.-C.; Kolc, J.; Rosenquist, N.R.; Tomioka, H. ibid, **1975,** *97,* 6586.
157. Laing, J.W.; Berry, R.S. *J. Am. Chem. Soc.* **1976,** *98,* 660.
158. Coulson, C.A. *Chem. Soc. Symp.* (Bristol) *Spec. Publ.* **1958,** *12,* 85.
159. Haselbach, E. *Helv. Chim. Acta* **1971,** *54,* 1981.
160. Hoffmann, R.; Imamura, A.; Hehre, W.J. *J. Am. Chem. Soc.* **1968,** *90,* 1499.
161. Dewar, M.J.S.; Ford, G.P.; Reynolds, C.H. *J. Am. Chem. Soc.* **1983,** *105,* 3162.

162. Bock, C.W.; George, P.; Trachtman, M. *J. Phys. Chem.* **1984,** *88,* 1467.
163. Noell, J.O.; Newton, M.D. *J. Am. Chem. Soc.* **1979,** *101,* 51.
164. Leopold, D.G.; Miller A.E.S.; Lineberger, W.C., Submitted for publication.
165. Bergman, R.G. *Acc. Chem. Res.* **1973,** *6,* 25.
166. (a) Lockhart, T.P.; Comita, P.B.; Bergman, R.G. *J. Am. Chem. Soc.* **1981,** *103,* 4082. (b) Lockhart, T.P.; Bergman, R.G. ibid, **1981,** *103,* 4091.
167. Breslow, R.; Napierski, J.; Clarke, T.C. *J. Am. Chem. Soc.* **1975,** *97,* 6275. Breslow, R.; Khanna, P.L. *Tetrahedron Lett.* **1977,** 3429.
168. Angus, R.O., Jr.; Johnson, R.P. Unpublished work.

CHAPTER 4

Fluorinated Organic Molecules

Bruce E. Smart

E.I. du Pont de Nemours & Company, Inc., Wilmington, Delaware

CONTENTS

1. INTRODUCTION

In the preface to his classic paper on hybridization in fluorocarbons, W.A. Bernett stated, "The unique chemical reactivity and unusual physical properties of fluorocarbons, due to the high electronegativity of fluorine, places these compounds in a class of their own. A theoretical understanding of some of their behavior is still lacking, however."[1] Although theory has advanced and a large amount of new experimental data has accumulated since this statement was made in 1969, many fluorocarbon properties remain puzzling and still defy theoretical understanding. This chapter, which surveys the data that best illustrate the characteristic effects of fluorine substituents on molecular geometry, bonding, and stability, emphasizes fluorocarbon bond strengths and thermodynamic stabilities. The various theories that attempt to explain fluorocarbon properties are briefly discussed, but the reader who desires more thorough analysis should consult the cited literature.

The limited availability of reliable, quantitative experimental data restricts the coverage to simple saturated and unsaturated derivatives. Except for free radicals, the properties of fluorinated reactive intermediates are not discussed. The chemistry of fluorinated carbocations, carbanions, and carbenes has been recently reviewed.[2]

Note on units: All energies and enthalpies are given in kilocalories per mole (kcal/mol), and entropies are in calories per mole-Kelvin (cal/mol-°K). To convert to SI units, multiply by 4.184 J/cal.

2. SATURATED FLUOROCARBONS

A. Structure and Bonding

a. Carbon–Fluorine Bonds

The quintessential fluorine substitutent effect is the familiar variation of C−F bond lengths and bond strengths in fluorinated methanes (Tables 4-1, 4-2). The C−F bond lengths decrease and bond strengths increase progressively in going from CH_3F to CF_4. The C−F bond dissociation energy (BDE) varies from about 109 kcal/mol in CH_3F to an extraordinary 130.5 kcal/mol in CF_4. No such trend is observed for the other halomethanes; in fact, the C−X (X = Cl, Br, I) BDEs decrease progressively as the halogen content increases. The bonding trends in fluoroethanes parallel those in the fluoromethanes. The C−F bond lengths in CH_3CH_2F, CH_3CHF_2, and CH_3CF_3 are 1.398(5),[3] 1.343(1),[4] and 1.335(5) Å,[5] respectively. The C−F BDEs in CH_3CH_2F and CH_3CF_3 are 107.7 ± 1 and 124.8 ± 2 kcal/mol, respectively.[6]

A concept that nicely illustrates the effect of a substituent on thermody-

TABLE 4-1. Halomethane Bond Lengths

	r(C−X) (Å)			
X	CH_3X[a]	CH_2X_2[b]	CHX_3[b]	CX_4[c]
F	1.385(2)	1.357	1.332(1)	1.319(0)
Cl	1.781(2)	1.772(0)	1.758	1.767(3)
Br	1.939(2)	1.934	1.930	1.942

[a]Reference 7.
[b]References 8–10.
[c]References 7, 9, and 10.

TABLE 4-2. Halomethane Bond Dissociation Energies

	D°(C−X) (kcal/mol)[a]			
X	CH_3X	CH_2X_2	CHX_3	CX_4
F	109.9 ± 1	119.5[b]	127.5[b]	130.5 ± 3
Cl	84.6 ± 0.2	80.1	77.6	73.1 ± 1.8
Br	70.9 ± 0.3	64[c]	62[c]	56.2 ± 1.8
I	57.2 ± 0.3	51.3[c]	45.7[c]	—

[a]Data from Reference 6 unless noted otherwise.
[b]Reference 11.
[c]Reference 12.

namic stability is the so-called incremental geminal stabilization (IGSTAB). The IGSTAB value is defined as the increase in thermodynamic stability of a geminal-substituted system relative to the corresponding monosubstituted system.[13,14] The IGSTAB values are derived from isodesmic reactions such as those in Equations 4-1, 4-2, and 4-3.

	ΔH_r° (kcal/mol)	
$2CH_3F \rightarrow CH_4 + CH_2F_2$	−12.3 ± 4.3	(4-1)
$3CH_3F \rightarrow 2CH_4 + CHF_3$	−32.2 ± 6.5	(4-2)
$4CH_3F \rightarrow 3CH_4 + CF_4$	−49.6 ± 8.4	(4-3)

Thus, the IGSTABs for F in CH_2F_2, CHF_3, and CF_4 are about −6.2, −10.7, and −12.4 kcal/mol, respectively. Similarly, the IGSTABs for fluorine in CH_3CHF_2 and CH_3CF_3 are about −6.5 and −9.9 kcal/mol, respectively.[13] These values clearly indicate that geminal fluorination increases thermodynamic stability. By contrast, the IGSTABs for chlorine in CH_2Cl_2, $CHCl_3$, and CCl_4 are only about −0.8, −1.0, and −0.1 kcal/mol, respectively. For the ethanes, chlorine is actually destabilizing (IGSTAB CH_3CHCl_2 = 1.6; CH_3CCl_3 = 1.9 kcal/mol).[13]

The relative effects of geminal versus vicinal substitution on thermodynamic stability can be gleaned from heats of isomerizations for Equations 4-4 to 4-7. (The ΔH_f° of FCH_2CH_2F is not known experimentally. The −106.6 kcal/mol value calculated by Buckley and Rodgers[15] is probably very accurate; see also Reference 230.) These comparative ΔH_r°'s again reflect the special stabilization imparted by geminal fluorination and they further illustrate the dramatic difference between fluorine and chlorine substituent effects.

		ΔH_r° (kcal/mol)	
(gauche)	$FCH_2CH_2F \rightarrow CH_3CHF_2$ (−106.6)	−12.2 ± 2	(4-4)
	$ClCH_2CH_2Cl \rightarrow CH_3CHCl_2$	−0.5 ± 0.7	(4-5)
	$CH_2FCHF_2 \rightarrow CH_3CF_3$	−15.3 ± 2.1	(4-6)
	$CH_2ClCCl_2H \rightarrow CH_3CCl_3$	1.0 ± 0.4	(4-7)

The unusual effects of fluorine as a substituent are normally attributed to three of its special properties: (a) its high electronegativity, (b) the three nonbonding electron pairs on F, and (c) the excellent match between the F2s and 2p orbitals with the corresponding orbitals of other second-period elements. Several theories that are based on these properties have been advanced to explain the bonding trends in fluoroalkanes. The most widely accepted ones involve an interplay of σ–π resonance and σ-electronegativity effects.

Pauling ascribed the bond shortening in fluoromethanes to resonance among valence bond structures involving doubly bonded fluorine **(1)**.[16] Since the numbers of these interactions are 0, 2, 6, and 12 in CH_3F, CH_2F_2, CHF_3, and CF_4, respectively, the C–F bonds are predicted to shorten and strengthen as the fluorine content of the methane increases. Interestingly, if the lower limits for the ΔH_r°'s in Equations 4-1 to 4-3 are taken, they fall nearly in the ratio of 1:3:6, which is precisely the ratio of double-bond to no-bond resonance structures for CH_2F_2, CHF_3, and CF_4.

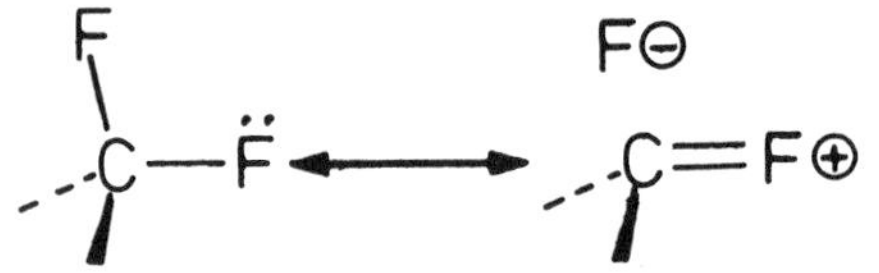

1 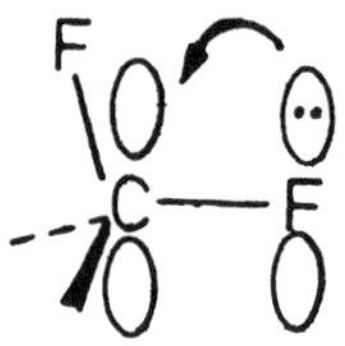 **2**

Ab initio calculations corroborate this valence bond description, wherein the principal stabilizing factor is back-donation of electron density from an F$p\pi$ lone pair into an antibonding σ*C–F orbital **(2)**.[17–19] Moreover, for XCH_2F derivatives σ^*–$p\pi$ electron transfer increases as the electronegativity of X increases. At the 4–31G level, for example, the C2p–F2p overlap population in CH_2F_2 is calculated to be about 26 times that in CH_3F.[17] More extensive 4–31G and 4–31G* calculations show that the total C–F bond overlap population increases continuously from CH_3F to CF_4 and parallels the progressive contraction of the C–F bonds.[19] By contrast, the C–Cl overlap populations decrease continuously from CH_3Cl to CCl_4.

The importance of electrostatics and bond ionicities also has been stressed. It is well known in the fluoromethane series that successive substitution of F on carbon increases the atomic charge on carbon, whereas the charge on F varies only slightly (Table 4-3). The C–F bonds therefore become progressively more ionic and stronger.

From CNDO/2 calculations, Oberhammer and colleagues[7,20,21] conclude that the C–F covalent bond orders in CH_3F and CF_4 differ by only about 3%, and they attribute the C–F bond shortening to a large increase in ionic bonding (p_{CF} = 0.034 in CH_3F; 0.160 in CF_4, where $p_{CF} = -q_F \cdot q_C$). They further claim that C–F bond lengths in CF_3X (X = H, F, Cl, Br, I) correlate only with the Pauling electronegativity of X, and that the difference in C–

TABLE 4-3. Atomic Charges and C_{1s} Binding Energy Shifts

	q_C (10^{-3}e)		q_F (10^{-3}e)		
Methane	STO−3G[a]	CNDO/2[b]	STO−3G[a]	CNDO/2[b]	ΔEC_{1s}[c]
CH_3F	169	180	−157	−190	2.8
CH_2F_2	383	400	−169	−190	5.48
CHF_3	532	610	−171	−200	8.28
$CClF_3$	—	630[d]	—	−180[d]	9.40[e]
$CBrF_3$	—	590[d]	—	−180[d]	8.52[e]
CF_4	674	810	−169	−200	11.05[e]

[a]Reference 22.
[b]Reference 23, unless noted otherwise.
[c]Shifts in electron-volts relative to CH_4. Reference 23.
[d]Reference 7.
[e]Reference 24.

X bond lengths in CH_3X versus CF_3X linearly correlates with the electronegativity of X, except for X = H. At the complete neglect of differential overlap (CNDO) level, the C−F bond orders are invariant in the CF_3X series, but C−X bond orders are not reported.

Obviously, these results conflict with the ab initio calculations that reveal the importance of both p−π interaction and electronegativity. The problems associated with comparing results from different quantum mechanical treatments and the possible inadequacy of semiempirical calculations for fluorocarbons are apparent.

The shifts in C_{1s} binding energies measured by photoelectron spectroscopy indicate systematic fluorine substituent effects (Table 4-3). The binding energies increase linearly with the number of fluorines bound to carbon and parallel the increase in atomic charge on carbon (q_C) in the methane series. Jolly and Bakke[24] have shown that the C_{1s} binding energy shifts, ΔEC_{1s}, for CF_3X derivatives (X = H, F, Cl, Br, I) are correlated by Equation 4-8, where F and R are field and resonance parameters for the substituents X (ρ = 0.948). This again supports the importance of both electronegativity and resonance in C−F bonding.

$$\Delta EC_{1s}(CF_3X) - \Delta EC_{1s}(CF_3H) = 3.76F + 0.323R \qquad (4\text{-}8)$$

Hybridization schemes based on valence bond logic are frequently used to rationalize fluorocarbon structure and bonding.[1,2,25,26,27a] Their basic postulate, which is derived from Walsh's rule, states that electronegative F bound to C causes the C atom to rehybridize and increase the amount of p character in its orbital directed toward the F. Thus by symmetry, the C−F bonds in CF_4 are taken to be sp^3 hybridized, but the C−F bond in CH_3F will have more p character (sp^n, $n > 3$) and the C−H bonds, more s character (sp^n, $n < 3$). Consequently, in agreement with experiment, the p-rich C−F bond in CH_3F is longer and weaker than the C−F bond in CF_4. This hybridization scheme also rationalizes the fluoromethane geometries. The

FCF bond angles in CF_4, CHF_3, and CH_2F_2 are respectively 109.5°,[7] 108.7°,[7] and 108.3°.[8] The hybridization argument would have the C−F bonds progressively increase in p character along this series of fluoromethanes and the FCF bond angles should correspondingly decrease, since bond angles between hybrid orbitals shrink as the orbitals acquire more p character. Conversely, the HCH angle in CH_2F_2 should be greater than tetrahedral, since the C orbitals directed toward the hydrogens have more s character than a normal sp^3 hybrid orbital. Experimentally, the angle HCH = 113.7° (111.9°) in CH_2F_2.[8,9]

Bernett has calculated that the s character of the carbon hybrid orbitals forming the fluoromethane C−H bonds increases in the order $CH_3F < CH_2F_2 < CHF_3$.[1] A parallel increase in C−H BDEs and a decrease in C−H bond lengths therefore are predicted. The C−H bond strengths in fact do increase in this order (cf Table 4-4), although the bond lengths (in angstroms) differ little experimentally: $r(CFH_2{-}H)$ = 1.100,[8] 1.095(10);[9] $r(CF_2H{-}H)$ = 1.092(3);[9] $r(CF_3{-}H)$ = 1.102(4).[7] This logic, however, fails to account for the comparatively strong C−H bonds in CH_4. The sp^3 $CH_3{-}H$ bond[7] (r = 1.099(2)Å) is actually stronger than the $CFH_2{-}H$ and $CF_2H{-}H$ bonds and only slightly weaker than the $CF_3{-}H$ bond. (Table 4-4).

Although this hybridization scheme qualitatively explains many fluorocarbon properties, there is some doubt about the proper description of s:p bonding ratios. Alternative schemes that do not rely on the perfect-pairing arguments of valence bond theory have been presented.[28,29] The validity of the Walsh–Bent hypothesis, which underlies the conventional hybridization picture outlined above, also has been questioned recently.[30] Exact sp^n hybridizations deduced from bond angles are now considered suspect.[30,31]

Among the various theories of fluorocarbon bonding, perhaps most intriguing are the highly original and controversial ones of Epiotis and coworkers. Based on perturbation molecular orbital (PMO) analyses, theories of nonbonded interactions, σ or geminal interactions, and bond ionicities are developed to explain substituent effects on molecular structure and stability. These theories have been discussed in a comprehensive monograph[32]

Table 4-4. Bond Dissociation Energies of Methanes

	$D°(C{-}X)$ (kcal/mol^a)			
X	$CH_3{-}X$	$CH_2F{-}X$	$CHF_2{-}X$	$CF_3{-}X$
H	105.1 ± 0.2	101.3 ± 1^b	103.2 ± 1^b	106.7 ± 1
F	109.9 ± 1	119.5^b	127.5^b	130.5 ± 3
Cl	84.6 ± 0.2	84.6^b	87.6	87.1^b
Br	70.9 ± 0.3	73^c	69 ± 2	70.6 ± 3
I	57.2 ± 0.3	—	—	55.0 ± 3

^{a}Reference 6, unless noted otherwise.
^{b}Reference 11.
^{c}Reference 12.

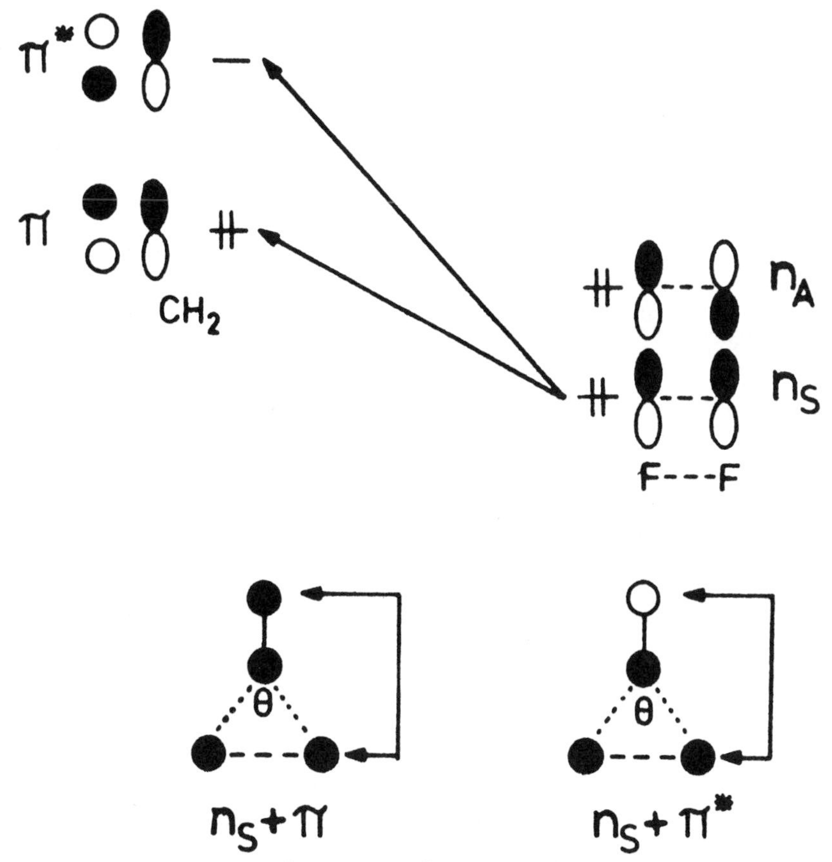

Figure 4-1. Interaction of π-type CH_2 fragment orbitals and F-group orbitals.

and recently expanded, modified, and integrated into a grandiose "Unified Valence Bond Theory of Electronic Structure."[33] To illustrate the principle of nonbonded attraction, the PMO arguments advanced to explain the FCF bond angle shrinkage in CH_2F_2 will be presented.[32]

The methylene π and π^* fragment orbitals and $F2p_z$ group orbitals that contribute to the highest molecular orbital (HOMO) of CH_2F_2 are shown in Figure 4-1. The phase relationships between the interacting fragment orbitals (only the symmetric F group molecular orbital interacts) reflect the effects of change in the FCF valence angle θ. As θ decreases, the four-electron $n_S + \pi$ destabilizing (repulsive) interaction decreases and the two-electron $n_S + \pi^*$ stabilizing interaction increases. These trends are apparent if one focuses on the secondary overlap (arrows) as θ varies. Quantitative ab initio MO calculations (STO−3G) of the energies for orbital interactions at θ = 120° versus 110° support these conclusions.[32,33] Accompanying these interactions is an increase in $F2p_z$−$F2p_z$ lone pair repulsion as θ decreases, but Epiotis contends that this repulsion is outweighed by the decrease in

($n_S + \pi$) overlap repulsion. The net effect of angle shrinkage is stabilization (ie, nonbonded or lone pair attraction). Complementary σ stabilization is also proposed.[32]

The nonbonded attraction proposal has been criticized by Kollman and others.[28,35] Supported by ab initio calculations, Kollman contends that Epiotis and co-workers underestimate the net F. . . F lone pair repulsions, and more important, his model calculations show that the source of the apparent attraction in an FCF fragment is an electronegativity effect and the lone pairs on F formally play no role. Epiotis, however, has attempted to repudiate these results.[32] The matter remains unresolved.

Although PMO analyses of structure and bonding can be very instructive, they often lead to considerable controversy. Their conclusions can differ markedly depending on the choice of fragment orbitals[34] and on whether factors that result in maximum stabilization or minimum destabilization are emphasized (see discussion of work by Epiotis in the section on alkene thermochemistry). This arbitrariness is unsettling and points to the need for reliable, quantitative assessments of the stabilizing and destabilizing factors that contribute to total molecular energy. For fluorocarbons, however, this is a formidable challenge (eg, see discussion in the section on alkene thermochemistry).

b. Carbon–Hydrogen and Carbon–Heteroatom Bonds

The dramatic effect of α-fluorination on C—F bond lengths and strengths does not occur with other C—X bonds. A comparison of the C—X (X = H, Cl, Br, I) BDEs in methanes shows that fluorination has no significant effect on C—Br and C—I bonds, only a slight effect on the C—Cl bond, and a small, irregular effect on the C—H bond (Table 4-4). This is also true for ethanes (Table 4-5). The striking exception is the C—F bond, whose BDE increases by about 20 kcal/mol on perfluorination.

The difference between CH_3—X and CF_3—X bond lengths is substantial for X = F and progressively decreases for the heavier halogens. Specifically,

TABLE 4-5. Bond Dissociation Energies of Ethanes

	$D°(C-X)$ (kcal/mol)[a]			
X	CH_3CH_2-X	CH_3CF_2-X	CF_3CH_2-X	CF_3CF_2-X
H	98.2 ± 1	99.5 ± 1[b]	106.7 ± 1	102.7 ± 0.5
F	107.7 ± 1	124.8 ± 2	109.4	126.8 ± 1.8
Cl	79.9 ± 1	—	—	82.7 ± 1.7
Br	67.8 ± 1	68.6 ± 1[c]	—	68.7 ± 1.5
I	53.4 ± 1	52.1 ± 1	56.3 ± 1	51.2 ± 1

[a]Reference 6, unless noted otherwise.
[b]Reference 36.
[c]Reference 37.

$r(CH_3{-}X) - r(CF_3{-}X) = 0.066(2), 0.029\ (5), 0.016(7)$, and $0.001(5)$ Å for X = F, Cl, Br, and I, respectively.[7]

The reasons for these effects are not well understood. As mentioned earlier, Oberhammer claims that $CF_3{-}X$ and $CH_3{-}X$ bond lengths strictly correlate with the electronegativity of X (except for X = H).[21] Bond ionicities indeed seem important, since α-fluorination noticeably affects the C–X BDEs only for the more electronegative F and Cl atoms. The bond ionicities from CNDO/2 calculations increase from $p_{CF} = 0.034$ in CH_3F to 0.160 in CF_4, but only from $p_{CCl} = 0.001$ in CH_3Cl to 0.057 in CF_3Cl.[7,20,21] However, any estimates or comparisons of covalent and ionic bond orders from approximate calculations at the CNDO/2 level should be viewed with some skepticism.

It is likely that resonance is also important. From considerations of orbital matching and π-donating ability, $p\pi{-}\sigma$ resonance **3** should increase in the order X = I $<$ Br $<$ Cl $\ll$ F, and the effect of α-fluorination on C–X BDEs should increase in the same order. Stated somewhat differently, for a given atom X in $CH_nF_{3-n}{-}X (X \neq F)$, the C–X BDEs do not significantly vary with n because the most important interactions between different C–F bonds occur to almost an equal extent in the radical and the molecule, which have equal numbers of fluorines.[19]

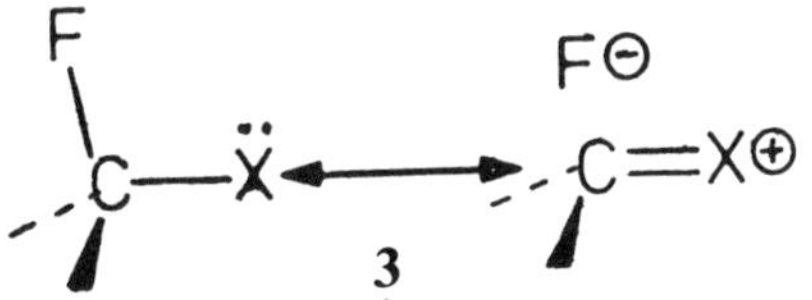

$$F{-}C{-}\ddot{X} \longleftrightarrow F^{\ominus}\ C{=}X^{\oplus}$$

3

More generally, the $p\pi{-}\sigma$ resonance depicted in **3** should be significant whenever X is a second-period atom with lone pairs (ie, N, O, or F). Ab initio calculations in fact indicate that this resonance increases in the order X = F $<$ O $<$ N.[17,38] Although the data are sparse, fluorination does appear to markedly affect C–O and C–N bonds, but not C–S or C–P bonds. The C–O BDEs in CF_3OCF_3 and CF_3OH are about 22 and 18 kcal/mol higher than those in CH_3OCH_3 and CH_3OH, respectively.[39] This compares to the approximate 20-kcal/mol difference in C–F bond strengths for CF_4 and CH_3F. Strengths of fluorinated C–N, C–S, or C–P bonds are not known experimentally, but their lengths are. The effects of CF_3 versus CH_3 substitution on geometry are shown below.[21,27]

	$O(CX_3)_2$	$N(CX_3)_3$	$S(CX_3)_2$	$P(CX_3)_3$
$r(C{-}A)_{X=F} - r(C{-}A)_{X=H} =$	−0.047 Å	−0.032 Å	0.014 Å	0.058 Å
$\angle(CAC)_{X=F} - \angle(CAC)_{X=H} =$	7.6°	7°	−1.7°	−1.4°

Fluorination appreciably shortens C–O and C–N bonds but actually lengthens C–S and C–P bonds. It also has been argued that electrostatic attraction between the highly positive charged C atom in the CF_3 group with the electronegative O and N atoms, but electrostatic repulsion with the electropositive S and P atoms, accounts for the observed differences in bond lengths.[21,27a]

The effects of CF_3 versus CH_3 substitution on bond angles are unusual. Normally the valence angle in AX_2 decreases as the electronegativity of X increases. Witness, for instance, the bond angles in FOF (103°) and FSF (98°) compared with those in ClOCl (111°) and ClSCl (102°).[21,27a] This has been rationalized by the familiar valence shell–electron pair repulsion (VSEPR) theory[40] and by MO–valence bond theory.[33] Since the instant CF_3 substituent effects cannot be explained by electronegativity differences, steric repulsion between CF_3 groups has been invoked to rationalize the unexpectedly large bond angles in $O(CF_3)_2$ and $N(CF_3)_3$.[21] It is equally plausible that $p\pi-\sigma$ resonance could be partly responsible for the opening of the valence angles.

The relative importance of various electronic effects to the bonding in fluoromethyl halides, ethers, and amines remains unresolved. Ab initio calculations that employ adequate basis sets to reliably treat these fluorocarbons have not been performed. To indicate the challenge, it has been shown that polarization functions at the double-zeta level are needed just to handle the comparatively simple $HOCH_2F$ and CH_3OCH_2F molecules.[41]

For a very good, more comprehensive review of fluorocarbon structure that covers the literature through 1974, the article by Yokozeki and Bauer[27a] is recommended. These authors extensively use hybridization and electronegativity arguments to explain the differences in the molecular structures of hydrocarbons and fluorocarbons.

β-Fluorination has its own peculiar effect on bond strengths. From the limited available data (Table 4-5), three β fluorines increase the comparative C–X bond strengths (CF_3CH_2X versus CH_3CH_2X) in the order X = F < I < H. Similarly, from the C–X BDEs in CF_3CF_2X versus CH_3CF_2X, the C–H bond is affected the most. The estimated BDEs given below suggest that the effect of β-fluorination is additive. It has been proposed that changes in electrostatic energy between CF_3CH_2X and $CF_3CH_2\cdot$ could in part account for the high C–X BDEs.[42,43]

	CH_3CHF-H	$CH_2FCHF-H$	CF_3CHF-H
$D°$ (kcal/mol)[44]	96.3 ± 1.5	98.8 ± 1.0	103.5 ± 1.0

β-Fluorination also increases 2 and 3° C–H bond strengths (Table 4-6). Unlike the progressive decrease in C–H BDEs for CH_3CH_2-H, $(CH_3)_2CH-H$, and $(CH_3)_3C-H$, the C–H BDEs increase in the fluorinated series. The $(CF_3)_3C-H$ BDE of 109 kcal/mol is an approximate value, but if correct, it is the strongest known C–H bond in a saturated organic compound. The comparative structures of $(CF_3)_3CH$[45] [r_g(C–H) = 1.110(10) Å; r_g(C–C) = 1.537(3) Å] and $(CH_3)_3CH$[46] [r_g(C–H) = 1.122(6) Å; r_g(C–C)

TABLE 4-6. Bond Dissociation Energies of Alkanes

R	$D°(R-H)$ (kcal/mol)[a]	R	$D°(R-H)$ (kcal/mol)[a]
CH_3	105.1 ± 0.2[b]	CF_3	106.7 ± 1
CH_3CH_2	98.2 ± 1[b]	CF_3CF_2	102.7 ± 0.5
$(CH_3)_2CH$	95.1 ± 1	$(CF_3)_2CF$	103.6 ± 0.6[c]
$(CH_3)_3C$	93.2 ± 1[b]	$(CF_3)_3C$	109[c]

[a]From Reference 6 unless noted otherwise.
[b]Reference 47 gives 104.4 (CH_4), 100.3 ± 1 (C_2H_6), 93.9 ± 1 (C_4H_{10}) kcal/mol.
[c]Reference 48.

= 1.535(1) Å] give no hint of an especially unusual C–H bond for the former.

The discussions of bond strengths throughout these first two sections have focused on substituent effects in the covalent structures but not in the associated radicals. It is fair to ask whether fluorination also affects the stability of radicals. Indeed, it is known that fluorination can profoundly affect radical geometry.[2,49,50,51] For example, methyl radicals become increasingly nonplanar with the successive replacement of H by F. β-Fluorination, however, has little or no effect on the geometry of ethyl radicals. From the BDE data for C–Br and C–I bonds in CH_3X versus CF_3X and CH_3CH_2X versus CF_3CF_2X (Tables 4-4, 4-5), it appears that fluorination has no significant effect, but this is somewhat circular reasoning.

The use of isodesmic reactions (Equations 4-9 to 4-12) has been recommended as a means to measure stabilization energies of substituted methyl or ethyl radicals with respect to $CH_3\cdot$ or $CH_3CH_2\cdot$.[52,53]

Reaction	ΔH_r°(kcal/mol)	
$FCH_2\cdot + CH_4 \rightarrow FCH_3 + CH_3\cdot$	3.9 ± 4.3	(4-9)
$F_2CH\cdot + CH_4 \rightarrow F_2CH_2 + CH_3\cdot$	4.0 ± 2.5	(4-10)
$F_3C\cdot + CH_4 \rightarrow F_3CH + CH_3\cdot$	−2.4 ± 4.2	(4-11)
$CH_3CF_2\cdot + CH_3CH_3 \rightarrow CF_3CH_3 + CH_3CH_2\cdot$	0.5 ± 4.0	(4-12)

From such data, it has been suggested that $F_2CH\cdot$ and $FCH_2\cdot$, for example, possess extra stability relative to $CH_3\cdot$.[52] However, owing to the sizable imprecision in the ΔH_f°'s for the fluororadicals, it is palpably meaningless to draw any conclusions from these isodesmic ΔH_r° values.

There does not appear to be any compelling evidence that fluorination affects the stability of free radicals.[54,55]

c. Carbon–Carbon Bonds

Aliphatic C–C bonds are normally strengthened by fluorination. Ethane C–C BDEs progressively increase with α-fluorination, and the C–C bond in CH_3CF_3 is about 11 kcal/mol stronger than that in CH_3CH_3 (Table 4-7).

TABLE 4-7. Ethane Bond Lengths and Bond Dissociation Energies

Ethane	r_a(C−C) (Å)[a]	$D°$(C−C) (kcal/mol)[b]
CH_3CH_3	1.532(1)	90.4 ± 0.2
CH_3CH_2F	1.502(3)	91.2[c]
CH_3CHF_2	1.498(4)	95.6 ± 2.5
CH_3CF_3	1.494(3)	101.2 ± 1.1[d]
CF_3CH_2F	1.501(4)	94.6 ± 4
CF_3CF_3	1.545(2)	98.7 ± 1.1[e]

[a]Reference 57.
[b]Reference 6, unless noted otherwise.
[c]Reference 12; 90.2 ± 2.6 from thermochemical data (Appendix I).
[d]Reference 58.
[e]Reference 59.

Further fluorination decreases the bond strengths relative to CH_3CF_3, but CF_3CH_2F and CF_3CF_3 still have stronger C−C bonds than CH_3CH_3. The difference between the CF_3–CF_3 and CH_3−CH_3 BDEs is consistent with the estimate that the C−C bond in polytetrafluoroethylene is some 8 kcal/mol stronger than that in polyethylene.[56]

The ΔH_r°'s for Equations 4-13 and 4-14 (IGSTAB ≅ −6.5 and −9.9 kcal/mol) show that geminal fluorination increases the overall thermodynamic stabilities of ethanes. This arises from increased stability of both C−C and C−F bonds. The effect of vicinal fluorination on thermodynamic stability and C−C or C−F bond strengths is difficult to assess owing to the lack of thermochemical data. The ΔH_r° of Equation 4-15 reflects more the special stability of CH_3CF_3 than any vicinal effect. The ΔH_r° of Equation 4-16, however, suggests that "incremental vicinal stabilization" (IVSTAB) for F is negligible.

	ΔH_r° (kcal/mol)	
$2CH_3CH_2F \rightarrow CH_3CHF_2 + CH_3CH_3$	−13.1 ± 2.9	(4-13)
$3CH_3CH_2F \rightarrow CH_3CF_3 + 2CH_3CH_3$	−29.7 ± 6.8	(4-14)

	ΔH_r° (kcal/mol)	
$2CH_3CF_3 \rightarrow CF_3CF_3 + CH_3CH_3$	13.2 ± 1.3	(4-15)
$2CH_3CH_2F \rightarrow FCH_2CH_2F + CH_3CH_3$ (−106.6)[15]	−0.9 ± 0.8	(4-16)

The trends in ethane C−C bond lengths versus bond strengths are puzzling in that the expected correspondence breaks down. For instance, CF_3CF_3 has a longer but much stronger C−C bond than CH_3CH_3, and whereas CH_3CF_3 and CH_3CH_2F have similar bond lengths; CH_3CF_3 has a 10-kcal/mol stronger C−C bond.

Neither the increased C−C bond strengths nor the anomalous bond lengths in fluoroethanes have been adequately explained. Simple hybridization arguments would predict CF_3CF_3 to have the shortest, strongest C—C bond because its bond should have the most s character. The unusually strong CH_3—CF_3 bond has been attributed to its relatively high ionic character.[58] From CNDO/2 calculations of atomic charges,[23] $-q_{C_2} \cdot q_{C_1} = 0.008$, 0.033, and 0.066 for CH_3CH_2F, CH_3CHF_2, and CH_3CF_3, respectively. This increase in electrostatic attraction parallels the increase in C—C bond strengths for these fluoroethanes. For CF_3CF_3, however the electrostatic effect is the largest ($-q_C \cdot q_C = -0.314$) and is repulsive, but its C—C bond is still some 8 kcal stronger than that in CH_3CH_3.

The relative kinetic reactivities of fluorinated cyclobutanes are often cited to illustrate the increased stability of fluorinated C—C bonds.[60,61] In general, fluorination kinetically stabilizes a cyclobutane toward thermal fragmentation, culminating with perfluorocyclobutane, which is stabilized over cyclobutane by almost 12 kcal/mol (Equations 4-17 and 4-18).

Reaction	log A	E_a (kcal/mol)	
□ → $2CH_2{=}CH_2$	15.6	62.5[62]	(4-17)
F_2□F_2 (F_2, F_2) → $2CF_2{=}CF_2$	15.97	74.2[63]	(4-18)

The fluorinated cyclobutane fragmentations fall into two categories, which presumably reflect the strengths of the individual C—C bonds being broken.[60] The lower energy pathways ($E_a \cong 69$ kcal/mol, Equations 4-19 and 4-20) involve cleavage of CH_2—CH_2 bonds whose energies are comparable to those in cyclobutane,[64] and unusually strong CH_2—CF_2 or CF_2—CF_2 bonds. The higher energy pathways ($E_a \cong 74$ kcal/mol; Equations 4-18 and 4-21) involve cleavage of two unusually strong bonds. [Again, it is curious that cyclo-C_4F_8 has longer—1.566(8) Å versus 1.548(3) Å[65]—yet stronger C—C bonds than c-C_4H_8). This analysis presumes that the transition states are late enough to feel both cyclobutane bonds being broken, and furthermore, it ignores any differences in cyclobutane strain energies. This assumption may not be valid (see the section titled, "Cyclopropane and Cyclobutane Strain Energies").

Reaction	log A	E_a (kcal/mol)[60]	
□F_2 → $CH_2{=}CH_2 + CH_2{=}CF_2$	15.61	69.2	(4-19)
□F_2, F_2 → $2CH_2{=}CF_2$	15.35	69.8	(4-20)
□F_2, F_2 → $CH_2{=}CH_2 + CF_2{=}CF_2$	15.27	73.6	(4-21)

Fluorine substitution has a uniquely different effect on the C—C bonds and stabilities of cyclopropanes. In comparison with cyclopropane (c-C_3H_6), hexafluorocyclopropane (c-C_3F_6) has slightly shorter C—C bonds [1.505(3) Å[66] vs 1.510(2) Å[67]], but it is thermally much less stable and it readily extrudes difluorocarbene at 190°C.[68] Various estimates of the strain energy in c-C_3F_6 place it 41–53 kcal/mol above the strain in c-C_3H_6,[1,66,69] but these estimates may be too high (see the section titled "Cyclopropane and Cyclobutane Strain Energies").

The cause of the added strain in c-C_3F_6 has not been settled, but local hybridization changes seem to be important.[1,70,71] Substitution of H by F increases the p character in the C orbitals directed toward F, as evidenced by the angles FCF (112.2°)[66] versus HCH (115.1°)[67] in c-C_3H_6. To compensate for this rehybridization, the C—C bonds in c-C_3F_6 increase in s character and the CCC valence angle opens; that is, the C—C bonds become more bent and are weakened. Bernett[1] has estimated that the C—C bonds in c-C_3F_6 are bent more than those in c-C_3H_6 by about 4°. The hybridization arguments also rationalize the unusually large differences in bond lengths of 1,1-difluorocyclopropane (c-$C_3H_4F_2$), wherein the C_2—C_3 bond opposite the CF_2 group is lengthened [1.533(1) Å] but the adjacent C_1—C_2 bond is shortened [1.464(2) Å][72] relative to the 1.510 Å C—C bonds in c-C_3H_6. The alternation of bond lengths in c-$C_3H_4F_2$ was anticipated by Hoffmann[73,74] from a qualitative MO analysis of σ and π substituent effects on cyclopropane bond lengths.

Ab initio STO—3G,[75a] 4–31G,[75b] and DZ quality[70,76] calculations have been performed on c-C_3H_5F and c-$C_3H_4F_2$. They correctly reproduce the bond alternation in the latter, but there is disagreement about the specific fluorine electronic effects. The DZ calculations predict a longer C_1—C_2 bond (1.494–1.497 Å) and a shorter C_2—C_3 bond (1.527–1.528 Å) in c-C_3H_5F compared to the corresponding bonds in c-$C_3H_4F_2$. The structure of c-C_3H_5F is not known experimentally.

The thermochemistry of geminally fluorinated cyclopropanes has been studied extensively and it is firmly established that a pair of geminal fluorines raises the strain energy of a cyclopropane ring by 12–14 kcal/mol.[69,77] For example, the heats of hydrogenation of **4–6** are respectively 14.2, 13.8, and 12.1 kcal/mol more exothermic than those for the corresponding hydrocarbons.[78] Ab initio 4–31G calculations of heats of hydrogenation for c-$C_3H_4F_2$ and c-C_3H_6 show a similar 11.7-kcal/mol extra destabilization in the former.[69] These results compare to O'Neal and Benson's[54] earlier estimate that the strain energy in a cyclopropane will increase by 4.5–5.0 kcal per F substituent.

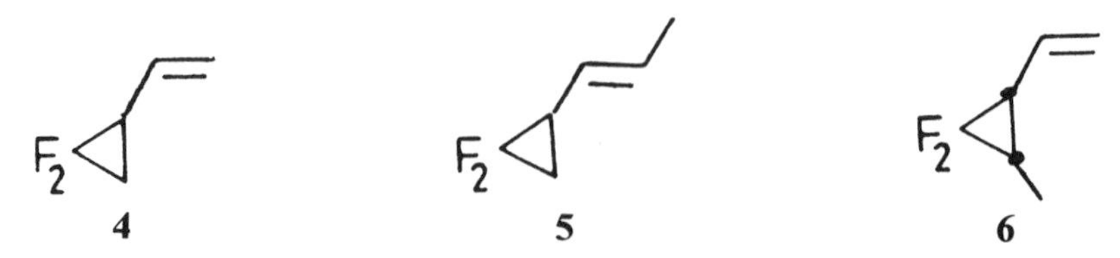

The kinetic effect of the increased strain in a *gem*-difluorocyclopropane is a specific weakening of the C—C bond opposite the CF_2 group by 8–10 kcal/mol, but the adjacent bond is affected only slightly (0–3 kcal/mol).[77] For example, the E_a's for the cis–trans isomerization of 1,1-difluoro-2,3-dimethylcyclopropane[79] and the rearrangement of 2,2-difluoro-1-vinylcyclopropane[80] are about 9.7 and 9.4 kcal/mol lower than those for their respective hydrocarbons. By contrast, the comparable E_a values for the rearrangements in Equations 4-22 and 4-23 and those in 4-25 reflect little perturbation of the bond adjacent to the CF_2 group.

	log A	E_a (kcal/mol)	
D_2, D_2 → D_2, D_2	14.3	41.6[81]	(4-22)
F_2 → F_2	13.2(2)	38.3(4)[82]	(4-23)
F_2, F_2 → F_2, F_2	12.6	29.6(5)[83]	(4-24)

A second pair of geminal fluorines causes dramatic, nonadditive kinetic destabilization (Equations 4-23 vs 4-24; 4-25 vs 4-26). Whereas the first CF_2 group reduces E_a in the methylenecyclopropane rearrangement by about 3 kcal/mol ($\Delta\Delta G\ddagger \cong$ 1.7 kcal/mol),[82] the second reduces it by an additional 8.7 kcal/mol ($\Delta\Delta G\ddagger \cong$ 5.8 kcal/mol).[83] The effect is equally pronounced in the spiropentane rearrangement, wherein the first pair of fluorines does not significantly affect E_a, but the second lowers it by nearly 13 kcal/mol ($\Delta\Delta G\ddagger$ $\cong$ 6.0 kcal/mol).[84] Again, however, the CF_2—CF_2 bond lengths in tetrafluorocyclopropanes do not reflect their low bond strengths. The C_1—C_2 bonds in 1,1,2,2-tetrafluorocyclopropane (1.47 Å)[85] and perfluorospiropentane (1.487(6) Å)[86] are 0.03–0.04 Å *shorter* than those in their hydrocarbon analogues, as opposed to the case for CF_3—CF_3 versus CH_3—CH_3.

	log A	E_a (kcal/mol)[84]	
X_2 → X_2 + X_2	15.9	57.6 (X = H)	(4-25)
	16.1(2)	58.0(5) (X = F)	
F_2, F_2 → F_2, F_2	13.8(1)	45.4(3)	(4-26)

Several additional examples that illustrate the kinetic effects of *gem*-difluorosubstitution on cyclopropane,[77] vinylcyclopropane,[78,87,88] and spiropentane[89] thermochemistry have been reported. By comparison, little is known about monofluorocyclopropanes. An ab initio STO−3G calculation predicted a single F substituent on a cyclopropane to be slightly stabilizing (0.8 kcal/mol),[90] but later 4–31G calculations[69,75b] indicated c-C_3H_5F to be destabilized by 4.6–6.5 kcal/mol relative to c-C_3H_6, in reasonable agreement with O'Neal and Benson's estimate.[54] To date, only two experimental results have been reported. Paquette and associates[91] found that the semibullvalene isomer **7a** was more stable than **7b**; Dolbier and Burkholder[92] also found from the fluoromethylenecyclopropane equilibrium (4-27) that F prefers not to be on a cyclopropane ring. These results, however, reflect only the *relative* effect of F on a cyclopropane ring versus a double bond.

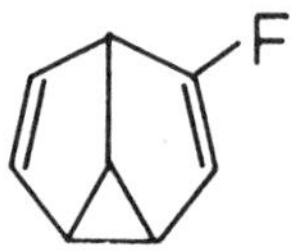

7a  **7b**

ΔH°_r (kcal/mol)
-2.60 ± 0.06 (4-27)

For a broader discussion of substituent effects on cyclopropanes and other strained systems, see Chapter 5 in this volume.

B. Conformation and Molecular Dynamics

Among their many perplexing phenomena, the conformational preferences and dynamic behavior of fluorocarbons are probably the least well understood. Fluorinated ethanes have been the most extensively investigated systems, and the discussion here is limited primarily to these "simple" molecules. Experimental and theoretical studies of more complex systems, including fluorinated haloalkanes, alkenes, and carbonyl compounds, have been reviewed elsewhere.[27a,32,33,93,94]

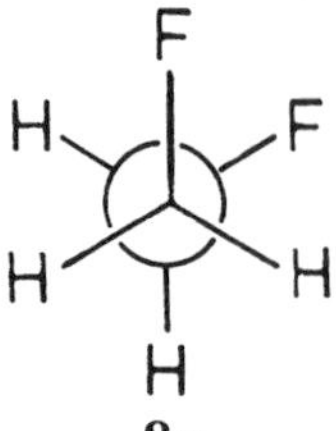

8g 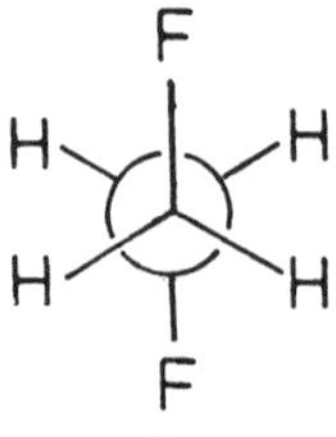 **8a** 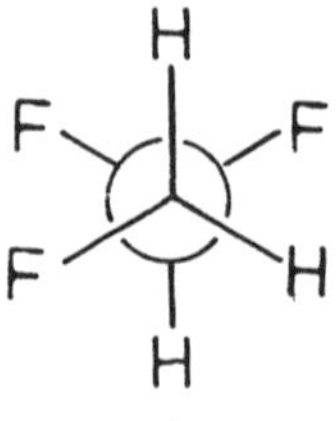 **9** 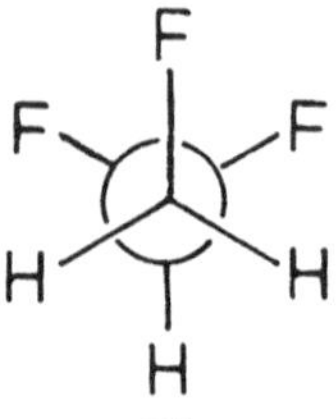 **10**

The gauche conformation of CH_2FCH_2F **(8g)** is more stable than the anti-form **(8a)** by 1.98 ± 0.08[95] (1.76 ± 0.51)[96] kcal/mol in the gas phase. For all other 1,2-dihaloethanes, including CH_2FCH_2Cl and CH_2FCH_2Br, the anti-form is favored by 0.5–2 kcal/mol.[93] For CH_2FCHF_2 in the gas phase, however, the conformation with fewer gauche F. . . F interactions **(9)** is preferred over **10** by 1.06 ± 0.40 kcal/mol.[97] (The $CH_2ClCHCl_2$ and $CH_2BrCHBr_2$ derivatives also prefer the analogous conformation, but by 1.8–2.0 kcal/mol.) Interestingly, **10** is more stable than **9** by 1.03 ± 0.06 kcal/mol in the liquid state, and only **10** exists in the solid state.[97] For CHF_2CHF_2, conformation **11a** predominates in all physical states. The energy difference between **11g** and **11a** in the gas phase (1.18 ± 0.10 kcal/mol)[97] is greater than the corresponding differences reported for $CHCl_2CHCl_2$ (0.5–0.8 kcal/mol) and $CHBr_2CHBr_2$ (0.6–1.1 kcal/mol).[93]

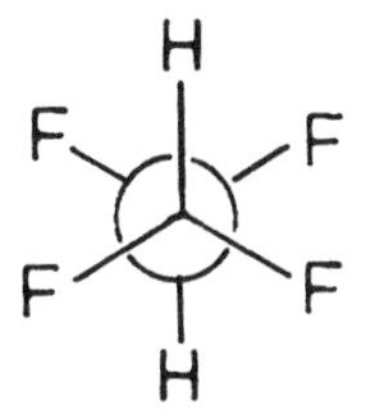

11a 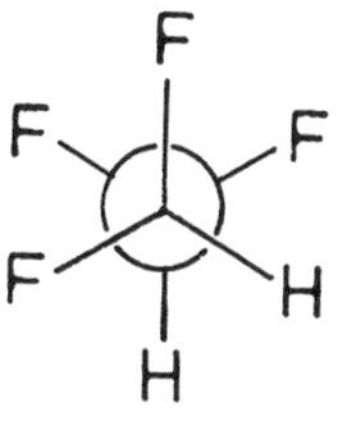 **11g**

The stability of gauche relative to anti-CH_2FCH_2F is unexpected from considerations of steric or dipole–dipole repulsion. A PMO analysis based on lone pair attraction and $\sigma_{CH}-\sigma^*_{CF}$ stabilization nicely rationalized the preference for the gauche form, but the same reasoning incorrectly predicted **10** and **11g** to be more stable.[32] An alternative PMO argument[35] and a hyperconjugation hypothesis[98–100] have been proposed, but neither scheme accounts for the different favored conformations in the series CH_2FCH_2F, CH_2FCHF_2, and CHF_2CHF_2.

Several semiempirical and ab initio calculations on fluoroethanes have been reported, but they often fail to reproduce the experimental results, no less clarify the factors that affect conformational stability. Semiempirical CNDO/2,[101] NNDO,[102] and MNDO[103] calculations correctly predict gauche CH_2FCH_2F to be more stable: E**(8a)** − E**(8g)** = 0.1, 0.3, and 1.3 kcal/mol, respectively, but CNDO/2 and NNDO give poor results for related fluoroethanes (cf Table 4-8). Both NNDO[102] and MNDO[104] badly underestimate the energy difference between **11a** and **11g**, and NNDO incorrectly predicts **10** to be more stable than **9:** E**(9)** − E **(10)** = 0.4 kcal/mol.

Ab initio calculations fail to correctly predict even the more stable CH_2FCH_2F conformer. A 4–31G calculation without geometry optimization gave E**(8g)** − E**(8a)** = 1.0 kcal/mol.[98] A higher quality (DZ + D_F) basis ((7,3)/[5,3] plus polarization functions on F) with geometry optimization gave worse results, with E**(8g)** − E**(8a)** = 1.32 kcal/mol,[105] and it was suggested that neglect of electron correlation may have caused the error. (The chronic problems encountered with CH_2FCH_2F conformers are similar to those found with CHF=CHF isomers; see section on olefin thermochemistry). The more stable conformer of CHF_2CHF_2 is correctly predicted by ab initio calculations, but the quantitative results are unsatisfying. A STO-3G//STO-3G basis (STO-3G-optimized geometries), gave $E(\mathbf{11g}) - E(\mathbf{11a}) =$ 0.54 kcal/mol, but 4–31G//STO-3G gave 2.4 kcal/mol.[104] Thus, depending on the choice of basis sets, the experimental difference of about 1.2 kcal/mol is either underestimated or overestimated.

Empirical methods have been considerably more successful. Meyer[93,106,107] and Abraham and Stölevik[108] have produced molecular mechanics force fields that calculate the conformation energies and rotational barriers of most fluorinated ethanes and propanes within 10% of experiment. The only exception is CH_2FCH_2F, which is correctly predicted to prefer the gauche form, but the gauche–anti energy difference is underestimated by 1–1.5 kcal/mol. A more thorough study of the reasons for this discrepancy may identify the factors that account for the special stability of gauche CH_2FCH_2F.[93]

The fluoroethanes exhibit peculiar trends in their barriers to internal rotation (Table 4-8). Not surprisingly, CH_3CH_2F has a higher barrier (3.30 kcal/mol) than CH_3CH_3 (2.93 kcal/mol). This compares to the barrier in CH_3CH_2Cl (3.72 kcal/mol),[109] and the increase in rotational barriers for CH_3CH_2X (X = H, F, Cl) parallels the increase in van der Waals radii of X (H = 1.20, F = 1.47, Cl = 1.75 Å^3).[110] Further fluorination has an irregular,

TABLE 4-8. Ethane Barriers to Rotation

Ethane	Barrier (kcal/mol)			
	CNDO/2[a]	NNDO[b]	4–31G[c]	Experiment
CH_3CH_3	2.18	1.9	3.26	2.928(25)[d]
CH_3CH_2F	2.00	1.5	3.63	3.30(3)[e]
CH_3CF_2H	1.88	1.8	3.39	3.21[f]
CH_3CF_3	1.76	1.5	3.40	3.25(20)[g]
CH_2FCH_2F				
g–a	—	—	3.22	2.36(58)[h]
g–g	—	—	7.74	4.52(72)[h]
CF_3CH_2F	1.46	1.9	—	4.58[i]
CF_3CHF_2	1.22	1.6	—	3.51[j]
CF_3CF_3	1.07	1.6	—	3.88[k]

[a]Reference 101.
[b]Reference 102.
[c]References 18, 98.
[d]Reference 117.
[e]Reference 118.
[f]Reference 119.
[g]Reference 94.
[h]Reference 96.
[i]Reference 120.
[j]Reference 121.
[k]Reference 122.

nonadditive effect wherein it decreases the barrier to 3.21 kcal in CH_3CF_2H and 3.25 ± 0.2 kcal in CH_3CF_3. The observed barriers then follow in the order $CH_3CF_3 < CHF_2CF_3 < CF_3CF_3 < CH_2FCF_3$. The difference in rotational barriers for CH_3CH_3 and CF_3CF_3 is only about 1 kcal/mol, which contrasts with the 8-kcal/mol difference between CH_3CH_3 and CCl_3CCl_3.[94]

Several calculations of barrier heights in fluoroethanes have been reported, but most failed to reproduce the observed trends. Semiempirical CNDO/2 and NNDO methods gave completely erroneous results (Table 4-8). Extended Hückel gave gauche–anti (3.06 kcal/mol) and gauche–gauche (4.79 kcal/mol) barriers in good agreement with experiment for CH_2FCH_2F, but it incorrectly gave CH_3CH_2F a lower barrier than CH_3CH_3.[110] Minimum basis STO-3G calculations had the same fault and were generally unsatisfactory in describing the conformational behavior of fluoroethanes.[18,112] A DZ quality calculation predicted CH_3CH_2F and CH_3CH_3 to have the same barriers.[113] By contrast, the split-valence 4–31G basis predicted the barrier heights remarkably well (Table 4-8). The unusual decrease in barriers upon geminal fluorination has been attributed to $p\pi-\sigma$ resonance[18] and electrostatic effects.[114]

In contrast to hydrocarbons, there are few experimental data on the conformation and molecular dynamics of perfluorocarbons. The different possible isomers and even the existence of rotational isomerism in some perfluoroalkanes are still matters of controversy. Systematic studies of conformational equilibria in *n*-perfluoroalkanes are just beginning to appear.[115] The most studied perfluorocarbon is the polymer polytetrafluoroethylene (PTFE). Unlike its hydrocarbon analogue, which favors the planar trans conformation, PTFE exists as a slightly twisted, helical structure distorted about 17° from the trans structure.[116] (Actually, it has two different helical forms, but they differ by only ≅ 2° in dihedral angle.) Empirical calculations predict a trans–gauche energy difference of 1.1–4.8 kcal/mol and an extremely high barrier between the conformers of more than 50 kcal/mol.[116] It is still not certain whether the helical twist in PTFE is inherent in the $(CF_2CF_2)_n$ polymer backbone or whether it is a result of the polymer's crystalline environment. Model ab initio calculations on $CH_3CF_2CF_2CH_3$ show no evidence for a stable rotamer with a 163° dihedral angle,[104] whereas empirical calculations can reproduce the helical twist observed in the polymer.[116]

3. UNSATURATED FLUOROCARBONS

A. Alkenes and Related Structures

a. Structure and Bonding

The C—F bond lengths in fluoroalkenes are shorter than those in fluoroalkanes with the same number of vicinal or geminal fluorines. Vinyl C—F bond lengths are typically 1.34–1.35 Å (Table 4-9) compared with 1.39–1.40

TABLE 4-9. Fluoroethylene Geometries[a]

	$CH_2{=}CHF$		*cis*-$CHF{=}CHF$		*trans*-$CHF{=}CHF$	
r_g(C=C), Å:	1.333(1)	1.330(18)	1.331(4)	1.330(11)	1.329(4)	1.320(9)
r_g(C−F), Å:	1.348(4)	1.351(15)	1.335(2)	1.342(8)	1.344(2)	1.338(3)
∠HCH, °:	114.7(44)	120.9(34)	—	—	—	—
∠HCF,°:	111.3(9)	107.7(27)	114.7(2)	113.9(6)	111.4(11)[b]	115(1)
Reference:	123	125	123	126	123	127

	$CH_2{=}CF_2$		$CHF{=}CF_2$		$CF_2{=}CF_2$
r_g(C=C), Å:	1.316(6)	1.340(6)	1.309(6)	1.341(12)	1.311(7)
r_g(C−F), Å:	1.324(3)	1.315(3)	1.336(2)[c]	1.316(11)[d]	1.319(2)
∠HCH, °:	119.3(3)	122.0(8)	—	1.342(24)[e]	—
∠HCF, °:	—	—	114.0(27)	116.0(24)	—
∠FCF, °:	109.7(7)	110.6(6)	109.1(5)	112.9(21)	112.6(6)
Reference:	123	128	123	129	123

[a]For comparison, r_g(C=C) = 1.339 Å, ∠HCH = 117.8° in $CH_2{=}CH_2$.[8]
[b]∠HCH probably 3–4° too small.[130]
[c]C_1−F, C_2−F assumed equal.
[d]C_1−F.
[e]C_2−F.

Å for monofluoroalkane C−F bonds. Vicinal fluorination does not noticeably affect vinyl C−F bond lengths, but geminal fluorination does. The C−F bonds in $CH_2{=}CF_2$, $CHF{=}CF_2$, and $CF_2{=}CF_2$ are 0.02–0.03 Å shorter than those in $CH_2{=}CHF$ or $CHF{=}CHF$. Vinyl C−F bonds and the C−F bonds in CHF_3 or CH_3CF_3 should have about the same strengths (125–127 kcal/mol), whereas geminal vinylic C−F bonds and those in CF_4 should be comparable (130 kcal/mol), assuming that C−F bond lengths and bond strengths correlate. From ΔH_f° data for $CH_2{=}CH\cdot$, $F\cdot$, and $CH_2{=}CHF$, the C−F BDE in $CH_2{=}CHF$ is 120.5 kcal/mol, about 13 kcal/mol higher than the CH_3CH_2−F BDE. The C−F bond strength in $CH_2{=}CF_2$ cannot be estimated, since ΔH_f° of $CH_2{=}CF\cdot$ is not known.

The C=C bond lengths of fluorinated olefins reported by different laboratories do not fully agree (Table 4-9). The electron diffraction data of Carlos and co-workers[123] show that $CH_2{=}CHF$ and *cis*- or *trans*-$CFH{=}CHF$ have C=C bonds about 0.01 Å shorter than that in $CH_2{=}CH_2$, but $CH_2{=}CF_2$, $CHF{=}CF_2$, and $CF_2{=}CF_2$ have bonds that are 0.02–0.03 Å shorter. The C=C bond length in $CH_2{=}CF_2$ (1.315 Å) reported by Laurie and Pence[124] from microwave data agrees with that of the Carlos group [1.316(6) Å]. Mijlhoff and co-workers have redetermined the fluoroethylene structures, except for that of $CF_2{=}CF_2$, from combined electron diffraction data and rotational constants from microwave data.[128] Although their structures for $CH_2{=}CHF$ and $CHF{=}CHF$ are similar to those of Carlos, they report that the C=C bond lengths in $CH_2{=}CH_2$, $CH_2{=}CF_2$, and $CFH{=}CF_2$ are equal within experimental error. This 0.03-Å discrepancy between the data of Carlos and Mijlhoff, inter alia, creates a serious problem in the proper choice of input geometries for MO calculations of *gem*-difluoroolefins.

The geometry of $CH_2{=}CF_2$ has been fully optimized with two high-quality DZ basis sets. Bock and co-workers[132] reported $r_e(C{-}F) = 1.341$ Å, $r_e(C{=}C) = 1.300$ Å, ∠HCH = 121.6°, and ∠FCF = 109.2° with a (7,3)/[5,3] basis. This compares to $r_e(C{-}F) = 1.310$ Å, $r_e(C{=}C) = 1.307$ Å, ∠HCH = 120.6°, and ∠FCF = 109.2° from a (DZ + D_C) basis[133] (Dunning and Hay's (9,5)/[3,2] basis[134] with d-polarization functions on C.) Both calculations give C=C bond lengths that agree better with the experimental value of Carlos and co-workers. It is dangerous, however, to draw any firm conclusions, since these calculations neglect the effects of electron correlation. It has been shown, for example, that geometry optimization of CHF=CHF isomers with DZ basis sets that do not include correlation results in overcontracted double bonds.[131] A notable feature of the fluoroethylene geometries, on which both calculations and experiments agree, is the contracted FCF angle in *gem*-difluoroolefins.

The explanations for the molecular structures of fluoroethylenes are analogous to those used to rationalize fluoroalkane structures. Bernett[1] has argued that C_1 in $CH_2{=}CF_2$ is sp^3 rather than sp^2 hybridized and its C−F bond length, strength, and the FCF angle should be similar to those in CF_4. Kollman[28] has attributed the contracted FCF angle to electronegativity effects in accord with the prediction of VSEPR theory and has calculated that the C-hybrid orbitals of the CH_2 group have more s character than those in $CH_2{=}CH_2$. This is consistent with the unusually large HCH angle in $CH_2{=}CF_2$. Pappas[135] has estimated that the effect of rehybridization is to shorten the C=C bond in $CH_2{=}CF_2$ versus $CH_2{=}CH_2$ by 0.012 Å.

An opposing view championed by Epiotis and co-workers attributes the FCF angle contraction in $CH_2{=}CF_2$ to F. . . F nonbonded attraction.[32,33] The PMO arguments are entirely analogous to those presented for CH_2F_2[32] (see discussion of Figure 4-1, above). An alternative PMO analysis that favors different fragment orbitals has been presented by Whangbo and colleagues.[34]

Ab initio calculations[17,132,136,137] attribute the shortening of C−F bonds in vinyl fluorides to resonance of type **12a** ↔ **12b.** (This represents interaction of a F2pπ lone pair with the π_{CC} MO of the double bond. The F lone pair

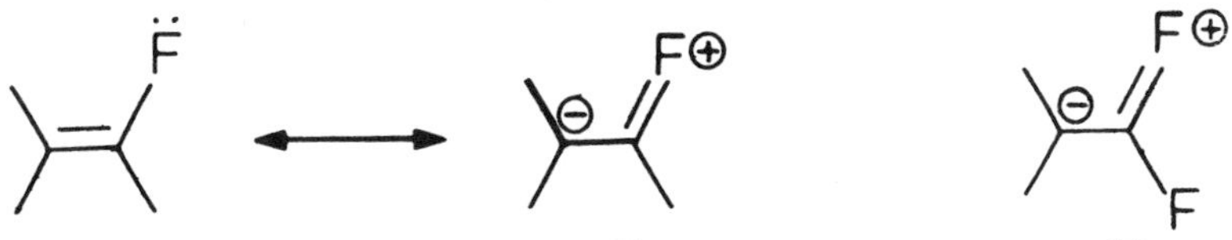

12a **12b** **12c**

does not mix appreciably with π^*_{CC} because their energy separation is too large.[138]) The atom charges (10^{-3} e) in $CH_2{=}CHF$ ($q_{C_1} = 156$, $q_{C_2} = -461$, $q_F = -313$) and $CH_2{=}CF_2$ ($q_{C_1} = 649$, $q_{C_2} = -533$, $q_F = -288$) calculated with a (9,5)/[5,4] basis set[132] show the expected $C_1{-}C_2$ charge alternation and indicate that **12c** is more important than **12b.** Thus, based on relative bond ionicities, both the C$-$F and C$=$C bonds in $CH_2=CF_2$ are predicted to be shorter and stronger than those in $CH_2=CHF$. It also is suggested that attraction between the fluorines in different electronic states **(12c)** might account for the small FCF angle in $CH_2=CF_2$.[132]

The effects of substitution in carbon–heteroatom double-bond versus single-bond systems differ markedly. The C$-$F bond lengths vary with the electronegativity of X in $(CF_3)_nX$ (see section titled "Carbon-Hydrogen and Carbon-Heteroatom Bonds") whereas they do not in $CF_2=X$ (X = O, S, Se). The identical geometry of the CF_2 groups in $CF_2=X$ is striking (Table 4-10). Ab initio 4–21G calculations[139] on $CF_2=O$ and $CF_2=S$ show that their gross atomic charges on C differ by about 0.7e, but the charge distributions in their bonds about C are nearly identical. The calculated dipole moments of $CF_2=S$ and $CF_2=O$ agree with experiment, but the former is predicted to be polarized $^-CF_2=S^+$ (0.09 debye), whereas the latter is polarized $^+CF_2=O^-$ (0.90 debye).

The replacement of H by F shortens the C$=$O and C$=$S bonds by 0.036 and 0.022 Å, respectively (Table 4-10). This contrasts with the shortening of the C$-$O bond in $(CH_3)_2O$ by 0.047 Å but the *lengthening* of the C$-$S bond by 0.014 Å in $(CH_3)_2S$ upon fluorination. The C$=$N bond lengths in $CH_2=NH$ and $CF_2=NF$ are equal within experimental error, even though their electronic structures are quite different. (An analogous similarity between $CH_2=CH_2$ and $CF_2=CHF$ exists, provided the Mijlhoff structure for $CF_2=CHF$ is accepted.) The atom charges from 4–21G calculations[140] are q_C = 0.13 and q_N = -0.76 in $CH_2=NH$, but are 1.33 and -0.44, respectively, in $CF_2=NF$. From Mulliken overlap populations, the π contributions to the C$=$N double bonds in the two molecules do not differ significantly, but the σ contribution in $CF_2=NF$ is much smaller. Thus it appears that the weak σ bonding offsets the strong electrostatic attraction in $CF_2=NF$, and the overall result is an unperturbed C$=$N bond.[140]

The structural data for higher fluoroolefins are limited principally to CF_3-substituted derivatives. Substitution of H by CF_3 normally shortens C$=$C bonds by about 0.04 Å, or about 0.03 Å for CF_3 versus CH_3 substitution. For example, r(C$=$C) = 1.318(8) Å in $CF_3CH=CH_2$ versus 1.341(2) Å in $CH_3CH=CH_2$.[146] The CCC angle is unaffected, 125.8(11)° versus 124.3(4)°, and the C$-CF_3$ bond [1.495(6) Å] is a tad shorter than the C$-CH_3$ bond.

TABLE 4-10. Geometries of $X_2C=Y$ Molecules

Molecule	r(C=Y) Å	r(C−F) Å	XCX Angle (°)	Ref.
$CF_2=O$	1.171(1)	1.315(0)	107.71(8)	141
$CH_2=O$	1.207(1)	—	116.2(1)	142
$CF_2=S$	1.589(2)	1.316(2)	107.1(2)	143
$CH_2=S$	1.611(1)	—	116.8(2)	144
$CF_2=Se$	1.743(3)	1.314(2)	107.5(4)	143
$CF_2=NF$	1.274(6)	1.300(3)	112.5(2)	140
$CH_2=NH$	1.273	—	117.0	145

TABLE 4-11. Geometries of $(CX_3)_2C{=}Y$ Molecules

Molecule	r_g(C=Y) (Å)	r_g(C−C) (Å)	CCC Angle (°)	Ref.
$(CF_3)_2C{=}O$	1.246(14)	1.549(8)	121.4(4)	148
$(CH_3)_2C{=}O$	1.210(3)	1.507(2)	116.7(3)	148
$(CF_3)_2C{=}NH$	1.294(29)	1.549(7)	121.6(4)	148
$(CF_3)_2C{=}CH_2$	1.373(13)	1.533(6)	123.6(3)	148
$(CH_3)_2C{=}CH_2$	1.342(3)	1.505	115.8(6)	149

A comparison of the structures of *cis*- and *trans*-$CF_3CH{=}CHCF_3$ with the $CH_3CH{=}CHCH_3$ isomers reveals similar features.[147] Surprisingly, the structures of *cis*- and *trans*-$CF_3CH{=}CHCF_3$ are nearly identical within experimental error. This contrasts with the CHF=CHF isomers wherein the FCC angle is about 3° larger in the cis isomer.[123,126,127]

Substitution of CH_3 by CF_3 shortens the N=N bond by about 0.02 Å to 1.235(10) Å in $CF_3N{=}NCF_3$.[147] The CNN angle is not affected by fluorination. Substitution of H by F has a similar effect wherein *trans*-FN=NF has a N=N bond 0.02 Å shorter than *trans*-HN=NH [1.231(10) vs 1.252(2)Å].[27]

Although substitution of H by CF_3 generally shortens double bonds, geminal CF_3 substitution has the opposite effect. The C=C and C−C bonds in $(CF_3)_2C{=}CH_2$ are appreciably longer than those in $(CH_3)_2C{=}CH_2$ (Table 4-11). The C=O and C−C bonds in $(CF_3)_2C{=}O$ are similarly lengthened relative to those in $(CH_3)_2C{=}O$. The abnormally large CCC angles, 123.6° and 121.4° in $(CF_3)_2C{=}CH_2$ and $(CF_3)_3C{=}O$, respectively, suggest that steric effects may be important.

Very few structures of perfluorinated olefins have been determined experimentally. The structure of $CF_3CF{=}CF_2$ is just partially solved, and only bond angles are known accurately.[150] Its FCF angle [112.2(28)°] is similar to that in $CF_2{=}CF_2$, and its CCC angle [127.8(7)°] is about 2° larger than that in $CF_3CH{=}CH_2$. Perfluorocyclobutene and cyclobutene have nearly identical C=C and C−C bond lengths.[27a,65] Hexafluorobutadiene is interesting because it exists preferentially in a gauche conformation with a CCCC dihedral angle of about 47°[151] and a barrier to planarity of about 2.85 kcal/mol.[152] By contrast, butadiene and 1,1,4,4-tetrafluorobutadiene are known experimentally to prefer the s-trans conformation,[27] and 2,3-difluorobutadiene is calculated to also favor the s-trans form.[138] A recent molecular mechanics calculation predicts the gauche conformer with a 48° dihedral angle to be more stable than the s-trans form of hexafluorobutadiene by 1.6 kcal/mol.[153] Hexafluorobutadiene is obviously an interesting molecule for further theoretical study.

b. Thermochemistry

The overall effect of fluorination on thermodynamic stability and its specific effect on C=C bond strength are often misunderstood. Experimentally, addition reactions of *gem*-difluoroolefins are all more exothermic than the

TABLE 4-12. Cyclobutene Ring Openings

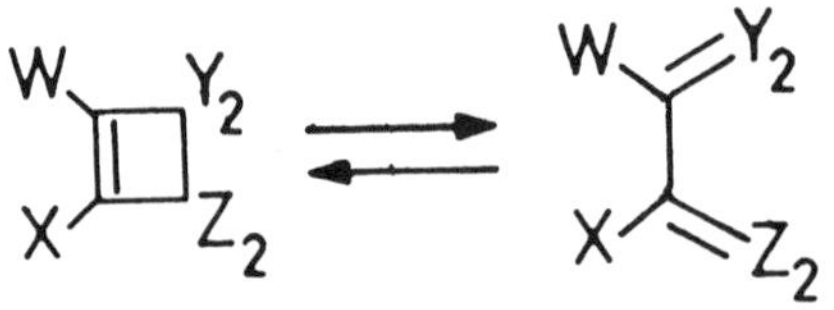

W	X	Y	Z	K_{eq} (315°C)	$\Delta H°$ (kcal/mol)	$\Delta S°$ (eu)	E_a (kcal/mol)	Ref.
H	H	H	H (**13**)	9×10^3	−8	4.5	32.5	155
F	F	F	F (**14**)	5.6×10^{-3}	11.7	9.6	47.1	156
F	H	F	H (**15**)	3.3	2.5	6.65	—	157
H	H	F	F (**16**)	77.5	—	—	47.9	158
CH_3	H	F	H (**17**)	16.0	1.1	7.4	—	157
CH_3	H	F	F (**18**)	0.5	—	—	—	158
CF_3	CF_3	F	F (**19**)	8.4	0.4	4.9	46.0	159

corresponding hydrocarbon reactions. For example, the heats of hydrogenation of $CH_2{=}CF_2$ (4-30) and $CHF{=}CF_2$ (4-31) are about 6 and 13 kcal/mol, respectively, greater than that for $CH_2{=}CH_2$ (4-28). From ΔH_f° data, the heats of Br_2, Cl_2, and HBr additions to $CF_2{=}CF_2$ are 10–15 kcal/mol more exothermic, the heat of polymerization of $CF_2{=}CF_2$ is 17 kcal/mol greater,[154] and the heat of $CF_2{=}CF_2$ cyclodimerization is 35 kcal/mol greater than the respective reactions of $CH_2{=}CH_2$. The $\Delta H°$ values for the c-C_4H_6 (**13**) and c-C_4F_6 (**14**) equilibria, −8 and 11.7 kcal/mol, respectively (Table 4-12), are often cited to exemplify the dramatic difference between the stabilities of fluorinated and nonfluorinated double bonds.

	ΔH_r° (kcal/mol)	
$CH_2{=}CH_2 + H_2 \rightarrow CH_3CH_3$	-32.6 ± 0.3	(4-28)
$CH_2{=}CHF + H_2 \rightarrow CH_3CH_2F$	-29.7 ± 0.8	(4-29)
$CH_2{=}CF_2 + H_2 \rightarrow CH_3CHF_2$	-38.8 ± 2.8	(4-30)
$CHF{=}CF_2 + H_2 \rightarrow CH_2FCF_2H$	-45.7 ± 5.5	(4-31)

It has been commonly accepted that fluorination destabilizes C=C bonds, but several results are inconsistent with this conclusion. The relative ΔH_r°'s for Equations 4-28 and 4-31 and the comparative cyclobutene equilibria in Table 4-12 indicate that three fluorines indeed thermodynamically destabilize double bonds. (Cyclobutenes **14, 16,** and **19** rearrange at similar rates and their equilibria involve the same changes in vinyl and allyl C—F bonds, but since their K_{eq}'s differ considerably, hexafluorobutadiene must be destabilized.) It is difficult, however, to reconcile the comparative ΔH_r°'s of Equations 4-28 and 4-30, which could be taken to mean $CH_2{=}CF_2$ is destabilized by about 6 kcal/mol relative to $CH_2{=}CH_2$, and the greater thermodynamic stability of 1,1,4,4-tetrafluorobutadiene relative to **16.** Subtle substituent changes also can have unusually pronounced effects on the cyclobutene equilibria, namely, substitution of H by CH_3 reduces K_{eq} by a factor of more than 150 (**16** vs **18**). Other equilibrium data seemingly indicate that *gem*-difluoro groups both destabilize (4-32 and 4-33) and stabilize (4-34) double bonds.

	ΔH_r°(kcal/mol)	
$CF_2{=}CHCH_3 \rightarrow CH_2{=}CHCF_2H$	-2.5[13]	(4-32)
	-5.1 ± 0.6[13]	(4-33)
	1.9 ± 0.1[82]	(4-34)

Fortunately, much of the confusion about the thermodynamic stability of fluorinated double bonds has been resolved by the studies of the Rodgers and Dolbier groups. Wu and Rodgers[56] have shown that the π-bond dissociation energy (D_π°) of $CF_2{=}CF_2$ is 52.3 ± 2 kcal/mol, about 7 kcal/mol less than that for $CH_2{=}CH_2$, where $D_\pi^\circ = D^\circ$ (CC$-$X) $-$ D° ($\cdot$CC$-$X). For $CH_2{=}CF_2$, $D_\pi^\circ = 62.8 \pm 2$ kcal/mol, which is 3.7 kcal/mol *greater* than D_π° ($CH_2{=}CH_2$) at 59.1 ± 2 kcal/mol.[160] The difference between the D_π° values of $CF_2{=}CF_2$ and $CH_2{=}CF_2$ agrees well with other thermochemical data. For example, the ΔH_r° for the isodesmic equation (4-35) indicates that $CF_2{=}CF_2$ is thermodynamically less stable than $CF_2{=}CH_2$. Equation 4-36 suggests that this is entirely a consequence of the relative π-bond strengths. (There seems to be no reason a priori for Equation 4-36 to equal ΔH_r° of Equation 4-35.)

	ΔH_r° (kcal/mol)	
$CF_2{=}CF_2 + CH_2{=}CH_2 \rightarrow 2CF_2{=}CH_2$	-14.6 ± 1.5	(4-35)

$$D^\circ_\pi(CF_2{=}CF_2) + D^\circ_\pi(CH_2{=}CH_2) - 2D^\circ_\pi(CF_2{=}CH_2) \cong -14.2 \text{ kcal/mol} \qquad (4\text{-}36)$$

The well-known [2+2] thermal cyclodimerization and cycloadditions of $CF_2{=}CF_2$ have been attributed in part to its weak π bond.[2,161] By contrast, the failure of $CH_2{=}CF_2$ to cyclodimerize is consistent with its relatively strong π bond. The activation energy for the dimerization of $CF_2{=}CF_2$ is 25 ± 1 kcal/mol,[63] which compares with an estimated value of greater than 44 kcal/mol for $CH_2{=}CF_2$.[160]

To properly analyze the thermodynamics of addition and rearrangement reactions of *gem*-difluoroolefins, it is necessary to assess the relative thermodynamic stabilities of vinylic CF_2 versus saturated CF_2 groups. A comparison of IGSTAB values has been recommended for this purpose.[13] Equation 4-37 gives an IGSTAB value for vinylic fluorine of only about -0.5

kcal/mol. (A value of −2.5 kcal/mol is obtained from ΔH_f°'s calculated from group values.)

		ΔH_r° (kcal/mol)	
$2CH_2{=}CHF \rightarrow$	$CH_2{=}CF_2 + CH_2{=}CH_2$	-1.1 ± 1.5	(4-37)
(−32.1)	(−81.7) (12.5)	(−5)	
$2CH_2{=}CHCH_2F \rightarrow$	$CH_2{=}CHCHF_2 + CH_2{=}CHCH_3$	−13.4	(4-38)
(−32.7)	(−92.6)		

The IGSTAB values for alkyl F (−6.5 kcal/mol, as noted in connection with Equations 4-1 to 4-3) and allylic F (−6.7 kcal/mol; Equation 4-38) are considerably larger. Thus, the driving force to convert vinyl CF_2 to alkyl (4-30) or allyl CF_2 groups (4-32, 4-33) is the greater stability of the latter two groups (ie, larger IGSTAB), not any inherent destabilization of the 1,1-difluoroolefins.[13]

The IGSTAB values should be used with caution, however, because by definition they are system dependent. For instance, it would be incorrect to assume similar values for the quite different allylic fluorines in $CH_2{=}CHCF_2H$ and those in cyclobutene **(16).** (This relates to the restrictions imposed on the use of group values to estimate ΔH_f°—see section titled "Heats of Formation.") The experimental result for the **16** equilibration (Table 4-12) in fact indicates that the IGSTAB for allylic fluorine in a $C{=}\underline{C}CF_2CF_2-$ system is less than that for vinyl CF_2.

Thermochemical data uniformly show that a single F stabilizes a double bond. The heat of hydrogenation of $CH_2{=}CHF$ (4-29) and the I_2-catalyzed equilibration of $FCH_2CH{=}CH_2$ to *cis*-$CHF{=}CHCH_3$ ($\Delta H_r^\circ = -3.34$ kcal/mol)[13] indicate about 3 kcal/mol added stability in a vinyl fluoride over its hydrocarbon analogue. Equation 4-39, which is useful for calculating substituent effects relative to CH_3,[90] is thermoneutral and implies that F and CH_3 have identical effects. A CH_3 group is considered to stabilize a double bond by 3.2 kcal/mol.[162] The congruence of these data casts doubt on an ab initio result indicating that F stabilizes a C=C bond considerably more than CH_3.[163]

$$(CH_3)_2CHF + CH_2{=}CHCH_3 \rightarrow (CH_3)_3CH + CH_2{=}CHF \qquad \Delta H_r^\circ\ (\text{kcal/mol}) = 0.0 \pm 1.0 \qquad (4\text{-}39)$$

Thermodynamic and kinetic data indicate that vicinal fluorination destabilizes a double bond. From the ΔH_r° for Equation 4-40, the IVSTAB for *cis*-CHF=CHF is about 4 kcal/mol (cf IVSTAB for CH_2FCH_2F; Equation 4-16). This destabilization is also reflected in the thermal cis–trans isomerizations of $CF_3CF{=}CFCF_3$[164] and CHF=CHF,[165] whose activation energies are about 6.4 and 4.7 kcal/mol, respectively, lower than those for $CH_3CH{=}CHCH_3$ and CDH=CDH isomerizations. As expected, *cis*-CHF=CHF is much less stable than $CH_2{=}CF_2$ (4-41).

	ΔH_r° (kcal/mol)	
$2CH_2{=}CHF \rightarrow CH_2{=}CH_2 + cis\text{-}CHF{=}CHF$	8.1 ± 3.8	(4-40)
$cis\text{-}CHF{=}CHF \rightarrow CH_2{=}CF_2$	-9.2 ± 3.1	(4-41)

Theoretical calculations generally agree with these experimental data. Ab initio DZ calculations for Equation 4-40 give ΔE = 6.65 to 8.80 kcal/mol, depending on the choice of basis set and input geometry.[132,133] The ΔE for Equation 4-41 is -11.7 kcal/mol by MINDO/3,[227] -9.6 kcal/mol by STO−3G//STO−3G,[34] and -6.3 to -16.0 kcal/mol by various DZ calculations.[132] A recent (DZ + D_C) calculation with complete geometry optimization gives -8.8 kcal/mol.[133] The ΔE also can be estimated from D°_π values. From the kinetic result cited above, assume $D^\circ_\pi(CH_2{=}CH_2) - D^\circ_\pi(cis\text{-}CHF{=}CHF) \cong 4.7$ kcal/mol. Since $D^\circ_\pi(CH_2{=}CF_2) - D^\circ(CH_2{=}CH_2) \cong 3.7$ kcal/mol, $D^\circ_\pi(CH_2{=}CF_2) - D^\circ_\pi(cis\text{-}CHF{=}CHF) \cong 8.4$ kcal/mol, which agrees remarkably with the (DZ + D_C) ΔE value.

Thermochemical data on fluoroalkylated olefins are scarce, but the following data indicate that a CF_3 group *destabilizes* a double bond by about 3 kcal/mol. The heat of hydrogenation of $CF_3CH{=}CH_2$ is calculated to be about 6 kcal/mol more exothermic than that for $CH_3CH{=}CH_2$ (4-42, 4-43). Since the IGSTABs in CH_3CF_3 (-9.8 kcal/mol) and $CH_2{=}CHCF_3$(-10.1 kcal/mol) are nearly equivalent, the comparative heats of hydrogenation correctly reflect relative thermodynamic stabilities of the olefins. Equation 4-44 similarly indicates 6-kcal/mol destabilization in $CF_3CH{=}CH_2$ relative to $CH_3CH{=}CH_2$. Since CH_3 stabilizes a double bond by 3.2 kcal/mol, the CF_3 group is destabilizing by about 3 kcal/mol.

		ΔH_r° (kcal/mol)	
$CH_2{=}CHCF_3 + H_2 \rightarrow$	$CH_3CH_2CF_3$ (-182.8)	-36.0	(4-42)
$CH_2{=}CHCH_3 + H_2 \rightarrow$	$CH_3CH_2CH_3$	-29.8 ± 0.2	(4-43)
$CH_3CF_3 + CH_2{=}CHCH_3 \rightarrow$	$CH_3CH_3 + CH_2{=}CHCF_3$	6.3 ± 2.2	(4-44)

The effects of fluorine substitution on C=O bonds are much larger but parallel those for C=C bonds. The comparative ΔH_r° values for Equations 4-45 and 4-46 and Equation 4-47 imply that *gem*-difluoro groups stabilize a C=O bond much more than a C=C bond.

From the approximation ΔH_r° (4-47) $\cong D^\circ_\pi(CF_2{=}O) + D^\circ_\pi(CH_2{=}CH_2) - D^\circ_\pi(CH_2{=}CF_2) - D^\circ_\pi(CH_2{=}O)$ and the D°_π values for $CH_2{=}O$ (71 ± 2 kcal/mol)[166] and the olefins, $D^\circ_\pi(CF_2{=}O)$ calculates to be about 109 kcal/mol, which is 38 kcal/mol higher than that for $CH_2{=}O$. This contradicts the prediction of Pickard and Rodgers that $CF_2{=}O$ should have a lower π-bond energy.[167]

		ΔH_r° (kcal/mol)	
$CF_2{=}O + HF \rightarrow$	CF_3OH	3.6	(4-45)
$CF_2{=}CH_2 + HF \rightarrow$	CF_3CH_3	-33.8	(4-46)
$CF_2{=}O + CH_2{=}CH_2 \rightarrow$	$CF_2{=}CH_2 + CH_2{=}O$	34.4 ± 1.5	(4-47)

The ΔH_r°'s of Equations 4-48 and 4-49 indicate that a single fluorine stabilizes a C=O bond by 19–21 kcal/mol, which compares with its modest 3-kcal/mol stabilization of a C=C bond.

	ΔH_r° (kcal/mol)	
$(CH_3)_2CHF + CH_2{=}O \rightarrow CHF{=}O + (CH_3)_2CH_2$	-18.9 ± 0.7	(4-48)
$CHF{=}O + CH_2{=}CH_2 \rightarrow CH_2{=}O + CH_2{=}CHF$	18.3 ± 0.9	(4-49)

The π-bond energy in $(CF_3)_2C{=}O$ is much lower than that in $(CH_3)_2C{=}O$. From experimental heats of hydration, Rogers and Rapiejko[168] estimated $\Delta H_r^\circ = -21$ kcal/mol for Equation 4-50, although there was considerable uncertainty about heat of hydration of $(CH_3)_2C{=}O$. Since two CH_3 groups stabilize a C=O bond by about 9 kcal/mol,[168] it was concluded that $(CF_3)_2C{=}O$ is destabilized by about 12 kcal/mol.

$$(CF_3)_2C{=}O + (CH_3)_2C(OH)_2 \rightarrow (CH_3)_2C{=}O + (CF_3)_2C(OH)_2 \qquad (4\text{-}50)$$

The exothermicity of Equation 4-51 indicates considerably greater destabilization. After correcting for the effects of CF_3 and CH_3 on a double bond, $(CF_3)_2C{=}O$ calculates to be about 42 kcal/mol less stable than $(CH_3)_2C{=}O$; that is, two CF_3's destabilize a C=O bond by 33 kcal/mol. Although the magnitude of the effect is debatable, two CF_3 groups clearly destabilize a C=O bond, and this agrees with the remarkable reactivity of $(CF_3)_2C{=}O$ in carbonyl addition reactions.[169]

	ΔH_r° (kcal/mol)	
$(CF_3)_2C{=}O + 2CH_2{=}CHCH_3 \rightarrow$ $(CH_3)_2C{=}O + 2CH_2{=}CHCF_3$	-29.9 ± 3.5	(4-51)

In summary, one or a pair of geminal fluorine substituents stabilize C=C and C=O bonds. Trifluoromethyl and presumably other perfluoroalkyl groups destabilize both C=C and C=O bonds. The lengths and strengths of fluorinated C=O bonds correlate; that is, shorter bonds are stronger bonds. By contrast, there is no correlation whatever between the lengths and strengths of fluorinated C=C bonds.

An interesting property of 1,2-difluoroethylenes, which has attracted considerable theoretical attention, is the greater thermodynamic stability of the cis over the trans isomer. It is now well established experimentally that this "cis effect" generally obtains for ethylenes having electronegative vicinal substituents.[170,171] This is contrary to expectation based on simple steric or electrostatic considerations. From a point charge model, it has been estimated that the Coulomb repulsion between the fluorines is about 1.2 kcal/mol higher in *cis*-CHF=CHF.[172] Coupled with the experimental difference between *cis*- and *trans*-CHF=CHF of 0.93 kcal/mol (ΔH_{298}°),[171] the extra stabilization in the cis isomer actually amounts to about 2 kcal/mol.

Several proposals including resonance stabilization,[171,173,174] conjugative destabilization,[35] nonbonded attraction,[32] and other electronic arguments[170] have been advanced to qualitatively explain the cis–trans CHF=CHF energy difference. The conflicting proposals of Epiotis and co-workers[32] and Bingham[35] have received the most attention.

Epiotis argues that p–π and p–σ attraction stabilizes *cis*-CHF=CHF over the trans isomer. A PMO analysis[32] of the π interactions reveals that the antisymmetric combination of F2p orbitals **(20)** can interact with the vacant antibonding π^*_{CC} orbital, but the symmetric combination **(21)** cannot. Consequently, **20** is depopulated relative to **21,** and pπ F. . . F overlap (lone pair attraction) results. Similarly for the σ system, since only the occupied orbital **23** can interact with the unoccupied σ^*_{CC} orbital, **22** is favored, and further attractive F. . . F overlap results. Although this treatment is oversimplified and has been modified, the importance of nonbonded attraction in *cis*-CHF=CHF is still evident from the more comprehensive PMO analysis.[34,175]

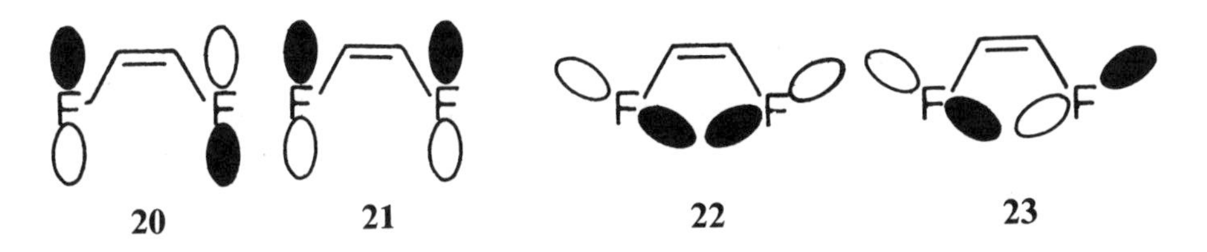

Bingham[35] also uses PMO arguments to support his claim that *cis*-CHF=CHF has no extra stability, but that the trans isomer is destabilized. Based on Eyring's principle of minimum bending of delocalized orbitals,[176] Bingham contends that electron delocalization will favor trans geometry only if antibonding orbitals are unoccupied. Since CHF=CHF can be considered to be a four-center, six-electron system isoelectronic to the butadiene dianion, the trans form will be more destabilized than the cis.

A raft of MO calculations has been performed on CHF=CHF isomers to attempt to reproduce their energy difference and to ascertain whether stabilizing or destabilizing effects are more important. Semiempirical[107] and ab initio[34,177] results have been interpreted to support Epiotis's hypothesis, whereas other ab initio results have been claimed to verify Bingham's hypothesis.[132,178] Since these cited ab initio calculations all incorrectly predict *trans*-CHF=CHF to be the more stable, however, the conclusions drawn from them appear rather dubious.

More extensive ab initio calculations that use polarized basis sets or include CI correctly predict *cis*-CHF=CHF to be the more stable. A 6–311G*//4–31G calculation by Binkley and Pople[172] gave ΔE_{tc} [E(trans) − E(cis)] = 0.26 kcal/mol, and an overlap population analysis indicated some F. . . . F π attraction, but no σ attraction, in *cis*-CHF=CHF. The majority of the extra stabilization, however, was attributed to correlation energy in the cis isomer. Cremer's correlation corrected (RSMP) ab initio calculations with 6–31G*, 6–311G*, and 6–311G** basis sets and experimental geometries gave ΔE_{tc} = 0.9–1.3 kcal/mol.[179] More than 80% of the energy difference arose from correlation corrections, but Cremer found some evidence for both π- and σ-lone pair attraction in *cis*-CHF=CHF. To complicate matters, a systematic survey of the role of electron correlation and polarization functions found that correlation effects were important for accurately reproducing the molecular structures of the cis and trans isomers, but once the geometry had been properly optimized, electron correlation did not contribute to ΔE_{tc}.[131]

These represent some of the highest quality calculations performed on fluorocarbons and they disagree on the specific electronic effects in the CHF=CHF isomers. The source of the small but significant 0.9-kcal/mol difference in energy between the isomers has generated an inordinate amount of research and controversy, but it seems likely that this contentious issue will remain unsettled for some time to come.

B. Acetylenes and Allenes

Among the shortest C−F bonds are those in HC≡CF (1.279 Å)[8] and FC≡CF (1.28 Å).[8] Only the C−F bond in FCN (1.26 Å)[180] is shorter. From ΔH_f°'s of HC≡C·, F·, and C_2HF (25.5 kcal/mol; see below), the C−F BDE in C_2HF is about 115 kcal/mol, some 5 kcal/mol less than that in CH_2=CHF. Fluorination does not significantly affect C≡C bond lengths, but it dramatically reduces their BDEs (Table 4-13). The C≡C bond in C_2F_2 is 115 kcal/mol weaker than that in C_2H_2!

Fluorine substitution has been predicted to destabilize a C≡C bond based on qualitative MO arguments.[90] Its quantitative effect is difficult to evaluate because of uncertainties in the thermochemical data. For C_2HF, experimental ΔH_f° values of 30 ± 15[9] and 25.5[181] kcal/mol have been reported, but for C_2F_2, disparate values of 5 ± 5[9] and −45 ± 6 kcal/mol[182] have appeared.

Several good-quality ab initio calculations on acetylenes have been published recently. From total E data, heats of hydrogenation (Table 4-14) and

Table 4-13. Acetylene (XC≡CY) Bond Lengths and Bond Strengths

X,Y	r(C≡C) (Å)[a]	$D°$(C≡C) (kcal/mol)[b]
H,H	1.207	228
H,F	1.198	178
H,Cl	1.204	166
H,CF_3	1.201	—
H,CH_3	1.206	—
F,F	1.20[c]	113[c]
Cl,Cl	1.195[c]	125
CF_3,CF_3	1.199[d]	—
CH_3,CH_3	1.213	—

[a]Reference 8, unless noted otherwise.
[b]Reference 181.
[c]Reference 9.
[d]Reference 151.

TABLE 4-14. Acetylene Heats of Hydrogenation

Acetylene	ΔE (kcal/mol)[a] 4–31G[b]	6–31G* MP2[c]	DZ + D_C[d]
HC≡CH	−42.0	−42.0	−42.0
HC≡CF	−58.7[e]	−55.9	−59.2
$HC≡CCH_3$	−36.7[f]	−38.2[f]	—
$HC≡CCF_3$	−48.2	—	—
FC≡CF	−77.7[g]	—	−74.1

[a]All values scaled to $\Delta H°$ value for C_2H_2.
[b]Total E's at standard geometries from Reference 183; actual value for C_2H_2 = −52.2 kcal/mol.
[c]Total E's from Reference 184; actual value for C_2H_2 = −46.3 kcal/mol.
[d](9,5)/[3,2] + polarization functions on C; actual value for C_2H_2 = −55.1 kcal/mol from Reference 133.
[e]4–31G//4–31G gives identical value; total E's from References 183 and 185.
[f]$\Delta H°$ = −39.8 ± 0.6 kcal/mol (experimental).
[g]4–31G//4–31G value, total E's of C_2F_4 and *cis*-$C_2H_2F_2$ from References 185 and 172.

thus ΔH_f°'s can be calculated. [The ΔE values in Table 4-14 are scaled to the experimental value for C_2H_2 and are not corrected for zero-point energy (zpe) differences. Scaling, however, reduces the error due to neglect of zpe to < 0.5 kcal/mol,[133] and therefore, $\Delta E \cong \Delta H_{298}^{\circ}$.] For C_2HF, $\Delta H_f^{\circ} \cong 25.5$ (4–31G) and 22.7 kcal/mol (6–31G*MP2), which agree well with the experimental value of 25.5 kcal/mol.[181] Similarly for C_2F_2, $\Delta H_f^{\circ} = 6.9$(4–31G//4–31G) and 3.3 kcal/mol (DZ + D_C), which are in accord with the 5 ± 5 kcal/mol experimental value.[9] In the following thermochemical calculations, the 4–31G calculated ΔH_f°'s are used.

The heat of hydrogenation of C_2HF is 16.7 kcal/mol more exothermic than that for C_2H_2. After correcting for the 3-kcal/mol added stability in $CH_2{=}CHF$ versus $CH_2{=}CH_2$, C_2HF is found to be thermodynamically destabilized relative to C_2H_2 by about 13.7 kcal/mol. The ΔH_r° for Equation 4-52 indicates greater destabilization (~ 21 kcal/mol).

$$\underset{(-82.2)}{(CH_3)_3CF} + HC{\equiv}CH \rightarrow (CH_3)_3CH + \underset{(25.5)}{HC{\equiv}CF} \qquad \frac{\Delta H_r^{\circ}\ (\text{kcal/mol})}{21.1} \qquad (4\text{-}52)$$

From the difference between the heats of hydrogenation of C_2H_2 and C_2F_2 (35.7 kcal/mol) and an estimated 4.7-kcal/mol destabilization in *cis*-CFH=CFH relative to $CH_2{=}CH_2$, C_2F_2 is calculated to be destabilized relative to C_2H_2 by about 40.4 kcal/mol, or about twice the destabilization in C_2HF. Both C_2HF and C_2F_2 are dangerously explosive and extraordinarily reactive.[2] Their thermodynamic instabilities undoubtedly account in large part for their chemical reactivities.

A CF_3 group also has been predicted to destabilize a triple bond.[90] The differences in heats of hydrogenation of $HC{\equiv}CCF_3$ and $HC{\equiv}CCH_3$ (8.4 kcal/mol) and the ΔH_r° of Equation 4-53 (ΔH_f $HC{\equiv}CCF_3$ = −98.6 kcal/mol from 4–31G calculations; Table 4-14) imply that $HC{\equiv}CCF_3$ is destabilized by 14.4 kcal/mol relative to $HC{\equiv}CCH_3$, or about 12.2 kcal/mol relative to C_2H_2 (CH_3 stabilizes a triple bond by 2.2 kcal/mol).[162] This compares with the modest 3-kcal/mol destabilization of a C=C bond by a CF_3 group. The unusual reactivities of $HC{\equiv}CCF_3$ and $CF_3C{\equiv}CCF_3$, especially in [4+2] cycloadditions,[2] likely result from their thermodynamic instabilities as well as their kinetic reactivities.

$$\begin{array}{llr} & & \Delta H_r^\circ \text{ (kcal/mol)} \\ (CH_3)_4C + \underset{(-98.6)}{HC{\equiv}CCF_3} \rightarrow \underset{(-197.3)}{(CH_3)_3CCF_3} + HC{\equiv}CCH_3 & -14.4 & (4\text{-}53) \end{array}$$

Only the molecular structures of $CFH{=}C{=}CHF$[186] and $CH_2{=}C{=}CF_2$[187] have been determined experimentally. The C=C bond lengths in $C_3H_2F_2$ are about 0.007 Å shorter than those in allene itself. The $C_1{-}C_2$ bond in $CH_2{=}C{=}CF_2$ (1.302 Å) is 0.009 Å shorter, and its FCF angle (110.2°) is nearly identical to that in $CH_2{=}CF_2$. There are no experimental thermodynamic data for fluorinated allenes.

The heats of hydrogenation of allenes, Equations 4-54 to 4-58, display differences similar to those observed for ethylenes. (The fluorinated allene ΔH_f°'s were calculated from Benson and Dolbier group values.) The data indicate mono- and *gem*-difluorination stabilize an allene bond about 1 kcal/mol less than an ethylene.

$$\begin{array}{llll} & & \Delta H_r^\circ \text{ (kcal/mol)} & \\ CH_2{=}C{=}CH_2 + H_2 & \rightarrow CH_2{=}CHCH_3 & -40.8 \pm 0.4 & (4\text{-}54) \\ CH_2{=}C{=}CHF + H_2 & \rightarrow CH_2{=}CHCH_2F & -38.5 & (4\text{-}55) \\ (1.3) & \quad (-37.2) & & \\ CH_2{=}C{=}CF_2 + H_2 & \rightarrow CH_2{=}CHCHF_2 & -44.5 & (4\text{-}56) \\ (-48.3) & \quad (-92.6) & & \\ CH_2{=}C{=}CF_2 + H_2 & \rightarrow CF_2{=}CHCH_3 & -41.8 & (4\text{-}57) \\ & \quad (-90.1) & & \\ CF_2{=}C{=}CF_2 + H_2 & \rightarrow CF_2{=}CHCHF_2 & -45.0 & (4\text{-}58) \\ (-142.6) & \quad (-187.6) & & \end{array}$$

The comparable ΔH_r°'s for Equations 4-54 and 4-57 show that geminal fluorination has little effect on the nonfluorinated double bond in $CH_2{=}C{=}CF_2$. The nearly identical heats of hydrogenation of $CH_2{=}C{=}CF_2$ (4-56) and $CF_2{=}C{=}CF_2$ imply that a second pair of geminal fluorines does not appreciably affect allene stability. This contrasts with the marked difrerence in stability between $CH_2{=}CF_2$ and $CF_2{=}CF_2$.

A fascinating property of $CH_2{=}C{=}CHF$ and $CH_2{=}C{=}CF_2$ is the relative reactivity of their double bonds in cycloaddition reactions. Biradical [2+2] additions occur predominantly on their fluorinated double bonds and appear to be thermodynamically controlled, whereas concerted [4+2] additions occur exclusively on their nonfluorinated double bonds.[188,189,190] This regiospecificity has been rationalized by a Frontier MO analysis of HOMO–LUMO control in the fluoroallene cycloadditions.[191]

Propyne is slightly more stable than allene (1 ± 0.8 kcal/mol). By contrast, as expected, $CH_3C{\equiv}CF$ (ΔH_f°(calcd) = 15.6 kcal/mol) is much less stable than $CH_2{=}C{=}CHF$ (ΔH_f°(calcd) = 1.3 kcal/mol). Although ΔH_f° of $CF_3C{\equiv}CF$ is not known experimentally and cannot be reliably estimated from group values, it is probably at least 26 kcal/mol higher than that for $CF_2{=}C{=}CF_2$, based on the estimated destabilizations in $HC{\equiv}CF$ and $HC{\equiv}CCF_3$. Experimentally, it is known that $CF_3C{\equiv}CF$ can be catalytically isomerized, apparently irreversibly, to $CF_2{=}C{=}CF_2$.[192,193]

C. Benzenes

The C–F bond lengths in monofluoro- and 1,2-difluorobenzenes are slightly shorter than those in monofluoro- and 1,2-difluoroalkenes. The more highly fluorinated benzenes have shorter C–F bonds than C_6H_5F, but any systematic trend is difficult to identify experimentally owing to the different methods of structural analysis (Table 4-15).

Several ab initio calculations on fluorobenzenes have been reported recently[194–197] and they reproduce the experimental structures (bond angles, r_0 bond lengths) reasonably well. An ab initio gradient calculation with a 4–

TABLE 4-15. Fluorobenzene Structures[a]

Fluorobenzene	r(C–F) (Å)	Angle CC(F)C (°)
C_6H_5F **(24)**	r_o 1.354[b]	123.4(5)
1,2-$C_6H_4F_2$ **(25)**	r_o 1.306(12)	119.4(3)
1,3-$C_6H_4F_2$ **(26)**	r_o 1.304(4)	
	r_g 1.324(2)	120.9(3)
1,4-$C_6H_4F_2$ **(27)**	r_a 1.354(4)	123.5(1)
1,3,5-$C_6H_3F_3$ **(28)**	r_o 1.304	123.7(1)[c]
	r_g 1.305(10)	
1,2,4,5-$C_6H_2F_4$ **(29)**	r_a 1.343(4)[d]	120.6(3)[d]
C_6HF_5 **(30)**	r_o 1.328	120.4[e]
C_6F_6 **(31)**	r_o 1.321	
	r_g 1.324(6)	120.0
	r_a 1.327(7)	

[a]Data from Reference 27a and references cited therein, except where noted.
[b]$r_o \cong 1.34$; Reference 197.
[c]Reference 198.
[d]Reference 199.
[e]Ab initio value, Reference 197.

21G basis set for a series of fluorinated benzenes shows that the C−F bond lengths decrease uniformly with an increase in ring fluorination.[197] The trend is reminiscent of that seen with fluoroalkanes.

One structural peculiarity is the deviation of some fluorobenzenes (**24, 27, 28**) from D_{6h} ring symmetry, but not others. The internal CCC bond angles for **25, 26,** and **29–31** are all within 1° of 120°. Ab initio calculations correctly reproduce these structural differences,[194,197] but the electronic factors responsible for the ring dissymmetry in **24, 27** and **28** have not been elucidated.

The C−F BDE in C_6H_5F is 125.7 ± 2 kcal/mol,[6] 5 kcal/mol higher than that in $CH_2{=}CHF$. The reason for the stronger aryl fluorine bond is not obvious. The degree of π-electron transfer from F to the π system, which produces some C−F double-bond character, is virtually identical in C_6H_5F and C_2H_3F.[183]. (This is also true for other π-donating substituents.) It is likely therefore that the C−F BDEs reflect the difference between the stabilities of the phenyl and vinyl radicals rather than any differences in C−F bonding between C_6H_5F and C_2H_3F. Indeed, it has been estimated that $C_6H_5\cdot$ is inherently less stable than $C_2H_3\cdot$ by 4.1 kcal/mol.[53] Moreover, this logic would suggest that the difference between the C−X bond strengths in C_6H_5X and C_2H_3X should be approximately constant for π donors X.

From experimental ΔH_f° data (ΔH_f° $C_6H_5\cdot$ = 79 ± 2 kcal/mol; $C_2H_3\cdot$ = 68.4 kcal/mol)[6,53] and Equation 4-59, $D°(C_6H_5{-}X) - D°(C_2H_3{-}X)$ = 3.3 ± 2.4, 7.2 ± 3.2, 4.7 ± 3.2, and 1.2 ± 2.2 kcal/mol for X = H, Cl, Br, and OC_2H_5, respectively. Thus, within experimental error, the differences between phenyl and vinyl C−X bond strengths are similar, except perhaps for the $C{-}OC_2H_5$ bonds. (For additional discussion of this approach, see Chapter 8, Volume 2 of this series, and Chapter 6, Volume 4.)

$$D°(A{-}X) - D°(B{-}X) = \Delta H_f^\circ(BX) - \Delta H_f^\circ(AX) + \Delta H_f^\circ(A\cdot) - \Delta H_f^\circ(B\cdot) \qquad (4\text{-}59)$$

From electron diffraction data on $C_6H_5CF_3$,[200] $r_g(C{-}F)$ = 1.345(3) Å, $r_g(C{-}CF_3)$ = 1.504(4) Å, and FCF angle = 106.6(1)°. These values are nearly identical to those reported for $CH_2{=}CHCF_3$.[146] There is no significant deviation of the ring in $C_6H_5CF_3$ from D_{6h} symmetry [$C_1C_2C_6$ angle = 120.4°(3)]. From thermochemical data, the $C_6H_5{-}CF_3$ BDE is 7.0 ± 3.8 kcal/mol higher than that in $CH_2{=}CH{-}CF_3$.

There is doubt about the $C_6F_5{-}X$ bond strengths owing to different estimates of $\Delta H_f^\circ(g)$ for $C_6F_5\cdot$. Choo, Golden, and Benson[201] favor a value of −130.9 ± 2 kcal/mol, whereas Krech, Price, and Sapiano[202a] report −92.6 ± 2.9 kcal/mol. With the former value, the $C_6F_5{-}X$ BDEs calculate to be 116.3 and 91.7 kcal/mol for X = F and Cl, respectively; with the latter, 154.6 and 130.0 kcal/mol, respectively. Since the higher BDE values reportedly have been confirmed by appearance potential measurements of $C_6F_5\oplus$ from C_6F_5X,[202a] the ΔH_f° value of Krech and co-workers for $C_6F_5\cdot$ is used for the following BDE calculations.

The effect of ring fluorination on bond strengths is remarkable. The calculated differences in C−X BDEs between C_6F_5X and C_6H_5X, assuming $\Delta H_f^\circ\ C_6F_5\cdot = -92.6 \pm 2.9$ kcal/mol, are shown below.

	X (kcal/mol)			
	H	F	Cl	Br
$D°(C_6F_5{-}X) - D°(C_6H_5{-}X) =$	40.8 ± 6.6	29.0 ± 5.5	34.4 ± 7.8	23.4 ± 9.6

As with alkyl fluorides, perfluorination increases the C−F bond strength; however, unlike the saturated systems, it also increases the phenyl C−H, C−Cl, and C−Br bond strengths by more than 20 kcal/mol. The magnitude of this difference is extraordinary; the cause, unknown.

The calculated $C_6F_5{-}CF_3$ BDE is 11.6 ± 7.1 kcal/mol *lower* than that in $C_6H_5{-}CF_3$. It will be shown that $C_6F_5CF_3$ is also thermodynamically much less stable than $C_6H_5CF_3$.

The ΔH_r° values for Equations 4-60 and 4-61 indicate that an F substituent stabilizes a benzene ring slightly more than a double bond (cf Equation 4-39).

	ΔH_r° (kcal/mol)	
$C_6H_5F + CH_2{=}CHCH_3 \rightarrow C_6H_5CH_3 + CH_2{=}CHF$	1.7 ± 0.9	(4-60)
$C_6H_5F + CH_2{=}CHCH_3 \rightarrow C_6H_6 + CH_2{=}CFCH_3$ (−41.4)	1.7	(4-61)

The CF_3 substituent has about the same effect as it does on a double bond (4-62, 4-63); that is, it destabilizes a benzene ring by about 3 kcal/mol.

	ΔH_r° (kcal/mol)	
$C_6H_5CF_3 + CH_2{=}CHCH_3 \rightarrow C_6H_6 + CH_2{=}C(CH_3)CF_3$ (−152.9)	−1.0	(4-62)
$C_6H_5CF_3 + CH_2{=}C(CH_3)_2 \rightarrow C_6H_5CH_3 + CH_2{=}C(CH_3)CF_3$	0.0	(4-63)

Curiously, these thermodynamic effects are not reflected in the heats of hydrogenation of C_6H_5X to $C_6H_{11}X$. From experimental and calculated ΔH_f° values, the heats of hydrogenation of C_6H_5X (X = H, F, Cl, Br, CH_3, CF_3) in fact all fall within 1.4 kcal of −50 kcal/mol.

Multiple fluorination thermodynamically destabilizes a benzene ring.[202b,c] For Equation 4-64, ΔH_r° equals 1.3 ± 1.0 and 1.9 ± 1.0 kcal/mol for 1,3- and 1,4-$C_6H_4F_2$, respectively, but it equals 5.1 ± 2 kcal/mol for 1,2-$C_6H_4F_2$ (IVSTAB = 2.5 kcal/mol).

$$2C_6H_5F \rightarrow C_6H_6 + C_6H_4F_2 \qquad (4\text{-}64)$$

Perfluorination results in substantial overall destabilization[202b,d] (Equation 4-65; IVSTAB ≅ 6 kcal/mol).

	ΔH_r° (kcal/mol)	
$6C_6H_5F \rightarrow C_6F_6 + 5C_6H_6$	36.9 ± 2.6	(4-65)

The comparative effects of CF_3 versus CH_3 substitution on the thermodynamic stabilities of benzene and pentafluorobenzene rings are astonishing. Methyl substitution has about the same effect on both rings (4-66). By contrast, CF_3 is highly destabilizing on a pentafluorophenyl ring. Since CF_3 destabilizes a benzene ring by about 6 kcal/mol relative to CH_3, the ΔH_r° for Equation 4-67 indicates that $C_6F_5CF_3$ is thermodynamically less stable than $C_6F_5CH_3$ by about 60 kcal/mol! (This startling result might make one question the $C_6F_5CF_3$ and $C_6F_5CH_3$ ΔH_f° values.) The ΔH_r° for Equation 4-68 shows a similarly large destabilization relative to C_6F_5H.

$$C_6F_5H + C_6H_5CH_3 \rightarrow C_6F_5CH_3 + C_6H_6 \qquad \Delta H_r^\circ \text{ (kcal/mol)} = -1.1 \pm 2.2 \tag{4-66}$$

$$C_6F_5CF_3 + C_6H_5CH_3 \rightarrow C_6F_5CH_3 + C_6H_5CF_3 \qquad \Delta H_r^\circ \text{ (kcal/mol)} = -53.5 \pm 2.7 \tag{4-67}$$

$$C_6F_5CF_3 + C_6H_6 \rightarrow C_6F_5H + C_6H_5CF_3 \qquad -52.4 \pm 3.9 \tag{4-68}$$

There are no experimental ΔH_f° data for polyperfluoroalkyl benzenes. However, the molecular structure of $(CCF_3)_6$ suggests considerable internal steric strain and thermodynamic instability. From X-ray crystal analysis, the benzene ring in $(CCF_3)_6$ is seen to be slightly distorted to a chair structure (~ 8° out-of-plane bending), and the CF_3 substituents alternate about 12° above and below the pseudo-plane of the ring.[203] This contrasts with $(CCH_3)_6$, which has a planar ring with the CH_3 groups bent out of plane by 10°.[204]

The special influence of perfluoroalkyl groups on the thermal stabilities of aromatics and their valence bond isomers are discussed next.

D. The Perfluoroalkyl Effect

Perfluoroalkyl groups confer unusual thermal stabilities to strained molecules. The extraordinary stabilities of **32** ($t_{1/2} > 2$ h at 360°C),[205] **33** ($t_{1/2} = 9$ h at 170°C),[206], and **34** ($t_{1/2} = 5.1$ h at 160°C),[207], compared with their hydro-

$(CF_3)_6$

32

$-(CF_3)_6$

33

$-(CF_3)_4$ S

34

carbon analogues, are often-cited examples. Other unusually stable molecules are **35**,[208] **36** ($t_{1/2} \cong 20$ min at 95°C),[209] and **37** ($t_{1/2} = 75$ min at 198°C),[210] which are the only isolable representatives of these ring systems.

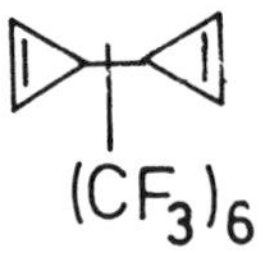

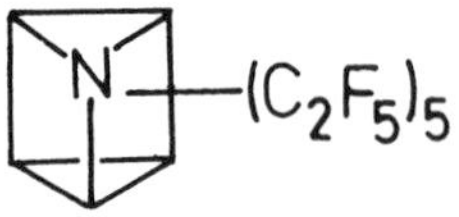

N $-(C_2F_5)_5$

35

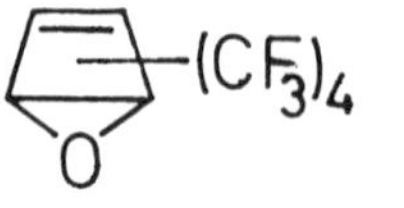

$-(CF_3)_4$ O

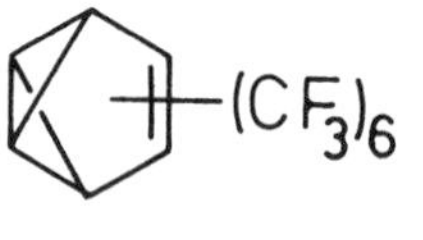

36

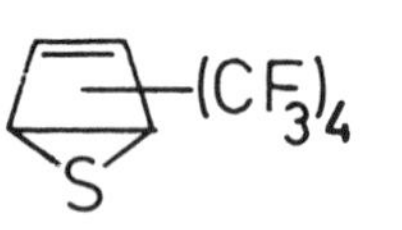

P P $-(CF_3)_4$

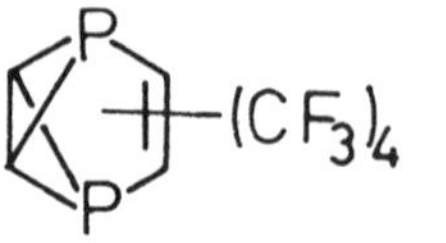

37

TABLE 4-16. Dewar Benzene Isomerizations

$$(X)_6\text{-Dewar benzene} \longrightarrow (X)_6\text{-benzene}$$

X	log A	E_a (kcal/mol)	$\Delta H°$ (kcal/mol)	Ref.
CH_3	15.03	37.2^a	-59.5	215
CF_3	12.59 ± 0.5	38.4 ± 1.2^b	-28.0 ± 1.4	213
C_2F_5	12.99 ± 0.16	36.3 ± 0.4^c	-9.0 ± 0.1^d	214
F	13.22	27.56	-51.0	216,217

$^a\Delta H\ddagger = 36.4$ kcal/mol, $\Delta S\ddagger = 7.5$ eu.
$^b\Delta H\ddagger = 37.4 \pm 1.2$ kcal/mol, $\Delta S\ddagger = -1.9 \pm 2$ eu.
$^c\Delta H\ddagger = 35.3$ kcal/mol, $\Delta S\ddagger = -8.7 \pm 3$ eu.
$^d\Delta S° = -16.3 \pm 0.1$ eu.

These are only a few examples of the many strained systems that are thermally stabilized by perfluoroalkyl groups. The synthesis and chemistry of fluorinated aromatic and heteroaromatic valence bond isomers have proved to be very fruitful. Several reviews of this fascinating chemistry are available.[2,207,211,212]

Another manifestation of the influence of perfluoroalkyl groups is their dramatic effect on the relative energies of isomers. A classic example is the 30-kcal/mol stabilization of Dewar-$(CCF_3)_6$ relative to its aromatic isomer in comparison with its $(CCH_3)_6$ analogues[213] (Table 4-16). Even more spectacular is the difference between Dewar-$(CC_2F_5)_6$ and Dewar-$(CCH_3)_6$, where $\Delta\Delta H°$ (isomerization) amounts to 50 kcal/mol.[214]

Lemal and Dunlap[213] dubbed the composite stabilizing influence of perfluoroalkyl groups on strained molecules "the perfluoroalkyl effect." The relative thermodynamic stability of Dewar-$(CCF_3)_6$ with respect to its aromatic isomer was attributed to the instability of the latter caused by steric strain. The steric strain in $(CC_2F_5)_6$ was shown to be especially pronounced, and thus its formation is much less favorable.[214] The relatively small $\Delta H°$ of aromatization for Dewar-$(CC_2F_5)_6$ coupled with its large negative $\Delta S°$ of aromatization (-16.3 eu) explains why Dewar-$(CC_2F_5)_6$ is actually more stable than its aromatic isomer at temperatures above 280°C.[214]

It is interesting that although Dewar-$(CCF_3)_6$ and Dewar-$(CC_2F_5)_6$ thermally isomerize more slowly than does Dewar-$(CCH_3)_6$, the kinetic E_a and $\Delta H\ddagger$ activation parameters for all three Dewar isomers are very similar. The lower reactivities of the Dewar isomers in fact result only from their relatively low activation entropies (Table 4-16, notes *a–c*). The observed thermal stabilities of perfluoroalkyl Dewar benzenes thus do not imply any inherent thermodynamic stabilization. To the contrary, the Dewar isomers are undoubtedly thermodynamically less stable than their hydrocarbon counterparts, since perfluoroalkyl groups destabilize double bonds as well as strained rings.[90,218,219] The perfluoroalkyl effect in Dewar benzenes therefore is most certainly a kinetic effect whose specific origins have been discussed in detail elsewhere.[214,218] (For a more general discussion of the effects of perfluoroalkyl groups and other substituents on the stability of valence isomers and related strained molecules, see Chapter 5 in this volume.)

The thermochemical calculations in the preceding section predict that a single CF_3 group should have a dramatic effect on comparative $\Delta H°$ of aromatization for Dewar-$C_6F_5CF_3$ versus Dewar-$C_6F_5CH_3$ or Dewar-C_6F_5H. (Recall that $C_6F_5CF_3$ is calculated to be thermodynamically destabilized relative to $C_6F_5CH_3$ by 60 kcal/mol.) Although the activation parameters for the aromatization of **38** (X = CF_3; log A = 14.20 ± 0.07, E_a = 30.6 ± 0.1 kcal/mol; X = CH_3: log A = 13.83 ± 0.06, E_a = 28.2 ± 0.1 kcal/mol; X = H: log A = 12.46 ± 0.14, E_a = 25.5 ± 0.2 kcal/mol) have been determined,[216] their enthalpies of isomerization have not. This strange prediction awaits experimental verification.

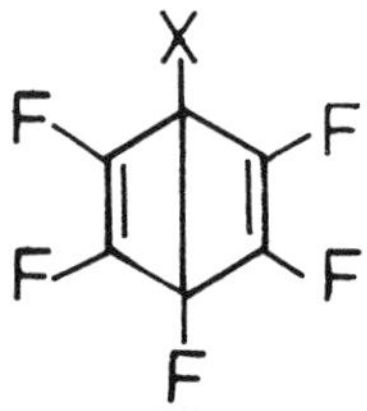

38

4. HEATS OF FORMATION

The heats of formation of fluorocarbons are particularly difficult to measure by the usual calorimetric methods.[220] Consequently, compared with hydrocarbons, relatively few precise thermochemical data are available for fluorocarbons. (Most of the published ΔH_f°'s of fluorohydrocarbons and perfluorocarbons are included in Appendix I.) There are several popular ways to estimate the heats of formation of hydrocarbons. The adequacy of these methods for fluorocarbons is briefly surveyed in this section.

Ab initio theory at the self-consistent-field (SCF) level can accurately predict ΔH_f°'s of small hydrocarbon molecules when applied to isodesmic,[221] or preferably, homodesmic[222] reactions. Full geometry optimization at a split-valence or DZ level is usually needed for best results.[223–225] For calculations on isodesmic or homodesmic reactions, all molecules in the reaction should be calculated at the same level of theory. This presumably will help to minimize errors due to basis set deficiencies. Moreover, to strictly compare the theoretical ΔE with experiment, corrections for differences in zero-point energies between reactants and products must be made, and the experimental $\Delta H°$ value at 0°K should be used.

Unfortunately, other than for simple fluorinated methanes and ethylenes, there are few published ab initio calculations on fluorocarbons, and even

TABLE 4-17. Calculated Versus Experimental Heats of Isodesmic Reactions

	ΔE_r (kcal/mol)			ΔH_r° (kcal/mol)		
Equation	STO−3G[a]	4-31G[a]	DZ + D_C[b]	MNDO[c]	MINDO/3[d]	Experimental
4-1	−22.9	−9.8	−10.1	−1.9	−9.9	−12.4 ± 4.3
4-69	−5.2	3.0	3.1	1.7	1.8	2.5 ± 0.9
4-13	—	−9.5[e]	—	−3.0	−9.6	−13.1 ± 2.9
4-37	−4.1	−1.2[f]	−1.3	0.3	−7.6	−1.1 ± 1.5

[a]Optimized geometries from Reference 19, unless noted otherwise.
[b](9,5)/[3,2] basis with d-polarization functions on C; optimized geometries, Reference 133.
[c]References 103 and 226.
[d]Reference 227.
[e]Standard geometries, References 17 and 228.
[f]Optimized geometries, Reference 132.

fewer systematic studies of the adequacy of different basis sets. In addition, the calculated ΔE's are usually compared directly with experimental ΔH_{298}° values, since data required to make the necessary zpe corrections for fluorocarbons are normally not available. Nonetheless, from published results it can be deduced what level of SCF theory is at least necessary, but not sufficient, to reliably calculate fluorocarbon energies.

The experimental and calculated heats of some isodesmic reactions at various levels of theory are presented in Table 4-17. The STO−3G basis set and Dewar's semiempirical MNDO and MINDO/3 methods are clearly inadequate. (The MINDO/3 results for reactions 4-1, 4-69, and 4-13 are deceptive owing to fortuitous cancellations of errors. The actual MINDO/3 ΔH_f°'s of CH_4, CH_3F, CH_2F_2, CH_3CH_2F, and CH_3CHF_2 are in error by about 11.5, 5.5, 2.0, 3.0, and 6.0 kcal/mol, respectively.) By contrast, the 4–31G and DZ basis sets give comparable, reasonably accurate results. The 4–31G basis, but not STO−3G, also has been shown to accurately calculate the *relative* heats of hydrogenation of $CH_2{=}CH_2$, $CH_2{=}CHF$, and $CH_2{=}CF_2$.[69] The split-valence 4–31G basis therefore appears to be the minimum basis set required for reliable fluorocarbon calculations.

$$2CHF_3 \rightarrow CH_2F_2 + CF_4 \qquad (4\text{-}69)$$

It should be cautioned, however, that even higher quality basis sets often yield significantly different results for a given isodesmic reaction. For instance, using the Roos and Siegbahn (7,3)/[5,3] basis and fully optimized geometries, Bock and co-workers have calculated reaction 4-37 to be 4.5 kcal/mol more exothermic than that with the 4–31G or (9,5)/[3,2] basis.[132] Similarly, the calculated exothermicities of the $C_2H_2F_2$ isomerization (4-41) differ by about 10 kcal/mol for the (7,3,1)/[4,2] and (9,5)/[4,2] basis sets.[132] A particularly pathological case is the homodesmic reaction, $2HC{\equiv}CF \rightarrow HC{\equiv}CH + FC{\equiv}CF$, for which ΔE is 5.6–5.7 kcal/mol from 4–31G//4–31G and 6–31G**//4–31G calculations, but is 50.9–55.5 kcal/mol from 4–

31G CI//4–31G calculations with and without corrections for higher excitations.[185] Obviously, ab initio calculations of thermochemical processes involving fluorocarbons at any level of Hartree–Fock/SCF theory should be treated with circumspection.

There are several empirical or semiempirical group-additivity and bond-additivity schemes that can quite accurately predict ΔH_f°'s of fluorocarbons, when used properly. The familiar group-additivity method of Benson[229], which works very well for hydrocarbons, fails for fluorocarbons, especially polar fluorocarbons. For instance, the Benson group values adequately predict ΔH_f° of CF_3CF_3, but the estimate of CF_3CH_3 deviates 7 kcal/mol from experiment. Dolbier and associates have derived alternative group value contributions to ΔH_f° for several fluorocarbon groups (Appendix II). When the Dolbier fluorocarbon and the Benson hydrocarbon values are used, the ΔH_f°'s of most fluorinated alkanes and alkenes are estimated quite accurately. *The Dolbier values, however, should not be used for molecules that have neighboring fluorine substituents.* From the Dolbier $C-(F)_3(C)$ value, for example, the estimated ΔH_f° of CF_3CF_3 is in error by 13 kcal/mol.

Two bond-additivity schemes that accurately predict ΔH_f°'s of perfluorocarbons, fluorohydrocarbons, and halofluorocarbons have been published. Buckley and Rodgers[15,230] have developed a semiempirical model as an extension to bond additivity whose results for one- and two-carbon saturated systems are normally well within 1 kcal/mol of experiment. Unfortunately, this method employs a heavily parameterized, complex equation that requires knowledge of molecular geometry. Problems with local minima in the calculations also can arise. This electrostatic model has not been extended to unsaturated systems, and it is liable to quickly become unwieldy for fluorocarbons much larger than fluoroethanes.

Joshi[231] has introduced a generalized bond energy, group contribution scheme that is probably less well-known than it should be. The procedure accurately calculates ΔH_f°'s for most fluorinated and other halogenated alkanes, alkenes, and benzenes. Heats of polymerization of halogenated monomers are also estimated remarkably well. Its main drawbacks are its complicated nomenclature for bond energy terms and its extensive corrections for various nonbonded interactions, ring strain, and so on. Notwithstanding, for estimating heats of formation of relatively large, polyfluorinated molecules for which reliable ab initio calculations would be prohibitively expensive, Joshi's method is probably the best one available.

5. CYCLOPROPANE AND CYCLOBUTANE STRAIN ENERGIES

The ring strain (E_s) of a cyclic molecule is conventionally defined as the difference between its experimental ΔH_f° and that for its corresponding hypothetically strain-free ring, calculated from thermochemical bond energies or group values. By this definition, the E_s values of cyclopropane (c-C_3H_6) and cyclobutane (c-C_4H_8) are 27.5 and 26.5 kcal/mol, respectively.[220,229] Greenberg reported E_s of perfluorocyclopropane (c-C_3F_6) to be 80.9 kcal/mol based on the difference between its ΔH_f° and the sum of three Dolbier $C(F)_2(C)_2$ group values. (Bernett[1] had previously estimated E_s = 68.6 kcal/mol using a group value of −98.1 kcal/mol). The ΔH_r° of Equation 4-70, which evaluates the E_s of c-C_3F_6 with reference to acyclic molecules,[69,222] also was cited by Greenberg to corroborate the E_s above of c-C_3F_6.

$$3CF_3CF_2CF_3 \rightarrow c\text{-}C_3F_6 + 3CF_3CF_3 \qquad \frac{\Delta H_r^\circ\ (\text{kcal/mol})}{76.7 \pm 7.6} \tag{4-70}$$

These estimates, however, misuse Dolbier group values and employ inappropriate reference molecules. The Dolbier group values cannot be used to estimate ΔH_f° of perfluorocarbons (see discussion near the end of the "Heats of Formation" section). The ΔH_f° of $CF_3(CF_2)_5CF_3$ estimated by such group values gives an outlandish error of 50.9 kcal/mol, for example. Also, a comparison of the experimental ΔH_f° of perfluorocyclohexane (c-C_6F_{12}) with that estimated from Dolbier $C(F)_2(C)_2$ group values leads to the absurd conclusion that c-C_6F_{12} has a strain energy of some 63 kcal/mol. That something is also amiss with Equation 4-70 becomes evident when an analogous equation (4-71) is used to evaluate E_s of c-C_6F_{12}.

$$6CF_3CF_2CF_3 \rightarrow c\text{-}C_6F_{12} + 6CF_3CF_3 \qquad \frac{\Delta H_r^\circ\ (\text{kcal/mol})}{54.8 \pm 15.0} \tag{4-71}$$

The choice of an appropriate $C(F)_2(C)_2$ group value for a $(CF_2)_n$ ring or chain is obviously crucial. A logical approach would be to assume that c-C_6F_{12} is strain free and to use a value of −94.4 ± 0.3 kcal/mol (1/6 · ΔH_f° of c-C_6F_{12}). This value can be tested in a calculation of the ΔH_f° of $CF_3(CF_2)_5CF_3$ [ie, ΔH_f° ($CF_3CF_2CF_3$) − (−94.4 ± 0.3) + 5(−94.4 ± 0.3) = −803.8 ± 3.5 kcal/mol]. The agreement with experiment (−809.2 ± 0.4 kcal/mol) is reasonable.

With a perfluoro-$C(F)_2(C)_2$ group value of −94.4 kcal/mol, E_s of c-C_3F_6 calculates to be 49.4 kcal/mol, which is considerably lower than previous estimates. Equation 4-72, which uses more suitable acyclic reference compounds than those in Equation 4-70, gives a similar E_s of 53.5 ± 3.6 kcal/mol.

$$3CF_3(CF_2)_5CF_3 \rightarrow 4c\text{-}C_3F_6 + 3CF_3CF_2CF_3 \qquad \frac{\Delta H_r^\circ\ (\text{kcal/mol})}{213.8 \pm 14.3} \tag{4-72}$$

It is instructive to compare this new E_s value with that of 1,1-difluorocyclopropane, c-$C_3H_4F_2$. The ΔH_f° of c-$C_3H_4F_2$ is not known experimentally, but it can be estimated to be −80.4 kcal/mol from its calculated heat of hydrogenation (−49.4 kcal/mol)[69] and the ΔH_f° of $CH_3CF_2CH_3$ (−129.8 kcal/mol). Since the Dolbier and Benson group values give −114.8 kcal/mol for c-$C_3H_4F_2$, its E_s is about 34.4 kcal/mol. The strain energies in c-C_3F_6 and c-$C_3H_4F_2$ differ by about 15 kcal/mol.

The difference in strain energies of c-C_3F_6 and c-$C_3H_4F_2$ is nearly the same as the difference in activation energies for their thermal decompositions (17.8 kcal/mol; Equations 4-73 and 4-74).[77]

		log A	E_a (kcal/mol)	
▷F_2 →	$CH_2{=}CH_2 + CF_2$:	14.1	56.4	(4-73)
F_2, F_2 ▷F_2 →	$CF_2{=}CF_2 + CF_2$:	13.25	38.6	(4-74)

Thus, the relative ease of CF_2: extrusion from c-C_3F_6 and c-$C_3H_4F_2$ is almost entirely a manifestation of their relative strain energies. This contrasts with a previous explanation used to reconcile the ΔE_a of 17.8 kcal/mol with the ΔE_s of more than 40 kcal/mol, based on an E_s of 76.7 kcal/mol for c-C_3F_6.[69]

A completely unexpected result is the reevaluated E_s of perfluorocyclobutane (c-C_4F_8). From its ΔH_f° (−368.7 kcal/mol) and the sum of four perfluoro-C(F)$_2$(C)$_2$ group values (−377.6 kcal/mol), the E_s of c-C_4F_8 is only about 8.9 kcal/mol; that is, *perfluorocyclobutane is about 17.6 kcal/mol less strained than cyclobutane.* Equation 4-75 indicates a somewhat higher E_s but still less than that for c-C_4H_8 itself. (The analogous equation for c-C_4H_8 gives 26.6 ± 0.3 kcal/mol.) These estimates contrast with earlier ones that variously placed E_s of c-C_4F_8 at 24,[69] 18.8,[69] and 5.5[1] kcal/mol above that in c-C_4H_8.

$$CF_3(CF_2)_5CF_3 \rightarrow \text{c-}C_4F_8 + CF_3CF_2CF_3 \qquad \Delta H_r^\circ \text{ (kcal/mol)} \; 14.3 \pm 4.6 \qquad (4\text{-}75)$$

The ΔH_f° of 1,1-difluorocyclobutane, c-$C_4H_6F_2$, is not known experimentally. A (DZ + D_C) calculation[133] gives ΔE = 33.4 kcal/mol for Equation 4-19, from which the ΔH_f° of c-$C_4H_6F_2$ is estimated to be −100.9 kcal/mol. (The ΔE for Equation 4-17 calculated with this basis set is 19.2 kcal/mol, which closely corresponds to the actual ΔH_r°, 18.6 ± 0.3 kcal/mol.) The difference between this estimated ΔH_f° and the sum of one Dolbier C(F)$_2$(C)$_2$

and three Benson $C(H)_2(C)_2$ groups (−119.7 kcal/mol) gives 18.8 kcal/mol for the E_s of c-$C_4H_6F_2$, which is about 10 kcal/mol greater than that in c-C_4F_8, but 7.7 kcal/mol less than the E_s in c-C_4H_8.

These relative strain energies roughly parallel the activation energies for c-C_4H_8, c-$C_4H_6F_2$, and c-C_4F_8 thermal ring fragmentations. For c-C_4H_8 versus c-$C_4H_4F_2$ (4-17, 4-19), ΔE_a = 6.7 kcal/mol and ΔE_s = −7.7 kcal/mol; for c-C_4H_8 versus c-C_4F_8 (4-17, 4-18), ΔE_a = 11.7 kcal/mol and ΔE_s = −17.6 kcal/mol. (If the E_s of 14.3 kcal/mol for c-C_4F_8 from Equation 4-75 is used, ΔE_s = −12.2 kcal/mol.) These comparisons suggest that the differences in cyclobutane reactivities can be attributed largely to differences in strain energies (ie, the more strained cyclobutane reacts faster). The need to explain why c-C_4F_8 has an E_a to fragmentation some 11.7 kcal/mol higher than that for c-C_4H_8, yet has up to 24 kcal/mol more ring strain, disappears with the new estimates of E_s for c-C_4F_8.

Finally, recall that the ΔE_a for the electrocyclic ring openings of hexafluorocyclobutene **(14)** and cyclobutene **(13)** is 14.6 kcal/mol (Table 4-12). Can this also be attributed to differences in E_s? Since there are no satisfactory group values to directly estimate E_s of **14,** it must be evaluated indirectly. The ΔH_f° of perfluorocyclohexene minus two perfluoro-$C(F)_2(C)_2$ group values (−266.9 ± 1.7 kcal/mol) seems a reasonable estimate. Since ΔH_f° of **14** is −235.9 ± 1 kcal/mol, its E_s is calculated to be 31 ± 2.8 kcal/mol. This is identical to the E_s of **13** (30 kcal/mol),[220] within experimental error. The ΔE_a (14.6 kcal/mol) for the cyclobutene ring openings therefore appears to be largely due to the difference in reaction enthalpies ($\Delta\Delta H_r^\circ$ = 19.7 kcal/mol).

Perhaps most important, the discussion in this section shows how different choices of reference systems can dramatically alter views of thermodynamic stabilities and interpretations of kinetic reactivities. Is perfluorocyclobutane really less strained than cyclobutane? It is, according to conventional definition, provided one accepts the premise that perfluorocyclohexane, or perfluoro-*n*-heptane and perfluoropropane, are no more strained than their hydrocarbon counterparts.

Appendix I. EXPERIMENTAL HEATS OF FORMATION

The heats of formation that were used throughout this chapter are listed in Tables I-1 to I-5. Unless noted otherwise, the values are taken from the compilation of internally consistent, computer-analyzed thermochemical data by Pedley and Rylance.[232] All values are in kilocalories per mole and refer to the standard state of the ideal gas at 1 atm and 25°C (298°K). Estimated values are enclosed in brackets.

TABLE I-1. ΔH_f° Values of Alkanes

Alkane	ΔH_f°	Alkane	ΔH_f°
CH_4	-17.8 ± 0.1	CH_2ClCH_2Cl	-31.0 ± 0.4
CH_3F	$[-56.8 \pm 2]$[a]	CH_2BrCH_2Br	-9.0 ± 0.4
CH_2F_2	-108.1 ± 0.2	CHF_2CF_2Br	-199.6 ± 0.9[d]
CHF_3	-167.0 ± 0.3	CF_2ClCF_2Cl	-215.2 ± 0.5
CF_4	-223.4 ± 0.1	CF_2BrCF_2Br	-189.0 ± 0.8
CH_3CH_3	-20.1 ± 0.0	$CH_3CH_2CH_3$	-25.0 ± 0.1
CH_3CH_2F	-62.9 ± 0.4[b]	c-C_3H_6	12.7 ± 0.1
CH_3CH_2Cl	-26.8 ± 0.1	CH_3CHFCH_3	-70.1 ± 0.4
CH_3CH_2Br	-14.9 ± 0.2	$CH_3CF_2CH_3$	-129.8 ± 3[e]
CH_3CHF_2	-118.8 ± 2.0	$CF_3CF_2CF_3$	-426.2 ± 1.7
CH_3CHCl_2	-30.5 ± 0.3	c-C_3F_6	-233.8 ± 2[f]
CH_2FCHF_2	-162.7 ± 1.7[c]	c-C_4F_8	-368.7 ± 2.5
$CH_2ClCHCl_2$	-35.6 ± 0.3	c-C_6F_{12}	-566.2 ± 1.8[g]
CH_3CF_3	-178.0 ± 0.4	$CF_3(CF_2)_5CF_3$	-809.2 ± 0.4
CH_3CCl_3	-34.6 ± 0.1		
CF_3CF_3	-322.7 ± 0.5		

[a]Reference 5.
[b]Reference 59.
[c]Reference 232 gives -178.8 ± 2.0 kcal/mol, which is in error. From $\Delta H_r^\circ = -63.98 \pm 0.50$ for $C_2F_3Cl(g) + 2H_2 \rightarrow C_2H_5F_3(g) + HCl(g)$ (Reference 233) and $\Delta H_f^\circ(CF_2CFCl) = -120.8 \pm 1.2$ (Reference 234a), $\Delta H_f^\circ(CH_2FCHF_2) = -162.7 \pm 1.7$ kcal/mol.
[d]Reference 220.
[e]Reference 234b.
[f]Reference 235.
[g]Reference 236.

TABLE I-2. ΔH_f° Values of Alkenes and Alkynes

Alkene	ΔH_f°	Alkyne	ΔH_f°
$CH_2{=}CH_2$	12.5 ± 0.3	$HC{\equiv}CH$	54.5 ± 0.2
$CH_2{=}CHF$	-33.2 ± 0.4	$HC{\equiv}CF$	$[25.5]$[d]
$CH_2{=}CHCl$	8.9 ± 0.3	$FC{\equiv}CF$	$[5 \pm 5]$[e]
$CH_2{=}CHBr$	19.0 ± 0.5	$HC{\equiv}CCH_3$	44.6 ± 0.5
$CH_2{=}CHOC_2H_5$	-33.7 ± 0.1		
$CH_2{=}CF_2$	-80.0 ± 0.8		
cis-$CHF{=}CHF$	-70.8 ± 3.1[a]		
trans-$CHF{=}CHF$	-69.4 ± 3.1[a]		
$CHF{=}CF_2$	-117.2 ± 2		
$CF_2{=}CF_2$	-157.9 ± 0.4		
$CH_2{=}CHCH_3$	4.8 ± 0.1		
$CH_2{=}CHCF_3$	-146.8 ± 1.6		
c-C_4F_6	-235.9 ± 1[b]		
c-C_6F_{10}	-455.7 ± 1.7[c]		

[a]Reference 241.
[b]Reference 237.
[c]Reference 236.
[d]Reference 181.
[e]Reference 9.

TABLE I-3. ΔH_f° Values of Benzenes

Benzene	ΔH_f°	Benzene	ΔH_f°
C_6H_6	19.8 ± 0.1	C_6F_5H	-192.6 ± 1.6
C_6H_5F	-27.7 ± 0.3	C_6F_6	-228.3 ± 0.3
C_6H_5Cl	12.3 ± 0.2	C_6F_5Cl	-193.7 ± 2.7
C_6H_5Br	24.9 ± 0.7	C_6F_5Br	-170.1 ± 4.0[a]
$C_6H_5OC_2H_5$	24.3 ± 0.1	$C_6H_5CH_3$	12.0 ± 0.1
1,2-$C_6H_4F_2$	-70.1 ± 0.2	$C_6H_5CF_3$	-143.2 ± 0.2
1,3-$C_6H_4F_2$	-73.9 ± 0.3	$C_6F_5CH_3$	-201.5 ± 0.4
1,4-$C_6H_4F_2$	-73.3 ± 0.3	$C_6F_5CF_3$	-303.2 ± 2.0[b]

[a]Reference 202a.
[b]Reference 238; Reference 232 gives -283.8 ± 1.8 kcal/mol which appears to be incorrect.

TABLE I-4. ΔH_f° Values of Carbonyl Compounds

Formula	ΔH_f°
$H_2C{=}O$	-26.0 ± 0.2
$HFC{=}O$	[−90.0][a]
$F_2C{=}O$	-152.9 ± 0.2
$(CH_3)_2C{=}O$	-51.9 ± 0.1
$(CF_3)_2C{=}O$	[−325.2][b]

[a]Reference 9.
[b]Reference 239.

TABLE I-5. ΔH_f° Values of Free Radicals[a]

Radical	ΔH_f°
$CH_3\cdot$	35.1 ± 0.2
$CH_2F\cdot$	-7.8 ± 2
$CHF_2\cdot$	-59.2 ± 2
$CF_3\cdot$	-111.7 ± 3.6
$CH_3CH_2\cdot$	25.9
$CH_3CF_2\cdot$	-72.3 ± 2
$CF_3CF_2\cdot$	-213.4 ± 1
$CH_2{=}CH\cdot$	68.4[b]
$C_6H_5\cdot$	79.0 ± 2
$C_6F_5\cdot$	-92.6 ± 2.9[c]

[a]Reference 6, unless noted otherwise.
[b]Reference 53.
[c]Reference 202.

Appendix II. DOLBIER GROUP VALUES

The Dolbier group contributions to $\Delta H_f^\circ(g)$ are listed in Table II-1. All values are in kilocalories per mole. The group values for C−(F)(H)$(C)_2$ and C−(F)(H)(C)(C_d) are not reported.

Table II-1. Dolbier Group Values[a]

Group	ΔH_f°
C−(F)$(H)_2$(C)	−52.7
C−$(F)_2$(H)(C)	−108.6
C−$(F)_2(C)_2$	−104.9
C−$(F)_3$(C)	−167.8
C−$(F)_3(C_d)$	−166.0
C−$(F)_2$(H)(C_d)	−107.6
C−$(F)_2$(C)(C_d)	−103.9
C−(F)$(H)_2(C_d)$	−52.2
C_d−(F)(H)	−38.4
C_d−$(F)_2$	−88.0

[a]References 13 and 240.

REFERENCES

1. Bernett, W.A. *J. Org. Chem.* **1969,** *34,* 1772.
2. Smart, B.E. In "The Chemistry of Functional Groups, Supplement D," Part 2, Patai, S. and Rappoport, Z. Eds.; Wiley: New York, 1983, Chapter 14, pp. 603–655.
3. Nygaard, L. *Spectrochim. Acta* **1966,** *22,* 1261.
4. Solimene, N.; Dailey, B. *J. Chem. Phys.* **1954,** *22,* 2042.
5. Rodgers, A.S.; Chao, J.; Wilhoit, R.C.; Zwolinski, B.J. *J. Phys. Chem. Ref. Data* **1974,** *3,* 117.
6. McMillen, D.F.; Golden, D.M. *Annu. Rev. Phys. Chem.* **1982,** *33,* 493.
7. Typke, V.; Dakkouri, M.; Oberhammer, H. *J. Mol. Struct.* **1978,** *44,* 85.
8. Harmony, M.D.; Laurie, V.W.; Kuczkowski, R.L.; Schwendeman, R.H.; Ramsay, D.A.; Lovas, F.J.; Lafferty, W.J.; Maki, A.G. *J. Phys. Chem. Ref. Data* **1979,** *8,* 619.
9. Stull, D.R.; Prophet, H. "JANAF Thermochemical Tables", 2nd ed.; NSRDS-NBS 37. Government Printing Office: Washington, D.C., 1971.
10. Landold-Börnstein, "Numerical Data and Functional Relationships in Science and Technology", Hellwege, K.-H.; and Hellwege, A.M. Eds.; Springer-Verlag: Berlin, new series, 11/7, 1976.
11. Pickard, J.M.; Rodgers, A.S. *Int. J. Chem. Kinet.* **1983,** *15,* 569.
12. Egger, K.W.; Cocks, A.T. *Helv. Chim. Acta* **1973,** *56,* 1516.
13. Dolbier, W.R., Jr.; Medinger, K.S.; Greenberg, A.; Liebman, J.F. *Tetrahedron* **1982,** *38,* 2415.
14. Dill, J.D.; Schleyer, P.v.R.; Pople, J.A. *J. Am. Chem. Soc.* **1976,** *98,* 1663.
15. Buckley, G.S.; Rodgers, A.S. *J. Phys. Chem.* **1983,** *87,* 126.
16. Pauling, L. "Nature of the Chemical Bond", 3rd ed.; Cornell University Press: Ithaca, N.Y., 1960, pp. 314–316.
17. Radom, L.; Hehre, W.J.; Pople, J.A. *J. Am. Chem. Soc.* **1971,** *93,* 289.
18. Radom, L.; Stiles, P.J. *Tetrahedron Lett.* **1975,** 789.
19. Baird, N.C. *Can. J. Chem.* **1983,** *61,* 1567.

20. Oberhammer, H. *J. Mol. Struct.* **1975,** *28,* 349.
21. Oberhammer, H. *J. Fluorine Chem.* **1983,** *23,* 147.
22. Hehre, W.J.; Pople, J.A. *J. Am. Chem. Soc.* **1970,** *92,* 2191.
23. Davis, D.W.; Banna, M.S.; Shirley, D.A. *J. Chem. Phys.* **1974,** *60,* 237.
24. Jolly, W.L.; Bakke, A.A. *J. Am. Chem. Soc.* **1976,** *98,* 6500.
25. Bent, H.A. *Chem. Rev.* **1961,** *61,* 275.
26. (a) Burdett, J.K. "Molecular Shapes", Wiley: New York, 1980. (b) Glidewell, C.; Meyer, A.Y. *J. Mol. Struct.* **1981,** *72,* 209.
27. (a) Yokozeki, A.; Bauer, S.H. *Top. Curr. Chem.* **1975,** *53,* 71. (b) Marsden, C.J.; Bartell, L.S. *Inorg. Chem.* **1976,** *15,* 2713.
28. Kollman, P.J. *J. Am. Chem. Soc.* **1974,** *96,* 4363.
29. Peters, D. *J. Chem. Phys.* **1963,** *38,* 561.
30. Magnusson, E. *J. Am. Chem. Soc.* **1984,** *106,* 1185.
31. Magnusson, E. *J. Am. Chem. Soc.* **1984,** *106,* 1177.
32. Epiotis, N.D.; Cherry, W.R.; Shaik, S.; Yates, R.L.; Bernardi, F. *Top. Curr. Chem.* **1977,** *70,* 1.
33. Epiotis, N.D. "Lecture Notes in Chemistry, No. 29. Unified Valence Bond Theory of Electronic Structure". Springer-Verlag: New York, 1982.
34. Whangbo, M.-H.; Mitchell, D.J.; Wolfe, S. *J. Am. Chem. Soc.* **1978,** *100,* 3698.
35. Bingham, R.C. *J. Am. Chem. Soc.* **1976,** *98,* 535.
36. Pickard, J.M.; Rodgers, A.S. *J. Am. Chem. Soc.* **1977,** *99,* 671.
37. Pickard, J.M.; Rodgers, A.S. *Int. J. Chem. Kinet.* **1977,** *9,* 759.
38. Radom, L.; Hehre, W.J.; Pople, J.A. *J. Am. Chem. Soc.* **1972,** *94,* 2371.
39. Batt, L.; Walsh, R. *Int. J. Chem. Kinet.* **1983,** *15,* 605.
40. Gillespie, R.J. *Angew. Chem. Int. Ed. Engl.* **1967,** *6,* 819.
41. Boggs, J.E.; Altman, M.; Cordell, F.R.; Dai, Y. *J. Mol. Struct.* **1983,** *94,* 373.
42. Pickard, J.M.; Rodgers, A.S. *J. Am. Chem. Soc.* **1976,** *99,* 691.
43. Wu, E.-C.; Rodgers, A.S. *J. Phys. Chem.* **1974,** *78,* 2315.
44. Martin, J.P.; Paraskevopoulos, G. *Can. J. Chem.* **1983,** *61,* 861.
45. Stølevik, R.; Thom, E. *Acta Chem. Scand.* **1971,** *25,* 3205.
46. Hildebrandt, R.L.; Wieser, J.D. *J. Mol. Struct.* **1973,** *15,* 27.
47. Castelhano, A.L.; Mariott, P.R.; Griller, D. *J. Am. Chem. Soc.* **1981,** *103.* 4263.
48. Evans, B.S.; Weeks, I.; Whittle E. *J. Chem. Soc. Faraday Trans. 1* **1983,** *79,* 147.
49. Kaplan, L. In "Free Radicals", Vol. II; Kochi, J.K., Ed.; Wiley: New York, 1978, Chapter 18, p. 361.
50. Stock, L.M.; Wasielewski, M.R. *Prog. Phys. Org. Chem.* **1981,** *13,* 253.
51. Molino, L.M.; Probiet, J.M.; Canadell, E. *J. Chem. Soc. Perkins Trans 2* **1982,** 1217.
52. Leroy, G.; Peters, D. *J. Mol. Struct.* **1981,** *85,* 133.
53. Leroy, G.; Peters, D.; Wilante, C. *J. Mol. Struct.* **1982,** *88,* 217.
54. O'Neal, H.E.; Benson, S.W. *J. Phys. Chem.* **1968,** *72,* 1866.
55. Dolbier, W.R., Jr.; Piedrahita, C.A.; Al-Sader, B.H. *Tetrahedron Lett.* **1979,** 2957.
56. Wu, E.-C.; Rodgers, A.S. *J. Am. Chem. Soc.* **1976,** *98,* 6112.
57. Al-Ajdah, G.N.D.; Beagley, B.; Jones, M.O. *J. Mol. Struct.* **1980,** *65,* 271.
58. Rodgers, A.S.; Ford, W.G.F. *Int. J. Chem. Kinet.* **1973,** *5,* 965.
59. Chen. S.S.; Rodgers, A.S.; Chao, J.; Wilhoit, R.C.; Zwolinski, B.J. *J. Phys. Chem. Ref. Data* **1975,** *4,* 441.
60. Conlin, R.T.; Frey, H.M. *J. Chem. Soc. Faraday Trans. 1* **1980,** *76,* 322.
61. Conlin, R.T.; Frey, H.M. *J. Chem. Soc. Faraday Trans. 1* **1979,** *75,* 2556.
62. Gerberich, H.R.; Walters, W.D. *J. Am. Chem. Soc.* **1961,** *83,* 4884.
63. Butler, J.N. *J. Am. Chem. Soc.* **1962,** *84,* 1393.
64. Dolbier, W.R., Jr.; Al-Fekri, D.M. *J. Am. Chem. Soc.* **1983,** *105,* 6349.
65. Chang, C.H.; Porter, R.F.; Bauer, S.H. *J. Mol. Struct* **1971,** *7,* 89.
66. Chiang, J.F.; Bernett, W.A. *Tetrahedron* **1971,** *27,* 975.
67. Bastiansen, O.; Fritsch, F.N.; Hedberg, K. *Acta Crystallogr.* **1964,** *17,* 583.
68. Birchall, M.J.; Fields, R.; Haszeldine, R.N.; McClean, R.J. *J. Fluorine Chem.* **1980,** *15,* 487.
69. Greenberg, A.; Liebman, J.F.; Dolbier, W.R., Jr.; Medinger, K.S.; Skancke, A. *Tetrahedron* **1983,** *39,* 1533.
70. Durmaz, S.; Kollmar, H. *J. Am. Chem. Soc.* **1980,** *102,* 6942.
71. Jason, M.E.; Ibers, J.A. *J. Am. Chem. Soc.* **1977,** *99,* 6012.

72. Paretta, A.T.; Laurie, V.W., Jr. *J. Chem. Phys.* **1975,** *62,* 2469.
73. Hoffmann, R. *Tetrahedron Lett.* **1970,** 2907.
74. Hoffmann, R.; Stohrer, W.P. *J. Am. Chem. Soc.* **1973,** *93,* 6941.
75. (a) Deakyne, C.A.; Allen, L.C.; Craig, N.C. *J. Am. Chem. Soc.* **1977,** *99,* 3895. (b) Clark, T.; Spitznagel, G.W.; Klose, R.; Schleyer, P.v.R. *J. Am. Chem. Soc.* **1984,** *106,* 4412.
76. Skancke, A.; Flood, E.; Boggs, J.E. *J. Mol. Struct.* **1977,** *40,* 263.
77. Dolbier, W.R., Jr. *Acc. Chem. Res.* **1981,** *14,* 195.
78. Roth, W.R.; Kirmse, W.; Hoffmann, W.; Lennartz, H.-W. *Chem. Ber.* **1982,** *115,* 2508.
79. Dolbier, W.R., Jr; Enoch, H.O. *J. Am. Chem. Soc.* **1977,** *99,* 4352.
80. Dolbier, W.R., Jr.; Sellers, S.F. *J. Am. Chem. Soc.* **1982,** *104,* 2494.
81. Slafer, W.D.; English, A.D.; Harris, D.O.; Shellhamer, D.F.; Meshishnek, M.J.; Aue, D.H. *J. Am. Chem. Soc.* **1975,** *97,* 6638.
82. Dolbier, W.R., Jr.; Fielder, T.H., Jr. *J. Am. Chem. Soc.* **1978,** *100,* 5577.
83. Dolbier, W.R., Jr.; Sellers, S.F.; Al-Sader, B.H.; Smart, B.E. *J. Am. Chem. Soc.* **1980,** *102,* 5398.
84. Dolbier, W.R., Jr.; Sellers, S.F.; Smart, B.E. *Tetrahedron Lett.* **1981,** *22,* 2953.
85. Stigliani, W.M.; Laurie, V.W. Private communication cited in Reference 72.
86. Dolbier, W.R., Jr.; Sellers, S.F.; Smart, B.E.; Oberhammer, H. *J. Mol. Struct.* **1983,** *101,* 193.
87. Dolbier, W.R., Jr.; Sellers, S.F. *J. Org. Chem.* **1982,** *47,* 1.
88. Sellers, S.F.; Dolbier, W.R., Jr.; Koroniak, H.; Al-Fekri, D.M. *J. Org. Chem.* **1984,** *49,* 1033.
89. Dolbier, W.R., Jr.; Sellers, S.F.; Al-Sader, B.H., Fielder, T.H., Jr.; Elsheimer, S.; Smart, B.E. *Is. J. Chem.* **1981,** *21,* 176.
90. Dill, J.D.; Greenberg, A.; Liebman, J.F. *J. Am. Chem. Soc.* **1979,** *101,* 6814.
91. James, D.R.; Birnberg, G.H.; Paquette, L.A. *J. Am. Chem. Soc.* **1974,** *96,* 7465.
92. Dolbier, W.R., Jr.; Burkholder, C.R. *Tetrahedron Lett.* **1983,** *24,* 1217.
93. Meyer, A.Y. In "The Chemistry of Functional Groups, Supplement D", Part 1, Patai, S. and Rappoport, Z. Eds.; Wiley: New York, 1983, Chapter 1, pp. 1–47.
94. Lowe, J.P. *Prog. Phys. Org. Chem.* **1968,** *6,* 1.
95. Harris, W.C.; Holtzclaw, J.R.; Kalasinsky, V.F. *J. Chem. Phys.* **1977,** *67,* 3330.
96. Friesen, D.; Hedberg, H. *J. Am. Chem. Soc.* **1980,** *102,* 3987.
97. Kalasinsky, V.F.; Anjaria, H.V.; Little, T.S. *J. Phys. Chem.* **1982,** *86,* 1351.
98. Radom, L.; Lathan, W.A.; Hehre, W.J.; Pople, J.A. *J. Am. Chem. Soc.* **1973,** *95,* 693.
99. Bartell, L.S. *J. Am. Chem. Soc.* **1977,** *99,* 3279.
100. Allinger, N.L.; Hindman, D.; Honig, H. *J. Am. Chem. Soc.* **1977,** *99,* 3282.
101. Gordon, M.S. *J. Am. Chem. Soc.* **1969,** *91,* 3122.
102. Kiss, A.I.; Lopata, A. *J. Mol. Struct.* **1983,** *104,* 411.
103. Dewar, M.J.S.; Rzepa, H.S. *J. Am. Chem. Soc.* **1978,** *100,* 58.
104. Scheiner, S. *J. Am. Chem. Soc.* **1980,** *102,* 3723.
105. Kveseth, K. *Acta Chem. Scand.* **1978,** *32A,* 51.
106. Meyer, A.Y. *J. Mol. Struct.* **1978,** *49,* 383.
107. Meyer, A.Y. *J. Comput. Chem.* **1980,** *1,* 111.
108. Abraham, R.J.; Stølevik, R. *Chem. Phys. Lett.* **1981,** *77,* 181.
109. Durrig, J.R.; Bucy, W.E.; Carriera, L.A.; Wurrey, C.J. *J. Chem. Phys.* **1974,** *60,* 1755.
110. Förster, H.; Vögtle, F. *Angew. Chem. Int. Ed. Engl.* **1977,** *16,* 429.
111. Pachler, K.G.R.; Tollenaere, J.P. *J. Mol. Struct.* **1971,** *8,* 83.
112. Palke, W.E. *Chem. Phys. Lett.* **1972,** *15,* 244.
113. Allen, L.C.; Basch, H. *J. Am. Chem. Soc.* **1971,** *93,* 6374.
114. Lowe, J.P.; Parr, R.G. *J. Chem. Phys.* **1966,** *44,* 3001.
115. Campos-Vallette, M.; Rey-Lafon, M. *J. Mol. Struct.* **1983,** *101,* 23.
116. Bates, T.W. In "Fluoropolymers", Wall, L.A. Ed.; Wiley-Interscience: New York, 1972, Chapter 13, pp. 451–474.
117. Weiss, S.; Leroi, G. *J. Chem. Phys.* **1968,** *48,* 962.
118. Flanagan, C.; Pierce, L. *J. Chem. Phys.* **1975,** *38,* 2963.
119. Angeno, H.; Raley, J.M.; Wollrab, J.E. *J. Mol. Struct.* **1974,** *52,* 163.
120. Ward, C.R.; Ward, C.H. *J. Mol. Spect.* **1964,** *12,* 289.
121. Tipton, A.B.; Britt, C.O.; Boggs, J.E. *J. Chem. Phys.* **1967,** *46,* 1606.
122. Gallaher, K.L.; Yokozeki, A.; Bauer, S.H. *J. Phys. Chem.* **1974,** *78,* 2389.
123. Carlos, J.L.; Karl, R.R.; Bauer, S.H. *J. Chem. Soc. Faraday Trans. 2* **1974,** *70,* 177.

124. Laurie, V.W.; Pence, D.T. *J. Chem. Phys.* **1963,** *38,* 2693.
125. Huisman, P.A.G.; Mijlhoff, F.C.; Renes, G.H. *J. Mol. Struct.* **1979,** *51,* 191.
126. Spelbos, A.; Huisman, P.A.G.; Mijlhoff, F.C.; Renes, G.H. *J. Mol. Struct.* **1978,** *44,* 159.
127. VanSchaick, E.J.M.; Mijlhoff, F.C.; Renes, G.H.; Geise, H.J. *J. Mol. Struct.* **1974,** *21,* 17.
128. Mijlhoff, F.C.; Renes, G.H.; Kohata, K,; Oyanagi, K.; Kuchitsu, K. *J. Mol. Struct.* **1977,** *39,* 241.
129. Mom, V.; Huisman, P.A.G.; Mijlhoff, F.C.; Renes, G.H. *J. Mol. Struct.* **1980,** *62,* 95.
130. Cremer, D. Private communication cited in Reference 131.
131. Gandhi, S.R.; Benzel, M.A.; Dykstra, C.E.; Fukunaga, T. *J. Phys. Chem.* **1982,** *86,* 3121.
132. Bock, C.W.; George, P.; Mains, G.J.; Trachtman, M. *J. Chem. Soc. Perkin Trans. 2* **1979,** 814.
133. Dixon, D.A.; Smart, B.E. Unpublished results.
134. Dunning, T.H., Jr.; Hay, J.P. In "Methods of Electronic Structure Theory", Schaefer, H.F., III, Ed.; Plenum Press: New York, 1977, Chapter 1, pp. 1–27.
135. Pappas, J.A. *J. Mol. Struct.* **1974,** *20,* 197.
136. Facelli, J.C.; Contreras, R.H. *Z. Naturforsch.* **1980,** *35A,* 1350.
137. Bak, B.; Kierkegaard, C.; Pappas, J.; Skaarup, S. *Acta Chem. Scand.* **1973,** *27,* 363.
138. Devaquet, A.J.; Townshend, R.E.; Hehre, W.J. *J. Am. Chem. Soc.* **1976,** *98,* 4068.
139. Oberhammer, H.; Boggs, J.E. *J. Mol. Struct.* **1979,** *55,* 283.
140. Christen, D.; Oberhammer, H.; Hammaker, R.M.; Chang, S.-C.; DesMarteau, D.D. *J. Am. Chem. Soc.* **1982,** *104,* 6186.
141. Nakata, M.; Kohata, K.; Fukuyama, T.; Wilkins, C.J.; Kuchitsu, K. *J. Mol. Struct.* **1980,** *68,* 271.
142. Duncan, J.L. *Mol. Phys.* **1974,** *28,* 1177.
143. Christen, D.; Oberhammer, H.; Zeil, W. *J. Mol. Struct.* **1980,** *66,* 203.
144. Johnson, D.R.; Powell, F.X.; Kirchoff, W.H. *J. Mol. Spectrosc.* **1977,** *39,* 136.
145. Pearson, R.; Lovas, F.J. *J. Chem. Phys.* **1977,** *66,* 4149.
146. Tokue, I.; Fukuyama, T.; Kuchitsu, K. *J. Mol. Struct.* **1973,** *17,* 207.
147. Bürger, H.; Pawelke, G.; Oberhammer, H. *J. Mol. Struct.* **1982,** *84,* 49.
148. Hilderbrandt, R.L.; Andreassen, A.L.; Bauer, S.H. *J. Phys. Chem.* **1970,** *74,* 1586.
149. Tokue, I.; Fukuyama, T.; Kuchitsu, K. *J. Mol. Struct.* **1974,** *23,* 33.
150. Lowrey, A.H.; George, C.; D'Antonio, P.; Karle, J. *J. Mol. Struct.* **1979,** *53,* 189.
151. Chang. C.H.; Andreassen, A.L.; Bauer, S.H. *J. Org. Chem.* **1971,** *36,* 920.
152. Wurrey, C.J.; Bucy, W.E.; Durig, J.R. *J. Chem. Phys.* **1977,** *67,* 2765.
153. Stølevik, R.; Thingstad, Ø. *J. Mol. Struct.* **1984,** *106,* 333.
154. Ben-Yehuda, M.; Katz, M.G.; Rajbenbach, L.A. *J. Chem. Soc. Faraday Trans. 1* **1974,** *70,* 908.
155. Frey, H.M. *Trans. Faraday Soc.* **1963,** *59,* 619.
156. Schlag, E.W.; Peatman, W.B. *J. Am. Chem. Soc.* **1964,** *86,* 1676.
157. Dolbier, W.R., Jr. Unpublished results.
158. Dolbier, W.R., Jr.; Al-Fekri, D.M. *Tetrahedron Lett.* **1983,** *24,* 4047.
159. Chesick, J.P. *J. Am. Chem. Soc.* **1966,** *88,* 4800.
160. Pickard, J.M.; Rodgers, A.S. *J. Am. Chem. Soc.* **1977,** *99,* 695.
161. Bartlett, P.D.; Wheland, R.C. *J. Am. Chem. Soc.* **1972,** *94,* 2145.
162. Hine, J. "Structural Effects on Equilibria in Organic Chemistry". Wiley: New York, 1975, pp. 270–276.
163. Cremer, D. *J. Am. Chem. Soc.* **1981,** *103,* 3633.
164. Schlag, E.W.; Kaiser, E.W., Jr. *J. Am. Chem. Soc.* **1965,** *87,* 1171.
165. Jeffers, P.M. *J. Phys. Chem.* **1974,** *78,* 1469.
166. Cruickshank, F.R.; Benson, S.W. *J. Phys. Chem.* **1969,** *73,* 733.
167. Pickard, J.M.; Rodgers, A.S. *J. Am. Chem. Soc.* **1976,** *98,* 6116.
168. Rogers, F.E.; Rapiejko, R.J. *J. Am. Chem. Soc.* **1971,** *93,* 4596.
169. Krespan, C.G.; Middleton, W.J. *Fluorine Chem. Rev.* **1967,** *1,* 145.
170. Hollein, H.C., Snyder, W.H. *J. Mol. Struct.* **1982,** *84,* 83.
171. Craig, N.C.; Piper, L.G.; Wheeler, V.L. *J. Phys. Chem.* **1971,** *75,* 1453.
172. Binkley, J.S.; Pople, J.A. *Chem. Phys. Lett.* **1977,** *45,* 197.
173. Viehe, H.G. *Chem. Ber.* **1963,** *93,* 953.
174. Pitzer, K.S.; Hollenberg, J.L. *J. Am. Chem. Soc.* **1954,** *76,* 1493.
175. Epiotis, N.D.; Yates, R.L. *J. Am. Chem. Soc.* **1976,** *98,* 461.
176. Eyring, H.; Stewart, G.H.; Smith, R.P. *Proc. Natl. Acad. Sci. USA* **1958,** *44,* 259.

177. Bernardi, F.; Bottani, A.; Epiotis, N.D.; Guerra, M. *J. Am. Chem. Soc.* **1976,** *100,* 6018.
178. Skancke, A.; Boggs, J.E. *J. Am. Chem. Soc.* **1979,** *101,* 4063.
179. Cremer, D. *Chem. Phys. Lett.* **1981,** *81,* 481.
180. Tyler, J.K.; Sheridan, J. *Trans. Faraday Soc.* **1973,** *59,* 2661.
181. Kloster-Jensen, E.; Pascual, C.; Vogt, J. *Helv. Chim. Acta* **1970,** *53,* 2109.
182. Ehlert, T.C. *J. Phys. Chem.* **1969,** *73,* 949.
183. Marriott, S.; Topsom, R.D. *J. Mol. Struct.* **1984,** *106,* 277.
184. Cremer, D. *J. Comput. Chem.* **1982,** *3,* 154.
185. Goddard, J.D. *Chem. Phys. Lett.* **1981,** *83,* 312.
186. Ellis, P.D.; Li, Y.S.; Tong, C.C.; Zens, A.P.; Durig, J.R. *J. Chem. Phys.* **1975,** *62,* 1311.
187. Durig, J.R.; Li, Y.S.; Tong, C.C.; Zens, A.P.; Ellis, P.O. *J. Am. Chem. Soc.* **1974,** *96,* 3805.
188. Dolbier, W.R., Jr.; Burkholder, C.R.; Piedrahita, C.A. *J. Fluorine Chem.* **1982,** *20,* 637.
189. Dolbier, W.R., Jr.; Burkholder, C.R. *Tetrahedron Lett.* **1980,** *21,* 785.
190. Dolbier, W.R., Jr.; Piedrahita, C.A.; Houk, K.N.; Strosier, R.W.; Gandour, R.W. *Tetrahedron Lett.* **1978,** 2231.
191. Domelsmith, L.N.; Houk, K.N.; Piedrahita, C.; Dolbier, W.R., Jr. *J. Am. Chem. Soc.* **1978,** *100,* 6908.
192. Banks, R.E.; Barlow, M.G.; Davies, W.D.; Haszeldine, R.N.; Taylor, D.R. *J. Chem. Soc.* **1969,** 1104.
193. Banks, R.E.; Barlow, M.G.; Mullen, K. *J. Chem. Soc. C* **1969,** 1131.
194. Schei, S.; Almenningen, A.; Almlöf, J. *J. Mol. Struct.* **1984,** *112,* 301.
195. Almlöf, J.; Faegri, K., Jr. *J. Chem. Phys.* **1983,** *79,* 2284.
196. Almlöf, J.; Faegri, K., Jr. *J. Am. Chem. Soc.* **1983,** *105,* 2965.
197. Boggs, J.E.; Pang, F.; Pulay, P. *J. Comput. Chem.* **1982,** *3,* 344.
198. Almenningen, A.; Hargittai, I.; Samdal, S. Private communication cited in Reference 199.
199. Schei, S.H.; Almenningen, A.; Almlöf, J. *J. Mol. Struct.* **1984,** *112,* 301.
200. Schultz, G.; Hargittai, I.; Seip, R. *Z. Naturforsch. A* **1981,** *36A,* 669.
201. Choo, K.Y.; Golden, D.M.; Benson, S.W. *Int. J. Chem. Kinet.* **1975,** *7,* 713.
202. (a) Krech, M.J.; Price, S.J.W.; Sapiano, H.J. *Can. J. Chem.* **1977,** *55,* 4222. (b) Harrop, D.; Head, A.J. *J. Chem. Thermodyn.* **1978,** *10,* 705. (c) Cox, J.D.; Gundry, H.A.; Head, A.J. *Trans. Faraday Soc.* **1964,** **60,** 653. (d) Cox, J.D.; Gundry, H.A.; Harrop, D.; Head, A.J. *J. Chem. Thermodyn.* **1969,** *1,* 77.
203. Couldwell, M.H.; Penfold, B.R. *J. Cryst. Mol. Struct.* **1976,** *6,* 59.
204. Karl, R.R., Jr.; Wang, Y.C.; Bauer, S.H. *J. Mol. Struct.* **1975,** *25,* 17.
205. Grayston, M.W.; Lemal, D.M. *J. Am. Chem. Soc.* **1976,** *98,* 1278.
206. Lemal, D.M.; Staros, J.V.; Austel, V. *J. Am. Chem. Soc.* **1969,** *91,* 3373.
207. Kobayashi, Y.; Kumadaki, I. *Acc. Chem. Res.* **1981,** *14,* 76.
208. Barlow, M.G.; Haszeldine, R.N.; Dingwall, J.G. *J. Chem. Soc. Perkin Trans. 1,* **1973,** 1542.
209. Wirth, D.; Lemal, D.M. *J. Am. Chem. Soc.* **1982,** *104,* 847.
210. Kobayashi, Y.; Fujino, S.; Hamana, H.; Hanzawa, Y.; Morita, S. *J. Org. Chem.* **1980,** *45,* 4683.
211. Kobayashi, Y.; Kumadaki, I. In "Advances in Heterocyclic Chemistry", Vol. 31; Katritzky A.R., Ed.; Academic Press: New York, 1983, p. 169.
212. (a) Zupan, M.; Sket, B. *Isr. J. Chem.* **1978,** *17,* 92. (b) Kobayashi, Y.; Kumadaki, I; *Top. Curr. Chem.* **1984,** *123,* 103.
213. Lemal, D.M.; Dunlap, L.H., Jr. *J. Am. Chem. Soc.* **1972,** *94,* 6562.
214. Dabbagh, A.-M.M.; Flowers, W.T.; Haszeldine, R.N.; Robinson, R.J. *J. Chem. Soc. Perkin Trans. 2* **1979,** 1407.
215. Oth, J.F.M. *Rec. Trav. Chim. Pays-Bas,* **1968,** *87,* 1185.
216. Cadman, P.; Ratajezak, E.; Trotman-Dickenson, A.F. *J. Chem. Soc. A* **1970,** 2109.
217. Haller, I. *J. Phys. Chem.* **1968,** *72,* 2882.
218. Greenberg, A.; Liebman, J.F.; Van Vechten, D. *Tetrahedron* **1980,** *36,* 1161.
219. Greenberg, A.; Liebman, J.F. "Strained Organic Molecules". Academic Press: New York, 1976.
220. Cox, J.D.; Pilcher, G. "Thermochemistry of Organic and Organometallic Compounds". Academic Press: New York, 1970.
221. Hehre, W.J.; Ditchfield, R.; Radom, L; Pople, J.A. *J. Am. Chem. Soc.* **1970,** *92,* 4796.
222. George, P.; Trachtman, M.; Bock, C.W.; Brett, A.M. *Tetrahedron* **1976,** *32,* 317.

223. Newton, M.D. In "Applications of Electronic Structure Theory", Schaefer, H.F., III, Ed.; Plenum Press: New York, 1977, Chapter 6, pp. 223–268.
224. Wiberg, K.B. *J. Am. Chem. Soc.* **1983,** *105,* 1227.
225. Schulman, J.M.; Disch, R.L. *J. Am. Chem. Soc.* **1984,** *106,* 1202.
226. Dewar, M.J.S.; Thiel, W. *J. Am. Chem. Soc.* **1977,** *99,* 4907.
227. Bingham, R.C.; Dewar, M.J.S.; Lo, D.H. *J. Am. Chem. Soc.* **1976,** *97,* 1294.
228. Latham, W.A.; Radom, L.; Hehre, W.J.; Pople, J.A. *J. Am. Chem. Soc.* **1973,** *95,* 699.
229. Benson, S.W. "Thermochemical Kinetics", 2nd ed.; Wiley: New York, 1976.
230. Buckley, G.S.; Rodgers, A.S. *J. Phys. Chem.* **1982,** *86,* 2059.
231. Joshi, R.M. *J. Macromol. Sci.-Chem.* **1974,** *A8,* 861.
232. Pedley, J.B.; Rylance, J. "Sussex-N.P.L. Computer-Analysed Thermochemical Data: Organic and Organometallic Compounds". University of Sussex: Brighton, UK, 1977.
233. Lacher, J.R.; Kianpour, A.; Oetting, F.; Park, J.D. *Trans. Faraday Soc.* **1956,** *52,* 1500.
234. (a) Erastov, P.A.; Kolesov, V.P. *J. Chem. Thermodyn.* **1982,** *14,* 103. (b) Williamson, A.D.; Le Breton, P.R.; Beauchamp, J.L. *J. Am. Chem. Soc.* **1976,** *98,* 2705.
235. Berman, D.W.; Bomse, D.S.; Beauchamp, J.L. *Int. J. Mass Spectrom. Ion Phys.* **1981,** *39,* 263.
236. Price, S.J.W.; Sapiano, H. *Can. J. Chem.* **1979,** *57,* 685.
237. Lacher, J.R.; Skinner, H.A. *J. Chem. Soc. A* **1968,** 1034.
238. Krech, M.J.; Price, S.J.W.; Yared, W.F. *Can. J. Chem.* **1973,** *51,* 3662.
239. Gordon, A.S. *Int. J. Chem. Kinet.* **1972,** *4,* 541.
240. Dolbier, W.R., Jr.; Medinger, K.S. *Tetrahedron* **1982,** *38,* 2411.
241. Jochims, H.W.; Lohr, W.; Baumgärtel, H. *Nouv. J. Chim.* **1979,** *3,* 109.

CHAPTER 5

Structures and Energies of Substituted Strained Organic Molecules

Arthur Greenberg and Tyler A. Stevenson

New Jersey Institute of Technology, Newark, New Jersey

CONTENTS

1. INTRODUCTION

The limits of stability of unsubstituted strained hydrocarbons have been substantially delineated over the course of the past three decades.[1] Lest we become too self-assured over progress in this area, it is wise to recall the very recent synthesis of [1.1.1]propellane[2] and its amazing stability despite

MOLECULAR STRUCTURE AND ENERGETICS, Vol. 3

predictions to the contrary.[1] Thus, it is likely that other surprises will occur and the reader is referred to Chapter 4 in Volume 2 of this series, which touches on the energetics of small propellanes and other exotic molecules.

Appropriate attachment of substituents can extend the range of accessible molecular systems, sometimes in a most spectacular manner. The only tetrahedrane to be even implicated with certainty as a viable molecule is the tetra-*tert*-butyl derivative, which melts at 130°C.[3] The origin of this remarkable stability is almost certainly kinetic rather than thermodynamic and arises from the steric encumbrance to exothermic dimerization, rearrangement, and decomposition to the acetylene monomers. We feel somewhat comfortable in explaining such steric effects because space-filling molecular models encourage us to believe that we understand them. That is why steric substituent effects are neglected in this chapter. Another striking example of the effects of substituents on stability is provided by a molecule such as hexakis (trifluoromethyl)-3,3′-bicyclopropenyl, which exhibits a half-life at 360°C of more than 2 h.[4] This is an illustration of the "perfluoroalkyl (R_f) effect" first noted by Lemal and Dunlap.[5] Substitution can be destabilizing as well, and in this regard we note increased lability of 1,1-difluorocyclopropanes[6] and the relative ease of carbene extrusion exhibited by hexafluorocyclopropane.[7] The energy relationships in fluorocarbons are dealt with at some length in Chapter 4 of this volume. Certain substituents can so dramatically transform strained and unsaturated molecules as to make them virtually unrecognizable compared to their hydrocarbon counterparts. Lithiocarbons are especially noteworthy in this regard.

We consider here the electronic effects of substituents on the molecular structures and energies of cyclopropanes, cyclobutanes, bicyclobutanes, cyclopropenes, and a variety of other cyclic species. Reference is made to comparison with unsaturated compounds including substituted ethylenes, acetylenes, and benzenes, and the relationship of energy and structure in cyclopropanes receives particular attention. This is also an area in which structures and energies give some fairly clear insights into reactivity. However, reactivity of substituted strained molecules must involve consideration of the structures and stabilities of activated complexes and intermediates and remains a story for a future chapter.

Throughout this chapter we report many calculational results based on ab initio quantum mechanical calculations. The reasons for this are threefold.

1. There are few published thermochemical data on substituted cyclopropanes and other strained molecules.
2. Calculations may be performed on molecules for which calorimetry is impractical (eg, vinylamine) or impossible (eg, cyclopropylcarbinyl cation), or on molecules or ions that do not actually exist (eg, the perpendicular conformer of cyclopropylcarbinyl cation) but find use as instructive models.
3. At a suitable level, calculational results well reproduce experimental data.

Appended to this chapter is a summary of 4–31G total energies and gas phase enthalpy of formation, $\Delta H_f^\circ(g)$, data for all monosubstituted compounds considered in this study. The calculational results correspond to gas phase ΔH_f° data and where ΔH_f° is employed, it is understood to mean gas phase.

2. SUBSTITUTED CYCLOPROPANES: STRUCTURES

Similarities between the vinyl and cyclopropyl groups have been evident since the early part of this century,[8–10] and they have been extensively reviewed.[11,12] Figure 5-1 depicts the molecular orbitals of ethylene, cyclopropane, and propane relevant to discussions in this chapter. Particularly striking is the ability of the cyclopropane ring to stabilize carbonium ions.[13–16] This is clearly a conjugative effect, as is clear from UV[17] and NMR spectral studies.[18–20] Specifically, dimethylcyclopropylcarbinyl cation (**1**) is most stable in the bisected conformation that allows conjugation involving the Walsh-type 3E′ orbital[18,19] (Figure 5-2*a*). The rotational barrier in this ion,

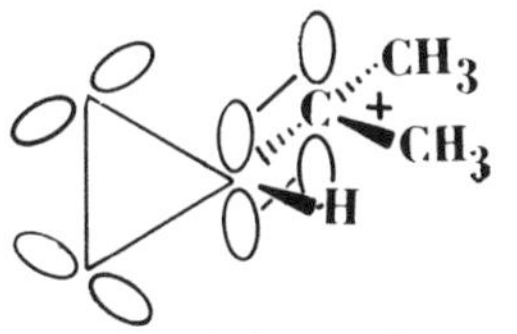

CH3 CH3 C + H   CH3 C + CH3 H

1 Bisected 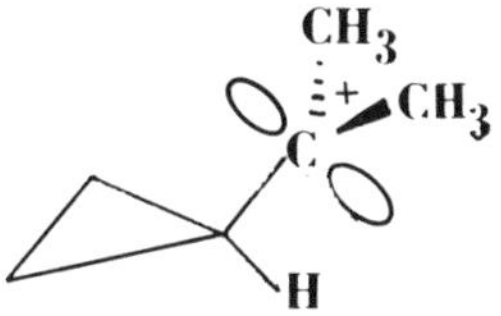   **1** Perpendicular

measured by dynamic NMR, corresponds to 13.7 kcal/mol,[20] indicating very substantial conjugation. Minimal basis set (STO-3G) ab initio molecular orbital (MO) calculations predict a rotational barrier of 26.2 kcal/mol for the unsubstituted cyclopropylcarbinyl cation, which is increased to 30.3 kcal/mol at the 4–31G level.[21] This is very similar to the calculated magnitude of the rotational barrier in allyl cation (STO-3G, 34.4 kcal/mol; 4–31G, 35.0 kcal/mol; 6–31G*, 34.8 kcal/mol).[21] These calculated barriers are likely to be upper bounds "due to the inability of single-determinant theory to properly describe the diradical character of the orthogonal structure"[21] and other closely related species. However, comparison between analogous ions should be meaningful. The STO-3G calculated rotational barrier in benzyl cation is 49 kcal/mol, about 13 kcal/mol higher than in cyclopropylcarbinyl cation.[22] This seems to support the view that phenyl is much more active than cyclopropyl in stabilizing a carbocationic center. It is consistent with an STO-3G calculation of the stabilization energy of reaction 5-1.[22]

$$C_6H_5CH_2^+ + c\text{-}C_3H_5CH_3 \xrightarrow{\Delta E_{STO-3G} = 17.0\ \text{kcal/mol}} C_6H_5CH_3 + c\text{-}C_3H_5CH_2^+ \quad (5\text{-}1)$$

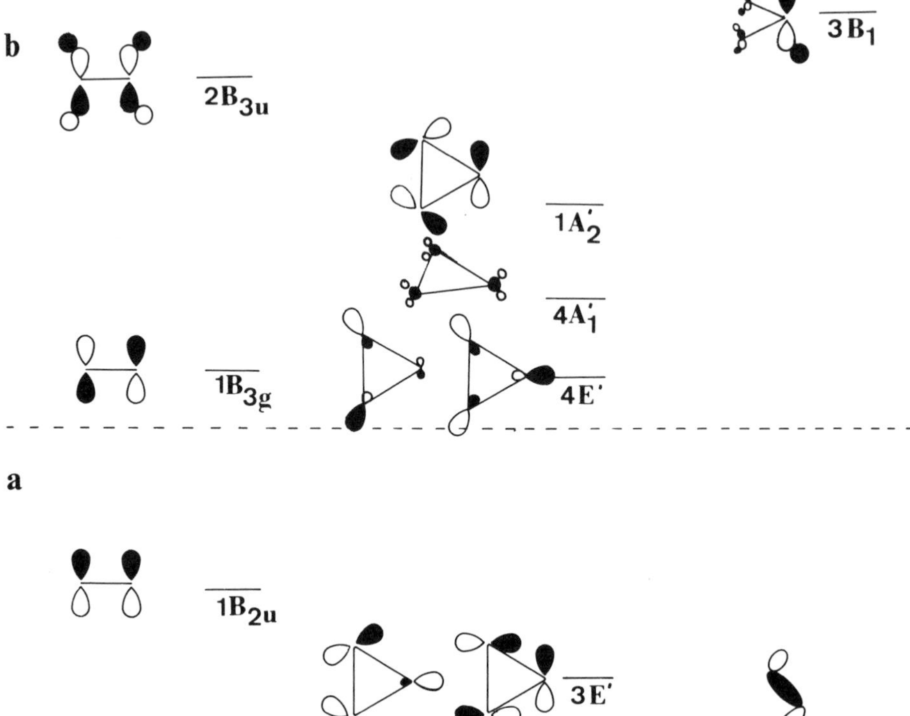

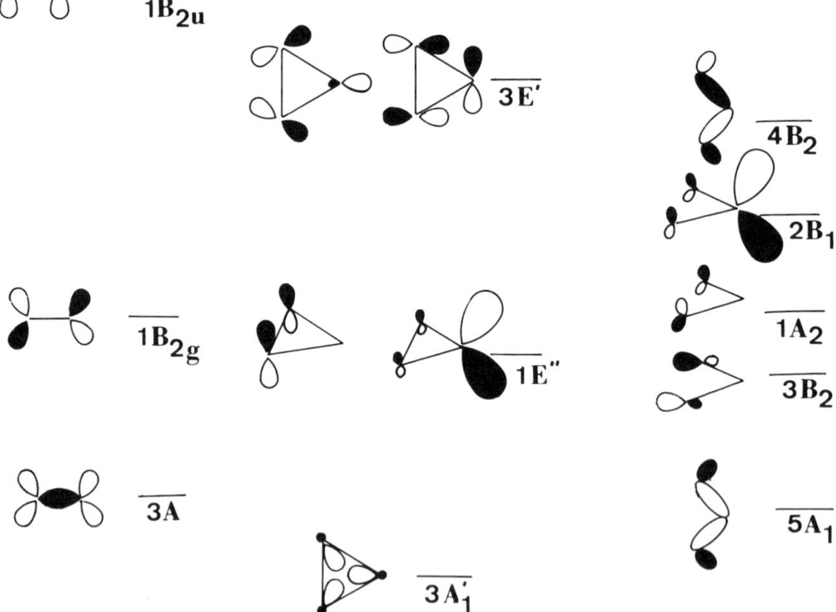

Figure 5-1. (*a*) High-lying occupied and (*b*) low-lying unoccupied molecular orbitals of ethylene, cyclopropane, and propane. (Adapted from Reference 176.)

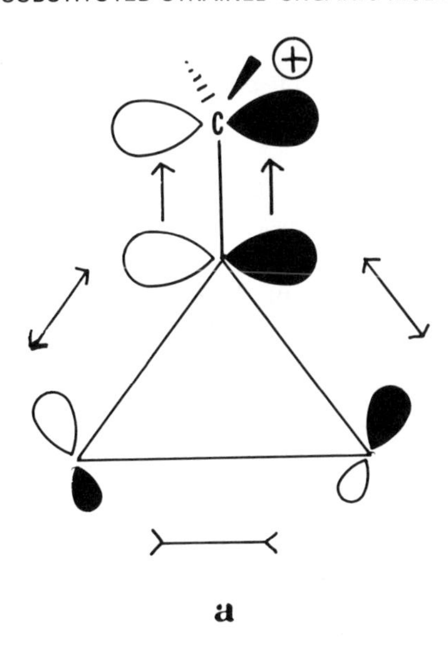

Figure 5-2. (*a*) Interaction of cyclopropane HOMO with vacant p orbital. (*b*) Structure of bisected (conjugated) cyclopropylcarbinyl cation (see Reference 26). (*c*) Structure of perpendicular cyclopropylcarbinyl cation (see Reference 26).

Certainly, part of this effect is due to the greater polarizability of the larger phenyl system rather than due solely to π-donor ability. When the analogous tertiary ions are compared (5-2), both the experimental [ion cyclotron resonance (ICR) spectroscopy] and the calculated (STO-3G) data indicate about equal stability for these cations. Solvolyses of the corresponding *p*-

nitrobenzoates has indicated greater stabilization by cyclopropyl than by phenyl,[23] and such trends have been noted by others.[24,25] It is interesting to compare allyl and cyclopropylcarbinyl cations in the manner of Equation 5-1. Equation 5-3 indicates that the stabilities are comparable. Recalling that 4–31G calculations indicate a rotational barrier in allyl cation that is 4.7 kcal/mol higher than in cyclopropylcarbinyl cation, the calculated 5.3-kcal/mol exothermicity for reaction 5-3 indicates stabilization of cyclopropylcarbinyl cation due to polarization by 10 kcal/mol relative to the smaller system.

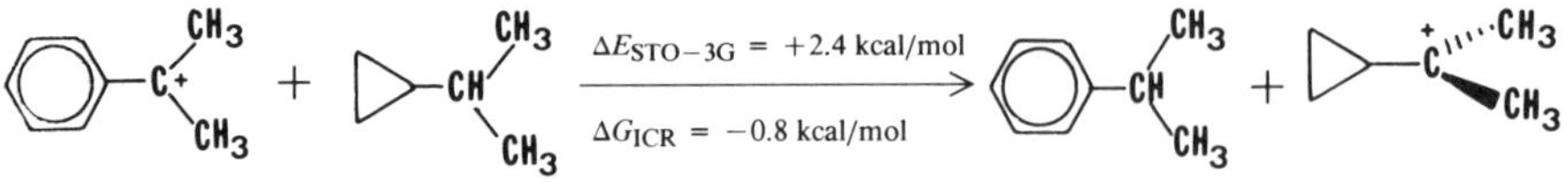

$\Delta E_{STO-3G} = +2.4$ kcal/mol

$\Delta G_{ICR} = -0.8$ kcal/mol

(5-2)

$\Delta E_{STO-3G} = +1.7$ kcal/mol

$\Delta E_{4-31G} = -5.3$ kcal/mol

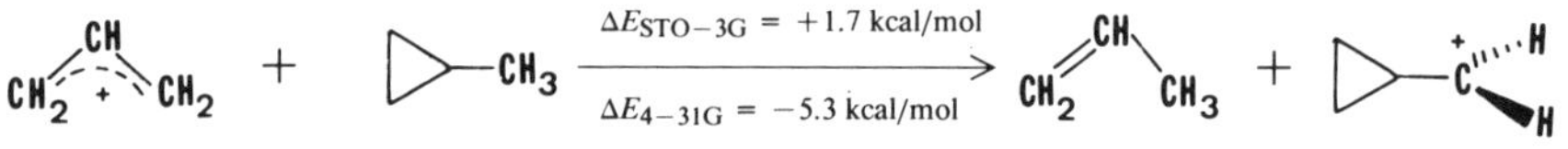

(5-3)

In contrast to the cation, double-zeta (DZ) calculations, which are essentially equivalent to 4–31G calculations, find an energy difference of only 1.1 kcal/mol favoring the bisected conformer of cyclopropylcarbinyl anion (**2**, planar CH_2^-) over the perpendicular conformer,[26] and this is consistent with

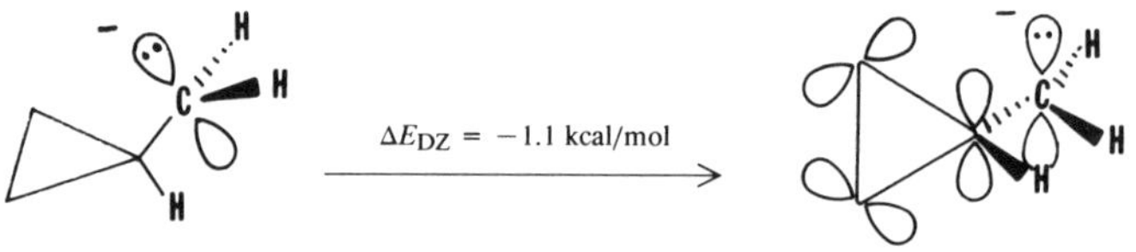

$\Delta E_{DZ} = -1.1$ kcal/mol

2 Perpendicular

2 Bisected

very little conjugative stabilization.[27,28] The most stabilizing interaction is between the lone pair and the $1A_2'$ orbital (Figure 5-3*a*). Thus, as these examples and our subsequent discussion will make clear, cyclopropane is a rather strong π donor and a very weak π acceptor.

Most of the available calculational and calorimetric data are for neutral molecules for which the magnitudes of substituent effects are smaller than in cyclopropylcarbinyl cations. Although the interpretation of stabilization in cyclopropylcarbinyl cation is typically made almost solely in terms of the large resonance stabilization in this ion, interpretations of the smaller effects in the neutral molecules are more subtle. These effects are, nevertheless, significant because they alter the geometries and energies of cyclopropane rings and alter the courses of their chemical reactions.

The bond asymmetries induced by substituents are of special interest because they may be associated with highly regioselective chemical reac-

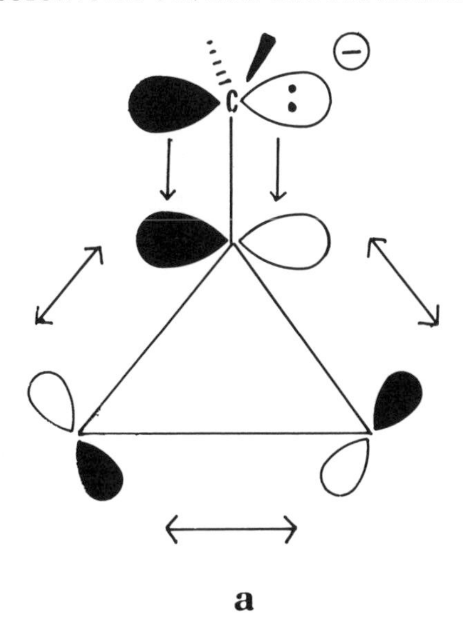

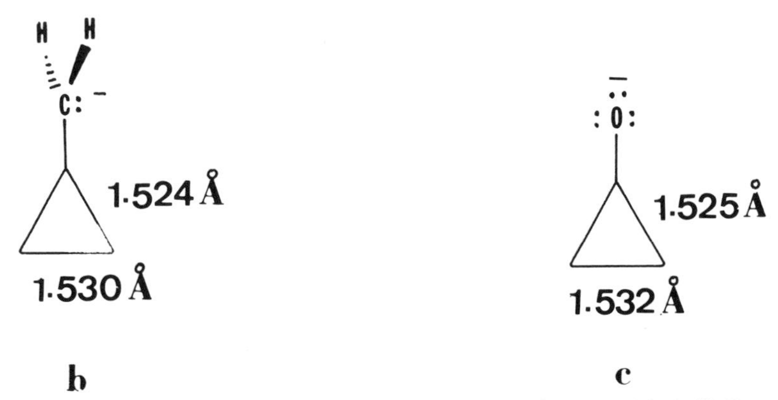

Figure 5-3. (*a*) Interaction of cyclopropane LUMO with filled p orbital. (*b*) Structure of bisected cyclopropylcarbinyl anion (see Reference 26). (*c*) Structure of cyclopropoxide anion (see Reference 26).

tions. For example, the mode of hydrogenation of cyclopropane rings is strongly dependent on the pattern of substitution (Scheme I).[29a,b] A dramatic effect on the geometric isomerization of *cis*-1,1-difluoro-2,3-dimethylcyclopropane **(3)** is evident when comparison is made with the corresponding isomerization of **4.**[6,30] The striking effects of geminal difluorosubstitution on

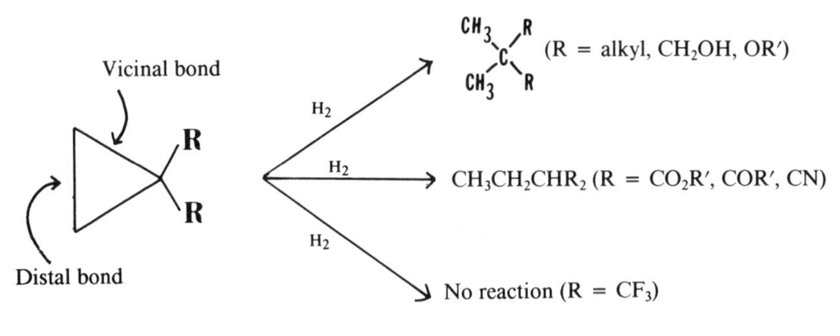

Scheme I

the ring geometry of 1,1-difluorocyclopropane (**5**)[31,32] appear to nicely correspond to the relatively facile opening of the distal bond in **3**.

F F CH3 CH3 **3** $\xrightarrow{E_a\ =\ 49.7\ \text{kcal/mol}}$ F F CH3 CH3 (5-4)

CH3 CH3 **4** $\xrightarrow{E_a\ =\ 59.4\ \text{kcal/mol}}$ CH3 CH3 (5-5)

F F 1.464 Å 1.553 Å **5**

Investigations into the role of π-accepting substituents on ring geometry received their main impetus from the predictions of Hoffmann and coworkers.[33–35] Assuming the dominant interaction to be between the cyclopropane 3E′ Walsh-type orbital and the vacant p- or π^* orbital on a substituent, they predicted lengthening of the vicinal (C_1C_2—C_1C_3) bonds and shortening of the distal (C_2C_3) bond. This is because transfer of electron density from the 3E′ orbital decreases the antibonding electron density in the distal bond and decreases the bonding electron density in the vicinal bonds (Figure 5-2*a*). This effect is well illustrated by the STO-3G-calculated geometry for the bisected cyclopropylcarbinyl cation (Figure 5-2*b*), particularly when it is compared to the perpendicular isomer (Figure 5-2*c*). Although minimal basis set ab initio calculations do not usually well reproduce geometries of substituted cyclopropanes,[36] the cyclopropylcarbinyl cation is dominated by the strong π interaction, which should be well reproduced at the STO-3G level.

Before proceeding to a discussion of bond length changes in cyclopropanes, recall that there are a number of different measures of bond lengths, and note that these are rather cavalierly mixed here. A discussion of these differences is to be found in Chapter 1 of Volume 2 in this series. For now, we note that it is the equilibrium bond length r_e, which is obtained by calculation and refers to a hypothetical vibrationless molecule at 0 K. The value r_s, obtained from microwave spectroscopy, is slightly larger, and r_g, obtained from electron diffraction is longer still. These values are comparable to calculated r_e only in the sense that they all refer to the gas phase. X-Ray crystallographic data furnish values that are not, strictly speaking, internuclear but rather are determined by the distance between centroids of electron density. Furthermore, they obviously involve crystalline material and a role for crystal packing in determining molecular coordinates must always be suspected. (See Chapter 1 of Volume 2 for further discussion of this issue.)

Hoffmann's predictions of the effects of π-acceptor substituents on cyclopropane ring geometries appear to be well borne out by experiment. In particular, a statistical study of X-ray crystallographic data from the Cambridge Crystallographic Database[37,38] indicated the universality of this effect, with a tendency toward shortening of the distal bond by twice the lengthening of each vicinal bond (see **6a**) when the substituent is in or close to the conju-

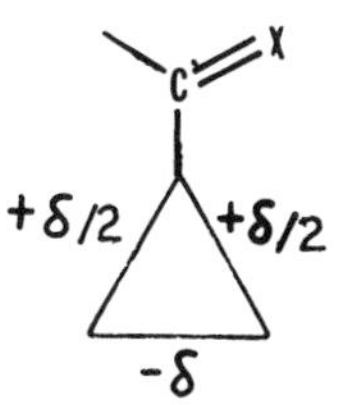

6a 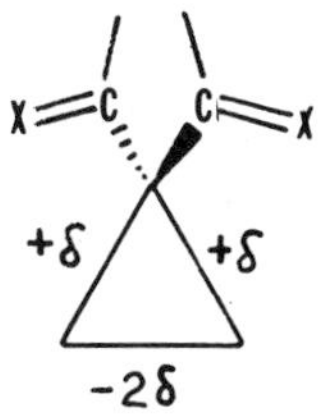 **6b** 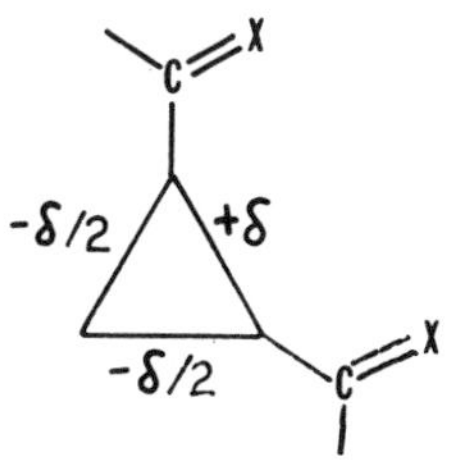 **6c**

gating bisected conformation. This is consistent with the finding of short distal bonds noted in an earlier study of cyclopropanes substituted by unsaturated substituents at C_1.[39] The magnitude of δ has been considered to be a measure of the extent of ring–substituent interaction.[38] Thus, for conjugated CH_2^+ (Figure 5-2*b*), the δ value (-0.06) is relatively large. (Note that the calculated structure for conjugated cyclopropylcarbinyl cation has vicinal bonds much greater in length than predicted from the Allen model.) On this basis alone, CN ($\delta = -0.017$) and phenyl ($\delta = -0.018$) are about equal in their interactions with a cyclopropane. Recall that consideration of phenyl as a π acceptor rather than a π donor is intuitively reasonable only when one considers cyclopropane's strong π-donor and weak π-acceptor properties. The carbonyl substituent ($\delta = -0.026$) presumably has a stronger interaction by this definition. Although CN and C_6H_5 appear to interact equally with cyclopropane by Allen's criterion,[38] calorimetric data discussed later yield a different conclusion. Allen[37,38] also provides evidence for structural additivity of substituent effects (see **6b** and **6c**) in support of an earlier suggestion.[35]

The shortening of the distal bonds in cyclopropanes has been advanced as the origin of stabilizations that render norcaradienes **7**,[40] **8**,[41] and **9**[42] observable by NMR in contrast to the parent hydrocarbon, which is much less stable than the isomeric 7-substituted cycloheptatriene (tropylidene). The relative stability of semibullvalene isomer **10**[43] has been similarly rationalized.

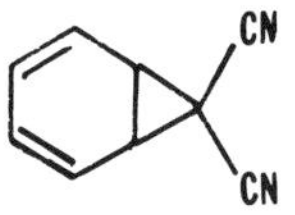

CN CN **7** CN CF_3 **8** COOH H **9** NC **10**

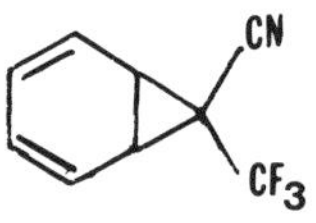

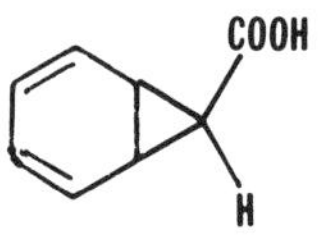

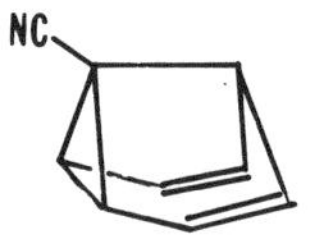

There are other important structural measures of the extent of conjugation of a cyclopropane ring with an unsaturated center. For example, the ring–substituent (ie, acceptor–donor) bond length (D_{ad}) has been employed as a structural criterion.[38] Allen has created a D_{ad} matrix in which cyclopropyl, vinyl, and alkyl derivatives are compared. The corresponding D_{ad} difference matrix is shown in Table 5-1. Those substituents which can adopt bisected conformations (torsional angle $\tau = 0 \pm 30°$; $180 \pm 30°$) or perpendicular conformations ($\tau = 90 \pm 30°$) are analyzed to separate bond length changes due to hybridization alone versus changes due to hybridization and conjugation. The difference is a measure of the conjugative interaction. Interpolation of D_{ad} for nonconjugated derivatives indicated $sp^{2.2}$ hybridization for the carbon orbital directed toward the substituent. The most interesting overall conclusion was that the cyclopropyl moiety is about 71% as effective a π donor as vinyl.[38] Another interesting point is the comparison between the conjugative part of bond shortening ("c" in Table 5-1) and Taft σ_R constants. One sees that vinyl, which is the poorest π acceptor has the lowest c value (-0.010). It is again worth remarking here that Allen[37,38] treats vinyl, attached to cyclopropane, as a π-acceptor substituent rather

TABLE 5-1. Donor–Acceptor Distance (D_{ad}) Difference Matrix[a]

1-Substituent	σ_R[c]	Interaction[d]	Distance (Å)[b] $D_{ad(\Delta)}$	$D_{ad(vi)}$	$D_{ad(\Delta)}/D_{ad(vi)}$ (%)
C_{sp3} (alkyl)		h	−0.019	−0.031	61
CH_2=CH— (vinyl)	−0.15	h	−0.027	−0.035	77
		h + c	−0.037	−0.049	76
		c	−0.010	−0.014	71
—COR (keto)	0.18	h	−0.023	−0.030	77
		h + c	−0.038	−0.052	73
		c	−0.015	−0.022	68
—COOH (acid)	0.11	h	−0.010	−0.017	59
		h + c	−0.030	−0.044	68
		c	−0.020	−0.027	74
—CN	0.08	h + c	−0.031	−0.044	71

[a]After Reference 38.
[b]$D_{ad(\Delta)}$ = donor–acceptor distance for cyclopropyl; $D_{ad(vi)}$ = donor–acceptor distance for vinyl.
[c]Values for Taft resonance parameter σ_R from: Ehrenson, S.; Brownlee, R.T.C.; Taft, R.W. *Prog. Phys. Org. Chem.* **1973,** *10,* 1–80.
[d]Interaction h = difference due to hybridization; interaction c = difference due to conjugation.

than a π donor, and this is reasonable in terms of cyclopropane's strong π-donor and weak π-acceptor behavior. The keto substituent, a much stronger π acceptor, has a larger c value (−0.015). This is also consistent with the relative δ values (vinyl, −0.022; carbonyl, −0.026). However, the COOH substituent is a poorer π acceptor than COR, using the σ_R criterion, yet has a higher c value. Also, CN is a stronger π acceptor than vinyl but has a lower D_{ad}. The criterion of D_{ad} shortening also implies some (p-d)–π delocalization in cyclopropylsilane, where D_{ad} (1.853 Å) is comparable to that in vinylsilane and the distal bond length (1.508 Å) is shorter than the vicinal bond (1.520 Å).[44]

Another structural measure of the π-donor ability of the cyclopropyl moiety is the tendency for π-acceptor substituents to adopt bisected conformations usually associated with measurable rotational barriers. This barrier was shown earlier to be quite high (13.7 kcal/mol) for dimethylcyclopropylcarbinyl cation **(1).** Hoffmann had predicted significant rotational barriers for carboxaldehyde derivatives of cyclopropane, ethylene, and benzene using extended Hückel calculations, which predicted the highest barrier for cyclopropanecarboxaldehyde.[45] A 4–21G calculated barrier for nitrocyclopropane (6.6 kcal/mol) is higher than experimental values obtained from microwave (3.3 ± 1.5 kcal/mol) and vibrational (4.7 ± 0.2 kcal/mol) spectroscopies, which are themselves similar to results for nitroethylene and nitrobenzene.[46] An interesting observation associated with these interactions is the remarkably high barrier to rotation about the C—C bond in 1, 1′-bis(1-nitrocyclopropyl) **(11),** in which conjugation between the nitro

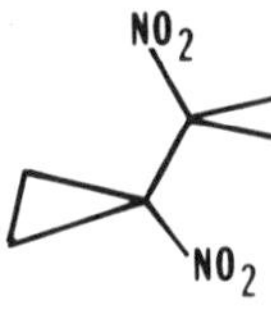

NO2

NO2

11

groups and cyclopropane rings must be diminished to allow C—C rotation.[47] The nitro group in 2-nitropropane rotates freely.[48] A 4–31G calculation on cyclopropylborane favored the bisected conformer over the perpendicular conformer by 7.3 kcal/mol with C—B bond lengths of 1.534 and 1.561 Å, respectively.[36] A 6–31G* calculational study[49] favored the s-cis (syn-periplanar) conformer of cyclopropanecarboxaldehyde by 0.3 kcal/mol over the s-trans (anti-periplanar) conformer and indicated a 5.8-kcal/mol rotational barrier, in good agreement with microwave results.[50] The experimental rotational barrier in cyclopropyldifluoroborane (4.2 kcal/mol) is very similar to the barriers in the vinyl and phenyl analogues.[51] The experimental solution phase barrier in cyclopropylcarbonyl fluoride (5.5 kcal/mol)[52] is similar to the gas phase barrier, comparable to that of the vinyl analogue, and about 1.5 kcal/mol lower than that in the phenyl compound.[53] Both cyclopropyldichloroborane[54] and cyclopropylcarbonyl chloride[55] are most stable in the bisected conformation with the latter favoring the s-cis structure. The s-trans conformer of nitrosocyclopropane has been calculated to be slightly more stable than the s-cis conformer,[56] although microwave data indicate only the presence of the s-trans conformer.[57]

Vinylcyclopropane is a particularly interesting study from a conformational point of view. The s-trans conformer **(12a)** is more stable by about 1.1 kcal/mol than the gauche conformer **(12b)**.[58] It is noteworthy that the gauche conformer is more stable than the s-cis structure, and this is probably

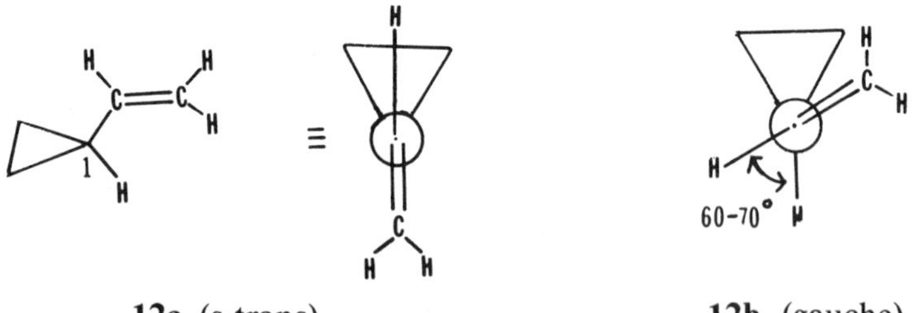

60–70°

12a (s-trans) **12b** (gauche)

due to steric effects. Calculational studies (4–21G)[59] indicate a distal bond length in **12a** of 1.510 Å, equal to that in the parent ring, and lengthening of the vicinal bonds (1.522 Å). The ring bonds in **12b** were about equal to those in **12a** despite the significant reduction in conjugation in the former. Differential conjugation in these conformers is indicated, however, by the differences in C_1C_4 distances: 1.482 Å in **12a** and 1.491 Å in **12b.** It is also interesting that the 1.1-kcal/mol difference between **12a** and **12b** is very similar to the free energy change in reaction 5-6.[60] This value, which may be roughly equated to the thermochemical stabilization energy of vinylcyclopropane, is lower than the 3.6-kcal/mol stabilization in 1,3-butadiene.[60] However, apportionment of stabilization between hybridization and conjugation effects is difficult. It has been noted that relatively weak conjugation in vinylcyclopropanes, compared to conjugated dienes, is manifested in small perturbations of the HOMO energies of the individual fragments, which explains their unexceptional reactivity in cycloaddition reactions.[61]

$$\xrightarrow[\leftarrow]{\text{LiN(CH}_3)_2} \quad \Delta G = -1.1 \text{ kcal/mol} \qquad (5\text{-}6)$$

It is worthwhile to briefly dwell on cyanocyclopropane, since there is very good agreement between calculational and experimental studies on its structure and it lacks the conformational complexity of other derivatives discussed earlier. Furthermore, discussion of its ΔH_f will be provided later in this chapter. Two microwave studies[62,63] are in excellent agreement and yield a distal bond length of 1.500 Å and vicinal bond lengths of 1.529 Å, numbers that indicate significant conjugation. The calculational study,[59] in excellent agreement, indicates a net flow of 0.18 e^- from ring to substituent in contrast to only 0.06 e^- for the vinyl substituent. This is consistent with CN (σ_R = 0.08) being a much better π withdrawer than vinyl (σ_R = -0.15), although it is not as good as many other substituents including —COR (σ_R = 0.18) and nitro (σ_R = 0.78). The ring perturbations are very slightly greater than those in vinylacetylene (σ_R, HCC—, -0.04).[63] Harmony and co-workers[63] compared π-electron affinities using σ_p^- rather than σ_R. A particularly striking conclusion from their work is that the cyclopropyl moiety is a better π donor than vinyl, at least for the CN and HCC substituents.[63] The basis for this conclusion is the comparison of bond lengths of, for example, C_2H_5—CN (1.459 Å), C_2H_3—CN (1.426–1.438 Å), and cyclo-C_3H_5—CN (1.420 Å).[63] This conclusion is contrary to Allen's conclusion,[38] noted earlier, in which cyclopropyl is considered to be only 71% as effective a π donor as vinyl. De Meijere[12] also notes that "... (the) ability of the cyclopropyl group to stabilize a center of positive charge is comparable with, or even greater than, that of the vinyl and phenyl group." This question is reexamined later in the context of stabilization energies. Harmony and Staley and co-workers also obtained the structure of cyclopropane isonitrile.[64]

It is reemphasized that cyclopropyl is a strong π donor. This is due to the importance of overlap of the 3E′ orbital with the vacant p or π* orbital of a conjugating substituent. The contribution of this interaction makes the ring geometries of cyclopropanes substituted by π acceptors readily understandable.

Cyclopropanes substituted by π-donor substituents have adopted structures not readily explicable in terms of Hoffmann's theory. Specifically, Hoffmann and co-workers[33–35] considered the main interaction to be one involving the A_2' Walsh-type antibonding orbital in cyclopropane. The interaction, depicted for the bisected conformation of CH_2^- in Figure 5-3*a*, would add electron density into the antibonding orbital and increase all three bond lengths. Durmaz and Kollmar[26] subsequently noted that this interaction is dominant only for substituents such as CH_2^- and O^-, which have very-high-energy occupied orbitals. Figures 5-3*b* and 5-3*c* show calculated (DZ) geometries for these two ions that clearly follow Hoffmann's predictions. Furthermore, these authors[26] observe shortened cyclo-C_3H_5—X bond lengths relative to CH_3—X for X = CH_2^- and O^- but not for other substituents. Aside from these two ionic substituents and similar species, simple explanations of molecular structure are elusive. Penn and Boggs[39] had noted the lengthening of distal bonds with substitution of π donors, but they lacked data on vicinal bonds. The structure of 1,1-difluorocyclopropane **(5)** is certainly inconsistent with Hoffmann's prediction. However, fluorine's electronic effects are overwhelmingly through the σ framework, and it appears to be a very poor and perhaps negligible π donor in this system.[32,36] Although structural data are lacking for fluorocyclopropane, a DZ calculational study indicates distortions similar to but smaller in magnitude than those in 1,1-difluorocyclopropane.[32]

Part of the difficulty in interpreting the effects of π donors has to do with inconsistencies in reported experimental data, as well as calculational data. For example, finding the precise structure of 1,1-dichlorocyclopropane has been a particularly elusive problem.[64,65] Initially, a microwave study[66] indicated long bonds (vicinal, 1.532 Å; distal, 1.534 Å) of virtually equal length, whereas an electron diffraction study[67] appeared to assume equal (and normal) cyclopropane bond lengths similar to those reported for chlorocyclopropane.[68] However, an NMR study[69] provided evidence for considerable ring dissymmetry (vicinal, 1.480 Å; distal, 1.544 Å), and this was strongly supported by small basis set DZ calculations.[70] The most recent structural determination of this molecule, by electron diffraction,[71] provides vicinal (1.495 Å) and distal (1.536 Å) bond lengths in good qualitative agreement with the NMR study and also in good agreement with 4–21G calculations.[72] The interpretation of the calculational results as well as the early study of 1,1-difluorocyclopropane indicates counterintuitively that π donation exists in the dichloro but not in the difluoro compound. However, π donation in 1,1-dichlorocyclopropane is said not to affect ring structure.[71]

Aminocyclopropane **(13)** is another interesting molecule that may well merit structural reinvestigation. Its conformation is one in which the amino group is pyramidal and the HNH angle is bisected by the molecule's plane of symmetry.[26,36,73] An early study[75,76] was corrected,[74] but it is apparent that there are significant discrepancies associated with the calculated structures.[26,36,73] Actually, two experimental approaches[39,74] and two of the calcu-

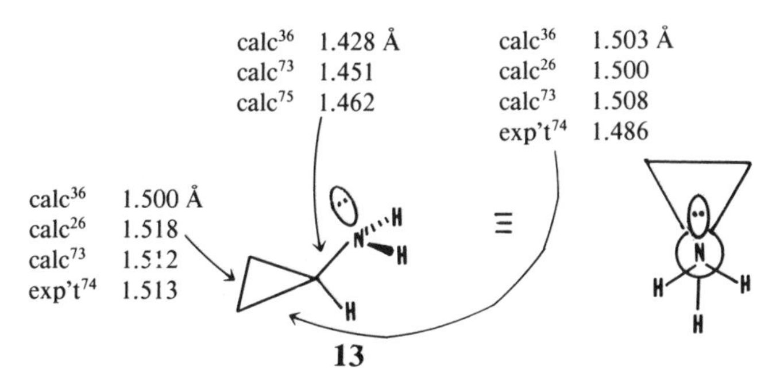

lational studies[26,73] agree quite well on the distal bond length. However, the discrepancy between calculated and experimental vicinal bond lengths is quite large. Although the stable conformation of aminocyclopropane can be described as a perpendicular conformation having a pyramidal amino group, the possibility for π conjugation still remains. The mechanism for conjugation in **13** entails overlap between the substituent lone pair orbital and the vacant 4E′ of cyclopropane said to be the most effective π-accepting MO[36] (see Figure 5-4).[36,76] Dominant π donation into this orbital would increase vicinal and decrease distal bond lengths.[26] This is not consistent with either the experimental or calculational results.

It appears, then, that conjugation effects in aminocyclopropane are small. This is consistent with the calculated charge on nitrogen (-0.70),[73] as well as the fact that the vertical ionization potential (IP_v) of aminocyclopropane (9.41 eV) is very similar to that of 2-aminopropane (9.31 eV).[77a] Both cases correspond to ionization of lone pair electrons and the similarity in band shapes is evidence for pyramidal amino groups in both molecules. This can be contrasted with photoelectron spectroscopy (PES) results for vinylamine, where the first vertical IP is 8.65 eV and the second π combination is at 11.90 eV; the 3.25-eV split indicates an extensive conjugative interaction.[77b] It is interesting that the tropylidene and norcaradiene isomers **14** and **15** are essentially equienergetic.[78]

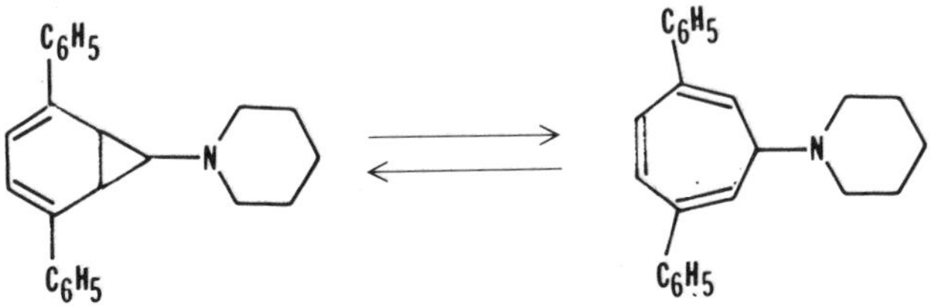

C_6H_5 C_6H_5

N   ⇌   N   (5-7)

C_6H_5 C_6H_5

14   **15**

In his summary of the effects of π-donor substituents on the geometries of cyclopropanes, F. H. Allen[37,38] notes a relative dearth of data compared to the π-acceptor derivatives. At the time of writing of that work, aminocyclopropane appeared to be an exception to the structural trend for these molecules, but the subsequent microwave investigation[74] corrected this.

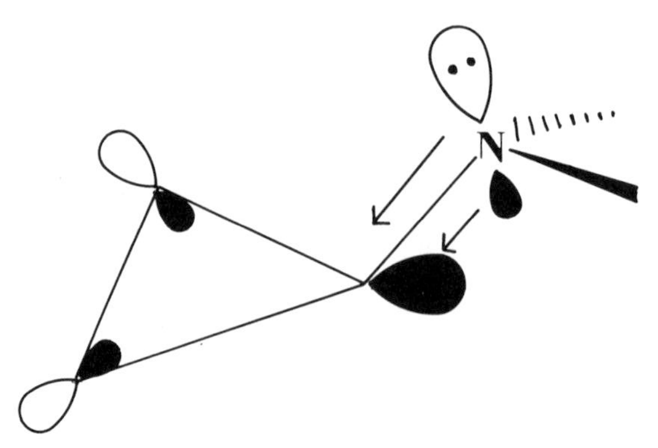

Figure 5-4. Possible conjugation mechanism in aminocyclopropane.

Allen concluded that the effects in cyclopropanes substituted by π donors were opposite those for the π acceptors and could be summarized by structure **16.** However, the experimental results for **5** and **13** do not appear to support the simple relationship depicted by **16.** A further discomforting

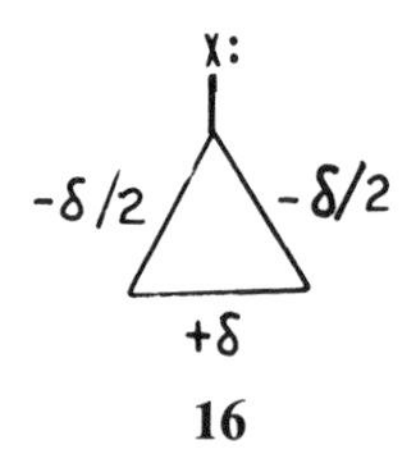

aspect of this analysis involves the predicted role of the phenyl substituent. Allen[37] noted that phenyl substituents adopting the bisected conformation tend to be associated with long vicinal and short distal bonds. Thus, by analogy to substituents such as COR, bisecting phenyls are considered to be π acceptors. However, the equality of ring bond lengths in *trans*-2,3-diphenyl-1,1-dibromocyclopropane and *trans*-2,3-bis(*p*-nitrophenyl)-1,1-dibromocyclopropane[79] does not appear to support this view, since the latter compound's nitrophenyl substituents should be considerably more effective π acceptors. Allen[37] noted further that most phenyl substituents tend to assume perpendicular conformations that are also associated with vicinal bond shortening and distal bond lengthening. This was explained by assuming that phenyl acts as a π donor through interaction with the appropriate cyclopropane 4E′ orbital. The prevalence of phenylcyclopropanes having perpendicular conformations[37] would even seem to suggest that the phenyl donor cyclopropyl acceptor interaction is more stabilizing than the phenyl acceptor–cyclopropyl donor interaction. In light of the very weak conjugation in aminocyclopropane and the great disparity between cyclopropane's ability to donate or accept π electrons, we are skeptical about this explanation. The role of steric effects in determining structures of phenylcyclopropanes must not be underestimated. Thus, the conformations in *gem*-diphenyl derivatives are the result of nonbonded repulsion, not an inherent electronic stabilization.[79,80] A study of the photoelectron spectra of phenylcyclopropanes also indicated little conjugation.[81]

The explanations of the effects on ring geometries produced by π-donor substituents that are all σ acceptors, aside from calculational subjects such as O^- and CH_2^-, have been subtle and not altogether convincing. Thus, the initial prediction by Hoffmann and co-workers of long vicinal and distal bonds, arising from π donation to the cyclopropane $1A_2'$ orbital, appears to be valid only for $X = CH_2^-$, O^-, and similar substituents. Similarly, the view that electronegative substituents such as Cl and Br simply withdraw electron density from bonding orbitals, thus weakening bonding and increasing bond length,[80] is also not supported. For example, the bond lengths in *syn*-1,2,3-trifluorocyclopropane **(17)**[82] and hexafluorocyclopropane **(18)**[83] are very similar to the cyclopropane bond length (1.510 Å). One would have expected significant bond lengthening on the basis of simple electron withdrawal from bonding orbitals and the absence of significant conjugative effects.[36]

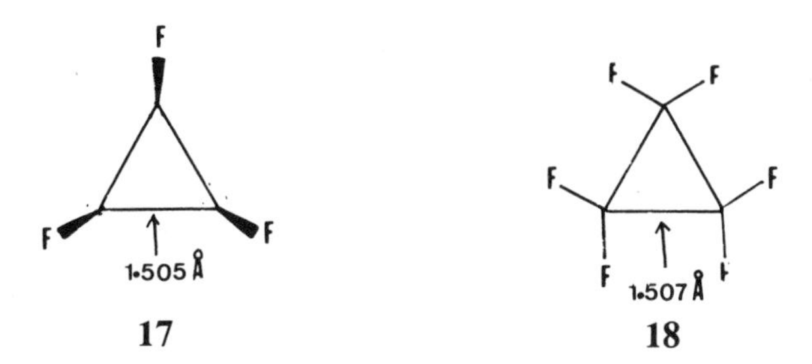

17 **18**

One way to explain the asymmetry in bond lengths in these π-donor-substituted cyclopropanes is to invoke hybridization effects. Although Durmaz and Kollmar[26] seem to attempt to separate the σ acceptor capability (electronegativity) properties of a substituent from effects on hybridization of the attached carbon, it seems impossible to do this in light of Bent's rules.[84] The bond dissymmetry in cyclopropanone **(19)**[85] could simply be explained in terms of sp^2 hybridization, which shortens C_1C_2 (C_1C_3). Furthermore, sp^2

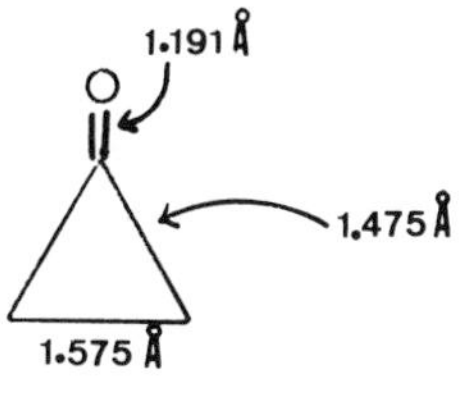

1.191 Å
O
1.475 Å
1.575 Å

19

hybridization tends to widen the $C_2C_1C_3$ angle, which helps explain the long distal bond in this molecule. Bernett[86] noted the unusual FCF angles in geminal difluoro compounds and postulated an enforced sp^3 hybridization that would tend to increase the s character of the carbon hybrid orbitals involved in ring bonding, thus shortening the vicinal bonds, expanding the $C_2C_1C_3$ angle, and presumably lengthening C_2C_3 (see **5**). Specific orbital interactions have been analyzed by L.C. Allen and co-workers.[87] They note changes in electron density primarily through σ interaction of the substituent with the cyclopropane ring, which tends to override the π-type overlap with the Walsh orbitals. However, the significance of π overlap is said to be the origin of the abnormally short C=O distance in this molecule.[87a]

Based on their calculational result that protonation of aminocyclopropane and hydroxycyclopropane does not significantly alter ring dimensions (see **13, 20–22**), Durmaz and Kollmar[26] claim that substituent electronegativity

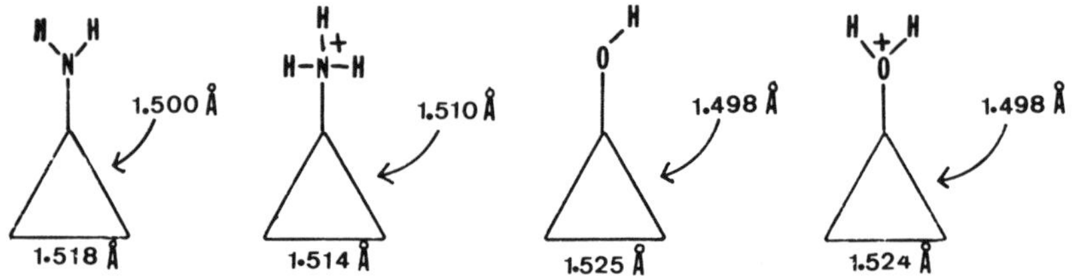

1.500 Å 1.518 Å **13** | 1.510 Å 1.514 Å **20** | 1.498 Å 1.525 Å **21** | 1.498 Å 1.524 Å **22**

has little effect on ring structure. However, they make a case that different substituents change local hybridization at the carbon to which they are attached. They show significant differences in bond angle distortion energies for each different substituent. Thus, they note that *gem*-difluoro substitution tends to widen the $C_2C_1C_3$ angle and that the associated ring strain increase could be minimized by shortened vicinal and lengthened distal bonds. However, Bent's rules predict that since NH_3^+ and OH_2^+ are more electronegative than NH_2 and OH, the s character in the carbon ring orbitals should increase, the vicinal bonds should be shortened, and the distal bonds should be lengthened in **20** and **22** relative to **13** and **21**, respectively.

Clark, Schleyer, and co-workers[36] have made a most interesting contribution to the discussion of substituent effects on cyclopropane structures. Figure 5-5 reproduces their plot of vicinal bond lengths in substituted cyclopropanes with the C—C bond lengths in the corresponding 2-propyl derivatives. There is a striking correspondence indicating similar substituent effects in cyclopropyl and isopropyl derivatives. They argue that the structures of the π-donor-substituted cyclopropanes normally encounterd are determined by inductive/electronegativity effects virtually identical to those in the unstrained isopropyl analogues. They indicate that substituent interactions with the appropriate $1E''$ molecular orbital in cyclopropane and the analogous $2B_1$ molecular orbital in propane are most important for these substituents. They note that these molecular orbitals have the largest coef-

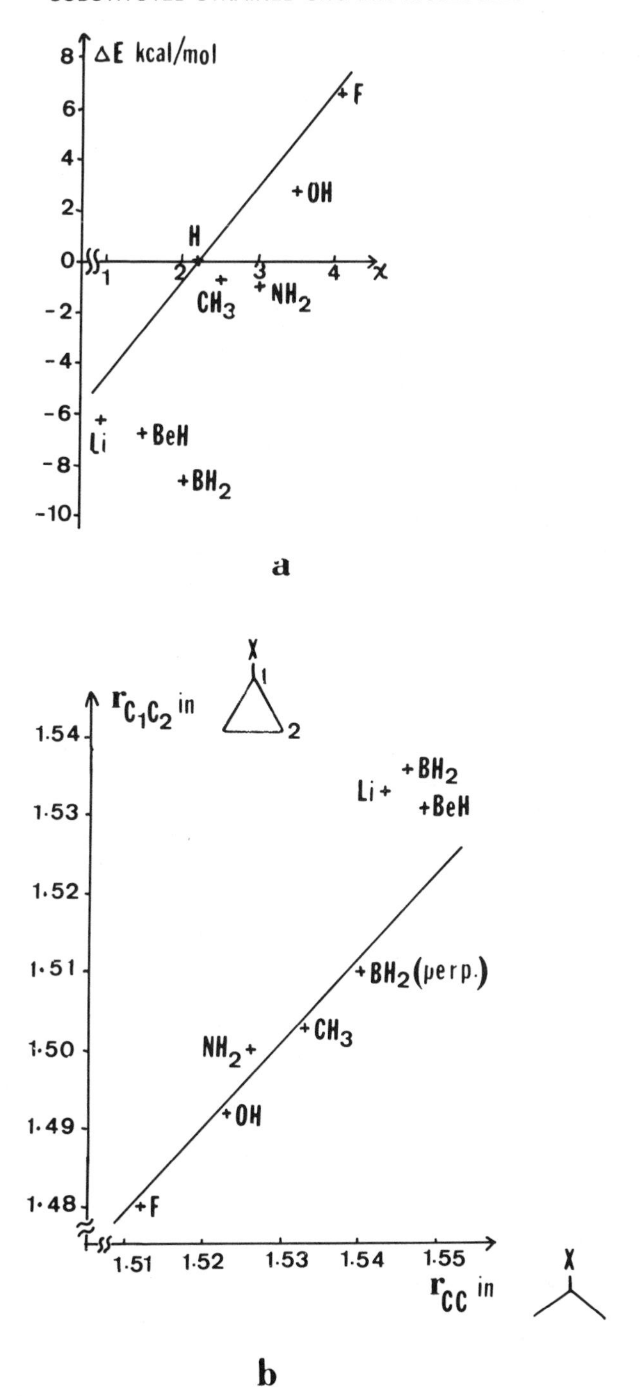

Figure 5-5. Relationship between (*a*) analogous bond lengths in cyclopropyl and isopropyl derivatives and (*b*) isopropyl stabilization energy of substituted cyclopropanes with electronegativity. (From Reference 36).

ficients of any MOs at the substituent-bearing carbons (see Figure 5-1). Departures from the line in Figure 5-5 (eg, BH_2, bisected), imply significant π interaction with the ring. According to this figure, fluorine has virtually no conjugative interaction and that of the amino group is small, consistent with the earlier discussion of aminocyclopropane. It is somewhat discomfiting that the π interaction of lithium appears to be about equal to that of bisected BH_2. This would suggest considerable covalency in the C—Li bond, and is inconsistent with the view of Clark, Schleyer and co-workers[36] that lithiocarbons closely mimic the corresponding carbanions.

Another point made by this group of workers is the very strong relationship between the structures of cyclopropyl versus 2-propyl derivatives (Figure 5-5*a*), which is similar to the relationship between isodesmic stabilization energies (Figure 5-5*b*). This type of relationship is the essence of the present series, "Molecular Structures and Energetics," and stabilization energies are considered in the next section of this chapter.

Before concluding our discussion of cyclopropane structures, a few simple observations are worthwhile. First, except for the predicted structures for cyclopropoxide and cyclopropylcarbinyl anion, all substituted cyclopropanes reported have either shortened vicinal bonds and lengthened distal bonds or vice versa. This is consistent with the relative frequencies of one of the E′ ring deformation modes (only one drawn) and the symmetric A_1'

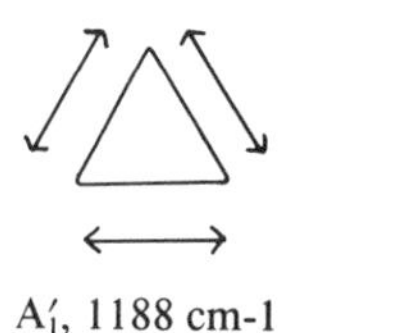

A_1', 1188 cm-1

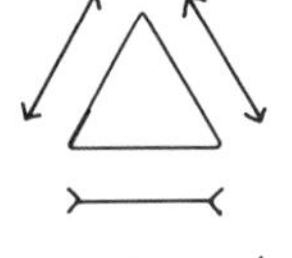

E′, 866 cm^{-1}

ring breathing mode.[88–91] Thus, the predominant ring distortion mode corresponds to the one that is energetically least demanding for the parent molecule. It is also important to point out that distortions in cyclopropanes tend to be much greater than in olefins. Thus, whereas the value for $+\delta$ (see **16**) in 1,1-difluorocyclopropane **(5)** is about 0.046 Å, the difference between comparable C=C bond lengths in ethylene (1.337 Å) and 1,1-difluoroethylene (1.316 Å) is only 0.021 Å.[92] In general, it has been noted that despite the anticipated resonance contributions of canonical structures such as **23** and **24,** the changes in the C=C lengths in substituted olefins are quite small.[93] The experimental C=C bond length in nitroethylene is less than 0.015 Å smaller than in ethylene (calculated difference 0.05 Å),[93] while calculations indicate that this bond is about 0.007 Å longer in aminoethylene.[93] Even 2-nitro-1-aminoethylene has a C=C bond length calculated to be only 0.021 Å longer than that in ethylene.[93] This resistance to C=C distortion reflects the higher frequency for the corresponding bond stretch (1623 cm^{-1})[91] compared to the ring deformation frequencies in cyclopropane.

Similar remarks can be made concerning substituted acetylenes. The

C≡C bond length change upon substitution of CH_3O for H is only about 0.001 Å, although F substitution appears to decrease bond length by about 0.01 Å.[94] Generally, there appear to be smaller deviations in alkynes compared to alkenes, reflecting the higher CC frequency (1974 cm^{-1})[91] in the former class.

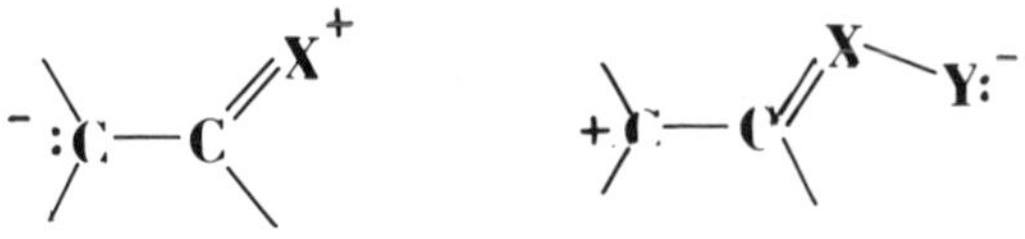

23 **24**

Also commentworthy is the statement that " . . . although many effects may be involved in bond-breaking reactions, it appears to be a good general rule that the longest ring bond is most easily cleaved, so that studies on the cyclopropanes themselves are useful in predicting reactivity."[36] We will briefly consider the effects of substitution on the mode of hydrogenolysis.[29] The reduction of 1,1-dimethoxycyclopropane (Scheme II) is consistent with the foregoing statement by Clark and co-workers. Since there is no reason to assume that this molecule does not follow the normal pattern for π donors, it is clear that the longer, weaker distal bond is broken. However, it is also apparent that the thermodynamically most stable isomer is formed. If one employs $\Delta H_f^\circ(g)$ data[95] for 2,2-dimethoxypropane and 1,1-dimethoxyethane and assumes that the enthalpy increment between the latter and 1,1-dimethoxypropane is the same as between methoxyethane and 1-methoxypropane, then the 2,2-isomer is 3.9 kcal/mol more stable than the 1,1-isomer. An even larger enthalpy difference can be obtained from published enthalpies of hydrolysis of 2,2-dimethoxypropane and 1,1-dimethoxypropane[96] if one assumes that differential enthalpies of solution and vaporization are small. In observing the course of hydrogenation of 1,1-dimethylcyclopropane (Scheme III) it is interesting to remark that the only reported measure of bond lengths (an STO-3G calculation)[36] indicates very slightly longer vicinal than distal bonds. Even if we assume that the two are of equal length, no obvious structural prediction concerning the regioselectivity of

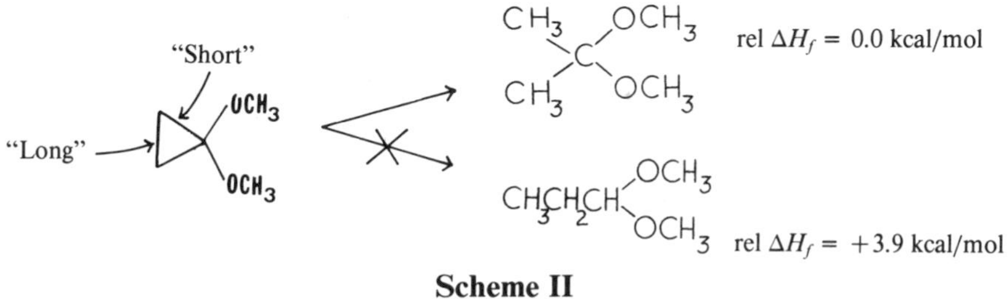

"Short"
"Long"
OCH_3
OCH_3
$(CH_3)_2C(OCH_3)_2$ rel ΔH_f = 0.0 kcal/mol
$CH_3CH_2CH(OCH_3)_2$ rel ΔH_f = +3.9 kcal/mol

Scheme II

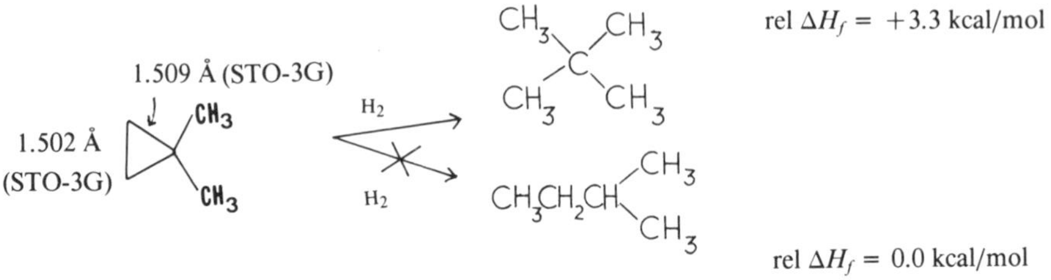

1.509 Å (STO-3G)

1.502 Å (STO-3G)

H_2 → $(CH_3)_4C$ rel ΔH_f = +3.3 kcal/mol

H_2 ↛ $CH_3CH_2CH(CH_3)_2$ rel ΔH_f = 0.0 kcal/mol

Scheme III

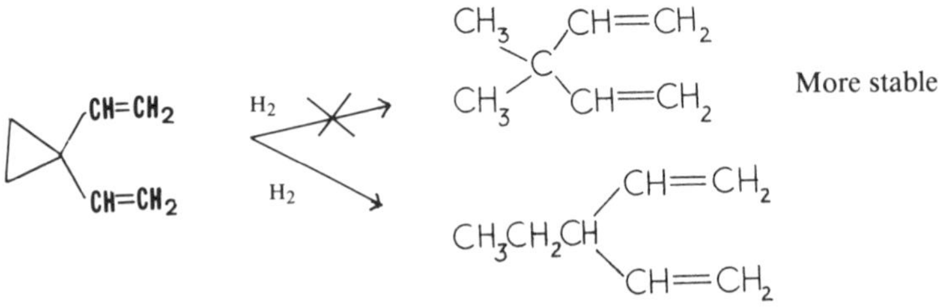

H_2 ↛ $(CH_3)_2C(CH{=}CH_2)_2$ More stable

H_2 → $CH_3CH_2CH(CH{=}CH_2)_2$

Scheme IV

hydrogenation is apparent. The course of hydrogenation[29] again clearly favors the most stable isomer. It is unfortunate that few data exist[95] allowing comparison of the stabilities of 1,1- and 2,2-disubstituted propanes. Liquid phase ΔH_f values indicate that 2,2-dinitropropane is 3.3 kcal/mol more stable than the 1,1-isomer.[95] Since the boiling points differ by only 1.5°C, it appears reasonable to assume that the 2,2-isomer is more stable in the gas phase as well. By analogy to the results for other cyclopropanes substituted by conjugating substituents at C_1, one would expect to reduce the vicinal bond to generate the contrathermodynamic product 1,1-dinitropropane. Similarly, one would anticipate 2,2-dimethyl-1,4-pentadiene to be more stable than 2-ethyl-1,4-pentadiene by an amount similar to the difference between neopentane and isopentane. The product (see Scheme IV) is clearly contrathermodynamic. A very interesting case could be 1,1-dichlorocyclopropane, where cleavage of the long distal bond would produce the contrathermodynamic product (Scheme V). It is assumed that $\Delta H_f^\circ(l)$[97] for the isomeric dichloropropanes is similar to $\Delta H_f^\circ(g)$, since the boiling points differ by only 19°C. The view of these hydrogenations is that they are kinetically

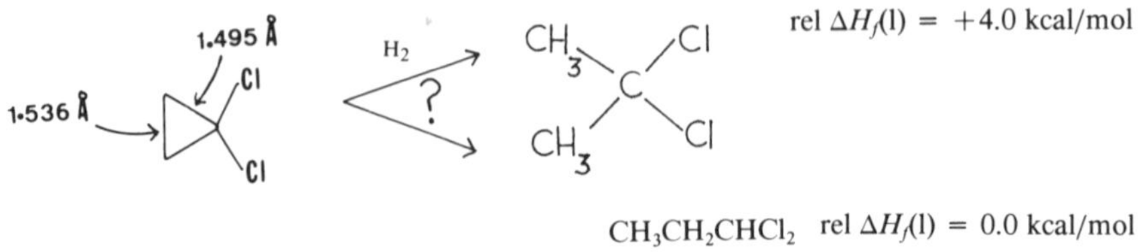

1.495 Å

1.536 Å

H_2 → ? $(CH_3)_2CCl_2$ rel $\Delta H_f(l)$ = +4.0 kcal/mol

$CH_3CH_2CHCl_2$ rel $\Delta H_f(l)$ = 0.0 kcal/mol

Scheme V

controlled,[29c] and we see that these products are often the thermodynamic ones as well.

3. SUBSTITUTED CYCLOPROPANES: STABILIZATION ENERGIES

There is only a small amount of published $\Delta H_f^\circ(g)$ data for substituted cyclopropanes.[98,99] Fuchs and co-workers[99] redetermined ΔH_v° values for phenylcyclopropane and cyanocyclopropane and employed these numbers in combination with heats of the isodesmic reaction 5-8 (Scheme VI) to obtain stabilization enthalpies. The values indicated that, of the four substituted compounds investigated, only phenylcyclopropane shows measurable stabilization (~2 kcal/mol, after correction for steric repulsion in isopropylbenzene). This is an interesting conclusion when one considers the earlier noted observations in which (a) photoelectron spectroscopic data indicated little conjugation between cyclopropyl and phenyl groups, (b) the conformational freedom of phenyl attached to cyclopropyl appeared to favor perpendicular (nonconjugated) conformations, and (c) Ibers's conclusions[79,80] concerning the similar effects of phenyl and *p*-nitrophenyl substituents on ring structure. It is also interesting that the cyano substituent produces significant changes on the geometry of the ring as well as the ring–substituent distance. Thus, conjugative interactions appear to be greater when Y = CN than when Y = phenyl. Perhaps this dichotomy will be better understood when electronegativity and inductive effects are considered along with resonance effects. The amino group would appear to very slightly destabilize a cyclopropane ring according to the data in Scheme VI. The case was made earlier that there is negligible conjugation between the cyclopropane ring and the amino group. Thus, inductive and hybridization effects may play the deciding role in these molecules. However, one must never lose sight of how small these enthalpy differences are compared to the magnitude of enthalpy of combustion measurements (see the discussion in Chapter 4 of Volume 2 in this series). Although Fuchs and co-workers[99] redetermined ΔH_v°, they accepted the published heat of combustion measurements, which are a source of larger potential errors.

$$c\text{-}C_3H_4XY \xrightarrow{\Delta H_r =} 2C_3H_8 + (CH_3)_2CXY \qquad (5\text{-}8)$$

X	Y	ΔH_r (kcal/mol)
H	H	−27.4
H	CN	−27.5
CH_3	CH_3	−27.8
H	NH_2	−28.2
H	C_6H_5	−24.8

Scheme VI

M	N	X	Y	ΔH_r (kcal/mol)	Ref.
H	H	CH_3	H	−0.5	100
H	H	CH_3	CH_3	−1.2	101
H	H	F	H	−2.6	102, 103
H	H	F	F	−1.9	103
H	H	OC_2H_5	H	−2.6 (est)	104
CH_3	CH_3	CH_3	CH_3	−5.2	101

Scheme VII

It is sometimes instructive to employ the ethylene framework for comparison of the stabilization energies of cyclopropanes. One example of such an appealingly simple comparison is provided by the equilibration of methylenecyclopropanes (Scheme VII). Clearly, the substituents examined above prefer vinyl to cyclopropyl sites. Methyl substituent effects appear to be additive, whereas the fluorine substituent effects are not. A related issue is the tendency to favor vinyl substitution in bullvalene isomers substituted

25 (A = Cl; B = Br) **26**

by Cl and Br **(25)**, while fluorine is found at the alkyl position **(26)**.[105] Hine[106] noted the importance of interaction terms that render such comparisons difficult to interpret simply.

Hine[106] has defined a double-bond stabilization parameter (termed D by Hine, but we shall employ "DBSP" here) according to equilibrium equation 5-9. This "reaction" does not really occur, but the DBSP values are actually obtained from equilibria of the type represented by Equation 5-10, where a statistical fit allows

$$CH_2{=}CHX + CH_2{=}CHCH_3 \rightarrow CH_2{=}CH_2 + CH_2{=}CHCH_2X \qquad (5\text{-}9)$$

$$XCH_2(H)C{=}C(H)Y \rightleftharpoons X(H)C{=}C(H)CH_2Y \qquad (5\text{-}10)$$

one to obtain interaction terms (T_V) that must be subtracted from the experimental ΔG.[106] Entropy changes and differences in solvation are assumed to be negligible. In the present work, a cyclopropyl stabilization parameter

(CSP) is defined according to Equation 5-11, which is analogous to Equation 5-9.

$$\text{c-C}_3\text{H}_5\text{–X} + \text{c-C}_3\text{H}_5\text{–CH}_3 \longrightarrow \text{c-C}_3\text{H}_6 + \text{c-C}_3\text{H}_5\text{–CH}_2\text{X} \quad (5\text{-}11)$$

Direct values for enthalpy changes in Equations 5-9 and 5-11 can be obtained through use of $\Delta H_f^\circ(g)$[95] for the four molecular components of each equation. Similarly, ab initio MO-calculated total energies may be employed to obtain total energies of reaction for Equations 5-9 and 5-11. For these purposes and especially for comparisons made later in this chapter, $\Delta H_f^\circ(g)$ and total energy (4–31G) data for vinyl and cyclopropyl compounds as well as ethynyl, phenyl, methyl, ethyl, and isopropyl analogues have been gathered into the appendix. These values are, at this writing, the best available at the 4–31G level and experimentally. Values for $\Delta H_f^\circ(g)$ and total energy (4–31G) for $CH_2{=}CHCH_2X$ are, by and large, absent and have been estimated using Equation 5-12. Similarly, corresponding values for cyclopropyl-CH_2X have been estimated using Equation 5-13. It is fair to be skeptical about very finely tuned quantitative interpretations of theoretical data, especially when the magnitude of stabilization energies for these neutral substituents typically is around 0–3 kcal/mol. However, use of isodesmic[107] and homodesmotic equations [108] (see Chapter 6 in Volume 4 of this series) tends to cancel errors arising from basis set inadequacies, lack of electron correlation, zero point energy (zpe) corrections, and so on. The 4–31G basis set has been shown to give good agreement with experiment,[108] even in some instances where the minimal basis set (STO-3G) fails badly. For example, the STO-3G basis set predicts a heat of hydrogenation (relative to that of cyclopropane) for 1,1-difluoroethylene that is too positive by 18.6 kcal/mol, whereas the 4–31G value is too positive by 1.1 kcal/mol.[109] Later in this section we will evaluate the agreement between 4–31G-calculated stabilization energies and data obtained from $\Delta H_f^\circ(g)$. Although it may seem presumptuous to state this, one should note that interpretation of 0–3 kcal/mol effects also places one close to the threshold of experimental noise in combustion calorimetry. Enthalpies of reaction involve much smaller numbers and can be more finely interpreted (see Chapter 4, Volume 2). The 4–31G basis set gave good agreement with the experimental isomerization noted in Scheme VI for X = Y = F.[109] Table 5-2 lists DBSP

$$\text{est}(\Delta H_f \text{ or } \Delta E_{4\text{-}31G})CH_2{=}CHCH_2X = CH_2{=}CHCH_3 + C_2H_5X - C_2H_6 \quad (5\text{-}12)$$

$$\text{c-C}_3\text{H}_5\text{–CH}_2\text{X} = \text{c-C}_3\text{H}_5\text{–CH}_3 + C_2H_5X - C_2H_6 \quad (5\text{-}13)$$

values using Hine's data[106] as well as calculational (4–31G) total energies and $\Delta H_f^\circ(g)$ data from the appendix. It is clear that CH_3O is about twice as strong a vinyl stabilizing substituent as methyl, whereas nitro and cyano are

TABLE 5-2. Double-Bond Stabilization Parameters (DBSP), Cyclopropyl Stabilization Parameters (CSP), and Isomerization Energies (IE)[a]

	DBSP (kcal/mol) according to Equation 5-9			CSP (kcal/mol) according to Equation 5-11		IE (kcal/mol) according to Equation 5-14
Substituent	From equilibration data[b]	From 4–31G total energies	From $\Delta H_f^\circ(g)$ data	From 4–31G data	From $\Delta H_f^\circ(g)$ data (see text)	From 4–31G data
H	0[c]	0[c]	0[c]	0[c]	0[c]	−7.9[d]
CH_3	3.2	3.2	2.8	1.6	1.6	−6.3
F	3.3	−0.01	3.3	−1.1	—	−9.0
CH_3O	5.2	6.1	6.9	0.5	—	−7.4
OH	—	6.5	6.3	1.9	—	−6.0
C_6H_5	—					
$CH{=}CH_2$	—	8.0	6.2	2.9	(−3.2)[e]	−5.0
CCH	—	3.3	3.1	1.0	—	−6.9
CO_2CH_3	3.2	2.4	7.1	—	—	−2.9
NO_2	2.9	1.1	−0.8	0.8	—	−7.1
CH_2OCH_3	2.6	—	—	—	—	—
$CH_2CO_2CH_3$	2.1	—	—	—	—	—
NH_2	—	11.2	—	4.4	3.0	−3.5

CN	2.3	2.3	1.7	1.8	1.9	−6.1
NC	—	2.8	—	0.5	—	−7.5
CHO	—	4.5	4.0	—	—	—
$COCH_3$	—	2.4	7.1	5.0	3.6	−2.9
CF_3	—	−2.5	—	—	—	—
SCH_3	3.2	—	—	—	—	—
Cl	1.8	—	—	—	—	—
$SOCH_3$	0.7	—	—	—	—	—
Br	0.3	—	—	—	—	—
SO_2CH_3	−0.4	—	—	—	—	—
NH_3^+	—	−3.9	−1.6	1.0	1.0	−7.0
O^-	—	35.3	22.4	8.7	—	0.8
CH_2^-	—	35.0	—	—	—	—
CH_2^+	—	23.3	—	27.1	—	19.2

[a]Note that IE = CSP − 7.9 kcal/mol (ie, IE and CSP are redundant).
[b]Reference 106.
[c]By definition.
[d]Add 5.4 kcal/mol to 4–31G isomerization energy to yield experimental value (−7.9) for X = H.
[e]Error in $\Delta H_f^\circ(l)$ of vinylcyclopropane suspected.

weaker stabilizers. Discrepancies with ΔH_f°(g)-derived data are serious for X = CO_2CH_3 and $COCH_3$, and it would appear that additional combustion calorimetry should be performed for these compounds. The calculational results for X = F do not appear to correctly predict the significant stabilizations experimentally observed. Strong stabilizations are predicted for X = O^-, CH_2^-, and CH_2^+ among charged substituents and X = NH_2 for neutrals. Significant destabilizations are calculated for σ-withdrawing substituents such as NH_3^+ and CF_3 in line with Hine's views.[106]

Table 5-2 also lists CSP according to Equation 5-11. With the exception of X = $COCH_3$, cyclopropyl stabilizations (CSP) are smaller than vinyl stabilizations (DBSP). It is interesting that X = NH_3^+ is a very slight stabilizer in contrast to its destabilizing effect when attached to vinyl. This may be understood in terms of the higher s-character orbital in ethylene compared to cyclopropane, thus accentuating destabilization induced by electronegative groups. The larger cyclopropyl group also is more polarizable than vinyl, providing another explanation of its enhanced stability in this case. Another interesting comparison is furnished by reaction 5-14, which represents isomerization to the 3-substituted propene (allyl) isomer, which is less stable than the 1- and 2-substituted isomers. This provides a direct comparison between the cyclopropyl and a primary position. Results for Equation 5-14 are also included in Table 5-2. The 4–31G basis set overestimates the isomerization energy by 5.4 kcal/mol. Table 5-2 corrects all isomerization energies by this amount. Only where X substantially stabilizes the ring and possibly destabilizes the primary position are the cyclopropyl isomers predicted to be more stable than the 1-allyl isomers.

$$c\text{-}C_3H_5X \longrightarrow CH_2{=}CH{-}CH_2X \qquad (5\text{-}14)$$

There have been few published studies on systematic investigations of the effects of substituents on stabilization energies of cyclopropanes substituted by neutral groups. Our earlier work on this subject, using the "first-row element sweep" pioneered by Schleyer and co-workers,[110] was based on STO-3G data and idealized geometries and was necessarily qualitative.[98] The conclusion that cyclopropanes act as mitigated ethylenes is consistent with the data presented above. The recent work of Clark, Schleyer and co-workers[36] indicates that π-donor substituents such as NH_2 produce essentially the same effects in isopropyl derivatives that they do in cyclopropyl derivatives. That is to say, for these species no special olefinlike stabilization effects are said to occur. This is graphically depicted in Figure 5-5*a* where significant deviations from the line occur for strong π-acceptor substituents such as BH_2, thus implying olefinlike conjugative behavior. These conclusions are, of course, consistent with the earlier noted views of cyclopropane as an exceedingly weak π acceptor and a strong π donor. However, that is not the entire story. If we accept Allen's estimate of $sp^{2.2}$ character for the exo-ring carbon orbitals on cyclopropane,[32] we would expect inductive destabilization by electronegative substituents, as Hine noted[106] occurs for olefins.

To obtain a quantitative feeling for substituent effects on cyclopropanes and closely related molecules, we have employed the techniques of correlation analysis. The Hammett equation and its modifications consider the interaction between a substituent X and reaction center Y through a carbon framework X–G–Y. The substituent constants are not meant explicitly for the X–G system. However, we will employ such substituent constants, specifically σ_R and σ_I, those utilized in the Taft dual-substituent parameter (DSP) equation.[111] It will become clear that useful information can be obtained by this approach. Moreover, a series of calculational substituent constants have been developed by Topsom:[112–114] σ_X, σ_F, and σ_{R°. These are, in fact, directly applicable to X–G systems (they were defined for G = vinyl). Discussion of the derivation of these constants and their meaning is the subject of Topsom's chapter in Volume 4 of this series. Another series of constants potentially useful for our analysis has been developed by Charton and is based on photoelectron spectral data.[115] Although it is not employed in the present analysis, the reader can review it in Chapter 9 also in Volume 4.

Isodesmic stabilization energies termed "methyl" and "ethyl stabilization energies" are calculated from reactions 5-15 and 5-16 respectively. Equation 5-17 is an isodesmic equation furnishing the "isopropyl stabilization energy." Calculated 4–31G total energies or $\Delta H_f^\circ(g)$ data may be inserted into these three equations. The identities of R examined in this work are cyclopropyl, vinyl, ethynyl, and phenyl. These data, tabulated in the appendix, are the best calculational and experimental data in hand at the writing of this chapter. Unfortunately, most of the 4–31G phenyl data are based on idealized ring geometries and therefore are the least optimized of the data sets. When comparing methyl, ethyl, and isopropyl stabilization energies, it is generally agreed that the last of these is most relevant for cyclopropyl[36,98,116] and possibly even vinyl.[98,116] However, fewer data, especially 4–31G-optimized total energies, are available for isopropyl derivatives than for methyl and ethyl compounds.

$$RX + CH_4 = CH_3X + RH \qquad (5\text{-}15)$$

$$RX + C_2H_6 = C_2H_5X + RH \qquad (5\text{-}16)$$

$$RX + C_3H_8 = (CH_3)_2CH{-}X + RH \qquad (5\text{-}17)$$

Tables 5-3 to 5-6 list calculated and experimental stabilization energies of substituted ethylenes, cyclopropanes, acetylenes, and benzenes. As

Table 5-3. Calculated and Experimental Stabilization Energies (kcal/mol) of Substituted Ethylenes

	Vinyl—X					
Stabilization energy model:	Methyl		Ethyl		Isopropyl	
X	ΔE_{stab}[a]	ΔH_{stab}[b]	ΔE_{stab}[a]	ΔH_{stab}[b]	ΔE_{stab}[a]	ΔH_{stab}[b]
H	0.0	0.0	0.0	0.0	0.0	0.0
Cl	12.8	5.7		0.8		−2.2
F	6.4	6.7	0.0	3.3	−5.1	0.6
CH_3	4.3	5.4	3.2	2.8	2.6	0.6
OMe	10.9	12.3	6.1	6.9		3.3
OH	10.6	11.9	6.6	6.3	1.9	2.2
NH_2	13.3	13.3	11.2	9.7	8.0	6.2
OAc		7.6[c]		1.8[c]		−2.3[c]
CH_2^-	39.6		35.0			
O^-	38.6	30.1	35.3	22.4	28.9	17.0
NH_3^+	1.5	4.9	−3.9	−1.6	−7.8	−7.7
CN	3.3	4.8	2.4	1.7		0.2
NC	5.8		2.8			
CHO	6.4	6.9[d]	4.5	4.0[d]		2.2[d]
COMe	3.9	10.5	2.4	7.2		6.9
CO_2Me	8.0	11.9	2.0	9.5		8.0
NO_2	4.7	3.4[e]	1.1	−0.8[e]		−4.7[e]
CF_3	−0.2	−1.7	−2.5			
Li	4.8		9.8		13.1	
BH_2	5.9		9.2		8.7	
CH_2^+	30.0	23.3	23.3	14.6		10.5
Vinyl	7.8	8.8	8.0	6.2		4.7
HCC—	3.9	5.9	3.3	3.1		1.1

[a]Calculated from 4–31G energies.
[b]From experimental heats of formation.
[c]All $\Delta H_f^\circ(g)$ values from Reference 95.
[d]If $\Delta H_f^\circ(g)$ of vi-CHO estimated by Hegedus (Hegedus, V.J.; Harrison, A.G. *Int. J. Mass Spectrom. Ion Phys.* **1979**, *30,* 293–306) were used, stabilization energy would be overestimated as compared to value listed.
[e]Used $\Delta H_f^\circ(g)$ for vi-NO_2 that was estimated here.

expected, the largest stabilizations are seen for charged substituents such as CH_2^-, O^-, and CH_2^+, which can act as strong conjugators. As mentioned previously, the NH_3^+ and CF_3 groups are seen to be destabilizers on vinyl at least for ethyl and isopropyl comparisons. In comparing the three different stabilization models, it is generally clear that they follow the following order: methyl > ethyl > isopropyl. In fact, some apparent methyl stabilizations (eg, nitroethylene) correspond to isopropyl destabilizations. This is essentially a restatement of the observation made by Clark, Schleyer, and co-workers[36] of significant methyl stabilization for isopropyl derivatives (ie,

TABLE 5-4. Calculated and Experimental Stabilization Energies (kcal/mol) of Substituted Cyclopropanes

Stabilization energy model:	Cyclopropyl—X					
	Methyl		Ethyl		Isopropyl	
X	ΔE_{stab}[a]	ΔH_{stab}[b]	ΔE_{stab}[a]	ΔH_{stab}[b]	ΔE_{stab}[a]	ΔH_{stab}[b]
H	0.0	0.0	0.0	0.0	0.0	0.0
Cl						
F[c]	5.3		−1.1		−6.2	
CH_3	2.7	4.2	1.6	1.6	1.0	−0.6
OMe	5.3		0.5			
OH	5.9		1.9		−2.7	
NH_2	6.5	6.6	4.4	3.0	1.2	−0.7
OAc						
CH_2^-						
O^-	12.0		8.7		2.3	
NH_3^+	6.3	7.5	1.0	1.0	−3.0	−5.1
CN[c]	2.8	4.5	1.8	1.3		−0.1
NC[c]	3.4		0.5			
CHO						
COMe	6.6	7.0	5.0	3.6		3.4
CO_2Me		6.1		3.9		2.2
NO_2	4.5		0.8			
CF_3						
Li	−2.1		2.9		6.2	
BH_2	6.1		9.4		8.9	
CH_2^+	33.7		27.1			
Vinyl	2.7	(−0.6)[d]	2.9	(−3.2)[d]		
HCC—	1.7		1.0			

[a]Calculated from 4-31G energies.
[b]From experimental heats of formation.
[c]Note that Hopkinson et al (Hopkinson, A.C.; McKinney, M.A.; Lien, M.H. *J. Comput. Chem.* **1983,** *4,* 513–523) found the strain energies of cyclopropane, cyclopropyl, cyclopropyl-CN, and cyclopropyl-NC at the 3-21G//3-21G level, using the isopropyl model. The corresponding stabilization energies by this model are 0, −5.1, 1.0, and −2.2 kcal/mol. Note that the stabilization for cyclopropyl-F in this work is in good agreement, and that the same order of stabilization appears for this work using the ethyl model.
[d]Did not include this value in correlations for reason explained in the appendix: we do not believe that conjugation of vinyl and cyclopropyl is destabilizing.

Equation 5-17, R = isopropyl). This is attributable to nonconjugative effects probably originating in the greater polarizability of the larger isopropyl framework. The calculated stabilizations of phenyl derivatives in Table 5-6 are clearly underestimated. This is seen by comparison with the experimental data and is clearly an artifact of the idealized phenyl structures employed. Table 5-7 lists linear regression results for comparison between theoretical and experimental stabilization energies. By and large the agree-

TABLE 5-5. Calculated and Experimental Stabilization (kcal/mol) Energies of Substituted Acetylenes

Stabilization energy model:	Ethynyl—X					
	Methyl		Ethyl		Isopropyl	
X	ΔE_{stab}[a]	ΔH_{stab}[b]	ΔE_{stab}[a]	ΔH_{stab}[b]	ΔE_{stab}[a]	ΔH_{stab}[b]
H	0.0	0.0	0.0	0.0	0.0	0.0
Cl	2.3					
F	−12.8		−19.2		−24.2	
CH_3	8.6	7.6	7.6	5.0	7.0	2.8
OMe	0.3		−4.5			
OH	0.7		−3.4		−8.0	
NH_2	11.5		9.4		6.2	
OAc						
CH_2^-	50.6		46.0			
O^-	55.0		51.7		45.3	
NH_3^+	−17.5		−22.8		−26.8	
CN	2.1		1.2			
NC	−1.2		−4.2			
CHO	1.7		−0.1			
COMe	2.0		0.5			
CO_2Me	2.3		−3.7			
NO_2	−18.6		−22.2			
CF_3	−9.1		−11.4			
Li	31.2		36.2		39.5	
BH_2						
CH_2^+	14.2	10.3	7.6	1.6		−2.5
Vinyl	8.3	8.1	8.6	5.5		4.0
HCC—	6.8	8.1	6.2	5.3		3.3

[a]Calculated from 4-31G energies.
[b]From experimental heats of formation.

ment is quite good, with the largest discrepancies seen for charged substituents having the largest magnitude effects. For the neutral substituents the discrepancy between calculational and experimental stabilization energies is usually less than 2 kcal/mol. Where occasional large discrepancies are found (eg, $CH_2{=}CHCOCH_3$ and $CH_2{=}CHCO_2CH_3$ in Table 5-3), we feel that $\Delta H_f^\circ(g)$ values are at fault.

Table 5-8 lists the substituent constants employed in the present work. Values for the Taft DSP parameters are taken from a recent compendium,[117] and values for σ_X, σ_F, and σ_{R° are taken from Topsom's work.[112–114] There is an excellent correlation between Topsom's σ_{R° and Taft's σ_R according to Equation 5-18. This allows calculation of one value from the other if one is lacking. Similarly, a plot of Taft's σ_I[117] against Topsom's σ_F[112–114] furnishes

TABLE 5-6 Calculated and Experimental Stabilization Energies (kcal/mol) of Substituted Benzenes

Stabilization energy model:	Phenyl—X					
	Methyl		Ethyl		Isopropyl	
X	ΔE_{stab}[a]	ΔH_{stab}[b]	ΔE_{stab}[a]	ΔH_{stab}[b]	ΔE_{stab}[a]	ΔH_{stab}[b]
H	0.0	0.0	0.0	0.0	0.0	0.0
Cl		5.7		0.8		−2.2
F	7.8	8.5	1.4	5.1	−3.7	2.4
CH_3	1.3	5.5	0.2	2.9	−0.4	0.7
OMe	6.2	9.8	1.4	4.4		0.8
OH	7.5	12.4	3.4	6.8	−1.2	2.7
NH_2	9.8	11.3	7.7	7.7	4.5	4.0
OAc						
CH_2^-	42.7		38.2			
O^-						
NH_3^+		8.7		2.2		−3.9
CN	1.2	3.7	0.3	0.6		−0.9
NC						
CHO	5.3	6.8	3.5	3.9		2.1
COMe		6.4		3.0		2.8
CO_2Me		8.4		6.0		
NO_2	2.5	3.5	−1.2	−0.7		−4.6
CF_3		2.0				
Li						
BH_2						
CH_2^+	43.3		36.6			
Vinyl	−1.3[c]	7.1	−1.0[c]	4.5		3.0
HCC—	3.1	7.0	2.5	4.2		2.2

[a]Calculated from 4–31G energies.
[b]From experimental heats of formation.
[c]If the energy of styrene were to be better optimized, the stabilization energy would be greater.

Equation 5-19, which can be similarly employed for calculations of missing values. It is apparent that the "inductive effect," as measured by σ_I, is identical to the field effect as measured by σ_F.

$$\sigma_{R^\circ} = 0.74\sigma_R + 0.03 \qquad n = 10;\ R^2 = 0.98 \tag{5-18}$$

$$\sigma_I = 1.00\sigma_F + 0.03 \qquad n = 19;\ R^2 = 0.94 \tag{5-19}$$

In attempting to employ isodesmic stabilization energies or enthalpies to obtain correlations with substituent constants, it is assumed that isodesmic entropies of reaction are negligible. This is true, with some readily understandable exceptions.[118] Since both π donors and π acceptors cause stabili-

TABLE 5-7. Linear Regression of 4–31G-Calculated and Experimental Stabilization Energies:

$$\Delta E_{stab} = m\,\Delta H_{stab(exp)} + b$$

where m = slope, n = number of points, b = the y-intercept, and R = correlation coefficient

Data set	Framework	n	R^2	R	Standard error	m	b
	A. Methyl Stabilization						
1	Vinyl[a]	16	0.95	0.97	2.53	1.27	−2.38
2	Vinyl[b]	13	0.92	0.96	1.19	0.83	0.41
3	Phenyl[c]	10	0.81	0.90	1.54	0.78	−0.86
4	Cyclopropyl[d]	6	0.93	0.96	0.81	0.95	−0.55
5	Ethynyl[e]	5	0.88	0.94	2.03	1.21	−0.65
6	All four[f]	34	0.91	0.95	2.34	1.26	−2.82
7	All four[g]	29	0.80	0.90	1.52	0.83	−0.12
	B. Ethyl Stabilization						
8	Vinyl[h]	16	0.85	0.92	3.91	1.48	−2.21
9	Vinyl[i]	10	0.94	0.97	0.90	1.00	0.67
10	Phenyl[j]	10	0.72	0.85	1.43	0.76	−0.75
11	Cyclopropyl[k]	6	0.98	0.99	0.34	1.49	−0.31
12	Ethynyl[e]	5	0.55	0.74	2.64	1.02	2.44
13	All four[f]	34	0.79	0.89	3.30	1.38	−1.45
14	All four[g]	30	0.46	0.68	2.28	0.74	0.54
	C. Isopropyl Stabilization						
15	Vinyl[l]	7	0.92	0.96	3.65	1.55	−0.06
16	Cyclopropyl[m]	4	0.83	0.91	0.98	0.76	1.04
17	All four[n]	15	0.82	0.90	3.80	1.42	−0.08

[a]Substituents: H, F, Me, OMe, OH, CN, NH_2, CHO, CO_2Me, NO_2, CF_3, CH_2^+, O^-, NH_3^+, vinyl, HCC— (omit Cl, COMe).
[b]Substituents: H, F, Me, OMe, OH, CN, NH_2, CHO, CO_2Me, NO_2, CF_3, vinyl, HCC— (omit Cl, COMe, charged substituents).
[c]Substituents: H, F, Me, OMe, OH, NH_2, CN, CHO, NO_2, HCC—.
[d]Substituents: H, Me, NH_2, CN, NH_3^+, COMe.
[e]Substitutents: H, Me, CH_2^+, vinyl, HCC—.
[f]Data sets 2–5 including H only once.
[g]Same as note *f*, but lacking charged substituents.
[h]Substituents: H, F, Me, OMe, OH, CN, NH_2, CHO, COMe, CO_2Me, NO_2, O^-, NH_3^+, vinyl, HCC—, CH_2^+.
[i]Same as set 8 less CO_2Me, F, COMe, and charged substituents.
[j]Substituents: H, F, Me, OMe, OH, NH_2, CN, CHO, NO_2, HCC.
[k]Substituents: H, Me, NH_2, CN, NH_3^+, COMe.
[l]Substituents: H, F, Me, OH, NH_2, O^-, NH_3^+.
[m]Substituents: H, Me, NH_2, NH_3^+.
[n]Includes phenyl, ethynyl; H used only once.

zation, they are treated in separate correlation equations. Since data sets are thus divided in two, there are between six and eight data points for each correlation. This admittedly diminishes statistical arguments based on regressions employing two or three independent parameters. However, the interpretation of the data will not need to rest on the significance of very small differences and the results will be interpreted semiquantitatively.

TABLE 5-8. Compilation of Taft DSP Constants (Reference 33) and Topsom Constants as well as Some Derived Constants (see Equations 5-18 and 5-19)

X	Substituent constants				
	σ_I[a]	σ_R[a]	σ_X[b]	σ_F[c]	σ_{R°[d]
H	0.0	0.0	0.0	0.0	0.0
F	0.54	−0.48	0.52	0.47	−0.29
Cl	0.47	−0.25	0.24[e]	0.44	−0.16[f]
Br	0.47	−0.25			
I	0.40	−0.16			
CH_3	−0.01	−0.16	0.17	−0.01	−0.09
OMe	0.30	−0.58	0.44	0.29	−0.42
OH	0.24	−0.62	0.43	0.30	−0.41
NH_2	0.17	−0.80	0.33[g]	0.15	−0.57
OAc	0.38	−0.23	0.46	0.41	−0.14[f]
Vinyl	0.11	−0.15	0.18	0.04	0.00
HCC—	0.29	−0.04	0.28	0.17	−0.02
n-Propyl	−0.01	−0.16	58		−0.09[f]
Phenyl	0.12	−0.11		0.06	−0.05[f]
OEt	0.28	−0.57			−0.39[f]
CH_2Cl	0.17	−0.08			−0.03[f]
CH_2Br	0.20	−0.10			−0.04[f]
CH_2I	0.17	−0.09			−0.03[f]
Ethyl	−0.01	−0.14			−0.07[f]
t-Bu	−0.01	−0.18		−0.01	−0.10[f]
OBu	0.28	−0.58			−0.40[f]
CN	0.57	0.08	0.31	0.45	0.08
NC	0.63	0.02[h]	0.43[g]	0.60[c]	0.05[f]
CHO	0.25[i]	0.20[f]	0.14	0.22	0.18
COMe	0.30	0.20	0.14	0.19	0.20
CO_2Me	0.32	0.11	0.19	0.25	0.11[f]
NO_2	0.67	0.10	0.40	0.66	0.18
CF_3	0.40	0.11	0.17	0.42	0.03
CO_2Et	0.30	0.11			0.11[f]
CO_2H	0.30	0.11	0.18	0.27	0.11[f]

[a]From Reference 33 unless noted otherwise.
[b]From Reference 21 unless noted otherwise.
[c]From Reference 22; note that there are also values listed in References 21 that are slightly different.
[d]From Reference 23 unless noted otherwise.
[e]6–31G*//3G result from R. Topsom, personal communication; see also References 21–23.
[f]Calculated by Equation 5-18.
[g]R. Topsom, personal communication; see References 21–23.
[h]Exner, O. In "Correlation Analysis in Chemistry", Chapman, N.B.; and Shorter, J., Eds.; Plenum Press: new York, 1978.
[i]Calculated from Equation 5-19.

TABLE 5-9. Correlation Analysis of Stabilization Energies by Taft DSP Equation:
$\Delta E = m_I\sigma_I + m_R\sigma_R + b$ (see Table 5-7)

	n	R^2	R	Standard error	m_I	m_R	b
A. Methyl Stabilization							
Vinyl—X							
π donors	6	0.97	0.99	1.07	−4.36	−17.62	0.73[a]
π acceptors	8	0.22	0.47	3.05	4.30	14.35	0.83[b]
π acceptors	5	0.83	0.91	1.37	3.25	21.38	0.02[c]
Cyclopropyl—X							
π donors	6	0.95	0.98	0.69	1.78	−7.36	0.71[a]
π acceptors	5	0.91	0.95	1.05	2.90	25.98	0.12[d]
Ethynyl—X							
π donors	6	0.94	0.97	2.68	−47.49	−20.87	2.02[a]
π acceptors	8	0.24	0.49	7.71	−16.08	7.80[e]	2.91[b]
Phenyl—X							
π donors	6	0.97	0.99	0.82	−4.40	−10.86	−0.25[a]
π acceptors	4	0.98	0.99	0.59	−0.92	28.10	−0.08[f]
B. Ethyl Stabilization							
Vinyl—X							
π donors	6	0.99	0.996	0.47	−15.27	−17.10	0.15[a]
π-acceptors	8	0.15	0.38	2.30	1.27	10.57	0.01[b]
Cyclopropyl—X							
π donors	6	0.94	0.97	0.63	−9.32	−6.87[g]	0.12[a]
π acceptors	5	0.90	0.95	0.90	−1.32	24.24	0.25[d]
Ethynyl—X							
π donors	6	0.96	0.98	2.78	−58.45	−20.28	1.48[a]
π acceptors	8	0.29	0.54	8.07	−19.15	4.28[b]	2.09[b]
Phenyl—X							
π donors	6	0.82	0.91	1.56	−6.48	−10.30	−0.84[a]
π acceptors	4	0.90	0.95	1.08	−4.23	21.97	0.14[f]
C. Isopropyl Stabilization[i]							
Vinyl—X	5	0.97	0.99	1.09	−22.32	−13.80	−0.01[j]
Cyclopropyl—X	5	0.93	0.97	1.16	−15.88	−4.00	−0.00[j]
Ethynyl—X	5	0.95	0.98	4.04	−65.00	−17.18	1.34[j]
Phenyl—X	5	0.78	0.88	1.97	−12.69	−7.68	−0.94[j]

[a]Substituents:H, F, Me, OMe, OH, NH_2.
[b]Substituents: H, CN, NC, CHO, COMe, CO_2Me, NO_2, CF_3.
[c]Substituents: H, CN, CHO, COMe, NO_2.
[d]Substituents: H, CN, NC, COMe, NO_2.
[e]Note inconsistency of sign of this value with that of m_{R° (Table 5-10).
[f]Substituents: H, CN, CHO, NO_2.
[g]Note that it is unusual for m_I to be greater than m_R.
[h]Note inconsistency of sign of this value with that of m_{R°.
[i]Only enough data to correlate π donors.
[j]Substituents: H, F, Me, OH, NH_2.

Table 5-9 lists correlations of methyl, ethyl, and isopropyl stabilization energies for neutral substituents using the Taft DSP approach. It is clear that correlations of π donors appear to be considerably better than for π acceptors. It is not immediately apparent why this should be so. However, if one examines the vinyl series, where this disparity is especially striking, it is clear that the magnitudes of the stabilizations of the π donors are larger than those for the π acceptors. This is not the case for the cyclopropyl series, since the hydrocarbon moiety is such a poor π acceptor (see below as well as earlier discussion). Another point, which this simple-minded approach ignores, is the interaction term between σ and π effects. Specifically, π donation and σ withdrawal should be mutually reenforcing, since in concert they tend to minimize charge distortion. On the other hand, σ withdrawal might tend to decrease the π-withdrawing effects of appropriate substituents, thus rendering comparison of the two types of correlation less straightforward.

Another interesting result is that acetylenes behave differently compared with ethylenes, cyclopropanes, and benzenes. Almost all the substituted acetylene correlations show larger sensitivities to inductive effects, in contrast to the other three classes of compounds where the reverse is true. Fuchs and co-workers[119] have also noted the extreme sensitivities of acetylene stabilities to inductive effects, and we have further analyzed these results using the "composite parameter approach."[120a] A third observation is that the m_I values become more negative for the π-donor class and less positive for the π-acceptor class in going from methyl to ethyl to isopropyl. This means that inductive stabilizations are decreased in this order, and this is in line with the observations in Tables 5-3 to 5-6 as well as the work of Clark, Schleyer, and co-workers.[36] Our correlations do not involve electropositive substituents such as Li and BH_2 where seemingly the reverse is true. These investigators provide rationale by asserting that isopropyl anion is less stable then methyl anion, but no proof is cited. In contrast to the slope of the line which measures sensitivity to inductive effects, m_I, m_R changes only slightly in comparing methyl, ethyl, and isopropyl stabilizations. It is also apparent that while methyl, ethyl, and isopropyl stabilization energy correlations give the same ordering of m_R for π donors, a different ordering may be apparent for π-acceptor substituents.

Table 5-10 lists correlations of stabilization energies using the Topsom parameters. The results are similar to those in Table 5-9, with generally similar values for m_R and m_{R° and the sum of m_X and m_F generally close to M_I.

It is important to note that the substituents vinyl, ethynyl, phenyl, and cyclopropyl have not been employed in the regression analyses in Tables 5-9 and 5-10. It is simple to rationalize these omissions by considering 1,3-butadiene as vinylethylene. It immediately becomes clear that the substituent is both donor and acceptor, and the actual stabilization energies are much different (discrepancy ~ 5 kcal/mol) from those predicted from regression equations of either π acceptors or π donors. A rather huge dis-

TABLE 5-10. Correlation of Stabilization Energies (kcal/mol) by Topsom Equation:

$$\Delta E = m_X\sigma_X + m_F\sigma_F + m_{R^\circ}\sigma_{R^\circ} + b$$

	n[a]	R^2	R	Standard error	m_X	m_F	m_{R°	b
				A. Methyl Stabilization				
Vinyl—X								
π donors	6	0.99	0.997	0.59	15.98	−14.55	−18.73	−0.10
π acceptors	8	0.67	0.82	2.22	37.15	−20.89	17.05	1.24
Cyclopropyl—X								
π donors	6	0.98	0.99	0.54	11.24	−5.31	−5.83	0.09
π acceptors	5	0.99	0.99	0.56	41.45	−26.23	29.10	−0.09
Ethynyl—X								
π donors	6	0.996	0.998	0.87	33.89	−77.77	−20.80	0.09
π acceptors	8	0.70	0.84	5.42	96.93	−80.51	−14.92	5.50
Phenyl—X								
π donors	6	0.95	0.98	1.32	−3.68	9.09	−15.18	0.14
π acceptors	4	—[b]	—[b]	—[b]	27.39	−22.55	35.70	—[d]
				B. Ethyl Stabilization				
Vinyl—X								
π donors	6	0.99	0.99	0.70	7.03	−20.19	−21.53	−0.05
π acceptors	8	0.80	0.89	1.25	32.47	−20.82	12.38	0.34
Cyclopropyl—X								
π donors	6	0.92	0.96	0.90	2.07	−11.18	−8.95	0.15
π acceptors	5	0.97	0.98	0.73	42.23	−31.15	24.78	0.12
Ethynyl—X								
π donors	6	0.99	0.997	1.23	25.55	−84.01	−23.35	0.14
π acceptors	8	0.71	0.85	5.77	94.69	−82.83	−18.54	5.00
Phenyl—X								
π donors	6	0.87	0.93	1.62	−12.66	3.44	−17.95	0.33
π acceptors	4	—[b]	—[b]	—[b]	63.23	−48.04	28.99	—[d]
				C. Isopropyl Stabilization[c]				
Vinyl—X	5	1.00	1.00	0.45	4.15	−6.71	−19.00	−0.03
Cyclopropyl—X	5	1.00	1.00	0.26	1.16	−18.64	−6.32	0.02
Ethynyl—X	5	1.00	1.00	0.31	25.15	−91.78	−20.70	−0.02
Phenyl—X	5	0.97	0.98	1.06	−11.96	−5.22	−15.38	0.07

[a]Same substituents employed for each correlation as in Table 5-9.
[b]Not valid, perfect line with four points.
[c]Only enough data for correlations with π donors.
[d]F-level or tolerance level not sufficient for program to caculate constant.

crepancy (~19 kcal/mol) is calculated for diacetylene using the same analysis.

Examination of Table 5-9 clearly indicates that for substituents that are π donors, the m_R values for cyclopropanes are much smaller than those for ethylenes. Similar results are apparent in Table 5-10. This clearly supports data cited earlier, which indicate that cyclopropane is a poor π acceptor. In contrast, for π-acceptor substituents, the m_R values are greater for cyclopropanes than for ethylenes. This implies that cyclopropane is a stronger π

TABLE 5-11. Relationships of Conjugation Sensitivity Slopes, m_R, to Ionization Potentials and Electron Affinities of Parent Hydrocarbons

		π Acceptors	
		Hydrocarbon IP (eV)	
Parent hydrocarbons	m_R	Adiabatic	Vertical
Vinyl—X	14.4	10.51[a]	10.51[b]
Ethynyl—X	7.8	11.40[a]	11.40[b]
Cyclopropyl—X	26.0	9.86[c]	10.60, 11.30[d]; avg 10.95
Phenyl—X	28.1	9.25[a]	9.25[b]

		π Donors
		Hydrocarbon EA (eV)
Parent hydrocarbon	m_R	Vertical[e]
Vinyl—X	−17.6	−1.78
Ethynyl—X	−20.9	−2.6
Cyclopropyl—X	−7.4	−5.29[f]
Phenyl—X	−10.9	−1.15

[a]Kimura, K.; Katsumata, S.; Achiba, Y.; Yamazaki, T.; Iwata, S. "Handbook of HeI Photoelectron Spectra of Fundamental Molecules". Japan Scientific Society Press, Halsted Press: New York, 1981.
[b]Levin, R.D.; Lias, S.G. "Ionization Potential and Appearance Potential Measurements, 1971–1981", NSRDS-NBS 71; National Bureau of Standards, U.S. Department of Commerce: Washington, D.C., 1982.
[c]Lias, S.G.; Buckley, P.J. *Int. J. Mass Spectrom. Ion Proc.* **1984,** *56,* 123–137.
[d]These are vertical IPs corresponding to two peaks in first band attributed to degenerate HOMOs (see note *a*); their average is taken as a crude approximation of the "inherent" vertical IP.
[e]Values obtained from Jordan, K.D.; Burrow, P.D. *Acc. Chem. Res.* **1978,** *11,* 341–348; *J. Am. Chem. Soc.* **1980,** *102,* 6882.
[f]Howard, A.E.; Staley, S.W. In American Chemical Society Symposium Series, No. 263, "Resonances in Electron-Molecule Scattering Van der Waals Complexes and Reactive Chemical Dynamics", Truhlar, D.G., Ed.; ACS, Washington, D.C.: 1984, pp. 183–192.

donor than ethylene, a view contrary to that espoused by Allen[37,38] but in line with the conclusions of Harmony and associates.[63] The study by Clark, Schleyer, and co-workers[36] emphasized π-donor substituents. However, these investigators did clearly show significant conjugation in cyclopropylborane. If one compares the isopropyl stabilization energies for the conjugated conformers of vinylborane (Table 5-3) and cyclopropylborane (Table 5-4), it is clear that they are both just under 9 kcal/mol. Alternatively, using the data in the appendix, the calculated rotational barrier for vinylborane (8.0 kcal/mol) is only slightly greater than that for cyclopropylborane (7.3 kcal/mol). These results indicate nearly equal π-donor capabilities for vinyl and cyclopropyl. In considering the magnitudes of m_R for the correlation employing π-acceptor substituents, it is interesting to take note of their relationship with the adiabatic ionization potentials of the parent hydrocarbons as noted in Table 5-11. Although two of these parent molecules, cyclopropane and benzene, exhibit Jahn–Teller distortion upon loss of an electron and are thus not quite comparable with the other two molecules, the relationship is intuitively satisfying. Benzene, the easiest to ionize, has the highest m_R, with π acceptors, and acetylene, with the highest IP, has the lowest m_R. Table 5-11 also lists values of m_R for π-donor substituents along with vertical electron affinities that correspond to resonances the interpretation of which is not so straightforward as IP data. A discussion of gas phase electron affinities is beyond the scope of this chapter. The only intuitively comfortable point is that the lowest m_R belongs to cyclopropyl, which has the least affinity for electrons. The foregoing approach has been discussed in greater detail in the literature.[120b]

One might ask how effective cyclopropane is at transmitting conjugation between vicinal substituents. It would seem logical that for cyclopropane to effectively connect such groups conjugatively, it must be simultaneously a good donor and a good acceptor, as vinyl is. The fact that it is a good π donor but a poor π acceptor suggests that it should not be particularly effective in transmitting conjugation effects, in agreement with at least some published studies.[121]

To conclude this section on cyclopropane structures and stabilities, we consider very briefly fluorinated species. More detailed discussion is to be found in Chapter 4 in this volume. In particular, the strain energy of 1,1-difluorocyclopropane, based on a strainless increment derived by Dolbier and co-workers,[122] appears to be about 12 kcal/mol higher than for cyclopropane, and this appears to be the basis of the relative ease of cleavage of its distal bond.[109] This value is very close to the value estimated by O'Neal and Benson[123] (4.5–5 kcal/mol destabilization per F on cyclopropane), even though the decreased incremental geminal stabilization (IGSTAB) is not accounted for. A strainless group increment of −95.8 kcal/mol has been estimated for a $C(F)_2(CF_2)_2$ group by using the diagonal, asymptotically diagonal, and related approaches discussed in Chapter 8 of Volume 2 in this series. The diagonal approach, for example, simply involves taking one-sixth of ΔH_f°(g) of perfluorocyclohexane as the strainless increment. The total strain in hexafluorocyclopropane comes to about 54 kcal/mol, lower than our previous estimate[109] and in surprisingly excellent agreement with the Benson–O'Neal estimate.[123] This extra strain appears to manifest itself in the relatively facile extrusion of CF_2 from the three-membered ring although an important part of this is the anomalous stability of the singlet carbene (see Chapter 3, Volume 1 of this series). Interestingly, three separate calorimetric determinations using different reactions give very similar values for ΔH_f°(g) of octafluorocyclobutane. When the unstrained group increment detailed above is employed, the strain energy of this molecule appears to be 8.5–12 kcal/mol lower than the parent.[123] Chapter 4 of this volume also details some of these relationships, and we have benefited from exchanges of data and views on these topics with Dr. Smart as well as Professor Dolbier.

Whereas extra strain in 1,1-difluorocyclopropane and hexafluorocyclopropane appear to be part of increased reactivity in these systems, perfluoroalkyl systems appear to behave differently. Thus, it was noted earlier that perfluoroalkyl groups dramatically enhance the thermal stabilities of strained molecules and the phenomenon was dubbed the "perfluoroalkyl (R_f) effect."[4,5] It is also worth recalling that 1,1-bis (trifluoromethyl) cyclopropane resists catalytic hydrogenation.[29] A glance at Table 5-3 indicates that CF_3 groups thermodynamically destabilize vinyl groups and certainly do not thermodynamically stabilize strained rings. This is sensible, since the substituent is a σ withdrawer and should, by Hine's [106] description, destabilize double bonds. Thus the "perfluoroalkyl (R_f) effect" appears to be totally kinetic in origin.[124]

4. SUBSTITUTED CYCLOBUTANES: STRUCTURES AND STABILIZATION ENERGIES

The stereochemistry of cyclobutanes has been extensively reviewed.[125] Cyclobutane offers an interesting difference in its conjugative ability compared to cyclopropane.[126] The symmetries of the degenerate HOMOs (Figure 5-6*a*) are such that conjugation may occur with π-acceptor substituents in both the bisected and perpendicular conformations (eg, CH_2^+, see Figure 5-6*b*). Although the ionization potentials of cyclobutane and cyclopropane are similar,[91] the interactions are much smaller in the cyclobutyl than in cyclopropyl derivatives because overlap is poorer in the former. This is the main reason that the rotational barrier in cyclobutylcarbinyl cation is much lower than that in cyclopropylcarbinyl cation,[127a] although conjugation in the perpendicular conformation of the former (Figure 5-6*b*) may also play a role. The 4–31G stabilization energy of cyclobutylcarbinyl cation[127b] is 20 kcal/mol less than that of cyclopropylcarbinyl cation when both bisected conformers are compared (ie, see Equation 5-20, X = CH_2^+, ΔE_{rxn} = +20 kcal/mol). Another interesting aspect is the possibility that certain 1,1-disubstituted cyclobutanes, having π acceptors, may prefer the bisected–perpendicular conformation **(27)** instead of the bisected–bisected conformation **(28)**, which one would anticipate for the analogous cyclopropanes.

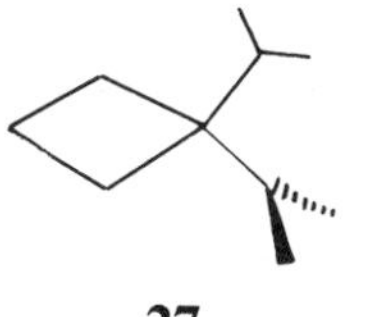

27 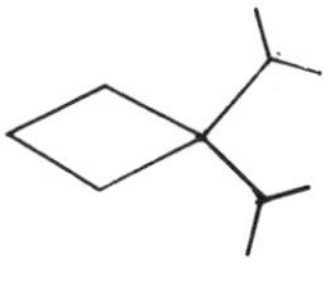 **28**

Another potential consequence of conjugation between π-acceptor substituents and the cyclobutane ring is depicted by structures **29** and **30.**[126] In the bisected conformation **(29)**, a short C_2C_4 (d_{24}) is predicted, while a short d_{13} is expected for conformation **30.** Allen[128] recently summarized the geome-

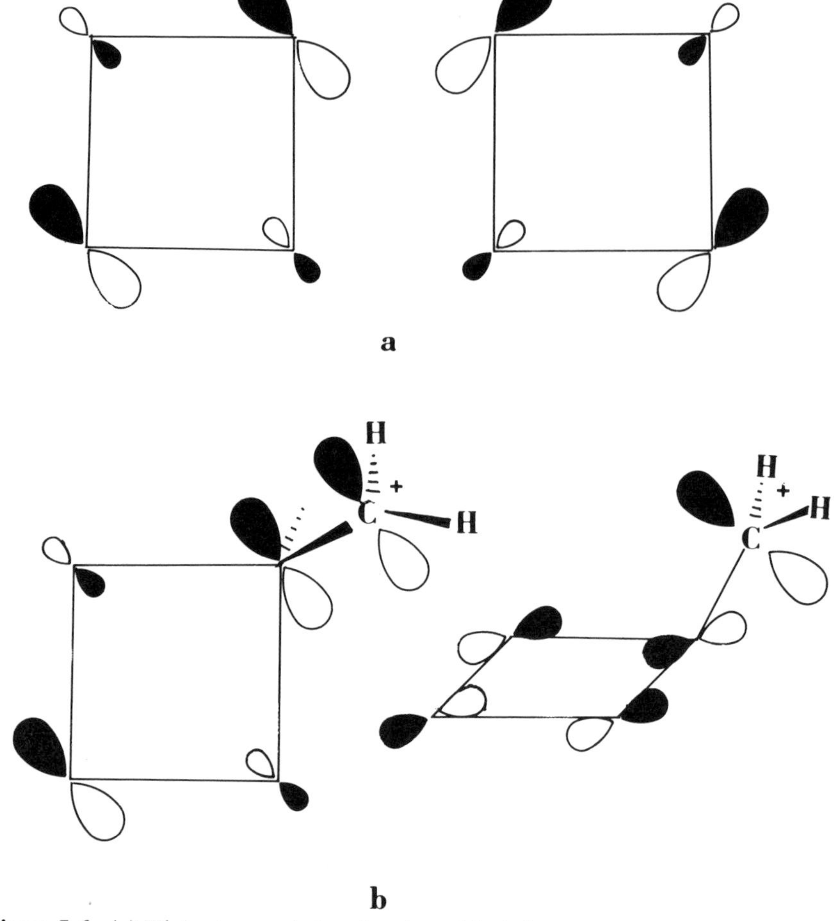

Figure 5-6. (*a*) Highest occupied molecular orbitals in cyclobutane. (*b*) Conjugation mechanisms in cyclobutylcarbinyl cation conformers (see Reference 126).

tries of substituted cyclobutanes and has examined 39 ring systems having only one carbonyl substituent. The seven structures classified as bisected (torsional angle, $\tau = -20° - +20°$; 160°–200°) had $d_{13} = 2.191$ Å and $d_{24} = 2.171$ Å, while the 12 structures classified as perpendicular ($\tau = 70 - 110°$) had $d_{13} = d_{24} = 2.173$ Å. Thus, while this very limited data set appears to support the predictions of Hoffmann and Davidson,[126] the results have "little statistical significance."[128] More to the point, there is no tendency to favor conjugated conformations over nonconjugated conformations as there

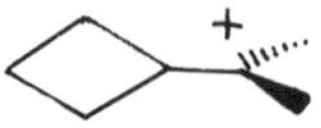

29 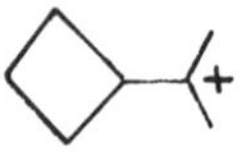 **30**

is for cyclopropanes having π-acceptor substitients.[128] Thus, resonance interactions are said to be "very small" by comparison to the cyclopropyl system.[128] A study of the structure of cyclobutylsilane[129] indicated that d_{12} (1.580 Å) was larger than d_{23} (1.564 Å), in contrast to the situation for cyclobutane rings substituted by the electronegative groups Cl, Br, and CN. This is similar to the case for cyclopropylsilane, where the vicinal bond lengths are longer than the distal bond lengths.[38] Both these observations make sense in light of the arguments advanced for the influence of substituent electronegativity in determining vicinal bond lengths.[36] However, another study[130] indicated very minor substituent effects on vicinal bond lengths in cyclobutanes.

The tendency to favor electropositive (σ-donor) groups in the axial position is emphasized by a calculated value of 2.8 kcal/mol favoring the "axial" cyclobutyl carbanion **(31)** over the "equatorial" stereoisomer and the favoring of axial lithiocyclobutane **(32)** by 0.4 kcal/mol.[130] In the case of cyclo-

31 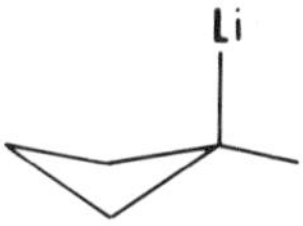 **32**

butylsilane, the equatorial isomer is favored by only 0.2 kcal/mol.[129] The C—Si bond length in cyclobutylsilane of 1.873 Å, is found to be considerably larger than the corresponding 1.839 Å bond length in cyclopropylsilane.[130] While this would seem to indicate negligible (p-d) π conjugation in the cyclobutyl derivative, there is some effect on d_{13} (2.14 Å) and d_{24} (2.20 Å) that might be invoked in favor of conjugation.

Allen[128] notes the wide range in average C—C bond lengths in substituted cyclobutanes (1.521–1.606 Å) and attributes these to nonbonded steric repulsions. It is also worthwhile recalling that the bond lengths in cyclobutanes, oxetanes, and thietanes are all longer than the bonds in the three-, five-, and six-membered ring compounds (Figure 5-7).[131] Furthermore, extremely long C—C bonds have been found in vicinal diphenylcyclobutane rings (eg, **33**), which are said to arise from conjugative interactions enhanced by the strained ring.[132,133] This interesting structural effect and its

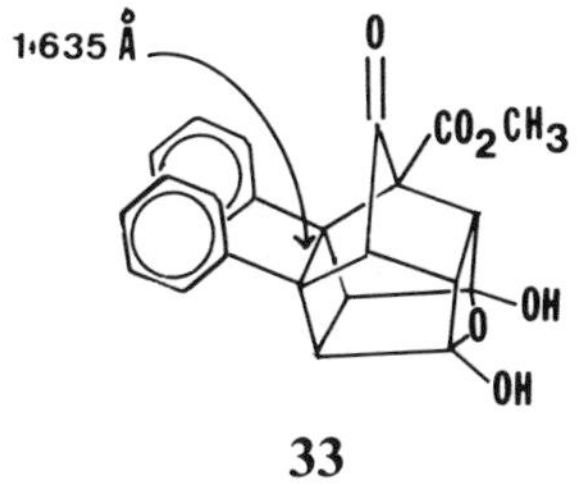

33

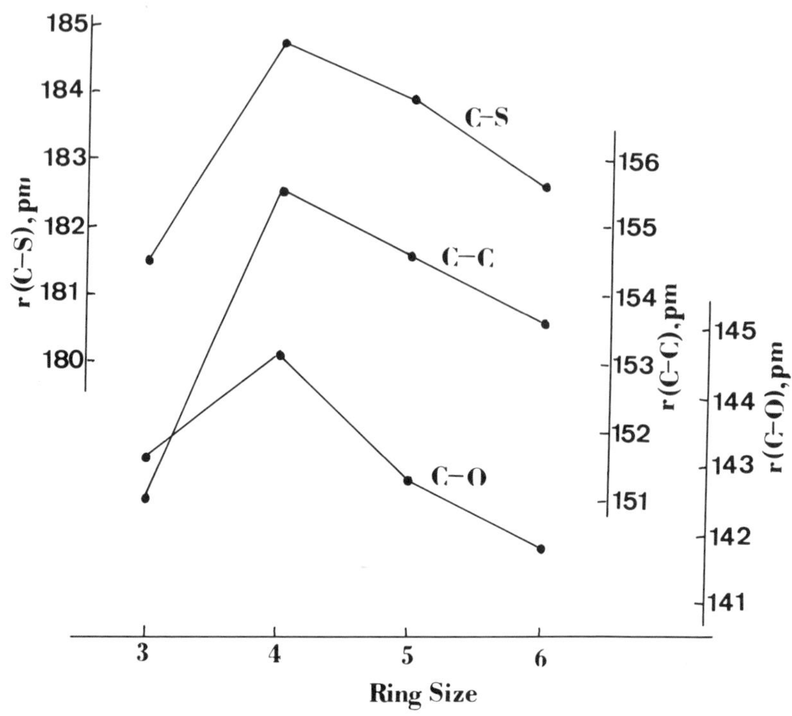

Figure 5-7. Relationships between ring size and bond length in three-, four-, five-, and six-membered rings (see Reference 131).

influence on the chemistry of these systems is discussed in Chapter 7 in this volume. One should recall that the in-plane ring deformation in cyclobutane is a lower energy process (750 cm^{-1}, 2.2 kcal/mol)[91] than that in cyclopropane.

Examination of the small body of published $\Delta H_f^\circ(g)$ data for monosubstituted cyclobutanes leads one to recognize an interesting trend first discussed by Fuchs and Hallman.[134] These data may be examined using homodesmotic equations 5-20 to 5-22 to examine stabilizations relative to cyclopropyl, cyclopentyl, and cyclohexyl derivatives (isopropyl stabilizations can be obtained by adding ΔE for Equation 5-20 to the isopropyl stabilization of the cyclopropane in this equation). The results are listed in Table 5-12. In all cases, the cyclobutyl derivative is found to be most stable thermodynamically. Fuchs and Hallman[134] examined this effect and very tentatively concluded that it could be an artifact of an incorrect $\Delta H_f^\circ(g)$ value for cyclobutane—a value too high by about 2 kcal/mol. However, we feel that this effect could be a real consequence of the considerable flexiblity of the cyclobutane ring, an "artful dodger" that can readily pucker, tilt methylene groups, and even stretch bonds to minimize nonbonded repulsions. This may be the source of the reduced strain in octafluorocyclobutane noted earlier and also discussed in Chapter 4.

$$\Delta E_{stab} = -\Delta E_{rxn} \quad (5\text{-}20)$$

$$\Delta E_{stab} = -\Delta E_{rxn} \quad (5\text{-}21)$$

$$\Delta E_{stab} = -\Delta E_{rxn} \quad (5\text{-}22)$$

One may discern tentative support for the Fuchs–Hallman argument in the comparison of 6–31G*/6–31G* calculated enthalpies of hydrogenation with experimental data.[135] The difference between ΔH_{hydrog} (obs) $-$ ΔH_{hydrog} (calc) was 5.9 ± 0.6 kcal/mol for cyclopropane, bicyclo[2.1.0]pentane, bicyclo[2.2.0]hexane, and *n*-butane. The value for cyclobutane is only 3.5 kcal/mol.[135] If the reported experimental value is indeed too exothermic by 2.5 kcal/mol, this would place cyclobutane in line with the other four values noted above, and it would also remove the apparent stabilizations of cyclobutyl derivatives listed in Table 5-12.

Support for our steric argument, however, can be obtained from a comparison of steric parameter data. Thus, the Taft E_s values are as follows: cyclobutyl (-0.06), ethyl (-0.07), isopropyl (-0.47), cyclopentyl (-0.51), and cyclohexyl (-0.79); no E_s value is published for cyclopropane.[136] However, Charton[137] has published a set of effective upsilon (υ_{eff}) values based on bond lengths and van der Waals radii: cyclobutyl (0.51), ethyl (0.56), cyclopentyl (0.71), isopropyl (0.76), cyclohexyl (0.87), and cyclopropyl (1.06). The cyclopropyl substituent is different from the others in that a confor-

TABLE 5-12. ΔH_f°(g) Values and Associated Stabilization Energies (see Equations 5-20 to 5-22) for Monosubstituted Cyclobutanes (c-C_4H_7—X)

		Stabilization energy (kcal/mol)		
X	ΔH_f°(g)[a]	Cyclopropyl	Cyclopentyl	Cyclohexyl
CH_3	-4.0[b]	4.3	4.4	3.3
C_2H_5	-6.6[c]	2.0[d]	1.8	1.6
NH_2	9.8	2.6	3.3	—
CN	35.2[c]	2.0	2.0[e]	1.8
CO_2CH_3	-83.7[c]	—	—	—

[a]Reference 95.
[b]ΔH_f°(l) = -10.6 kcal/mol (Reference 95); employed Trouton's rule ($\Delta H_v^\circ = 0.021\ T_b$) and boiling point (39.4°C, "CRC Handbook") to obtain ΔH_f°(g).
[c]See Reference 134.
[d]ΔH_f°(l) (ethylcyclopropane) = -5.93 kcal/mol; employed Trouton's rule with boiling point (72.5°C, "CRC Handbook") to obtain ΔH_f°(g)).
[e]ΔH_f°(g) cyanocyclopentane = 11.7 kcal/mol; see Reference 134.

mational dependence is factored in.[137a] Any apparent E_s parameter for cyclopropyl would have to include conjugative effects that would tend to make the value more negative than it otherwise would be. It appears that cyclobutyl can indeed accommodate larger substituents more comfortably than the other systems examined.

Wiberg has taken note of the short C_1C_3-nonbonded distances in cyclobutane and has speculated that this may be a significant or major contributor to its strain energy.[137b] In noting that bicyclo[1.1.1]pentane maintains a C_1C_3-nonbonded distance of only 1.845 Å, he examined the ability of electronegative substituents to decrease this distance and electropositive substituents to increase it.[137b] Perhaps such a decrease in nonbonded repulsion may be the reason for the reduction of strain in octafluororcyclobutane.

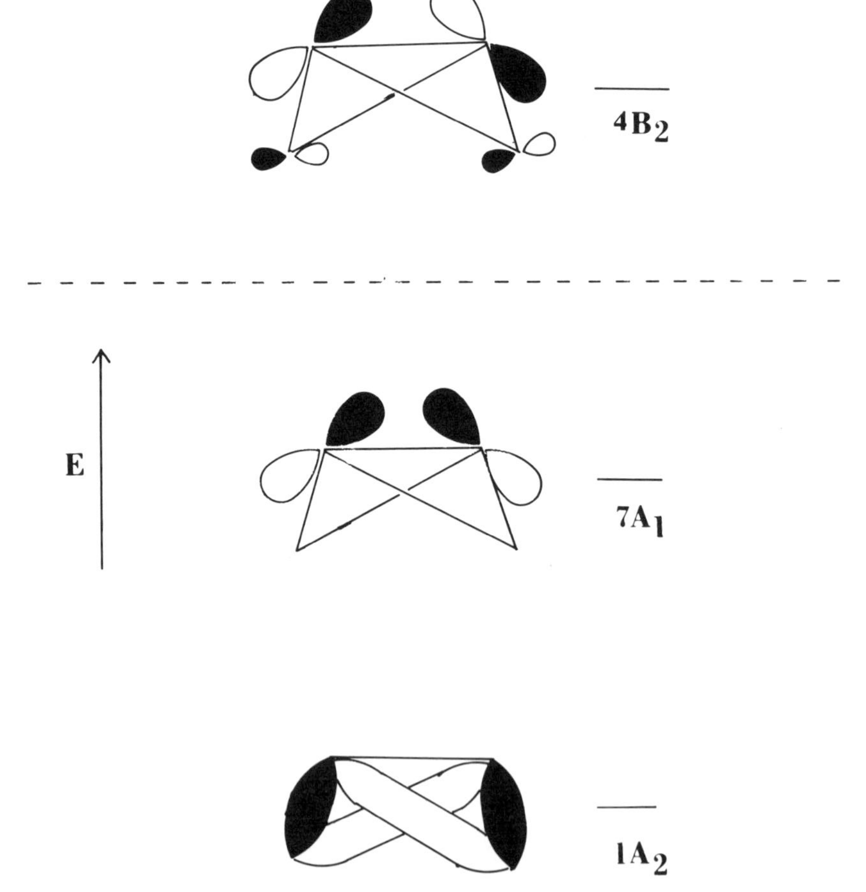

Figure 5-8. Two highest occupied and the lowest unoccupied orbitals of bicyclobutane. (Adapted from Reference 176.)

5. SUBSTITUTED BICYCLOBUTANES: STRUCTURES AND ENERGETICS

The strain energy of bicyclobutane is more than 10 kcal/mol higher than that of two cyclopropane rings.[1] The highest occupied molecular orbital in bicyclobutane (Figure 5-8) is almost entirely due to overlap of canted p orbitals between C_1 and C_3. The central bond's high energy and high p character ($\sim$91%)[138] should make this system particularly sensitive to conjugative effects, particularly with bridgehead-substituent π acceptors. Moreover, the high value for $^1J_{CH}$ (202 Hz)[139] corresponds to an orbital having 39% s character directed toward bridgehead substituents. Thus, just as we have shown earlier that acetylenes are particularly sensitive to inductive effects, so should 1-substituted bicyclobutanes be. Two measures of this sensitivity are provided in the recent literature: (a) substitution of methyl groups at the bridgehead positions lowers the oxidation potentials of bicyclobutanes much more than substitution at the 2 and 4 positions[140a,b] (see Scheme VIII, where the $E_{1/2}$ voltages are from Reference 140*a*), and (b) the sensitivity to inductive effects (ρ_I) of the pK_a of 3-substituted-1-bicyclobutanecarboxylic acids is another measure of this factor[141] (see Scheme XIX).

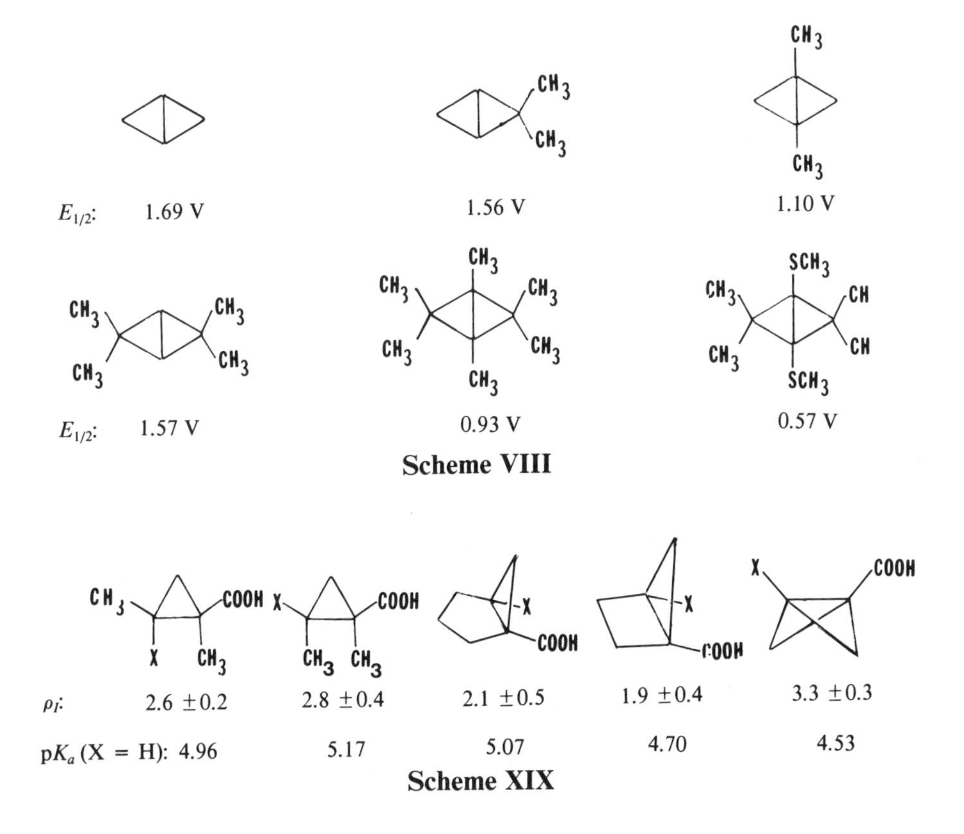

Scheme VIII

Scheme XIX

The flexibility in the bicyclobutane nucleus is manifested in extensive structural changes in the ring structure induced by substituents.[142–145] Thus, the parent hydrocarbon (**34**, see Figure 5-9) has approximately equal bond lengths and a flap angle (θ, angle between planes of the three-membered rings) of 121.7°. If the 2- and 4-positions are joined by a one- or two-carbon bridge, the flap angle decreases and the central bond (C_1C_3) length (d_{13}) decreases while the peripheral bond lengths (eg, d_{12}) increase (eg, **35** and **36** in Figure 5-9). The presence of phenyl groups at the bridgehead carbons sig-

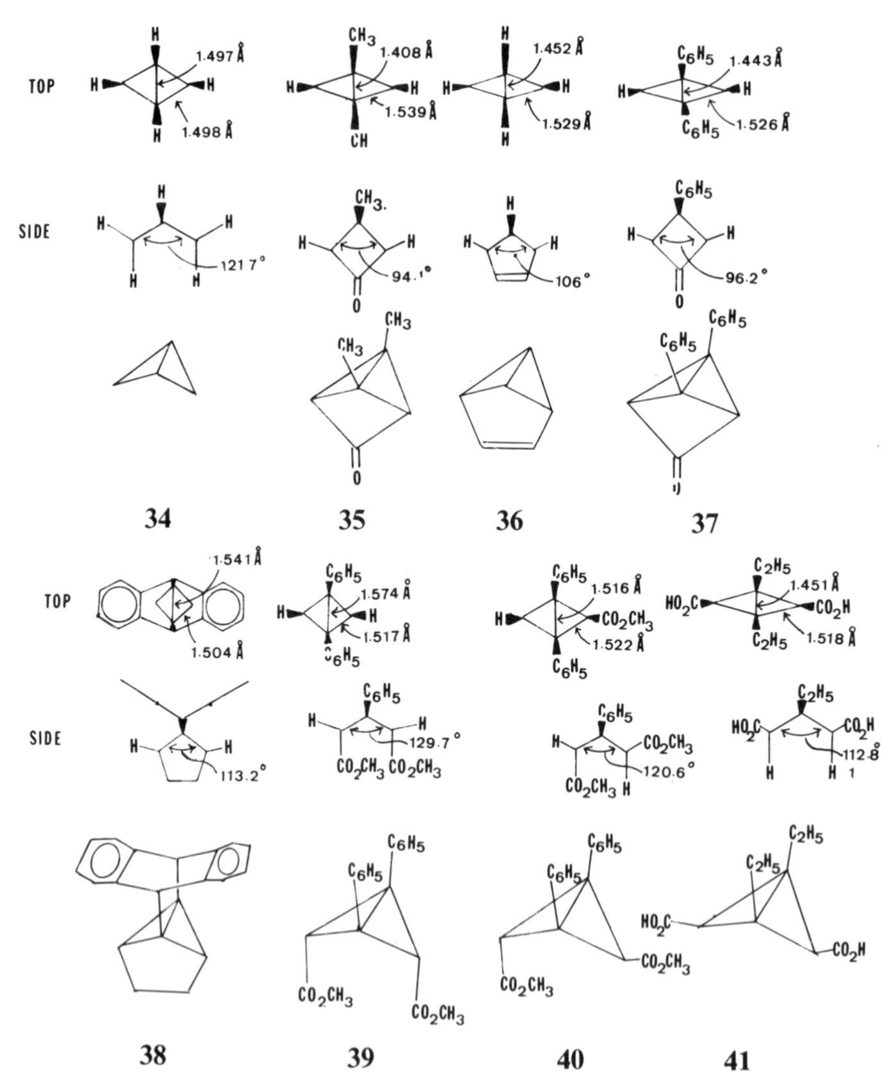

Figure 5-9. Top (T), side (S), and perspective (P) views of the structures of substituted bicyclobutanes (see Reference 145).

nificantly lengthens the central bond due to π conjugation, which is further evidenced by significant UV spectral shifts.[146] Thus, the central bond in **37** (Figure 5-9) is 0.035 Å longer than that in **35.** This conjugation, depicted by **37a,** is strongly reminiscent of the bond lengthening noted for 1,2-diphen-

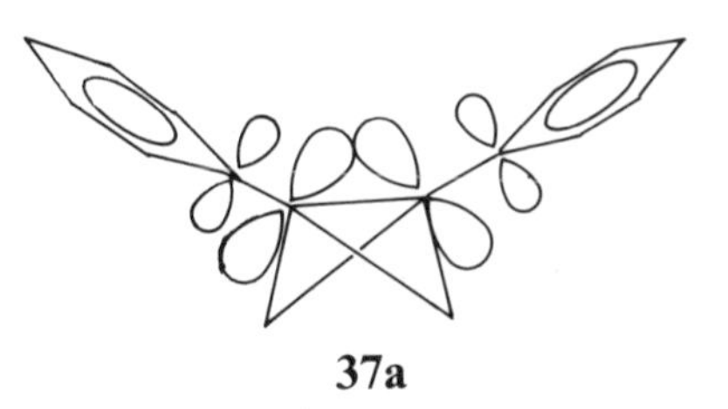

37a

ylcyclobutanes and discussed in Chapter 7. Bridging between the bridgehead carbons in **38** (Figure 5-9) has the effect of lengthening the central bond so that it is longer than the peripheral bond. Molecule **38** is an example of a [4.1.1]propellane. The geometries of smaller [n.1.1]propellanes ($n < 3$) are not discussed here, but their energetics are briefly presented in Chapter 4 of Volume 2 of this series. A third means of stretching the central bond is to widen the flap angle. This is seen in the endo–endo isomer of dimethyl-1,3-diphenylbicyclobutane-2,4-dicarboxylate **(39).** Steric repulsion of the two carbomethoxy groups widens the bond angle, which may be compared to that of the exo–endo isomer **(40),** which lacks such steric repulsion and has more normal bond lengths. The effects of phenyl conjugation on the central bond lengths are seen in these two structures. It would not be very surprising to observe significant phenyl conjugation in these compounds. However, one would not expect corresponding conjugation in cyclopropanes, since bicyclobutane is a much stronger π donor. An interesting contrast is furnished by **41** in Figure 5-9. In this molecule the π-accepting carboxyl groups are conjugated and shorten the C_1—C_3 bond distal to them in the same manner as in cyclopropanes. The flap angle is a small one, "driven" by the short central bond length that "originates" in the conjugation of the C_2 and C_4 substituents. This is an interesting contrast with **39,** wherein steric repulsion is the "origin" of the opening of the flap angle and "drives" the increase in length of the central bond or 2,4-bridging in **35,** where pinching the flap angle "drives" the decrease in bond length. It should be noted that although central bond lengths in substituted bicyclobutanes range from 0.89 Å below that of the parent to 0.77 Å above it,[145,147] all measured peripheral bond lengths exceed those of the parent hydrocarbon.[143] One should also note that the mean cyclopropane bond length, $D = (2d_{12} + d_{13})/3$, varies from 1.490 to 1.536 Å in substituted bicyclobutanes, excluding those containing or substituted by silyl groups.[145]

The covariation of structural parameters in substituted bicyclobutanes has been analyzed,[142-145,148a] and is depicted in Figure 5-10.[143] An interesting interpretation of the flap angle variation rests on the consideration of the

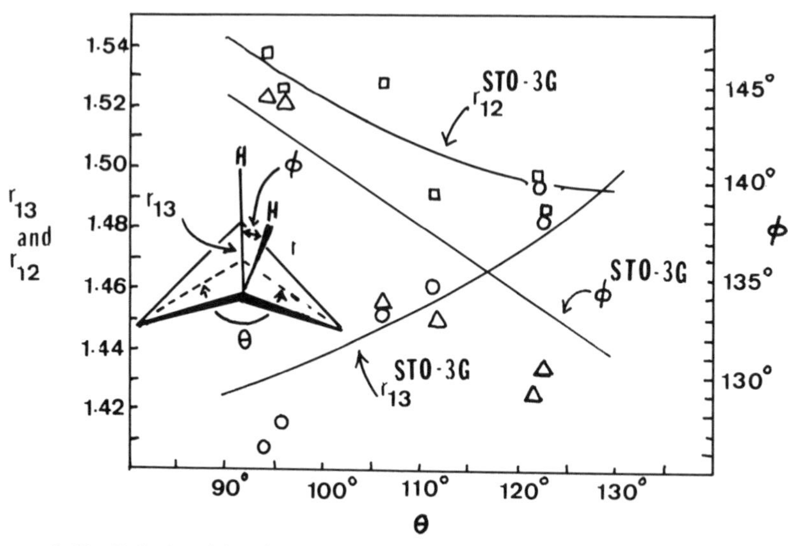

Figure 5-10. Relationships between structural parameters of bicyclobutanes (see Reference 143).

two bridgehead carbons as radical centers. In the parent hydrocarbon they are virtually planar. When the flap angle is decreased, there is a tendency for the bridgehead carbons to pyramidalize, increasing the s character and thus bonding in the central bond. This will also decrease d_{13}.

An interesting molecule for examination could be 2,2,4,4-tetrafluorobicyclobutane (**42,** X = H), in which one might anticipate a significantly lengthened central bond and an increase of the flap angle along with decreased sensitivity to acid, which frequently is the decomposition pathway. The calculated IP of 2,2,4,4-tetrafluorobicyclobutane is 10.6 eV, some 1.6 eV higher than that calculated for the parent hydrocarbon.[140b] Perhaps 1,3-diphenyl substitution (**42,** X = C_6H_5), which would increase acid sensitivity, might make ^{19}F monitoring of ring inversion feasible. Compound **42** (X = CF_3) has been made and is rather stable.[148b]

42 ⇄ ? **42**

There are not many thermochemical data for substituted bicyclobutanes. Comparison of the gas phase enthalpy of hydrogenation of bicyclobutane to cyclobutane (−45.1 kcal/mol)[95] with the experimental solution phase value for 1,3-dimethylbicyclobutane (−40.6 kcal/mol)[149] indicates at least 4.5 kcal/mol stabilization in the latter if solvent effects are neglected in the condensed phase reaction. Recall that the product *cis*-1,3-dimethylcyclobutane is also probably stabilized relative to acyclic model compounds. Thus, the stabilization in this derivative is fairly high, hence reminiscent of the acetylenes. Enthalpies of hydrogenation for 1-cyanobicyclobutane and methyl 1-bicyclobutanecarboxylate can be obtained from published experimental values of $\Delta H_f^\circ(g)$.[150] The value for the ester (−45.6 kcal/mol) is consistent with very slight destabilization in the bicyclobutyl derivative compared to the cyclobutyl derivative. The value for the nitrile (−38.6 kcal/mol) seems to suggest at least 6.5 kcal/mol of stabilization. This appears to be unreasonably high, considering that the cyano group's dominant effect should be inductive rather than conjugative ($\sigma_I = 0.57$; $\sigma_R = 0.08$) and that the inductive effective should be highly destabilizing. It appears that the enthalpy of combustion reported for 1-cyanobicyclobutane may be 6–8 kcal/mol too low, perhaps because of precombustion oligomerization. The value for the carbomethoxy group appears to be more reasonable in light of its parameters ($\sigma_I = 0.31$; $\sigma_R = 0.16$). Published STO-3G calculations[151] indicate very strong conjugative interactions with π acceptors at the bridgehead position, but higher level calculations and experimental data are required to check this prediction.

6. SUBSTITUTED CYCLOPROPENES

The cyclopropene ring is primed to interact strongly with substituents. Its very high strain energy (54.5 kcal/mol)[1] could, in principle, be significantly reduced through interaction with Walsh-type as well as π orbitals. The olefinic $^1J_{CH}$ coupling constant is 220 Hz,[152] indicating 44% s character for the carbon orbital at C_1. The HOMO of cyclopropene has particularly large coefficients at C_3, which are well suited for coupling substituents at this position to the π orbitals. This is the source of the significant split between the π orbitals in the anti conformer of 3,3′-bicyclopropenyl **(43),** calculated by 4-31G to be 1.14 eV.[153,154] This is manifested in the low IP (8.76 eV) observed in the 3,3′-dimethyl derivative **44,** which is most likely to be stable in the gauche conformation found for bicyclopropyl.[153] The IP for this molecule is almost 1 eV lower than that of the parent compound.[153,154]

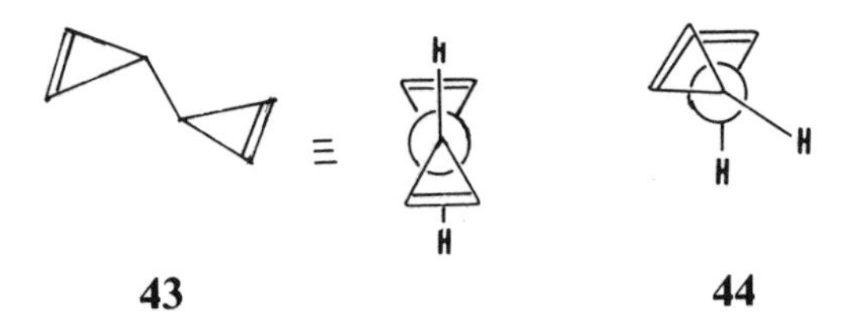

Some substituents at the 3-position can interact strongly enough with the double bond in cyclopropenes to produce what we may regard as a new entity rather than a perturbed cyclopropene. Certainly, cyclopropenium cations and cyclopropenones are the most extreme examples of this phenomenon. For example, diphenylcyclopropenone **(45)** is thermally stable to 130°C,[155] and its 21-kcal/mol ring resonance energy supports the widely held view that it is aromatic.[156–158] Recently, methylenecyclopropene **(46)** has been isolated, and its spectral properties suggest significant delocalization.[159,160] However, the present discussion considers only cyclopropenes having two formal sp^2-hybridized carbons, for which there are fewer data.

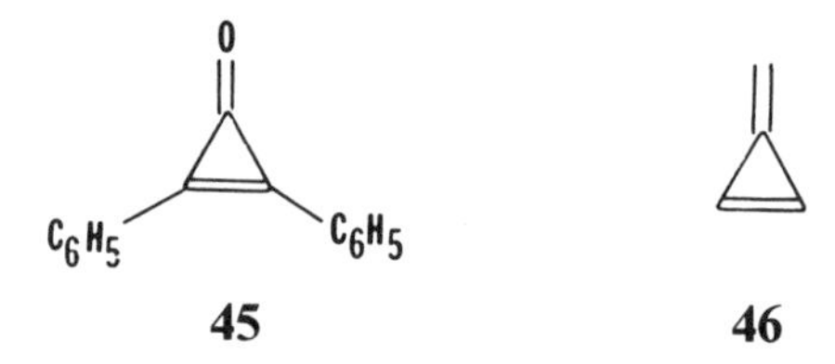

45 **46**

F.H. Allen[161] has reviewed the effects of substituents on the geometries of substituted cyclopropenes. Unfortunately, most of these are cyclopropenones, methylenecyclopropenes, and closely related molecules. Cyclopropene's C=C distance (1.296 Å) is shorter than that in ethylene and even shorter than in allene. Allen remarks that various 1-phenylcyclopropenes (eg, **47**) have short bonds to the aromatic substituent, which he estimates have about one-third double-bond character. By way of contrast, the C_1—C_1', bond length in biphenyl is 1.488 Å.[161]

$\tau = \pm 15°$

$d = 1.422$ Å

47

The ring structure of 3,3-difluorocyclopropene[162] **(48)** is reminiscent of 1,1-difluorocyclopropane[31,32] in that adjacent bonds are shortened and distal bonds lengthened relative to the parent hydrocarbon **(49).** Furthermore, lengthening of the C—F bonds in **48** is accompanied by compression of the FCF angle to 105.5°. This helps maintain the nonbonded F . . . F distance (2.173 Å) fairly close to that in 1,1-difluorocyclopropane (F . . . F, 2.197 Å, C—F, 1.355 Å; ∠FCF, 108.3°) in line with observations by Hargittai[163] (see also Chapter 1 in Volume 2 of this series). Cyclopropene **50**[164] has geminal π acceptors at C_3 and maintains a shortened distal bond and lengthened vicinal bonds just as in the related cyclopropanes. A calculational study of

F F 1.438 Å 1.321 Å ≡ 1.365 Å F 105.5° F 1.509 Å 1.296 Å

48 **49**

3-nitrosocyclopropene **(51)** indicates that the C_s anti form is the most stable structure.[165]

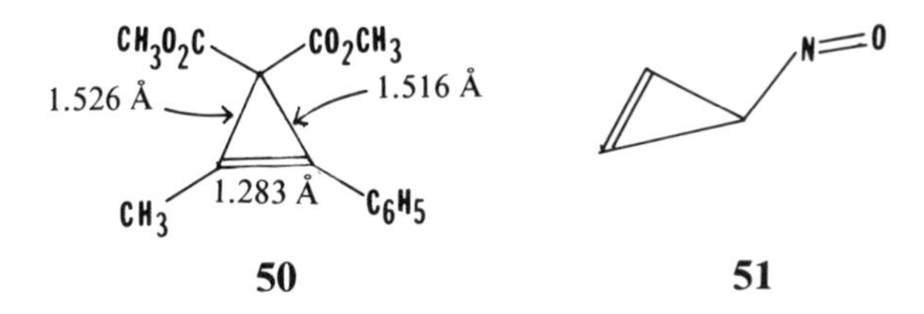

50 **51**

There are few thermochemical and good calculational data on substituted cyclopropenes. Gas phase enthalpies of hydrogenation of cyclopropene (Equation 5-23) and 1-methylcyclopropene (Equation 5-24) (ΔH_f°(l), 1-methylcyclopropene from Reference 95 combined with estimated ΔH_v° using Trouton's rule) yield −53.5 and −52.0 kcal/mol, respectively. The latter number appears to suggest a very modest stabilization energy, in contrast to what one expects for a position having some acetylenic properties and similarity to the 1-bicyclobutyl position. However, solution heats of hydrogenation[149] of 1,2-dimethylcyclopropenes (Equations 5-25 and 5-26) suggest very considerable stabilizations, perhaps even greater than we expect. Some interesting calculational (4–31G) comparisons are furnished by Equations 5-27 to 5-32.[109,166] The most interesting results involve the 3-lithio, 3-fluoro, and 3,3-difluoro derivatives. The extra stabilizations in the fluorinated derivatives (Equations 5-29 and 5-32) suggest an effect reminiscent of aromatic stabilization. The calculated stabilization in 3,3-difluorocyclopropene is about 40% that in cyclopropenone[109] or 8 kcal/mol. 3-Fluorocyclopropene can be studied at −50°C.[167] Its geminal J_{HF} appears to indicate an unusually large FCH angle, which could well enhance the molecule's stabilization through fluorine hyperconjugation. 3-Lithiocyclopropene is slightly destabilized (Equation 5-31), since it assumes some of the aspects of an antiaromatic ring. Its isomer, 1-lithiocyclopropene (Equation 5-28), is highly stabilized, since it resembles lithium acetylide. We will not discuss the highly unconventional geometries for polylithiated strained and unsaturated molecules except to point out that 3,3-dilithiocyclopropene has a low energy structure having planar tetracoordinate carbon.[168]

$$\Delta H = -53.5 \text{ kcal/mol} \qquad (5\text{-}23)$$

$$\Delta H = -52.0 \text{ kcal/mol} \qquad (5\text{-}24)$$

$$\Delta H = -43.3 \text{ kcal/mol} \qquad (5\text{-}25)$$

$$\Delta H = -40.2 \text{ kcal/mol} \qquad (5\text{-}26)$$

$$\Delta E_{4-31G} = -4.9 \text{ kcal/mol} \qquad (5\text{-}27)$$

$$\Delta E_{4-31G} = -27.0 \text{ kcal/mol} \qquad (5\text{-}28)$$

$$\Delta E_{4-31G} = -8.1 \text{ kcal/mol} \qquad (5\text{-}29)$$

$$\Delta E_{4-31G} = +0.2 \text{ kcal/mol} \qquad (5\text{-}30)$$

$$\Delta E_{4-31G} = +3.6 \text{ kcal/mol} \qquad (5\text{-}31)$$

$$\Delta E_{4-31G} = -9.6 \text{ kcal/mol} \qquad (5\text{-}32)$$

Interesting fluorinated derivatives of cyclopropene have been reported. We had earlier noted the amazing stability of hexakis(trifluoromethyl)-3,3′-bicyclopropenyl.[4] Perfluorocyclopropene **(52)**[169] is stable for 12 h at 105°C. However, it is flammable and explosive in air at room temperature and 1 atm. It reacts violently with triethylamine at −100°C.[169a] 1,3,3-Trifluorocyclopropene has also been synthesized.[169b] The trifluoromethyl derivatives **53–55** have been reported and survive the high temperatures of their generation.[170]

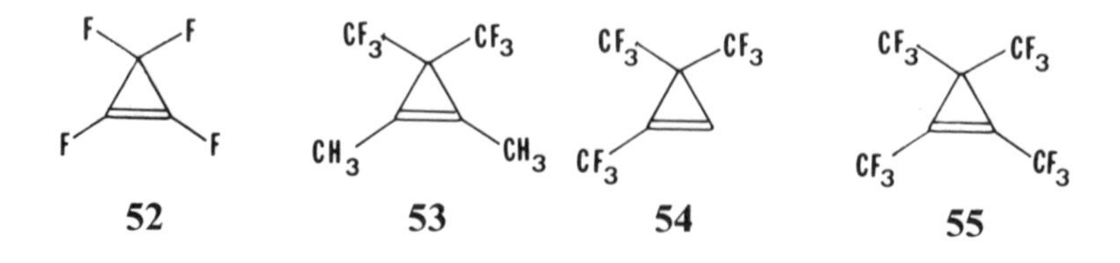

7. SUBSTITUTED SPIROPENTANES

Spiropentane **(56)** is an interesting molecule because its strain energy exceeds the sum of the strains of two cyclopropane rings by 10 kcal/mol.[1] Its molecular orbitals are depicted in Figure 5-11. One interesting aspect of this molecule is that its HOMO is higher in energy than that of cyclopropane, and this is reflected in its lower IP (9.45 eV, adiabatic),[171,172] which could be compared to the value for cyclopropane (9.8 eV)[173] except that the latter suffers Jahn–Teller distortion upon ionization. A more relevant comparison would be 1,1-dimethylcyclopropane, which has an adiabatic IP of 9.08 eV,[174] lower than that of spiropentane. This may explain why spiropentane is anodically stable while 1,1-dimethylcyclopropane is oxidized at > 2.5 V (vs SCE).[175] Of additional interest is the calculated prediction[176] that the difference between the Walsh orbitals and those having appreciable coefficients on hydrogen is calculated to be 0.092 au for cyclopropane and 0.118 au for spiropentane. This should indicate a difference in the balance of σ and π effects between the two molecular systems.

Few structural data are available on substituted spiropentanes. A study of chlorospiropentane[177] assumed equilateral triangular geometry for the three-membered rings. The electron diffraction structure of octafluorospiropentane **(57)** has been compared with the structure of spiropentane **(56)**.[178] The

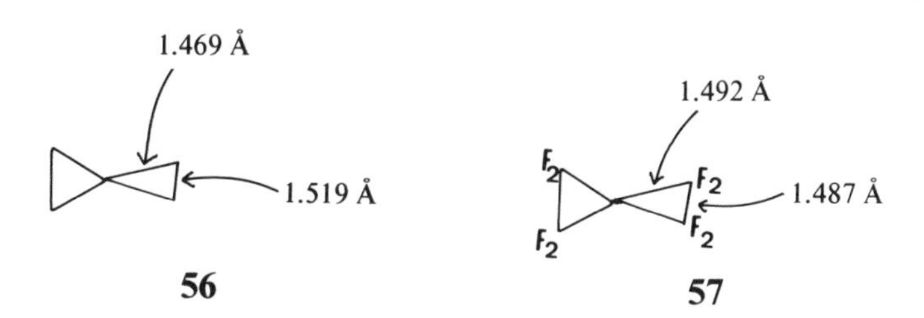

bond length variations in **57** are interesting in that they do not simply reflect additive relationships of parameters in 1,1-difluorocyclopropane **(5)**. In addition, the explanation of Clark, Schleyer, and co-workers[36] for the bond lengths in **5** does not seem to explain the structure of **57**. They explain the short vicinal and long distal lengths in **5** by citing inductive withdrawal from cyclopropane's 1E″ orbital (Figure 5-1*a*), thus decreasing the distal bonding and vicinal antibonding interactions. However, the cyclopropyl E″ pair is significantly split in spiropentane, and a glance at Figure 5-11 indicates that a more complex explanation is required to explain the structure of **57**. Thermolysis of **57** at 259°C produces exclusive extrusion of CF_2 in contrast to less substituted fluorospiropentanes, in which extrusion competes with isomerization or isomerization occurs exclusively.[179,180] The mode of reaction is highly dependent on the specific pattern of substitution.[179,180]

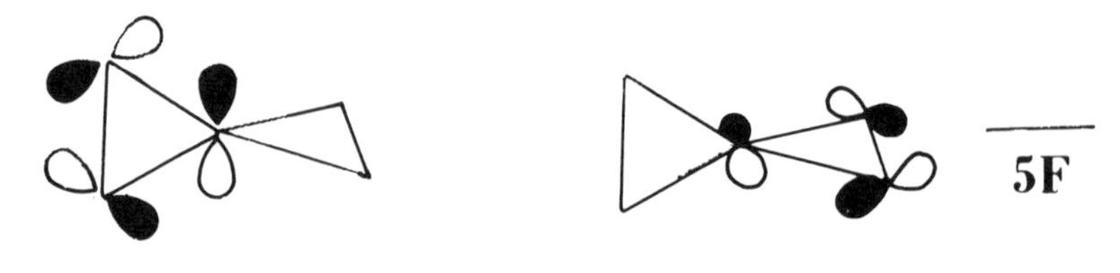

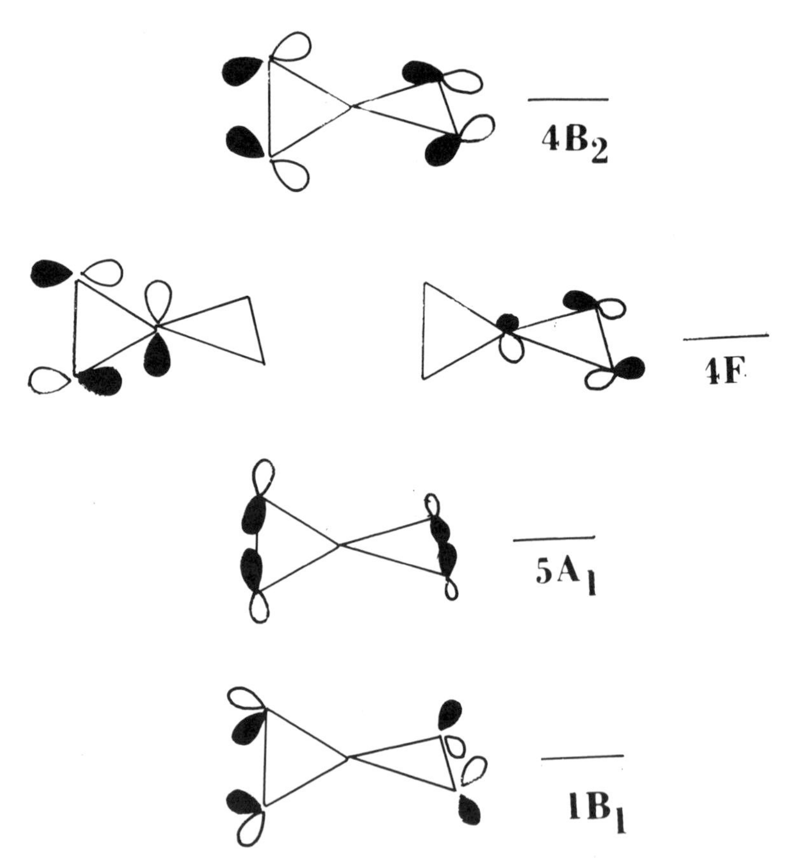

Figure 5-11. Highest-lying occupied and lowest unoccupied orbitals of spiropentane. (Adapted from Reference 176).

8. SUBSTITUTED POLYCYCLIC MOLECULES

It is worthwhile very briefly noting some polycyclic molecules that might provide very significant substrates for substituent effects. Quadricyclane (**58,** X = H) has an extremely low ionization potential (7.86 eV, adiabatic),[181,182] consistent with its electrochemical oxidation at only 0.91 V.[140] This molecule should therefore interact strongly with π-acceptor substituents such as CH_2^+, BH_2, and NO_2. Its highest occupied molecular orbital (Figure 5-12*a*) has nodes at the bridgehead positions joining the five-membered rings, thus indicating greatest conjugative stabilization by substituents placed at the other four cyclopropanoid positions. Placement of vicinal CF_3 groups at these positions raises $E_{1/2}$ by 1.28 V, whereas analogous substitution at the 2,3-positions of norbornadiene raises $E_{1/2}$ by only 0.95 V.[140a]

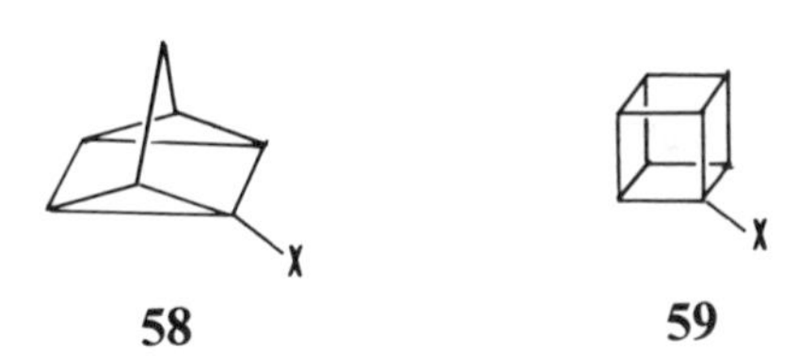

The highest occupied orbitals of cubane (**59,** X = H) are the triply degenerate $3T_{2g}$ set, which is followed by a triply degenerate subjacent $1T_{2u}$ set (see Figure 5-12*b*). The low IP of this molecule (9.0–9.1 eV, vertical;[183] 8.74 eV, adiabatic[171,172]) clearly indicates the presence of strain. A nice comparison molecule is adamantane, roughly the same size and polarizability as cubane, which also as a triply degenerate HOMO set. Its adiabatic IP is 9.25 eV.[171,172] The $3T_{2g}$ set has significant coefficients at hydrogen, which might indicate particular sensitivity to inductive effects, mitigated however by the polarizability of this eight-carbon polycycle. This is an interesting contrast to cyclopropane, where the Walsh orbitals almost exclusively involve carbons. Although the connection is not obvious, it is still worth remarking that cubane is 10^3 times more acidic than cyclopropane, even though $^1J_{CH}$ values indicate the same formal hybridization for the carbons.[184] The subjacent $1T_{2u}$ MO set is very similar to the Walsh-type orbitals in cyclopropane or cyclobutane. The IP and $E_{1/2}$ values of a number of other strained polycycles have been summarized by Gassman and Yamaguchi.[140a]

An interesting group of compounds for examination of substituent effects are the benzene valence isomers **43** and **60–62.** One might speculate about the ability to perturb the energy differences between these compounds as a function of location of the substituent.[185] In particular, Dewar benzene **(60)** and benzvalene **(61)** are quite close in energy. The vertical ionization potential of **60** is 9.40 eV[186] in fairly good agreement with MINDO/3 calculations[187] and this corresponds to ionization from the antisymmetric π-orbital combination.[188] Thus, substituents attached to the double bonds should exert the strongest effects. The HOMO for benzvalene is essentially π bonding,[188] and the vertical IP (8.54 eV)[186] also agrees fairly well with the MINDO/3 calculation.[187] The vertical IPs of prismane **(62)** and 3,3′-bicyclopropenyl **(43)** have been predicted to be 8.40 and 8.63 eV, respectively.[187] We have discussed the IP data available for 3,3′-bicyclopropenes.

Figure 5-12*c* depicts the HOMO for the recently synthesized[2]

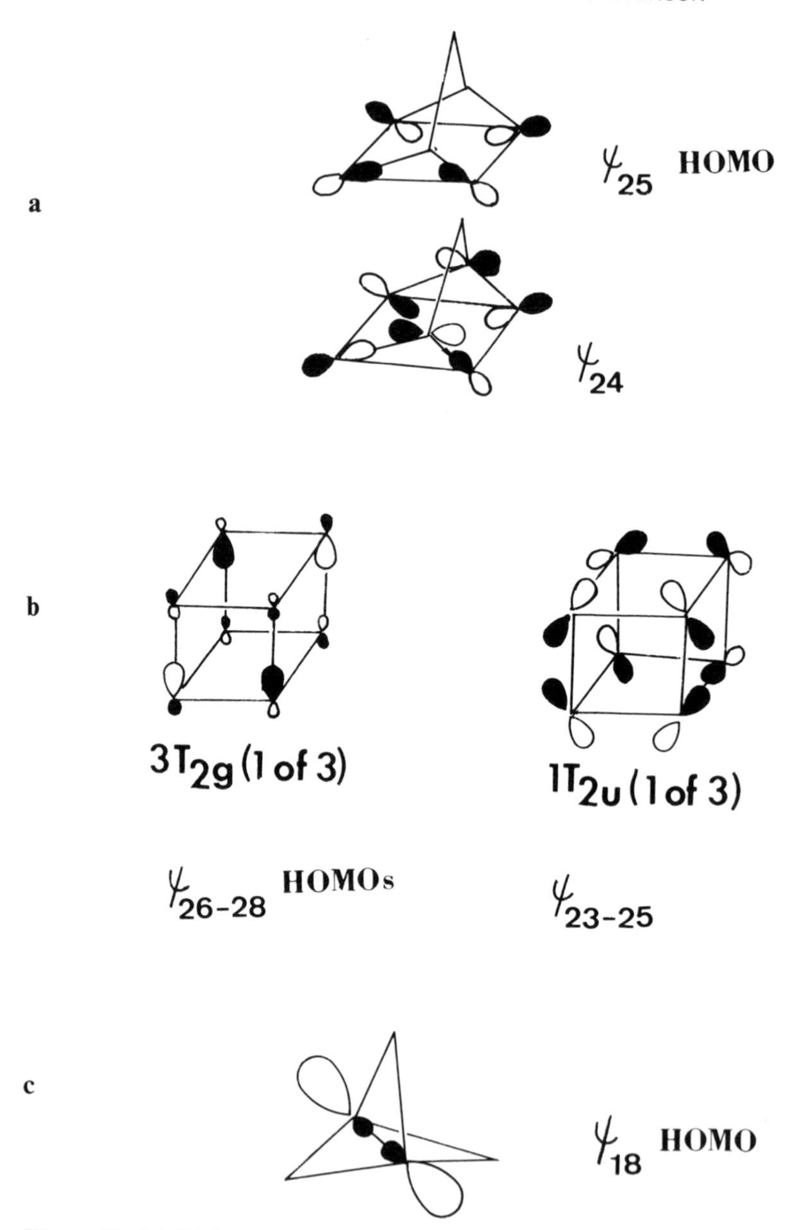

Figure 12. (*a*) Highest occupied orbitals of quadricyclane (see References 181, 182). (*b*) Highest occupied orbitals of cubane (see Reference 183). (*c*) Highest occupied orbital of [1.1.1]propellane (see Reference 189).

[1.1.1]propellane. It is interesting that this MO has no coefficients on substituent-bearing carbons.[189] The highly unusual bonding in this molecule, so reminiscent of electron-deficient species such as boron hydrides, makes this a highly interesting candidate for examination of substituent effects. Thus, we have come full circle. The previously unanticipated hydrocarbon mentioned at the beginning of this chapter may now furnish the most unique test of substituent effects.

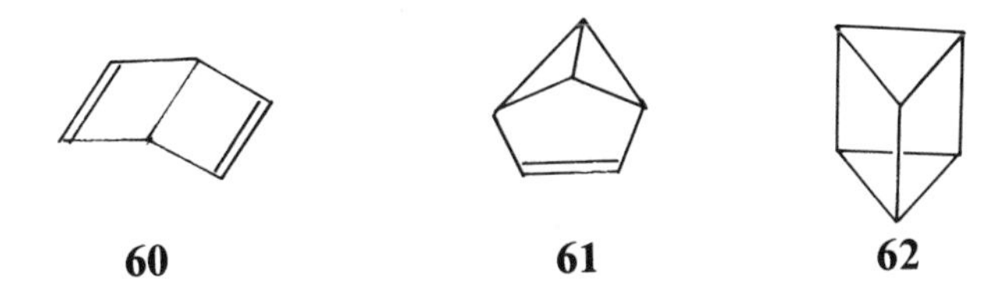

60 **61** **62**

REFERENCES

1. Greenberg, A.; Liebman, A. "Strained Organic Molecules". Academic Press: New York, 1978.
2. Wiberg, K.B.; Walker, F.H. *J. Am. Chem. Soc.* **1982,** *104,* 5239–5240.
3. (a) Maier, G.; Pfriem, S.; Schafer, U.; Matusch, R. *Angew. Chem. Int. Ed. Engl.* **1978,** *17,* 520–521. (b) X-ray structure: Irngartinger, H.; Goldmann, A.; Reiner, J.; Nixdorf, M.; Rodewald, H.; Maier, G.; Malsch, K.-D.; Emrich, R. *Angew. Chem. Int. Ed. Engl.* **1984,** *23,* 993–994.
4. Grayston, M.W.; Lemal, D.M. *J. Am. Chem. Soc.* **1976,** *98,* 1278–1280
5. Lemal, D.M.; Dunlap, L.H., Jr. *J. Am. Chem. Soc.* **1972,** *94,* 6562–6564.
6. Dolbier, W.R., Jr. *Acc.Chem. Res.* **1981,** *14,* 195–200.
7. Birchall, J.M.; Fields, R.; Haszeldine, R.N.; McClean, R.J. *J. Fluorine Chem.* **1980,** *15,* 487–495.
8. Demjanov, N.J. *Berichte* **1907,** *40,* 4393–4397, 4961–4963.
9. Kohler, E.P.; Conant, J.B. *J. Am. Chem. Soc.* **1917,** *39,* 1404–1420.
10. Carr, E.P.; Burt, C.P. *J. Am. Chem. Soc.* **1918,** *40,* 1590–1600.
11. Charton, M. In "The Chemistry of Alkenes", Vol. 2; Zabicky, J., Ed.; Wiley: London, 1970, pp. 511–610.
12. De Meijere, A. *Angew. Chem. Int. Ed. Engl.* **1979,** *18,* 809–826.
13. Hart, H.; Sandri, J.M. *J. Am. Chem. Soc.* **1959,** *81,* 320–326.
14. Schleyer, P.v.R.; Buss, V. *J. Am. Chem. Soc.* **1969,** *91,* 5880–5882.
15. Richey, H.G., Jr. In "Carbonium Ions", Vol. 3; Olah, G.A.; and Schleyer, P.v.R., Eds.; Wiley: New York, 1972, Chapter 25.
16. Wiberg, K.B.; Hess, B.A., Jr.; Ashe, A.J., III. In "Carbonium Ions", Vol. 3; Olah, G.A.; and Schleyer, P.v.R., Eds.; Wiley, New York, 1972, Chapter 26.
17. Deno, N.C.; Richey, H.G., Jr.; Liu, J.S.; Hodge, J.D.; Houser, J.J.; Wisotsky, M.J. *J. Am. Chem. Soc.* **1962,** *84,* 2016–2017.
18. Pittman, C.U., Jr.; Olah, G.A. *J. Am. Chem. Soc.* **1965,** *87,* 2998–3000.
19. Deno, N.C.; Liu, J.S.; Turner, J.O.; Lincoln, D.N.; Fruit, R.E., Jr. *J. Am. Chem. Soc.* **1965,** *87,* 3000–3002.
20. Kabakoff, D.S.; Namanworth, E. *J. Am. Chem. Soc.* **1970,** *92,* 3233–3234.
21. Hehre, W.J. In "Methods of Electronic Structure Theory", Vol. 4; Schaefer, H.F., III, Ed.; Plenum Press: New York, 1977, pp. 277–331.
22. Wolf, J.F.; Harch, P.G.; Taft, R.W.; Hehre, W.J. *J. Am. Chem. Soc.* **1975,** *97,* 2902–2904.
23. Brown, H.C.; Peters, E.N. *J. Am. Chem. Soc.* **1973,** *95,* 2400–2401.
24. Deno, N.C.; Richey, H.G., Jr.; Liu, J.S.; Lincoln, D.N.; Turner, J.O. *J. Am. Chem. Soc.* **1965,** *87,* 4533–4538.
25. Volz, H.; Shin, J.H.; Streicher, H.-J. *Tetrahedron Lett.* **1975,** 1297–1300.

26. Durmaz, S.; Kollmar, H. *J. Am. Chem. Soc.* **1980,** *102,* 6942–6945.
27. Streitweiser, A., Jr., 1982, see ref. 112 in Reference 5 above.
28. (a) Perkins, M.J.; Ward, P. *J. Chem. Soc. Chem. Commun.* **1971,** 1134–1135. (b) Perkins, M.J.; Ward, P. *J. Chem. Soc. Perkin Trans. 1* **1974,** 667–675.
29. (a) Groger, C.; Musso, H.; Rossnagel, I. *Chem. Ber.* **1980,** *113,* 3621–3628. (b) Musso, H. In "Current Trends in Organic Synthesis", Nozaki, H., Ed. (IUPAC); Pergamon Press: Oxford, 1983, pp. 371–378. (c) Musso, H. Personal communication to A. Greenberg.
30. Dolbier, W.R., Jr.; Enoch, H.O. *J. Am. Chem. Soc.* **1980,** *99,* 4532–4533.
31. Perretta, A.T.; Laurie, V.W. *J. Chem. Phys.* **1975,** *62,* 2469–2473.
32. Skancke, A.; Flood, E.; Boggs, J.E. *J. Mol. Struct.* **1977,** *40,* 263–270.
33. Hoffmann, R. *Tetrahedron Lett.* **1970,** 2907–2910.
34. Hoffmann, R. *Proc. R. Congr. Pure Appl. Chem.* **1971,** *2*(3), 233.
35. Hoffmann, R.; Stohrer, W.-D. *J. Am. Chem. Soc.* **1971,** *93,* 6941–6948.
36. Clark, T.; Spitznagel, G.W.; Klose, R.; Schleyer, P.v.R. *J. Am. Chem. Soc.* **1984,** *106,* 4412–4419.
37. Allen, F.H. *Acta Crystallogr.* **1980,** *B36,* 81–96.
38. Allen, F.H. *Acta Crystallogr.* **1981,** *B37,* 890–900.
39. Penn, R.E.; Boggs, J.E. *J. Chem. Soc. Chem. Commun.* **1972,** 666–667.
40. Ciganek, E. *J. Am. Chem. Soc.* **1965,** *87,* 652–653.
41. Ciganek, E. *J. Am. Chem. Soc.* **1965,** *87,* 1149–1150.
42. Wehner, R.; Gunther, H. *J. Am. Chem. Soc.* **1975,** *97,* 923–924.
43. Paquette, L.A.; Volz, W.E. *J. Am. Chem. Soc.* **1976,** *98,* 2910–2917.
44. Oberhammer, H.; Boggs, J.E. *J. Mol. Struct.* **1979,** *57,* 175–182.
45. Hoffmann, R. *Tetrahedron Lett.* **1965,** 3819–3824.
46. Skancke, A. *Acta Chem. Scand. A* **1982,** *36,* 637–639.
47. Kai, Y.; Knochel, P.; Kwiatkowski, S.; Dunitz, J.D.; Oth, J.F.M.; Seebach, D.; Kalinowski, H.O., *Helv. Chem. Acta* **1982,** *65,* 137–161.
48. Durig, J.R., Smooter-Smith, J.A.; Li, Y.S.; Wasacz, F.M., *J. Mol. Struct.* **1983,** *99,* 45–60.
49. De Mare, G.R.; Peterson, M.R. *J. Mol. Struct. (Theochem),* **1983,** *104,* 115–130.
50. Volltrauer, H.N.; Schwendeman, R.H. *J. Chem. Phys.* **1971,** *54,* 260–267.
51. Odum, J.D.; Szafran, Z.; Johnston, S.A.; Li, Y.S.; Durig, J.R. *J. Am. Chem. Soc.* **1980,** *102,* 7173–7180.
52. Noe, E.; Young, R.M. *J. Am. Chem. Soc.* **1982,** *104,* 6218–6220.
53. Martin, G.J.; Gouesnard, J.-P. *Tetrahedron Lett.* **1975,** 4251–4254.
54. Durig, J.R.; Trowell, P.L.; Szafran, Z.; Johnston, S.A.; Odom, J.D. *J. Mol. Struct.* **1981,** *74,* 85–95.
55. Durig, J.R.; Bist, H.D.; Saari, S.V.; Smooter-Smith, J.A.; Little, T.S. *J. Mol. Struct.* **1983,** *99,* 217–233.
56. Skancke, A.; Boggs, J.E. *Acta Chem. Scand.* **1978,** *A32,* 893–894.
57. Corkill, M.J.; Cox, A.P.; Norris, J.A. *J. Chem. Soc. Chem. Commun.* **1978,** 388–389.
58. De Meijere, A.; Lüttke, A. *Tetrahedron,* **1969,** *25,* 2047–2058.
59. Skancke, A.; Boggs, J.E. *J. Mol. Struct.* **1979,** *51,* 267–274.
60. Staley, S.W. *J. Am. Chem. Soc.* **1967,** *89,* 1532–1533.
61. Pasto, D.J.; Fehlner, T.P.; Schwartz, M.E.; Baney, J.F. *J. Am. Chem. Soc.* **1976,** *98,* 530–534.
62. Pearson, R., Jr.; Choplin, A.; Laurie, V.W. *J. Chem. Phys.* **1975,** *62,* 4859–4861.
63. Harmony, M.D.; Nandi, R.N.; Tietz, J.V.; Choe, J.-I.; Getty, S.J.; Staley, S.W. *J. Am. Chem. Soc.* **1983,** *105,* 3947–3951.
64. Taylor, W.H.; Harmony, M.D.; Cassada, D.A.; Staley, S.W. *J. Chem. Phys.* **1984,** *81,* 5379–5383.
65. Boggs, J.E. *J. Mol. Struct.* **1983,** *97,* 1–16.
66. Flygare, W.H.; Narath, A.; Gwinn, W.D. *J. Chem. Phys.* **1962,** *36,* 200–208.
67. Akelseev, N.V.; Barzdain, P.P.; Shostakovskii, M.V., *Zh. Strukt. Khim.* **1972,** *13,* 512–514.
68. Schwendeman, R.H.; Jacobs, J.D.; Krigas, T.M. *J. Chem. Phys.* **1964,** *40,* 1022–1028.
69. Cole, K.C.; Gilson, D.F.R. *J. Mol. Struct.* **1975,** *28,* 385–390.
70. Skancke, A. *J. Mol. Struct.* **1977,** *42,* 235–241.
71. Hedberg, L.; Hedberg, K.D.; Boggs, J.E. *J. Chem. Phys.* **1982,** *77,* 2996–3002.
72. Pulay, P.; Fogarasi, G.; Pang, F.; Boggs, J.E. *J. Am. Chem. Soc.* **1979,** *101,* 2550–2560.
73. Skancke, A.; Boggs, J.E. *J. Mol. Struct.* **1978,** *50,* 173–182.
74. Mathur, S.N.; Harmony, M.D. *J. Chem. Phys.,* **1978,** *69,* 4316–4318.

75. Hendricksen, D.K.; Harmony, M.D. *J. Chem. Phys.* **1969,** *51,* 700–705.
76. Harmony, M.D.; Bostrom, R.E.; Hendricksen, D.J. *J. Chem. Phys.* **1975,** *62,* 1599–1600.
77. (a) Kimura, K.; Katsumata, S.; Achiba, Y.; Yamazaki, T.; Iwata, S. "Handbook of HeI Photoelectron Spectra of Fundamental Organic Molecules". Japan Scientific Societies Press/Halsted Press: New York, 1981, pp. 117–119. (b) Albrecht, A.; Allan, M.; Haselbach, E.; Neuhas, L.; Carrupt, P.A. *Helv. Chim. Acta* **1984,** *67,* 220–223.
78. Staley, S.W.; Fox, M.A.; Cairncross, A. *J. Am. Chem. Soc.* **1977,** *99,* 4524–4526.
79. Jason, M.E.; Ibers, J.A. *J. Am. Chem. Soc.* **1977,** *99,* 6012–6021.
80. Lauher, J.W.; Ibers, J.A. *J. Am. Chem. Soc.* **1975,** *97,* 561–567.
81. Ghosh, P.K. In "Chemical Analysis", Vol. 67; Elving, P.J.; and Winefordner, J.D., Eds.; Wiley: New York, 1983, pp. 122–125.
82. Gillies, C.W. *J. Mol. Spectrosc.* **1976,** *59,* 482–492.
83. Chiang, J.F.; Bernett, W.A. *Tetrahedron* **1971,** *27,* 975–980.
84. Bent, H.A. *Chem. Rev.* **1961,** *61,* 275–311.
85. Pochan, J.M.; Baldwin, J.E.; Flygare, W.H. *J. Am. Chem. Soc.* **1969,** *91,* 1896–1898.
86. Bernett, W.A. *J. Org. Chem.* **1969,** *34,* 1772–1776.
87. (a) Deakyne, C.A.; Allen, L.C.; Laurie, V.W. *J. Am. Chem. Soc.* **1977,** *99,* 1343–1349. (b) Deakyne, C.A.; Allen, L.C.; Craig, N.C. *J. Am. Chem. Soc.* **1977,** *99,* 3895–3903.
88. Herzberg, G. "Molecular Spectra and Molecular Structure", Vol. 2; "Infrared and Raman Spectra of Polyatomic Molecules", Van Nostrand Press: Princeton, N.J., 1945, pp. 351–353.
89. Baker, A.W.; Lord, R.C. *J. Chem. Phys.* **1955,** *23,* 1636–1643.
90. Brecher, C.; Krikorian, E.; Blau, J.; Halford, R.S. *J. Chem. Phys.* **1961,** *35,* 1097–1108.
91. Shimanouchi, T. "Tables of Molecular Vibration Frequencies", Consolidated Volume 1; NSRDS-NBS 39. National Bureau of Standards: Washington, D.C., 1972.
92. Yokozeki, A.; Bauer, S.H. *Top. Curr. Chem.* **1975,** *53,* 71–119.
93. Marriot, S.; Topsom, R.D. *J. Mol. Struct. (Theochem)* **1984,** *109,* 305.
94. Hencher, J.L. In "The Chemistry of the Carbon–Carbon Triple Bond", Part 1; Patai, S., ed.; Wiley: London, 1978, pp. 57–67.
95. Pedley, J.B.; Rylance, J. "Sussex–N.P.L. Computer-Analysed Thermochemical Data: Organic and Organometallic Compounds", University of Sussex: Brighton, UK, 1977.
96. Wiberg, K.B.; Squires, R.R. *J. Am. Chem. Soc.* **1981,** *103,* 4473–4478.
97. Rozhnov, A.M.; Nesterova, T.N. *Russ. J. Phys. Chem.* **1978,** *52,* 1068–1069.
98. Dill, J.D.; Greenberg, A.; Liebman, J.F. *J. Am. Chem. Soc.* **1979,** *101,* 6814–6826.
99. Fuchs, R.; Hallman, J.H.; Perlman, M.O. *Can. J. Chem.,* **1982,** *60,* 1832–1835.
100. Chesick, J.P. *J. Am. Chem. Soc.* **1963,** *85,* 2720–2723.
101. Crawford, R.J.; Tokunaga, H. *Can. J. Chem.* **1974,** *52,* 4033–4039.
102. See also: Dolbier, W.R., Jr.; Medinger, K.S.; Greenberg, A.; Liebman, J.F. *Tetrahedron* **1982,** *38,* 2415–2420.
103. Dolbier, W.R., Jr.; Burkholder, C.R. *Tetrahedron Lett.* **1983,** 1217–1220.
104. Kirmse, W.; Murawski, H.-R. *J. Chem. Soc.* **1977,** 122–124.
105. Oth, J.F.M.; Merenyi. R.; Schroder, G. *Tetrahedron Lett.* **1968,** 3941–3946.
106. Hine, J. "Structural Effects on Equilibria in Organic Chemistry", Wiley: New York, 1975, pp. 270–276.
107. Hehre, W.J.; Ditchfield, R.; Radom, L.; Pople, J.A. *J. Am. Chem. Soc.* **1979,** *92,* 4796–4801.
108. George, P.; Trachtman, M.; Brett, A.M.; Bock, C.W. *J. Chem. Soc. Perkin Trans.* **1977,** *8,* 1036–1047.
109. Greenberg, A.; Liebman, J.F.; Dolbier, W.R., Jr.; Medinger, K.S.; Skancke, A. *Tetrahedron* **1983,** *39,* 1533–1538.
110 (a) Collins, J.B.; Dill, J.D.; Jemmis, E.D.; Apeloig, Y.; Schleyer, P.v.R.; Seeger, R.; Pople, J.A. *J. Am. Chem. Soc.* **1976,** *98,* 5419–5427. (b) Dill, J.D.; Schleyer, P.v.R.; Pople, J.A. *J. Am. Chem. Soc.* **1976,** *98,* 1663–1668.
111. Ehrenson, S.; Brownlee, R.T.C.; Taft, R.W. *Prog. Phys. Org. Chem.* **1973,** *10,* 1–80.
112. Topsom, R.D. *Acc. Chem. Res.* **1983,** *16,* 292–298.
113. Marriot, S.; Topsom, R.D. *J. Am. Chem. Soc.* **1984,** *106,* 7–10.
114. Marriot, S.; Topsom, R.D. *J. Mol. Struct. (Theochem)* **1984,** *106,* 277–286.
115. Charton, M. In "The Chemistry of the Functional Groups", Supplement C; Patai, S.; and Rappoport, Z., (eds.; Wiley: New York, 1983, Chapter 8.
116. Fuchs, R. *J. Chem. Educ.* **1984,** *61,* 133–136.
117. Charton, M. *Prog. Phys. Org. Chem.* **1981,** *13,* 119–251.

118. Liebman, J.F.; Hyman, A.S.; Ladon, L.H.; Stevenson, T.A.; Greenberg, A. Unpublished observations.
119. Furet, P.; Hallak, G.; Matcha, R.L.; Fuchs, R. Submitted for publication to *Canad. J. Chem.*; R. Fuchs, personal communication.
120. (a) Charton, M.; Greenberg, A.; Stevenson, T.A. *J. Org. Chem.* **1985,** *50,* 2643–2646. (b) Greenberg, A.; Stevenson, T.A. *J. Am. Chem. Soc.,* **1985,** *107,* 3488–3494.
121. (a) Trachtenberg, E.N.; Odian, G. *J. Am. Chem. Soc.* **1958,** *80,* 4018–4022. (b) Fuchs,. R.; Bloomfield, J.J. *J. Org. Chem.* **1963,** *28,* 910–912.
122. Dolbier, W.P., Jr.; Medinger, K.S.; Greenberg, A.; Liebman, J.F. *Tetrahedron* **1982,** *38,* 2415–2420.
123. O'Neal, H.E.; Benson, S.W. *J. Phys. Chem.* **1968,** *72,* 1866–1887.
124. Greenberg, A.; Liebman, J.F., Van Vechten, D. *Tetrahedron* **1980,** *36,* 1161–1166.
125. Moriarty, R.M. In "Topics in Stereochemistry", Vol. 8; Eliel, E.L.; and Allinger, N.L., Eds.; Wiley: New York, 1974, pp. 271–421.
126. Hoffmann, R.; Davidson, R.B. *J. Am. Chem. Soc.* **1971,** *93,* 5699–5705.
127 (a) Radom, L.; Pople, J.A.; Schleyer, P.v.R. *J. Am. Chem. Soc.* **1972,** *94,* 5935–5945. (b) Franke, W.; Schwarz, H.; Thies, H.; Chandrasekhar, J.; Schleyer, P.v.R.; Hehre, W.J.; Saunders, M.; Walker, G. *Chem. Ber.* **1981,** *114,* 2808–2824. (For 4–31G total energy of cyclobutane, see Reference 166, below.)
128. Allen, F.H. *Acta Crystallogr.* **1984,** *B40,* 64–72.
129. Dakkouri, M.; Oberhammer, H. *J. Mol. Struct.* **1983,** *102,* 315–324.
130. Jonvik, T.; Boggs, J.E. *J. Mol. Struct. (Theochem)* **1981,** 293–302.
131. Hargittai, I. "The Structure of Volatile Sulphur Compounds". D. Reidel: Dordrecht, Holland, 1985.
132. Harano, K.; Ban, T.; Yasuda, M.; Osawa, E.; Kanematsu, K. *J. Am. Chem. Soc.* **1981,** *103,* 2310–2317.
133. Osawa, E.; Ivanov, P.M.; Jaime, C. *J. Org. Chem.* **1983,** *48,* 3990–3993.
134. Fuchs, R.; Hallman, J. H. *Can. J. Chem.* **1983,** *61,* 503–505.
135. Wiberg, K.B. *J. Am. Chem. Soc.* **1983,** *105,* 1227–1233.
136. Taft, R.W. In "Steric Effects in Organic Chemistry", Newman. M.S., ed.; Wiley: New York, 1956.
137. (a) Charton, M. *Top. Curr. Chem.* **1983,** *114,* 57–91. (b) Wiberg, K.B. *Tetrahedron Lett.* **1985,** 599–602.
138. Pomerantz, M.; Fink, R.; Gray, G. *J. Am. Chem. Soc.* **1976,** *98,* 291–292.
139. Wilberg, K.B.; Lampman, G.M.; Ciula, R.P.; Conner, D.S.; Schertler, P.; Lavanish, J. *Tetrahedron* **1965,** *21,* 2749–2769.
140. (a) Gassman, P.G.; Yamaguchi, R. *Tetrahedron* **1982,** *38,* 1113–1122. (b) Richtsmeier, S.C.; Gassman, P.G., Dixon, D.A. *J. Org. Chem.* **1985,** *50,* 311–317.
141. McDonald, R.N.; Reitz, R.R., *J. Am. Chem. Soc.* **1976,** *98,* 8144–8155.
142. Irngartinger, H.; Lukas, K.L. *Angew. Chem. Int. Ed. Engl.* **1979,** *18,* 694–695.
143. Paddon-Row, M.N.; Houk, K.N.; Dowd, P.; Garner, P.; Schappert, R. *Tetrahedron Lett.* **1981,** 4799–4802.
144. Eisenstein, M.; Hirschfeld, F.L. *Acta Crystallogr.* **1983,** *B39,* 61–75.
145. Allen, F.H. *Acta Crystallogr.* **1984,** *B40,* 64–72.
146. Woodward, R.B.; Dalrymple, D.L. *J. Am. Chem. Soc.* **1969,** *91,* 4612–4613.
147. A d_{13} of 1.838 Å has been reported for 1,3-trimethylsilyl-2,4-dimethyl-2,4-disilabicyclobutane (see Reference 145).
148 (a) Newton, M.D.; Schulman, J.M. *J. Am. Chem. Soc.* **1972,** *94,* 767–773. (b) Mahler, W. *J. Am. Chem. Soc.* **1962,** *84,* 4600–4601.
149. Turner, R.B.,; Goebel, P.; Mallon, B.J.; Doering, W.v.E.; Coburn, J.F,.; Pomerantz, M. *J. Am. Chem. Soc.* **1968,** *90,* 4315–4322.
150. Hall, H.K., Jr.; Baldt, J.H. *J. Am. Chem. Soc.* **1971,** *93,* 140–145.
151. Greenberg, A. *Tetrahedron Lett.* **1978,** 3509–3512.
152. Closs, G. L. In "Advances in Alicyclic Chemistry", Vol. 1; Hart, H.; and Karabatsos, G.J., Eds.; Academic Press: New York, 1966, pp. 53–127.
153. Greenberg, A.; Liebman, J.F. *J. Am. Chem. Soc.* **1981,** *103,* 44–48.
154. Spanget-Larsen, J.; de Korswagen, C.; Eckert-Maksić, M.; Gleiter, R. *Helv. Chim. Acta* **1982,** *65,* 968–982.
155. Breslow, R.; Haynie, R.; Mirra, J. *J. Am. Chem. Soc.* **1959,** *81,* 247–248.
156. Greenberg, A.; Tomkins, R.P.T.; Dobrovolny, M.; Liebman, J.F. *J. Am. Chem. Soc.* **1983,** *105,* 6855–6858.

157. Grabowski, J.J.; Simon, J.D.; Peters, K.S. *J. Am. Chem. Soc.* **1984,** *106,* 4615–4616.
158. Steele, W.V.; Gammon, B.E.; Smith, N.K.; Chickos, J.S.; Greenberg, A.; Liebman, J.F. *J. Chem. Thermodyn.* **1985,** *17,* 505–511.
159. Billups, W.E.; Lin, L.-J.; Casserly, E.W. *J. Am. Chem. Soc.* **1984,** *106,* 3698–3699.
160. Staley, S.W.; Norden, T.D. *J. Am. Chem. Soc.* **1984,** *106,* 3699–3700.
161. Allen, F.H. *Tetrahedron* **1982,** *38,* 645–655.
162. Ramaprasad, K.R.; Laurie, V.W.; Craig, N.C. *J. Chem. Phys.* **1976,** *74,* 4832–4835.
163. Hargittai, I. *Top. Curr. Chem.* **1982,** *96,* 44–77.
164. Rømming, C.; Berg, A.S., *Acta Chem. Scand.* **1979,** *A33,* 271–274.
165. Poppinger, D. *Aust. J. Chem.* **1976,** *29,* 465–478.
166. Hehre, W.J.; Pople, J.A. *J. Am. Chem. Soc.* **1975,** *97,* 6941–6955.
167. Craig, N.C.; Sloan, K.L.; Sprague, J.R.; Stevens, P.S. *J. Org. Chem.* **1984,** *49,* 3847–3848.
168. Collins, J.B.; Dill, J.D.; Jemmis, E.D.; Apeloig, Y.; Schleyer, P.v.R.; Seeger, R.; Pople, J.A. *J. Am. Chem. Soc.* **1976,** *98,* 5419–5427.
169. (a) Sargeant, P.B.; Krespan, C.G. *J. Am. Chem. Soc.* **1969,** *91,* 415–419. (b) Craig, N.C.; Lai, R.K.-Y.; Penfield, K.W. *J. Phys. Chem.* **1980,** *84,* 899–906.
170. Cullen, W.R.; Waldman, M.C. *Can. J. Chem.* **1979,** *48,* 1885–1892.
171. Dewar, M.J.S.; Worley, S.D. *J. Chem. Phys.* **1969,** *50,* 654–667.
172. Bodor, N.; Dewar, M.J.S.; Worley, S.D. *J. Am. Chem. Soc.* **1979,** *92,* 19–24.
173. Bieri, G.; Burger, F.; Heilbronner, E.; Maier, J.P. *Helv. Chim. Acta* **1977,** *60,* 2213–2233.
174. Lossing, F.P. *Can. J. Chem.* **1972,** *50,* 3973–3981.
175. Shono, T.; Matsumura, Y. *Bull. Chem. Soc. Japan* **1975,** *48,* 2861–2864.
176. Jorgensen, W.L.; Salem, L. "The Organic Chemist's Book of Molecular Orbitals", Academic Press: New York, 1972.
177. Woerner, L.M.; Harmony, M.D. *J. Chem. Phys.* **1966,** *45,* 2339–2343.
178. Dolbier, W.R., Jr., Sellers, S.F.; Smart, B.E.; Oberhammer, H. *J. Mol. Struct.* **1983,** *101,* 193–197.
179. Dolbier, W.R., Jr.; Sellers, S.F.; Smart, B.E. *Tetrahedron Lett.* **1981,** 2953–2956.
180. Dolbier, W.R., Jr., Sellers, S.F.; Al-Sader, B.H.; Fielder, T.H., Jr. *J. Am. Chem. Soc.* **1981,** *103,* 717–718.
181. Martin, H.-D.; Heller, C.; Haselbach, E.; Lanyiova, Z. *Helv. Chim. Acta* **1974,** *57,* 465–472.
182. Haselbach, E.; Bally, T.; Lanyiova, Z.; Baertschi, P. *Helv. Chim. Acta* **1979,** *72,* 583–592.
183. Bischoff, R.; Eaton, P.E.; Gleiter, R.; Heilbronner, E.; Jones, T.B.; Musso, H.; Schmelzer, A.; Stober, R. *Helv. Chim. Acta* **1978,** *61,* 547–557.
184. Luh, T.-Y.; Stock, L.M. *J. Am. Chem. Soc.* **1974,** *96,* 3712–3713.
185. Greenberg, A.; Liebman, J.F. *Tetrahedron* **1979,** *35,* 2623–2627.
186. Levin, R.D.; Lias, S.G. "Ionization Potential and Appearance Potential Measurements, 1971–1981". NSRDS-NBS 71. National Bureau of Standards: Washington, D.C., 1982.
187. Bews, J.R.; Glidewell, C. *J. Mol. Struct. (Theochem)* **1982,** *86,* 197–207.
188. Newton, M.D.; Schulman, J.M.; Manus, M.M. *J. Am. Chem. Soc.* **1974,** *96,* 17–23.
189. Jackson, J.E.; Allen, L.C. *J. Am. Chem. Soc.* **1984,** *106,* 591–599.

APPENDIX

Ab initio (4–31G) total energies in atomic units (hartree, 1 hartree = 627.5 kcal/mol), level of geometry optimization, and gas phase enthalpies of formation [ΔH_f°(g), kcal/mol] for substituted methanes, ethanes, 2-propanes, acetylenes, ethylenes, cyclopropanes, and benzenes. Unless otherwise specified, ΔH_f°(g) values are from the Pedley and Rylance compendium.[95] Terms under geometry optimization (Geom) have the following meaning: 4–31G, 4–21G, STO-3G, DZ (double-zeta): optimized at respective calculational levels; EXP, based on experimental structure; *p*4–31G or STO-3G: partial optimization of these levels; Est, estimated; Std, standard structures. Superscript numbers designate Appendix References and Notes A1–A92.

Methanes (CH_3X)

X	4–31G Energy	Geom	ΔH_f°	X	4–31G Energy	Geom	ΔH_f°
H	40.13977	4–31G[1]	−17.8	NH_2	95.07166	4–31G[1]	−5.5
CH_3	79.11593	4–31G[1]	−20.1	NH_3^+	95.43739	pSTO-3G[12]	147.6[13]
CH_2^+	78.19496	STO-3G[2]	219[3]	NO_2	243.27451	4–31G[1]	−17.9
CH_2^-	78.38126	4–31G[1,4]		OCH_3	153.83622	Exp[14]	−44.0
	78.36268	p4–31G[5,6]		OH	114.87152	4–31G[1]	−48.2
CO_2CH_3	266.4298	STO-3G[7]	−98.0	O^-	114.21840	4–31G[1]	−38.5[15]
$COCH_3$	191.67699	Exp[8]	−51.9	F	138.85681	4–31G[1]	−56.8[16]
CHO	152.68653	4–31G[1]	−39.6	Cl	498.52260	Exp[8]	−19.6
$CH=CH_2$	116.90510	4–31G[1]	4.8	Li	46.96000	4–31G[1]	
CCH	115.70133	4–31G[1]	44.6	BH_2	65.34844	4–31G[1]	
CF_3	375.33319	4–31G[1]	−178.8				
CN	131.72827	4–31G[1]	17.7[9]				
NC	131.69409	p4–31G[10]	41.3[11]				

Ethanes (C_2H_5X)

X	4–31G Energy	Geom	ΔH_f°	X	4–31G Energy	Geom	ΔH_f°
H	79.11593	4–31G[1]	−20.1	NC	170.67496	p4–31G[17]	33.8[18]
CH_3	118.09381	4–31G[1]	−25.0	NH_2	134.05115	p4–31G[4,19]	−11.4
CH_2^+	117.18167	4–31G	208[3]		134.04795	Std[6,20]	
CH_2^-	117.36467	4–31G[1,4]		NH_3^+	134.42205	pSTO-3G[12]	138.8[13]
	117.34778	p4–31G[5,6]		NO_2	282.25646	4–31G[21]	−24.4
CO_2CH_3	305.41558	4–21G[24]	−102.5[23]	OCH_3	192.82002	Exp[24]	−51.7
$COCH_3$	230.65561	4–31G[25]	−57.6	OH	153.85411	Std[26]	−56.1
CHO	191.66572	4–21G[27]	−44.8	O^-	153.19990	pSTO-3G[12]	−48.5[15]
$CH{=}CH_2$	155.88080	Std[28]	−0.1	F	177.84496	4–31G[29]	−62.5[30]
CCH	154.67849	4–31G[31]	39.5	Cl			−26.8
CF_3	414.31301	4–31G[32]		Li			−15.2
CN	170.70592	4–31G[34]	12.3	BH_2	104.31919	Std[35]	

2-Propanes (CH_3CHXCH_3)

X	4–31G Energy	Geom	ΔH_f°	X	4–31G Energy	Geom	ΔH_f°
H	118.09381	4–31G[1]	−25.0	NC			
CH_3	157.07260	4–31G[36]	−32.1	NH_2	173.03415	4–31G[36]	−20.0
CH_2^+			199[3]	NH_3^+	173.40627	pSTO-3G[12]	127.8[13]
CH_2^-				NO_2			−33.2
CO_2CH_3				OCH_3			−60.2
$COCH_3$			−62.7	OH	192.83938[36]	4–31G[36]	−65.1
CHO			−51.5	O^-	192.18789	pSTO-3G[12]	−58.8[15]
$CH=CH_2$			−6.5	F	216.83100	4–31G[36]	−70.1
CCH			32.6[37]	Cl			−34.7
CF_3				Li	124.80046	4–31G[36]	
CN			5.9	BH_2	143.29581	4–31G[36]	

Acetylenes (HCCX)

X	4–31G Energy	Geom	ΔH_f°	X	4–31G Energy	Geom	ΔH_f°
H	76.71141	4–31G[38]	54.5	NC	168.26375	Est[39]	
CH_3	115.70133	4–31G[1]	44.6	NH_2	131.6616	4–31G[4,40]	
CH_2^+	114.78923	STO-3G[41]	281[42]	NH_3^+	131.98118	4–31G[43]	
CH_2^-	115.03353	4–31G[1]		NO_2	279.81649	Std[44]	
CO_2CH_3	303.00513	4–21,31G[45]		OCH_3	190.40837	Exp, 4–31G[46]	
$COCH_3$	228.25181	Std[44]		OH	151.44421	4–31G[47]	
CHO	189.26094	Std[44]		O^-	150.8778	4–31G[48]	
$CH=CH_2$	153.48995	STO-3G[28]	69[49]	F	175.4099	STO-3G[50]	
CCH	152.28374	STO-3G[38]	108.8[51]	Cl	535.0979	STO-3G[50]	
CF_3	411.89036	Exp[52]		Li			
CN	168.3033	STO-3G[50]		BH_2			

Ethylenes ($CH_2{=}CHX$)

X	4–31G Energy	Geom	ΔH_f°	X	4–31G Energy	Geom	ΔH_f°
H	77.92216	4–31G[38]	12.5	NC	169.48569	6–31G[53]	
CH_3	116.90510	4–31G[1]	4.8	NH_2	132.87521	4–31G[6,44]	11.5±4.1[91]
CH_2^+	116.02511	STO-3G[54]	226[42]		132.85890	Std[20,55]	
	115.96939	STO-3G[54,55]		NH_3^+	133.22214	STO-3G[65]	173[57]
CH_2^-	116.22676	4–31G[1,6]	29.0[58]	NO_2	281.06442	4–31G[44]	9.0[59]
	116.18251	4–31G[55,61]		OCH_3	191.63602	4–31G[61]	−26.0[62]
CO_2CH_3	304.22496	Exp,est[63]	−79.6	OH	152.67074	4–31G[64]	−29.8[65]
$COCH_3$	229.46562	Std[44]	−32.1[66]	O^-	152.0624	4–31G[7]	−38.3[62]
CHO	190.47909	4–31G[44]	−16.2[67]	F	176.65147	4–31G[68]	−33.2
$CH{=}CH_2$	154.69983	4–31G[69]	26.3	Cl	536.32548	STO-3G[70]	5.0[71]
CCH	153.48995	STO-3G[28]	69[49]	Li			
CF_3	413.11525	Exp[72]	−146.8	BH_2	103.14016	Std[6,20]	
CN	169.51592	4–31G[68]	43.2[73]		103.12736	Std[20,55]	

Cyclopropanes (cyclo-C_3H_5X)

X	4–31G Energy	Geom	ΔH_f°	X	4–31G Energy	Geom	ΔH_f°
H	116.88385	4–31G[74]	12.7	NC	208.44363	4–31G, 6–31G	
CH_3	155.86431	4–31G[36]	6.2[62]	NH_2	171.82615	4–31G[36]	18.4
CH_2^+	154.99277	STO-3G[41,76]		NH_3^+	172.19149	DZ[77]	170.6[13]
	154.94407	STO-3G[41,78]		NO_2	320.02569	Exp, 4–21G[76,79]	
CH_2^-					320.01871	4–21G[78,80]	
						4–31G,	
CO_2CH_3				OCH_3	230.58874	DZ$<$Exp[81]	
$COCH_3$	268.43155	4–31G[82]		OH	191.62506	4–31G[36]	
CHO				O^-	190.98167	DZ[83]	
$CH=CH_2$	193.65341	4–31G[84]	(35.9)[85]	F	215.61111	4–31G[36]	
CCH	192.44807	Exp[86]		Cl			
CF_3				Li	123.70079	4–31G[36]	
CN	208.47677	4–21G[87]	43.7[73]	BH_2	142.10216[36,76]		
					142.09053[36,78]		

Benzenes (C_6H_5X)

X	4–31G Energy	Geom	ΔH_f°	X	4–31G Energy	Geom	ΔH_f°
H	230.37750	4–31G[36]	19.8	NC			
CH_3	269.35568	Std[44]	12.0	NH_2	285.32500	Std[6,44]	20.8
CH_2^+	268.65231	STO-3G[41]			285.31795	Std[4,44]	
CH_2^-	268.68709	pSTO-3G[1]		NH_3^+			176.5[13]
CO_2CH_3			−68.8	NO_2	433.51617	Std[44]	16.2
$COCH_3$			−20.7	OCH_3	344.08385	pSTO-3G[88]	−16.2
CHO	342.93279	Std[44]	−8.8	OH	305.12115	Std[44]	−23.0
$CH=CH_2$	307.14078	Std[44]	35.3	O^-			
CCH	305.94397	Std[44]	75.2	F	329.10876	4–31G[90]	−27.7
CF_3			−143.2	Cl			12.3
CN	321.96792	Std[44]	51.6	Li	237.20279[36]		
				BH_2			

APPENDIX REFERENCES AND NOTES

A1. Pross, A.; DeFrees, D.J.; Levi, B.A.; Pollack, S.K.; Radom, L.; Hehre, W.J. *J. Org. Chem.* **1981,** *46,* 1693–1699.
A2. Cremer, D.; Binkley, J.S.; Pople, J.A.; Hehre, W.J. *J. Am. Chem. Soc.* **1974,** *96,* 6900–6903.
A3. Lossing, F.P.; Semeluk, G.P. *Can. J. Chem.* **1979,** *48,* 955.
A4. Pyramidal substituent.
A5. Pross, A.; Radom, L. *Aust. J. Chem.* **1980,** *33,* 241–248.
A6. Planar substituent; conjugated with π system if present.
A7. Hopkinson, A.C.; Lien, M.H. *Int. J. Quant. Chem.* **1980,** *18,* 1371–1391.
A8. Kollman, P.; Rothenberg, S. *J. Am. Chem. Soc.* **1977,** *99,* 1333–1342.
A9. An, X.-W.; Mansson, M. *J. Chem. Thermodyn.* **1983,** *15,* 287–293.
A10. This work; employed model structure using C—N, C=N same as C_2H_5NC. See discussion of this molecule, 4–31G CH_3 structure from Reference A1.
A11. Baghal-Vayjooee, M.H.; Collister, J.L.; Pritchard, H.O. *Can. J. Chem.* **1977,** *55,* 2634–2636.
A12. Radom, L. *Aust. J. Chem.* **1975,** *28,* 1–6.
A13. Bowers, M.T., Ed.; "Gas Phase Ion Chemistry", Vol. 2.; Academic Press: New York, 1979, Chapter 9.
A14. Tse, Y.C.; Newton, M.D.; Allen, R.C. *Chem. Phys. Lett.* **1980,** *75,* 350–356.
A15. Sullivan, S.A.; Beauchamp, J.L. *J. Am. Chem. Soc.* **1976,** *98,* 1106–1165.
A16. Average of two estimated values: −57.8 kcal/mol (Joshi, R.M. *J. Macromol. Sci. Chem.* **1974,** *8,* 861–885); −55.9 kcal/mol (Stull, D.R.; Westrum, E.F.; Sinke, G.C. "The Chemical Thermodynamics of Organic Compounds", Wiley: New York; 1969).
A17. This work; employed with NC = 1.167 Å from 6–31G study of C_2H_3NC (Moffat, J.G. *J. Phys. Chem.* **1977,** *81,* 82–86); C—N = 1.415 Å, interpolated from STO-3G studies of C_2H_5—CN and C_2H_5—NC (Moffat, J.B. *Int. J. Quant. Chem.* **1981,** *19,* 771–779) and 4–31G study of CH_3—CN (Reference A1).
A18. Calculated using ΔH_{isom} = 21.5 kcal/mol (Reference 11) and ΔH_f(g) of C_2H_5CN = 12.3 kcal/mol (Reference 95).
A19. This work; employed 4–31G NH_2 structure from CH_3NH_2 (Reference A1) with C—N = 1.450 Å), conformation from Radom, L.; Hehre, W.J.; Pople, J.A. *J. Am. Chem. Soc.* **1972,** *94,* 2371–2381.
A20. Apeloig, Y.; Schleyer, P.v.R.; Pople, J.A. *J. Am. Chem. Soc.* **1977,** *99,* 5901–5905.
A21. This work; employed with conformation from STO-3G study; 4–31G structure of NO_2 and C—N from CH_3NO_2 (Reference A1).
A22. Used geometry from Klimkowski, V.J.; Scarsdale, J.N.; Schafer, L., *J. Comput. Chem.* **1983,** *4,* 494–498.
A23. Estimated from: $CH_3CH_2CO_2CH_3$ = $CH_3CO_2CH_3$ − $CH_3CO_2C_2H_5$ + $CH_3CO_2C_2H_5$; all ΔH_f°(g) from Pedley and Rylance (Reference 95).
A24. Employed experimental geometry from: Harmony, M.D.; Laurie, V.W.; Kuczkowski, R.S.; Schwendeman, R.H.; Ramsay, D.A.; Lovas, F.J.; Lafferty, W.J.; Maki, A.G. *J. Phys. Chem. Ref. Data* **1979,** *8,* 619–721.
A25. This work; CO syn to C_2H_5; employed 4–31G structure of CH_3CO group of acetaldehyde (Bouma, W.J.; Radom, L.; Rodwell, W.R. *Theor. Chim. Acta* **1980,** *56,* 149–55) with C_2—C_3 = C_3—C_4 (1.494 Å), C_2C_3O = 124.2°; $C_2C_3C_4$ = 116°.
A26. Reference A20; energy lower than experimental and model calculations.
A27. Used geometry from: Klimkowski, V.J.; Van Nuffel, P.; Van den Enden, L.; Van Alsenoy, C.; Geise, H.J.; Scarsdale, J.N.; Schafer, L., *J. Comput. Chem.* **1984,** *5,* 122–128.
A28. Hehre, W.J.; Pople, J.A. *J. Am. Chem. Soc.* **1975,** *97,* 6941.
A29. Bach, R.D.; Badger, R.C.; Lang, T.J. *J. Am. Chem. Soc.* **1979,** *101,* 2845–2848.
A30. Average of three estimated values: −62.2 and −62.5 kcal/mol (see two References A16) and −62.9 kcal/mol (Chen, S.S.; Rodgers, A.S.; Chao, J.; Wilhoit, R.C.; Zwolinski, B. *J. Phys. Chem. Ref. Data* **1975,** *4,* 441).
A31. Employed CH_3—C_{sp} as well as C_{sp}—H and C_{sp}—C_{sp} equal to corresponding bonds in propyne (Reference A1).

A32. Employed 4–31G CF_3 group from CH_3CF_3 (Reference A1).
A33. No energies given unless fully optimized. This is because stabilization energies for the unsaturated species are available from Schleyer, P.v.R.; Chandrasekhar, J.; Kos, A.J.; Clark, T.; Spitznagel, G.W. *J. Chem. Soc. Chem. Commun.* **1981,** *17,* 882–884, and can be obtained by subtraction of one isodesmic equation from another.
A34. Employed C—CN = 1.455 Å, CN, 1.142 Å from 4–31G study of CH_3CN (Reference A1).
A35. Dill, J.D.; Schleyer, P.v.R.; Pople, J.A. *J. Am. Chem. Soc.* **1976,** *98,* 1663.
A36. Clark, T.; Spitznagel, G.W.; Klose, R.; Schleyer, P.v.R. *J. Am. Chem. Soc.* **1984,** *106,* 4412–4419.
A37. Stull, D.R.; Westrum, E.F.; Sinke, G.C. "The Chemical Thermodynamics of Organic Compounds". Wiley: New York, 1969.
A38. Lathan, W.A.; Hehre, W.J.; Pople, J.A. *J. Am. Chem. Soc.* **1971,** *93,* 808.
A39. Employed using geometry from: Wilson, S. *Astrophys. J.* **1978,** *220,* 363–365.
A40. See Reference A48; note, close to planar at full optimization.
A41. Hehre, W.J. In "Modern Theoretical Chemistry", Vol. 4; Schaefer, H.F., III, Ed.; Plenum Press: New York, 1977, pp. 277–331.
A42. Lossing, F.P. *Can. J. Chem.* **172,** *50,* 3973.
A43. Employed C—N (1.39 Å) interpolated from DZ cycopropylamine geometry (Skancke, A.; Boggs, J.E. *J. Mol. Struct.* **1978,** *50,* 173–182), cyclopropylammonium ion (Durmaz, S.; Killmar, H. *J. Am. Chem. Soc.* **1980,** *102,* 6942–6945), and ethynylamine (Reference A48); N—H (0.994 Å, Reference A1), assume tetrahedral NH_3.
A44. Marriot, S.; Topsom, R.D. *J. Mol. Struct. (Theochem).* **1984,** *15,* 277.
A45. This work; employed 4–21G geometry of CO_2CH_3 group as in Reference A22, C_{sp}—X bond taken as 1.473 Å; an interpolated value of 1.463 was suggested but did not converge, therefore lengthened to 1.473 Å.
A46. This work; triple bond, 1.186 Å, C_{sp}—H, 1.050 Å from HCC—OH, Reference A48 and CC—O, 1.31 Å, O—CH_3, 1.43 Å, ∠ COC, 113.4° from HCC—OCH_3; Lister, D.G.; Palmieri, P. *J. Mol. Struct.* **1976,** *32,* 355–363.
A47. Bouma, W.J.; Nobes, R.H.; Radom, L.; Woodward, C.E. *J. Org. Chem.* **1982,** *47,* 1869–1875.
A48. Hopkinson, A.C.; Lien, M.H.; Yates, K.; Mezey, P.G.; Csizmadia, I.G. *J. Chem. Phys.* **1977,** *67,* 517–523.
A49. Dewar, M.J.S.; Kohn, M.C. *J. Am. Chem. Soc.* **1972,** *94,* 2699.
A50. Powell, M.F.; Peterson, M.R.; Csizmadia, I.G. *J. Mol. Struct. (Theochem.)* **1982,** *9,* 323–325.
A51. This value is based on experimental gas phase enthalpy of hydrogenation of 3,5-octadiyne (−128.58 ±0.23 kcal/mol, unpublished data courtesy R. Fuchs) combined with published value for *n*-octane; this value is compared with twice the difference between 1-butyne and ethyne to obtain the $\Delta H_f^\circ(g)$. This value is used in place of that listed in Reference A37.
A52. Experimental geometry employed: Shoolery, J.N. *J. Chem. Phys.* **1951,** *19,* 1364.
A53. Model geometry employing 6–31G structure: Moffat, J.B. *J. Phys. Chem.* **1977,** *81,* 82–86.
A54. Radom, L.; Hariharan, P.C.; Pople, J.A.; Schleyer, P.v.R. *J. Am. Chem. Soc.* **1973,** *95,* 6531.
A55. Planar substituent, perpendicular to π system.
A56. Jordan, F. *J. Phys. Chem.* **1975,** *80,* 76.
A57. $\Delta H_f^\circ(g)$, $CH_3CHNH_2^+$ = 157 kcal/mol (ie, vinylamine is protonated at the beta carbon, not nitrogen); Lossing, F.P.; Lam, Y.T.; MacCall, A. *Can. J. Chem.* **1981,** *59,* 2228–2231. At the 4-31G level, $CH_2{=}CHNH_3^+$ is less stable than $CH_3CHNH_2^+$ by 16.2 kcal/mol; Mueller, K.; Brown, L.D. *Helv. Chim. Acta* **1978,** *61,* 1407–1418.
A58. MacKay, G.I.; Lien, M.H.; Hopkinson, A.C.; Bohme, D.K. *Can. J. Chem.* **1978,** *56,* 131–140. See also Reference A92.
A59. Estimated: $C_6H_5NO_2$ (16.2 kcal/mol) + CH_2CHCH_3 (4.8 kcal/mol) = C_6H_5—CH_3 (12.0 kcal/mol) + CH_2CHNO_2.
A60. Cremashi, P.; Morosi, G.; Simonetta, M. *Theochem* **1981,** *2,* 397–400.
A61. Bernardi, F.; Epiotis, N.D.; Yates, R.L.; Schlegel, H.B. *J. Am. Chem. Soc.* **1976,** *98,* 2385.
A62. Value estimated in: Dill, J.D.; Greenberg, A.; Liebman, J.F. *J. Am. Chem. Soc.* **1979,** *101,* 6814–6826.

A63. The CO-*cis* structure: employed model CO_2CH_3 group from HCO_2CH_3, see Harmony et al in Reference A24; For C_2—C_3, used 1.470 Å (Allen, F.H. *Acta Crystallogr* **1981,** *B37,* 890–900). The CO-*trans* conformer did not converge.
A64. OH-*cis* structure: Bouma, W.J.; Radom, L.; Rodwell, W.R. *Theor. Chim. Acta* **1980,** *56,* 149–155.
A65. Holmes, J.L.; Lossing, F.P. *J. Am. Chem. Soc.* **1982,** *104,* 2648–2649.
A66. Estimated value: Hegedus, V.J.; Harrison, A.G. *Int. J. Mass Spectrom. Ion Phys.* **1979,** *30,* 293–306.
A67. Schiess, P.,; Radimerski, P. *Helv. Chim. Acta* **1974,** *57,* 2583–2597. $\Delta H_f^\circ(g)$ = 20.6 kcal/mol, estimated in Reference A66.
A68. Marriot, S.; Topsom, R.D. *J. Mol. Struct. (Theochem.)* **1984,** *109,* 306.
A69. Modeled on data: Bock, C.W.; Trachtman, M.; George, P. *J. Mol. Spectrosc.* **1980,** *84,* 243–255.
A70. Bernardi, F.; Bottoni, A.; Epiotis, N.D. *J. Am. Chem. Soc.* **1978,** *100,* 7205–7209. Note typographical error in reference. See Kollman, P.; Nelson, S.; Rothenberg, S. *J. Phys. Chem.* **1978,** *82,* 1403–1406.
A71. Value from Reference 95 (Pedley–Rylance compendium) = 8.9 kcal/mol; The 5.0-kcal/mol value employed is derived from group contribution theory (Alfassi, Z.B.; Golden, D.M.; Benson, S.W. *J. Chem. Thermodyn.* **1973,** *5,* 411–420) as suggested by J.F. Liebman (see Kolesov, V.P.; Papina, T. S.; *Russ. Chem. Rev.* **1983,** *52,* 405).
A72. This work; experimental geometry (Tokue, I.; Fukuyama, T.; Kuchitsu, K. *J. Mol. Struct.* **1973,** *17,* 207) used except for C—F = 1.361 Å, 4–31G study of CH_3CF_3 (Reference A1) and angle HCH and C—H values from 4–31G study of ethylene (Reference A38).
A73. Fuchs, R.; Hallman, J.H.; Perlman, M.O. *Can. J. Chem.,* **1982,** *60,* 1932–1835.
A74. Dupuis, M.; Pacansky, J. *J. Chem. Phys.* **1982,** *76,* 2511.
A75. This work: employed NC = 1.167 Å from 6–31G study of CH_2CHNC, Moffat, ring—NC bond = 1.39 Å interpolated using 6–31G data.
A76. Substituent in bisected conformation.
A77. This work; DZ geometry (Durmaz, S.; Kollmar, H. *J. Am. Chem. Soc.* **102,** 6942–6945) employed with C—H = 1.071 Å, ∠ HCH = 114° from 4–31G study of cyclopropane (Reference A74); N–H = 0.994 Å, from 4–31G study of CH_3NH_2 (Reference A1).
A78. Substituent in perpendicular conformation.
A79. This work; a 4–21G structure (Skancke, A. *Acta Chem. Scand. A* **1982** *36,* 637–639) did not converge; therefore, used experimental ring structure of nitrocyclopropane (Mochal, A.R.; Britt, C.O.; Boggs, J.E. *J. Chem. Phys.* **1973,** *58,* 3221, and 4–21G nitro group of the Skancke study; C—N = 1.475 Å employed, which is identical to that for the perpendicular conformation, since shorter C—N did not converge for the bisected conformer.
A80. This work; employed 4–21G structure from Skancke study (Reference A79).
A81. This work; same ring structure and conformation as cyclopropanol DZ result (Durmaz; see Reference A77), OCH_3 structure from Harmony et al (Reference A24); manually optimized ring–substituent bond = 1.416 Å.
A82. This work; 4–31G structure of CH_3CO group of acetaldehyde (Bouma, W.J.; Radom, L.; Rodwell, W.R. *Theor. Chim. Acta* **1980,** *56,* 149–155) 4–31G angles around C=O from *trans*-acrolein (Bock, C.W.; Trachtman, M.; George, P. *J. Mol. Spectrosc.* **1979,** *78,* 298–308); exocyclic ring–carbon bond = 1.474 Å (Allen, F.H. *Acta Crystallogr.* **1981,** *B37,* 890–900) for this geometry, $E = -268.43151$. The distal bond was manually optimized (shortened by 0.013 Å) to arrive at the listed energy. The starting point was a distal bond shortening of 0.026 Å to 1.475 Å as recommended (Allen, F.H. *Acta Crystallogr.* **1980,** *B36,* 81–96).
A83. This work; DZ structure of Durmaz (Reference A77) used with C—H = 1.071 Å, ∠HCH = 114° from 4–31G study of cyclopropane (see Reference A74).
A84. Hehre, W.J. *J. Am. Chem. Soc.* **1972,** *94,* 6592.
A85. $\Delta H_f^\circ(l)$ (from Reference 95) = 29.3 kcal/mol; ΔH_v° from Trouton's rule (BP, 39.5–41.5°C, Overberger, C.G.; Borchert, A.E. *J. Am. Chem. Soc.* **1960,** *82,* 4896–4899) = 0.021 × 314.7 K = 6.6 kcal/mol; $\Delta H_f^\circ(g)$ = +35.9 kcal/mol. However, this value is not used because it is clearly too high, predicting 3 kcal/mol destabilization on conjugating vinyl and cyclopropyl groups.
A86. Experimental structure: Klein, A.W.; Schrumf, G. *Acta Chem. Scand. A* **1981,** *A35,* 431–435.

A87. This work; used 4–21G structure (Skancke, A.; Boggs, J.E. *J. Mol. Struct.* **1977,** *40,* 263–270.
A88. Klessinger, M.; Zywietz, A. *Theochem* **1982,** *7,* 341–350.
A89. This value based on experimental gas phase enthalpy of hydrogenation of phenylacetylene (−68.24 ± 0.24 kcal/mol); unpublished data courtesy of R. Fuchs.
A90. von Negy-Felsobuki, E.; Topsom, R.D.; Pollack, S.; Taft, R.W. *J. Mol. Struct. (Theochem)* **1982,** *88,* 255–263.
A91. Ellenberger, M.R.; Eades, R.A.; Thomsen, M.W.; Farneth, W.E.; Dixon, D.A. *J. Am. Chem. Soc.* **1979,** *101,* 7151–7154.
A92. ΔH_f°(g) allyl anion (29.5 ±0.9 kcal/mol) in good agreement with earlier value (Oakes, J.M.; Ellison, G.B. *J. Am. Chem. Soc.* **1984,** *106,* 7734–7741).

CHAPTER 6

Macroincrementation Reactions: A Holistic Estimation Approach for the Properties of Organic Compounds

Joel F. Liebman

University of Maryland Baltimore County, Catonsville, Maryland

CONTENTS

MOLECULAR STRUCTURE AND ENERGETICS, Vol. 3

Statements in angular brackets such as this, ≪ ≫, are intended to supplement the text by offering details of arguments, additional examples, and extensions of the ideas presented. They may be omitted in an initial reading without loss of continuity.

1. MACROINCREMENTATION REACTIONS: INTRODUCTION, METHODOLOGY AND PHILOSOPHY

How the physical properties of a species relate to the underlying molecular structure is a topic of seminal importance in chemistry. The physical property of greatest interest to the author has long been the heat of formation, a quantity that directly relates the energetics of molecules to their structure, both in comparison to idealized reference states (eg, the presence of aromaticity and/or strain) and with reference to the magnitudes of intramolecular and intermolecular interactions (eg, the presence of steric repulsion and/or hydrogen bonding). It must be emphasized that most of the conceptual models and calculational studies of chemical energetics ignore the explicit presence of solvent and indeed any molecule other than the one being investigated.

To the extent that the chemist's models reflect "reality," it is generally because the impact of the presence of these other molecules has been successfully intuited and incorporated. For example, the practicing chemist generally "knows" how to qualitatively, it not quantitatively, describe the effects due to solvent polarity and hydrogen bonding power. To enhance their reality, quantum chemical calculations are generally compared with the gaseous state. (Chapter 9 in Volume 1 of this series offers a discussion of what is seemingly ignored with benign neglect when one makes the casual equivalence of molecular energetics as derived from quantum chemical calculations and of ideal gases at standard temperature and pressure.)

To the extent that intermolecular interactions are sufficiently small, condensed phases (liquids, solids, solutions) mimic the gas. Though exceptions unequivocally exist, it is nonetheless rare at least among neutral, covalent compounds for the relative stability of a set of isomers to differ in the two phases. Likewise, it is rare among these species for the number, much less nature, of intramolecular bonds to change upon a phase change. The major exception is the number of hydrogen bonds, and this is a rather predictable effect. Thus the general success of our discussions of chemical energetics is attributable in part to this general weakness of *inter*molecular interactions for covalent and neutral compounds.

However, the energies associated with many interesting *intra*molecular phenomena (eg, many substituent effects) are also small. To allow for maximum and most successful use of experimentally and calculationally derived numbers, to provide for the most accurate test of our fundamental understanding and intuition, all species in this chapter, unless otherwise said, are taken as ideal gases at 1 atm and 298 K (standard temperature and pressure). Furthermore, the expression "heat of formation" implicitly refers to compounds in this well-defined, ideal and idyllic state. (Also, nearly all the heats of formation used in this chapter have been taken from References 1 or 2, and in the absence of an explicit literature citation, are to be assumed from these sources. (At the time of writing, the second edition of Reference 1 and Reference 2 were not in print. To facilitate the use of this chapter, therefore the primary references for heat of formation of all neutral species not found in Pedley and Rylance—Reference 1, first edition—are explicitly given.)

Several million organic compounds have been isolated and/or synthesized and their structures characterized.

≪We do not mean to imply that all (some, or even any) interatomic bond distances and angles have been determined but rather that it is known "what" is attached to "what," and so one can construct a molecular and/or conceptual model for the compound. Chapter 1 of Volume 2 in this series shows how to do better than merely "what" is attached to "what?" when comparing molecular structures in the gaseous and condensed phases.≫

Nevertheless there are data on the heats of formation (in some phase) of but several thousand. This phenomenal numerical disparity in the amount of thermochemical and "other-chemical" data arises from the dual difficulties of obtaining the desired species in adequate quantity and purity and the necessity of performing replicate, exacting measurements. Moreover, the numerical disparity is steadily worsening: there are about 300,000 new compounds reported each year, but fewer than 100 heats of formation are generally reported during that time. If we are to have any hopes of correlating molecular structure and energetics, we must attempt to understand, use, and develop estimation techniques in the study of chemical energetics. Indeed, organic chemists have already devised numerous methods of varying degrees of rigor, complexity, and applicability.

Space disallows discussing all or even many of these approaches, and the reader is deferred to References 3–8 for a rather thorough sampling. Likewise, Chapter 8 in Volume 2 of this series offers a conceptually thorough and mathematically rigorous discussion of the unifying threads of resonance and strain energies, heats of formation and reaction, and three estimation methods: the use of group increments, exchange increments, and macroincrements. The numerous and diverse strengths of group increments have been discussed at length by Benson in Reference 3. Simply saying "Benson" is probably adequate to encourage immediate recall of the method by many readers. The last of the three, also called macroincrementation reactions, has been done "informally" and implicitly for decades, although the word "macroincrementation" is rather new. [9–11] To preclude any unnecessary ambiguity, we now define this old/new approach, which is used extensively in this chapter.

The approach of macroincrementation reactions starts with the molecule of interest and attempts to describe it in terms of "whole" other molecules that are better understood. More precisely, given two sets of molecules, if the net numbers and types of atoms, bonds, groups (eg, CH_3—, —CH_2—), and more general structural features (eg, cyclopropane and benzene rings) are the same, it is postulated that the resulting values from adding and subtracting the heats of formation (or any other property of interest) for the two sets are the same. That is, all macroincrementation reactions are assumed to be thermoneutral. The list of general structural features is long, and it usually is argued that the discrepancies between estimation and experiment are due to some feature that was ignored. For example, consider the macroincrementation reactions relating the hydrocarbons ethylene, propane, and benzene and their monofluoro derivatives, vinyl fluoride, isopropyl fluoride, and fluorobenzene:

$$(CH_3)_2CHF + CH_2{=}CH_2 = C_3H_8 + CH_2{=}CHF \qquad (6\text{-}1)$$
$$(CH_3)_2CHF + C_6H_6 = C_3H_8 + C_6H_5F \qquad (6\text{-}2)$$
$$CH_2{=}CHF + C_6H_6 = CH_2{=}CH_2 + C_6H_5F \qquad (6\text{-}3)$$

While we might think all three reactions are thermoneutral, the discrepancies in fact are 0.4, 2.4, and 2.0 kcal/mol (2, 10, and 8 kJ/mol), respectively. (See Section 3 for a thorough discussion of the energetics of haloethylenes and Chapter 4 for a thorough discussion of fluorinated species.) Questions these discrepancies elicit include: Is it "really" correct to assume that the C—F and C—H bonds in all three sets of species are the same? Is it "really" correct to assume that as the C—H bond goes, so goes the C—F? Is it "really" correct to ignore whether hydrogen or fluorine is bonded to a tetrahedral, tetracoordinate, sp^3-hybridized carbon and the other to a trigonal, tricoordinate, sp^2-hybridized carbon? If any of the answers to these questions is "no" (see Section 3), our list of general structural features should distinguish between the two types of C—F bond. It is also fair to ask whether the answers to these questions are equally valid for the heavier halogens.

Likewise, recall that CH_3-containing species are often thermochemically anomalous[12] compared to homologues with longer alkyl chains. More precisely, consider the homologous series: HY, CH_3Y, CH_3CH_2Y,

$CH_3(CH_2)_2Y, \ldots, CH_3(CH_2)_{n-1}Y, CH_3(CH_2)_nY$. Although for large enough n there is a limiting difference for the heats of formation of subsequent members of this series that is independent of the nature of the arbitrary group Y (see Chapter 5, Volume 2), the differences between that the first two and between the second and third members and CH_3CH_2Y are not only different from that limit, they are also different from each other and depend on the nature of Y.

Indeed, consider the following three reactions for arbitrary groups Y and Z:

$$HY + CH_3Z = HZ + CH_3Y \tag{6-4}$$
$$CH_3Y + CH_3CH_2Z = CH_3Z + CH_3CH_2Y \tag{6-5}$$
$$CH_3CH_2Y + CH_3CH_2CH_2Z = CH_3CH_2Z + CH_3CH_2CH_2Y \tag{6-6}$$

The first of these, Equation (6-4), is often not a macroincrementation reaction because there are H—Y′ and C—Z′ bonds on the left that generally do not correspond to the H—Z′ and C—Y′ bonds on the right. (For example, HY may be HOH and CH_3Z may be CH_3CN, in which case Y′ = O and Z′ = C.) By contrast, the second and third reactions (6-5 and 6-6) always are macroincrementation reactions. Saying that CH_3-containing species are anomalous is meant to convey that reaction 6-6 is more likely to be thermoneutral than reaction 6-5 even though both are macroincrementation reactions. This will be referred to as the "methyl versus ethyl dichotomy" in this chapter.

It would appear that macroincrementation reactions involving methyl of the type

$$CH_3Y + RZ = CH_3Z + RY \tag{6-7}$$

are less reliable than analogous ones involving ethyl,

$$CH_3CH_2Y + RZ = CH_3CH_2Z + RY \tag{6-8}$$

Indeed, except for —Y and —Z both bonding through carbon (ie, Y′ = Z′ = C), use of the former reaction is a priori discouraged. Ethyl is so much more electronically "normal" than methyl, yet unlike larger R groups it is sterically so nearly innocuous that it may be considered to be a paradigmatic alkyl group, R_{pd}. Although it is important to identify paradigmatic groups for use in general macroincrementation reactions—these provide general structural features—we may not always have the choice to use them. That is, in the absence of data on some needed C_2H_5 derivative, we may have to make use of that of CH_3 or even of H.

All the macroincrementation reactions given above may be written schematically as follows.

$$A + B + \ldots = X + Y + \ldots \tag{6-9}$$

Alternatively, they may be written as follows.

$$A = X + Y + \ldots - B - \ldots \tag{6-10}$$

For example, macroincrementation reaction 6-3 may be rewritten as

$$CH_2{=}CHF = CH_2{=}CH_2 + C_6H_5F - C_6H_6 \qquad (6\text{-}11)$$

As such, the desired heat of formation is expressed as the arithmetic sum and difference of the heats of formation of other, better understood species. In either formulation, given that values for all the other species are known, simple arithmetic allows one to derive the value of interest. Since one's "whole" knowledge of chemical energetics and "whole" molecules are used in this approach, macroincrementation reactions may be referred to as a "holistic" estimation approach.

Other properties that have been studied using macroincrementation reactions include the entropy of formation for the ideal gas, the heat of vaporization and molar polarizability for the liquid, and molar volumes for both liquids and solids. In general, success for these have also been found.

≪Major co-workers for these other properties have been: A. Greenberg, A. S. Hyman, L. H. Ladon, and T. A. Stevenson (entropy); L. H. Ladon (heat of vaporization); T.-L. Dang, S. G. Lias, and P. J. Rybczynski (molar polarizability); S. G. Lias and M. W. Victor (molar volume); and D. T. Dao (zero-point energies)≫

Nevertheless, we have considered other properties here only when it is necessary; for example, to estimate the heat of vaporization of some liquid species in order to derive an input number for use in deriving the heat of formation for the "real" problem of interest, the heat of formation of the ideal gas. Our decision to limit discussion in this chapter to essentially one property was made by considering four interrelated reasons.

1. *Brevity.* Books have been written about molecular properties. We have but one chapter to describe the methodology and some successful applications of macroincrementation reactions.
2. *Availability.* Data are much more readily available on gaseous species than on liquids. We say "liquids" not "condensed phase" because it appears that few predictive approaches developed to date are at all adequate for solids. More precisely, although there are numerous formulas for estimating heats of vaporization, such formulas are absent for estimating heats of sublimation except for some well-defined classes of compounds. (See Chapter 3, Volume 2.)
3. *Applicability.* As noted above, most conceptual models developed in organic chemistry (eg, resonance and inductive effects) do not explicitly involve intermolecular interactions. As such, they are implicitly for isolated and unsolvated molecules. Molecules in an ideal gas are truly in such a state.
4. *Simplicity.* Where the condensed phase (pure liquid, solution, or pure solid) is needed for an effect to be significant, it is usually difficult to disentangle. (See, however, Chapter 7 in Volume 1 and Chapter 2, this

volume.) Moreover, we will make some "heroic" attempts at discussing solid glucose and aqueous glycine to invite the biochemist to use macroincrementation reactions. Indeed, wherever the condensed phase is not important, why should we bother discussing anything but the ideal gas?

These four reasons for limiting our attention to the heat of formation of gas phase species also provided limitations on what molecules would be discussed in this chapter. In particular, we opted for compounds of the general type $A_mB_{n-m}H_n$ where the atoms A and B were anything but hydrogen and B is not carbon. For example, the $n = 8$, eight-membered ring heterocyclic case of tetroxane, $(CH_2O)_4$, is included, whereas the $n = 8$, eight-membered carbocyclic case of cyclooctatetraene, C_8H_8, is not. Likewise, the $n = 8$ substituted benzene, phenylhydrazine ($C_6H_5NHHH_2$) is discussed, but another $n = 8$ substituted benzene, styrene ($C_6H_5CHCH_2$) is not. Expanding and explaining a little, our reasons are again:

1. *Brevity.* Had we included C_nH_n species as well as $A_mB_{n-m}H_n$, the number of compounds would have been multiplied excessively. And, had we considered general, hence arbitrary, organic compounds, it might have been hard to convince reader and author alike that there was not some bias in the choice of these compounds.
2. *Availability.* There are more than enough compounds of the type $A_mB_{n-m}H_n$ to illustrate the various points we wish to make about chemical energetics.
3. *Applicability.* Many of these $A_mB_{n-m}H_n$ species allow us to introduce some new feature of interest quite naturally. For example, the discussion of zwitterions arises in the context of the oxyallyl form of cyclopropanone, $n = 4$, and of aqueous solutions of glycine, $n = 5$, Sections 2 and 6, respectively, of this chapter.
4. *Simplicity.* Although most of the compounds of interest are comparatively low molecular weight (n rarely exceeds 8), they are nonetheless generally polyfunctional. They are often the simplest members of a given class of compounds and so provide useful insights into numerous species. Insights gleaned by the author are included in this chapter—those from readers will be explicitly welcomed.

Does the concept of macroincrementation reactions look familiar? It may. This chapter documents its systematic use in providing qualitative insights and interrelationships as well as quantitative results in the study of the energetics of organic compounds. Making the foregoing more explicit, and presaging the remainder of the chapter, experience has shown us that there are five major and admittedly interrelated uses of macroincrementation reactions:

1. To confirm some experimental value that is in doubt. If one is confident that one has not left out any important interactions, then theory and

experiment "should" be in close agreement. If they are not, either an important interaction has been left out or at least one of the experimental numbers is in error. Clearly, however, in such cases it is then useful to try to find some independent corroborative evidence for that error.

2. To predict the value of some property of the compound of interest. Ideally, one "should" use more than one "fundamentally distinct" macroincrementation reaction so that the predictions can be compared and averaged. This is often not possible, however. "Fundamentally distinct" is taken to mean differing in a substantive way and not, for example, gotten by replacing ethyl by propyl or hydroxy by methoxy in a given macroincrementation reaction. Rather regrettably, because of length limitations, this use as a predictor is generally ignored here. (The data compendium by Lias and her co-workers[2] also contain numerous predictions of the heats of formation of organic compounds using group and macroincrement methods.)
3. To compare some property of two species such as their relative resonance energies or strain energies. This may be done in two rather distinct ways. First, one may write two corresponding macroincrementation reactions for the two compounds and then compare the derived correction terms. (This is done in this volume in Chapter 5 for the strain energy of substituted cyclopropanes and in Chapter 2 for the resonance energy of [*n*]-annulenes, and in Chapter 6 of Volume 4 and Reference 11 for the resonance energy of amides and esters.) Alternatively, by using one molecule as an input quantity for a reaction predicting the property for the other, the discrepancy between the predicted and experimental values for the whole reaction allows directly the desired comparison.
4. To provide a generalization of isodesmic[13] reactions wherein, generally, quantum chemical approaches provide the "alphanumeric" correction terms. For example, substituent effects on a ring system of interest, *R*, have often been compared to those on methane; consider reaction 6-12.

$$RH + CH_3X = RX + CH_4 \qquad (6\text{-}12)$$

The calculated energy of this reaction, "the methyl stabilization energy," is then used as an "alphanumeric" correction term for either of the equivalent reactions 6-12 and 6-13.

$$RX = RH + CH_3X - CH_4 \qquad (6\text{-}13)$$

where now experimental heats of formation are used for those of *R*H, CH_3X, and CH_4. To correct for the "methyl versus ethyl dichotomy" (ie, the anomalous properties of methyl groups), the CH_3X and CH_4 are often replaced by other more suitable hydrocarbons and their derivatives. Along with analogous conceptual substitutions for other components (eg, appropriate olefins in lieu of just ethylene when discussing resonance energies), this refinement of isodesmic reactions goes under the names of "homodesmotic reactions"[14] and "group separation reactions."[15] Since quantum chemically "assisted" isodesmic, homodesmotic, and group separation reactions are discussed at length elsewhere in this series (most notably Chapters 4 and 5 of this volume, we have opted not to do so in this chapter.

5. To intentionally leave out some interaction (stabilizing or destabilizing molecular feature) in the macroincrementation reaction and define the difference of the experimental and predicted quantities as a new "alpha-numeric" correction term. The philosophy of correction terms is a common feature of macroincrementation and the commonly used group increment method. Indeed, one may derive correction terms for use in the latter approach by adopting the value derived from a suitable macroincrementation reaction. Such quantities include the resonance energy of an ester or amide (see Reference 11 and Chapter 6, Volume 4) and the strain energy of a cycloalkane. These correction terms may then be used in a description of the molecule and/or as a correction term for related species. For example, one need not always derive the strain energy for cyclopropane de novo, instead this quantity can be used as a benchmark for discussing the energetics of strained hydrocarbons and their substituted derivatives. (See, eg, Chapter 5 for an explicit discussion and Reference 10 on cyclopropanone, where this was done informally and implicitly.)

Having spoken mostly philosophically and pedagogically in this introductory section, we close with some practical comments on the units and notation used in this chapter. Most organic chemists are familiar with kilocalories per mole—many thermochemists, following official decree, use kilojoules per mole. In this chapter we use both sets of units, with the defined conversion factor 1 kcal $\equiv$ 4.184 kJ. To save space, the "per mole" is dropped, and although values in kilocalories often are given to the nearest tenth, kilojoules are given just to the nearest unit. Error bars (shown in parentheses) are not given unless they are particularly large and/or significant. Combining all the foregoing conventions, a value of 3.2 (0.5) kcal (13 (2) kJ) is to be taken to mean 3.2 $\pm$ 0.5 kcal/mol (or equivalently, 13 $\pm$ 2 kJ/mol).

2. THE ENERGETICS OF THE ISOMERIC n = 4 SPECIES, CYCLOPROPANONE, ALLENE OXIDE, AND THE ZWITTERIONIC OXYALLYL

Our first application of macroincrementation reactions in this chapter reviews and extends the first application the author made using this approach[10]: the energetics of the isomeric n = 4 species, cyclopropanone, **1,** allene oxide, **2,** and the zwitterionic oxyallyl, **3.** Starting with the simplest of the three, cyclopropanone, we recognized that it has an sp^2-hybridized,

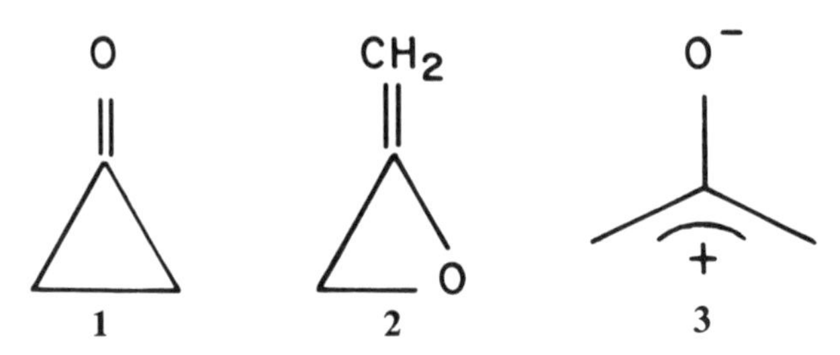

trigonally coordinated carbon and so is even more strained than the parent hydrocarbon, cyclopropane.

≪This "folklore" statement is most simply documented by comparing the strain energies of small cycloalkanes, cycloalkenes, and methylenecycloalkanes. For cyclopropane derivatives, these are 28.3, 54.5, and 41.7 kcal (118, 228, and 174 kJ); for cyclobutane derivatives, 27.4, 30.6, and 28.8 kcal (115, 128, and 120 kJ); and for cyclopentane derivatives, 7.3, 6.8, and 6.3 kcal (31, 28, and 26 kJ). (All these values were taken from Reference 16.)≫

Two macroincrementation reactions were suggested that explicitly build in this destabilizing structural feature, despite the possibility of potential problems with ion thermochemistry affecting the trustworthiness of one of our predictions:

$$\mathbf{1} \equiv \text{cyclo-}(CH_2)_2C{=}O = \text{cyclo-}(CH_2)_2C{=}CH_2 + (CH_3)_2C{=}O - (CH_3)_2C{=}CH_2 \qquad (6\text{-}14)$$

$$\mathbf{1} \equiv \text{cyclo-}(CH_2)_2C{=}O = \text{cyclo-}(CH_2)_2CH^+ + (CH_3)_2C{=}O - (CH_3)_2CH^+ \qquad (6\text{-}15)$$

These result in comparable values, −3 and 0 kcal (−13 and 0 kJ), and to our joy, they satisfactorily agreed with the experimentally determined value of +3.8 kcal (+ 16 kJ) reported some years afterwards.[17] Although the discrepancy of about 5 kcal (22 kJ) is *not small,* by comparison, the "leading" semiempirical method, MNDO (modified neglect of diatomic overlap), "predicted"[18] a value of 11.4 kcal (48 kJ) some years after the experiment was reported. This example of cyclopropanone is not, however, intended as self-congratulatory and we emphasize that one cannot count on being this "lucky." With more than a decade of additional awareness of chemical energetics, we now discuss the new insights that have been gleaned both about cyclopropanone and more general compounds from this study.

We start by recalling that CH_3-containing species are often thermochemically anomalous compared to homologues with longer alkyl chains. As such, a priori, it would have been preferable to have employed reaction 6-16 than the earlier one (6-14)

$$\mathbf{1} \equiv \text{cyclo-}(CH_2)_2C{=}O = \text{cyclo-}(CH_2)_2C{=}CH_2 + (C_2H_5)_2C{=}O - (C_2H_5)_2C{=}CH_2 \qquad (6\text{-}16)$$

In fact, although fortuitously, the predictions are numerically identical to within 0.2 kcal (1 kJ). We now acknowledge that the H case is generally even more anomalous than the CH_3 case, even when both are macroincrementation reactions. Yet thinking back, the avoidance of the seemingly related reaction 6-17 was made more "instinctively" than consciously because of the more well-articulated awareness of the governing principle, the "methyl versus ethyl dichotomy":

$$\mathbf{1} \equiv \text{cyclo-}(CH_2)_2C{=}O$$
$$= \text{cyclo-}(CH_2)_2C{=}CH_2 + CH_2{=}O - CH_2{=}CH_2 \qquad (6\text{-}17)$$

Using this expression, the "predicted" value would have been 9.6 kcal (40 kJ), and we would have seriously erred, had we used it.

The second awareness commences with the recollection that we anticipated macroincrementation reaction 6-14 to be preferable to 6-15 because "one would think that the 'bare' positive charge in the cyclopropyl and isopropyl cations would be extremely sensitive to the molecular environment."[19] Yet our two predictions, −3 and 0 kcal (−13 and 0 kJ) were so close. Indeed, we might think that the reaction using cyclopropyl cation "should" result in a higher heat of formation for cyclopropanone than that using acetone, and this was not found to be the case. We were surprised and relieved.

Was our happiness inappropriate, or at least premature? Old thermochemical data of ion energetics, much more than data for neutral species, appear to be all too often untrustworthy. As such, let us redo that latter macroincrementation reaction (6-15), using more recent ion data on cyclopropyl cation. Making use of the most recent value[20] for the heat of formation of cyclopropyl cation (ie, one obtained from evaluation of the proton affinity of cyclopropene), we find a new "predicted" heat of formation of cyclopropanone of −9 kcal (−37 kJ). This new result is surprising, if not outright disappointing, because we would have expected a *too-high* heat of formation of cyclopropanone. However, there is ambiguity about the very existence of the cyclopropyl cation that is employed in our macroincrementation reaction. More precisely, is the observed cation at all "opened" to its considerably more stable isomer, the allyl cation, and thus does "cyclopropyl cation not contain a cyclopropane ring at all? Quantum chemical calculations disagree[21,22] as to whether the rearrangement from cyclopropyl to allyl cation has a nonzero activation barrier. If there is such a barrier, it is no doubt small. This suggests that the disparate experimental values for the heat of formation of cyclopropyl cation arise from allyl cations with differing degrees of twisting and accompanying destabilization.

We may, however, derive a heat of formation of an untwisted and otherwise undistorted cyclopropyl cation. This is done by adding to the most trusted experimental heat of formation of the acyclic allyl cation[2] the energy difference of cyclopropyl and allyl cations that were calculated at the highest (ie, most rigorous) quantum chemical level given in Reference 22 for these two ions. Using this derived number for the heat of formation of cyclopropyl cation results in a new predicted heat of formation of cyclopropanone of nearly 19 kcal (79 kJ). This result is even higher than the value found experimentally but more in accord with our biases that assert that macroincrementation reaction 6-14 is to be trusted more than reaction 6-15.

We additionally note that *gem*-difluorocyclopropanes are considerably more strained than their unhalogenated analogues[23,24] There are highly successful literature comparisons of how the *gem*-difluorine and ketone groups increase strain energies.[23] In that *gem*-difluorocyclopropanes are more strained than the parent cyclopropane, we suggest that the first macroincrementation reaction (6-14)—the most unequivocal and simple—will in fact underestimate the strain energy of cyclopropanone. Equivalently, we are not surprised that the thermodynamic stability of cyclopropanone predicted by macroincrementation reactions 6-15 and 6-16 has been overestimated: our preferred estimate of the heat of formation of cyclopropanone is low. Self-consistency between macroincrementation reactions and experimental results for the energetics of cyclopropanone returns.

As part of the original study, we also endeavored to determine the relative energies of cyclopropanone and allene oxide. The following macroincrementation reactions were suggested for allene oxide:

$$\mathbf{2} \equiv \text{cyclo-}(\text{—}CH_2O\text{—})C{=}CH_2$$
$$= \text{cyclo-}(CH_2)_2C{=}CH_2 + \text{cyclo-}(CH_2)_2O - \text{cyclo-}(CH_2)_3 \tag{6-18}$$
$$\mathbf{2} \equiv \text{cyclo-}(\text{—}CH_2O\text{—})C{=}CH_2$$
$$= \text{cyclo-}(CH_2)_2C{=}CH_2 + CH_3OCH{=}CH_2 - CH_3CH_2\,CH{=}CH_2 \tag{6-19}$$

These resulted in predictions of the heat of formation of allene oxide of 23 and 20 kcal (96 and 84 kJ), and so about 23 kcal (96 kJ) above that first proposed for the heat of formation of cyclopropanone. This value is encouragingly close to that obtained from literature ab initio calculations,[25] 21.5 kcal (90 kJ). We now recognize that the "methyl versus ethyl dichotomy" naturally arises here—after all, *methyl* vinyl ether was used in Equation 6-19. The use of *ethyl* vinyl ether—that is, Equation 6-20—is inherently preferable.

$$\mathbf{2} \equiv \text{cyclo-}(\text{—}CH_2O\text{—})C{=}CH_2$$
$$= \text{cyclo-}(CH_2)_2C{=}CH_2 + CH_3CH_2OCH{=}CH_2 \tag{6-20}$$
$$- CH_3CH_2CH_2CH{=}CH_2$$

The new value for the heat of formation of allene oxide, 19.6 kcal (82 kJ), is nearly identical to that found before, suggesting that the "methyl versus ethyl dichotomy" is not too burdensome. We also note that an important conceptual difference between macroincrementation reaction 6-18 and either reaction 6-19 or 6-20 has been ignored: the resonance energy intrinsic in all vinyl ethers is absent in reaction 6-18 and overestimated in 6-19 and 6-20.

≪To the extent that vinyl ethers are stabilized by resonance structures containing positive oxygen and negative carbon, eg, $^{-}CH_2$—CH=O^+—CH_3, the presence of two sp^2 atoms in the three-membered ring of allene oxide decreases the resulting stabilization. However, in that the resonance energy of an acyclic vinyl ether is only about 3.5 kcal (14 kJ) as determined by using the following macroincrementation reaction:

$$CH_3CH_2—O—CH{=}CH_2 = CH_3CH_2CH_2CH{=}CH_2 + CH_3CH_2—O—CH_2CH_3 - CH_3CH_2CH_2CH_2CH_3$$

the underestimate in reaction 6-18 and overestimate in reactions 6-19 and 6-20 is but about 1–2 kcal ($\simeq$ 5–10 kJ).≫

We thus propose a heat of formation of allene oxide of 21 kcal (88 kJ), nearly identical to that taken before, 21.5 kcal (90 kJ). In any case, we now propose a difference of about 24 kcal (100 kJ) between the heats of formation of cyclopropanone and allene oxide: the new value remains comparable to that from the literature ab initio quantum chemical computations.

As an additional, albeit seminal part of the original study, we also attempted to determine the relative energies of cyclopropanone and the zwitterionic oxyallyl. For this latter species, the following macroincrementation reaction was suggested:

$$\mathbf{3} \equiv {}^{+}(CH_2)_2CO^{-} = (CH_2CHCH_2)^{+} + CH_3O^{-} - CH_4 + E_{Coul} \qquad (6\text{-}21)$$

where E_{Coul} is taken to be the Coulombic or electrostatic interaction of the positive and negative charges of the zwitterion. How do we compute the electrostatic interaction above? The following assumptions were explicitly made.

1. We employ Coulomb's law, Equation 6-22.

$$E_{Coul} = \sum_{i<j} \frac{q_i q_j}{\epsilon\, r_{ij}} \qquad (6\text{-}22)$$

   where we sum over the charges q_i and q_j at a separation of r_{ij}. When two equal and opposite charges of magnitude equaling that of the charge of the electron (or proton) are 1 Å apart, their attraction energy equals 14.4 eV, or 332 kcal (1390 kJ) in somewhat more common units.
2. The dielectric constant ϵ is unity: there is no "stuff" between the charged atoms. Furthermore, only point charge interactions have been considered.
3. All the unit negative charge is to be located on the oxygen.
4. The unit positive charge is distributed equally between the two terminal carbons: no extra self-repulsion in the cation between these two like-charged atoms is assumed. After all, whatever destabilization arises here is also found in the free allyl cation.
5. All the geometric parameters, the C—C and C—O bond lengths, and the

CCO and CCC angles, were taken from what had been the most recently reported ab initio molecular orbital calculation.[25] Using a then current compendium of experimental values[26] for the heats of formation of allyl cation, methoxide ion, and methane, and the assumptions above as to molecular geometry

≪With respect to molecular geometry, it may simply be shown that changing the C—C and C—O bond lengths by about 0.1 Å in opposite directions has no *qualitative* and little *quantitative* effect on our calculations of the relative stability of cyclopropanone and the zwitterionic oxyallyl.≫

and electrostatics, it was determined that the heat of formation of zwitterionic oxyallyl was 55 kcal (230 kJ) above that of cyclopropanone. This was compared with the result from the ab initio calculation, 83 kcal (347 kJ). Though these values were rather discrepant, it was noted that the more "orthodox" calculation was no doubt an upper bound—there was a low-lying empty orbital in zwitterionic oxyallyl, and so even a small configuration interaction (CI) study would no doubt find a preferential stabilization of the zwitterionic species. To be most meaningful, comparable CI calculations must be done on both the cyclopropanone and the zwitterion. Although, in fact, the requisite CI on both species remains seemingly unreported, a recent computational study[27] showed that a CI stabilizes oxyallyl in its zwitterionic form more than in its triplet diradical form. Is this stabilization "merely due to the poor description of *any* species with a negative charge,[28] be it bearing a total negative charge or be it a neutral zwitterion?

≪The values of -1 for the oxygen charge and of $+\frac{1}{2}$ for the terminal carbons of the allyl component are clearly "exaggerations"[29] (ie, they are larger than the "true" charges). Nevertheless these choices allow us to make a well-defined estimate of the energetics of the zwitterion. Indeed, we do not know how to find the "true" charge or how to compute the energy of a species such as zwitterionic oxyallyl bearing charges other than the whole and half-integral charges used here.≫

We do not know in the particular, although a recent literature study[30] of the ground, biradical, and zwitterionic forms of ethylene (the so-called X, T, and V states) noted that the energetics of the zwitterion was the most sensitive of the three to changes in calculational methodology. All these arguments suggest that the literature molecular orbital calculation gives an upper bound to the relative stability of cyclopropanone relative to the zwitterionic oxyallyl, and so our findings are in more *quantitative* as well as *qualitative* agreement than when we wrote that first article on these species.

Another conceptually significant change in the macroincrementation reaction for zwitterionic oxyallyl is implicit in the use of the modified reaction:

$$\mathbf{3} \equiv {}^+(CH_2)_2CO^- = CH_2CRCH_2^+ + R'O^- - RR' + E_{Coul} \quad (6\text{-}23)$$

In this reaction R and R′ are to be chosen to better mimic the intramolecular environment found in oxyallyl than does R = H and R′ = CH_3. That is, we idealize the use of groups more paradigmatic than H and CH_3. However, we find there is but one R other than R = H for which we have data, CH_3.

≪This additional exercise amply documents our assertion that when discussing the thermochemistry of ions, it is imperative to use the "best," and thus generally the most recent, data. For example, the heat of formation of allyl ion reported in Reference 26 differs by about 10 kcal (40 kJ) from that given in the most recent compendium, Reference 2.≫

Thus we use CH_3. The choice of R′ is much wider, since there has been considerable interest in the energetics of gas phase alkoxide ions, which have been among the most "popular" of anions (see Reference 2 for the accumulated thermochemistry and numerous associated primary literature citations for these species). We opt for R = $(CH_3)_2CH$—because it is of comparable size, polarizability, and composition to the group attached to the —O^- in oxyallyl. So doing reduces the predicted heat of formation of zwitterionic oxyallyl to about 46 kcal (192 kJ). This value is still sufficiently above that of cyclopropanone that it comes as no surprise that the zwitterion oxyallyl does not contribute in any significant way to the chemistry of the parent cyclopropanone. We are encouraged by the qualitative and quantitative consistency of our findings, both with experiment and with more "orthodox" theory. This gives us confidence in applying macroincrementation reactions to other species, whether they be strained or zwitterionic or "merely" of interest.

3. THE $n = 3$ SPECIES, THE VINYL HALIDES

We now turn to a set of $n = 3$ examples, the vinyl halides, $CH_2 = CH—X$, for X = F, Cl, and Br. Are these species stabilized relative to other organic halides? One might think so, since they have either two (for F) or three (for Cl and Br) resonance structures involving π bonding of the olefin with the halogen.

$$CH_2{=}CH\text{-}X \leftrightarrow {}^-CH_2CH{=}X^+ \leftrightarrow {}^+CH_2CH{=}X^- \quad \text{(F excepted)} \quad (6\text{-}24)$$

However, the same species might be expected to be destabilized because of σ-electron withdrawal from the unsaturated carbon framework. (See Reference 15 and Chapter 5 of this volume for a more thorough discussion of this latter effect.) Intuitively, both the π-stabilization and σ-destabilization should be maximized for fluorine and minimized for bromine—as noted in Chapter 4 of this volume, the orbitals of fluorine are more compatible with those of carbon than for any other halogen—while simultaneously account must be taken of the fact that fluorine is the most electronegative element. It is not intuitively obvious how to balance these two competing influences on molecular stability.

Notice that we have left vague the organic halides to which we are comparing these vinyl halides. The approach taken in References 15 and 31 employed isopropyl halides because in the cases of both vinyl and isopropyl derivatives, the carbon bearing the halogen is bound twice to carbon as well: in vinyl, the same carbon is bound twice to the halogen-bearing carbon, whereas in isopropyl two carbons are so bound. Macroincrementation reaction 6-25 is thus suggested:

$$CH_2{=}CH{-}X = CH_2{=}CH{-}R + (CH_3)_2CH{-}X - (CH_3)_2CH{-}R \quad (6\text{-}25)$$

What R do we choose as our paradigm? We have expressed caveats about R = H and CH_3, and preferences for R = C_2H_5. In the case of reaction 6-25, we might expect the differences between R = H, CH_3, and C_2H_5 to be rather small because (a) $(CH_3)_2CH$— and $CH_2{=}CH$— are rather small and so steric effects are quite minimal, and (b) H, CH_3, and C_2H_5 are rather electronically "innocuous," nonpolar hydrocarbon substituents and so their interactions with the nonpolar hydrocarbons $(CH_3)_2CH$— and $CH_2{=}CH$— are expected to be comparable. That our expectations are realized is shown by the fact that the differences of the heats of formation of $(CH_3)_2CH{-}R$ and $CH_2{=}CH{-}R$ are nearly independent of the choice among the three R groups at our disposal: for R = H, CH_3, and C_2H_5, we find −37.5, −37.0, and −36.8 kcal (−157, −155, and −154 kJ).

≪Had we chosen the large, but still quite electronically innocuous substituent $(CH_3)_3C$—, the difference between $(CH_3)_2CH{-}R$ and $CH_2CH{-}R$ would have been −34.4 kcal (−144 kJ). Likewise, had we chosen the rather small, but noninnocuous (interactive) substituent $—CH{=}CH_2$, the difference would have been −32.7 kcal (−137 kJ). Both new results corroborate our assertions that contrary to what was thought earlier, H, CH_3, and C_2H_5 are comparatively sterically and electronically equivalent when dealing with hydrocarbons.≫

We may thus return to Equation 6-25 using any of the three choices, R = H, CH_3, and C_2H_5. Rather than deciding, we simply choose a composite or averaged value, −37.1 kcal (−155 kJ). This results in the following predictions of the heats of formation of vinyl fluoride, chloride, and bromide: −32.9, 2.4, and 13.6 kcal (−138, 10, and 57 kJ). The value for the fluoride is almost indistinguishable from the experimental value in our archives −33.2 kcal (−139 kJ), whereas those for the heavier halogens differ markedly: for the chloride and bromide we find 8.8 and 18.9 kcal (37 and 79 kJ). It would appear that halogen substitution on ethylene results in no stabilization at best, and destabilization in general. What is also found, and quite perplexing at that, is that the destabilization follows no obvious or monotonic order relative to the position of the halogen in the periodic table: zero = no destabilization $\cong$ F $<$ Br = some destabilization $<$ Cl = even more destabilization. But are the experimental heats of formation correct?

To attempt to understand this, we make use of the following alternative macroincrementation reaction

$$CH_2{=}CH{-}X = CH_2{=}CH{-}R - C_6H_5{-}R + C_6H_5{-}X \qquad (6\text{-}26)$$

This reaction explicity compares substituted ethylenes and benzenes, generically species **4** and **5**:

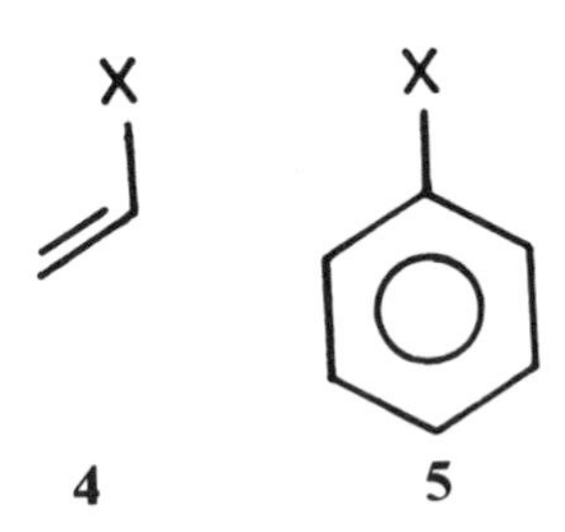

Again, we inquire what R is to be used, although from the above it is likely that R = H, CH_3, and C_2H_5 will not show significant differences. This is corroborated by experiment: the differences of the heats of formation of $CH_2{=}CH{-}R$ and $C_6H_5{-}R$ are 6.9, 7.1, and 7.3 kcal (29, 30, and 31 kJ) respectively. Again, although any of the three R groups could be used, we choose the average, 7.1 kcal (30 kJ), from which we predict the heats of formation of C_2H_3 (F), C_2H_3 (Cl), and C_2H_3 (Br) to be -34.9, 5.0, and 17.7 kcal (-146, 21, and 74 kJ). Fluorine and bromine show small and comparable destabilization energies, 1 kcal (5 kJ), but the error is much larger for chlorine, 3.8 kcal (16 kJ). This new order, F $\cong$ Br $<$ Cl is different from the previous one and likewise lacks any explanation based on the positions of the three halogens in the periodic table.

Since the halogen-bearing carbons in ethylene and benzene frameworks have comparable geometries, atomic hybridizations, and orbital energies, we predict that reaction 6-26 is nearly thermoneutral. (Also see Chapter 6, Volume 4.) While our expectation is confirmed for fluorine and bromine (as well as for CH_3 and C_2H_5), it is flagrantly denied for chlorine. Is this disparity for chlorine real? Are the experimental data for the vinyl halides indisputable?

We think not. It is interesting to note that an alternative value for the heat of formation of vinyl chloride has been suggested[32] recently: 5.5(5) kcal (23(2) kJ). Not only is this new value nearly indistinguishable from that predicted from macroincrementation reaction 6-26, but use of it in the earlier reaction (6-25) results in a monotonic trend for destabilizations: $F < Cl <$ Br. This monotonic trend is most encouraging.

4. THE $n = 4$ SPECIES, THE DIHALOETHANES, AND THE ENERGETICS OF GEMINAL AND VICINAL SUBSTITUTION

We now turn to a particularly instructive area, the $n = 4$ cases for the 1,1-dihaloethanes and 1,2-dihaloethanes, CH_3CHX_2 and CH_2XCH_2X, for $X =$ F, Cl, and Br. These species may be interrelated with the aforementioned vinyl halides by both formal and experimental combination with HX and indeed, these hydrohalogenation/dehalogenation reactions have served the calorimetrist well.

≪From a calorimetric, or even more general thermochemical, point of view, it is far better to consider the dehydrohalogenation reaction. Not only does the hydrohalogenation reaction have the a priori possibility of forming both dihaloethane isomers, but it is not obvious that the more stable isomer would even be formed in higher yield. That is, addition of HX by initial H^+ attack would result in CH_3CHX_2 via Markownikoff addition and the intermediacy of CH_3CHX^+, whereas Michael addition of X^- or free radical addition of $X^{\cdot}$ would eventually yield XCH_2CH_2X via the intermediacy of $(XCH_2CHX)^-$ or $(XCH_2CHX)^{\cdot}$.≫

In addition, dihaloethanes have been viewed as archetypes for the study of geminally (1,1-) and vicinally (1,2-) substituted species, and so provide examples for a particularly important type of structural isomerism.[33] Indeed, it has been argued[34] that logic applicable for dihaloethanes may be extended to other classes of molecules (eg, disubstituted ethylenes and disubstituted hydrazines). For all these classes of molecules, it appears that the *gem*- or 1,1-disubstituted species are usually more stable than the isomeric *vic*- or 1,2-species. This is a simple, general, and powerful rule of molecular energetics.

We recall another simple, general, and powerful rule: double substitution of methane is generally energetically preferred over single. That is, CH_2X_2 species are usually found to be more stable (see References 32 and 34–36 and Chapter 4, this volume) than would have been predicted from assuming that macroincrementation reaction 6-27 is thermoneutral. We note that XCH_2CH_2X has two singly substituted carbons whereas CH_3CHX_2 has one doubly substituted carbon. Does not CH_3CHX_2 thus gain stability over XCH_2CH_2X because of the differing energies of mono- and disubstitution? Macroincrementation reaction 6-28 explicitly tests this conjecture:

$$CH_2X_2 = 2CH_3X - CH_4 \tag{6-27}$$

$$CH_3CHX_2 = XCH_2CH_2X + CH_2X_2 + CH_4 - 2CH_3X \tag{6-28}$$

Thermoneutrality of this reaction would show the first-mentioned rule (the higher stability of 1,1-disubstituted ethanes than their corresponding 1,2-isomers) to be derivable from the second (the inherent stabilization of disubstituted methanes). Alternatively, if CH_3CHX_2 turns out to be experimentally more stable than predicted from macroincrementation reaction 6-28, assuming thermoneutrality (ie, if the experimentally determined heat of formation of 1,1-isomers be found to be lower than the theoretical), the rule about 1,1-disubstituted ethanes can be said to be distinct from the rule about disubstituted methanes and additionally gains credibility. There is, of course, a third possible result from the test above: CH_3CHX_2 may be less stable than predicted by macroincrementation reaction 6-27, in which case there is nothing "magical" about 1,1-disubstitution. Indeed, in that case, it could be cogently argued that geminal disubstitution is intrinsically destabilizing (except for so-substituted methanes) and not only for the cases of substituents that obviously sterically repel and thus stabilize the vicinal isomer.

What is found experimentally? Let us take data on all the relevant disubstituted methanes and ethanes from two specialized review articles.[32,35] So doing, using Equation 6-28, one finds for fluorine, chlorine, and bromine net destabilization of the 1,1-dihaloethane by 8.8, 2.2, and 1.2 kcal (37, 9, and 5 kJ). Even for the rather small and electronically innocuous substituent $X = CH_3$, the 1,1-disubstituted compound is destablilized by 1.2 kcal (5 kJ). We thus assert that 1,1-disubstituted species are generally intrinsically destabilized relative to their 1,2-disubstituted isomers. That is, the first rule should be replaced by its opposite.[37]

5. THE $n = 5$ SPECIES, BROMOACETONE AND IODOACETONE, AND THE ENERGETICS OF α-HALOKETONES AND RELATED SPECIES

The next species we discuss are also $n = 5$ compounds, bromoacetone and iodoacetone, CH_3COCH_2X (X = Br and I, respectively). It has long been known[38] that α-haloketones react more readily with nucleophiles than do the simple alkyl halides that lack the keto group.

≪We exclude from this discussion the basic nucleophiles, hydroxide and alkoxide ions, because they generally react by loss of a proton and of the halide ion to form transitory cyclopropanones (cf Section 2) in the Favorskii rearrangement[41] that converts α-haloketones into carboxylic acids or their esters (eg, 2-bromocyclohexanone reacts with $C_2H_5O^-$ to form ethyl cyclopentanecarboxylate.)≫

This relative reactivity has generally been explained[39] in terms of a stabilized (ie, energetically lower) transition state involving the keto carbon as well as the "ipso" carbon from which the halogen is displaced. However, MNDO calculations argue against this and suggest a normal S_N2 reaction path.[40] It is further suggested that the electronic environment of the central carbon in an S_N2 reaction mimics that of a carbanion and so is stabilized by the keto group. A simple alternative interpretation, apparently unmentioned in the literature, agrees that the transition state is comparably normal for S_N2 reactions and argues that the rapid rate results from destabilization of the ground state haloketone by the electrostatic repulsion of the bond dipoles of the keto and C—X bonds.

What insights can macroincrementation reactions provide? Macroincrementation reactions to date have not been particularly applicable to the energetics of transition states; nonetheless we may deduce the degree of destabilization in the substituted ketones. Macroincrementation reaction 6-29 documents the existence of significant destabilization for X = Br and I:

$$CH_3COCH_2CH_3 + CH_3CH_2CH_2X = CH_3CH_2CH_2CH_3 + CH_3COCH_2X \qquad (6\text{-}29)$$

The theoretically predicted heats of formation of gaseous bromoacetone and iodoacetone are −47.6 and −35.1 kcal (−199 and −147 kJ). By contrast, the experimental values are −43.3(1.9) and −31.3(1.2) kcal (−181(8) and −131(5) kJ), corresponding to about 4 kcal (17 kJ) destabilization in both cases. We may thus conclude that ground state destabilization makes a meaningful contribution to the high reactivity of α-haloketones with nucleophiles.

Indeed, this destabilization should have been anticipated given the long known, but more indirectly demonstrated, lessened stability of α-diketones.

≪Documentation[42] includes the high stability of the monoenol form of cyclic α-diketones relative to that of cyclic monoketones and the facility of base-catalyzed transformations such as the benzilic acid rearrangement of acyclic α-diketones to form α-hydroxycarboxylic acids.≫

These species are related to the α-haloketones but have two adjacent electronegative oxygens instead of one halogen and one oxygen. The destabilization of α-diketones may be quantitated by using macroincrementation reaction 6-30 to study the energetics of the $n = 6$ species butane-2,3-dione (biacetyl):

≪The data on this species come from our standard archive, Reference 1, as for most of the species of interest in this chapter. To alert the reader, we note that this interesting and important compound is misnamed as 3-oxobutanal in that source.≫

$$CH_3COCH_2CH_3 + CH_3CH_2COCH_3 = CH_3CH_2CH_2CH_3 + CH_3COCOCH_3 \qquad (6\text{-}30)$$

The predicted heat of formation is −85.1 kcal (−356 kJ), to be compared with the highly precise experimental value of −78.2(0.3) kcal (−327(1) kJ). This corresponds to a net destabilization of 7 kcal (29 kJ), certainly a nonnegligible quantity. We argue that the destabilization arises because of repulsion of the two keto groups and so would expect the net charges on the keto carbon and oxygen atoms to decrease in the diketone, and accordingly the heat of vaporization of butane-2,3-dione should be comparatively low. Using the same macroincrementation reaction 6-30 but now for predicting the heat of vaporization, one finds 11.7 kcal (49 kJ) to be contrasted with the experimental value of 9.3 kcal (39 kJ). This 2.4-kcal (10-kJ) discrepancy is in accord with our expectations. So far, so good.

Let us now consider the energetics of the isoelectronically related 2,3-dimethylbutadiene, wherein the two =O groups of the dione are replaced by two $=CH_2$ groups. Two predictions may be made: (a) net stabilization of the diene will be found, since this is almost the archetypal example of a conjugated species (ie, our prediction will be too positive, and (b) since most hydrocarbons are essentially nonpolar, no disparity in the heat of vaporization arises for this compound

≪The absence of a disparity can also be derived from the literature finding that the heat of vaporization of a hydrocarbon almost completely depends on the total number of carbons and is largely independent of its molecular structure. (See rule 1 for the estimation of heats of vaporization given in Reference 43.)≫

Macroincrementation reaction 6-31 tests both predictions:

$$CH_3C(CH_2)CH_2CH_3 + CH_3CH_2C(CH_2)CH_3 = CH_3CH_2CH_2CH_3 + CH_3C(CH_2)C(CH_2)CH_3 \qquad (6\text{-}31)$$

Both predictions are confirmed: the derived heats of formation and of vaporization are 13.2 and 7.4 kcal (55 and 31 kJ), to be compared with the experimental values of 10.5 and 7.4 kcal (44 and 31 kJ).

To compare even more explicitly 2,3-dimethylbutadiene and butane-2,3-dione, one should consider macroincrementation reaction 6-32.

$$CH_3COCOCH_3 + 2CH_3C(CH_2)CH_2CH_3 = CH_3C(CH_2)C(CH_2)CH_3 + 2CH_3CH_2COCH_3 \qquad (6\text{-}32)$$

We anticipate that (a) this new reaction will result in a heat of formation of 2,3-dimethylbutadiene "far" too high, since it incorporates the purported destabilization of 2,3-diketobutane and lacks the conjugation energy of the diene, and (b) the heat of vaporization will be low because that of the α-diketone is low. Both predictions are confirmed: the predicted heat of formation is 19.9 kcal (83 kJ), whereas the experimental value is 10.5 kcal (44 kJ), and the predicted heat of vaporization is 4.9 kcal (21 kJ), whereas the experimental value is 7.4 kcal (31 kJ). We are encouraged by the self-consistency of all our findings.

6. THE $n = 5$ SPECIES, GLYCINE, AND THE ENERGETICS OF α-AMINO ACIDS

In this section we discuss the energetics of amino acids, and commencing with the simplest one, the $n = 5$ species, glycine. It is well established[44,45] that as an isolated, gas phase molecule the preferred structure is amino acetic acid, H_2NCH_2COOH, whereas in the condensed phase, either in aqueous media or as a bulk solid, it exists as the zwitterion[45,46] $^+H_3NCH_2COO^-$. Whether one views glycine as a bifunctional molecule or as a zwitterion, it is not at all surprising that salts of glycine readily form in which a proton either has been lost to form glycinate anion $H_2NCH_2COO^-$ or has been gained to form glycinium ion $^+H_3NCH_2COOH$. It is perhaps presumptuous to attempt to predict the properties of aqueous solutions, but we nonetheless do so where we warn the reader (and ourselves) not to expect as accurate or unambiguous results as given elsewhere in this chapter for isolated molecules.

Of course it is necessary to convince ourselves we can derive the properties of the simpler, nonzwitterionic, "nonpolar" species in the gas phase first. Four interrelated macroincrementation reactions may immediately be written:

$$CH_3COOH + CH_3NH_2 - CH_4 = H_2NCH_2COOH \quad (6\text{-}33)$$
$$CH_3CH_2COOH + CH_3NH_2 - CH_3CH_3 = H_2NCH_2COOH \quad (6\text{-}34)$$
$$CH_3COOH + CH_3CH_2NH_2 - CH_3CH_3 = H_2NCH_2COOH \quad (6\text{-}35)$$
$$CH_3CH_2COOH + CH_3CH_2NH_2 - CH_3CH_2CH_3 = H_2NCH_2COOH \quad (6\text{-}36)$$

These result in "predictions" of the heat of formation of gas phase nonpolar glycine of −90.8, −92.5, −94.6, and −93.5 kcal (−380, −387, −396, and −391 kJ) respectively. From our earlier preferences of C_2H_5 over CH_3 and CH_3 over H, we would expect that the reliability of the four results should increase in the order: (6-33) $<$ (6-34) $\cong$ (6-35) $<$ (6-36). Encouragingly, the most recently reported experimental value[47] for this quantity is −93.5(1.2) kcal (−391(5) kJ), in complete agreement with our purportedly most reliable prediction.

What about the gas phase zwitterion? We opt to parallel both the macroincrementation reactions for the nonpolar species as well as our earlier successful treatment of the polar form of cyclopropanone, the zwitterionic oxyallyl. In the particular, one could consider the related following four macroincrementation reactions:

$$CH_3COO^- + CH_3NH_3^+ - CH_4 + E_{Coul} = {}^+H_3NCH_2COO^- \quad (6\text{-}37)$$
$$CH_3CH_2COO^- + CH_3NH_3^+ - CH_3CH_3 + E_{Coul} = {}^+H_3NCH_2COO^- \quad (6\text{-}38)$$
$$CH_3COO^- + CH_3CH_2NH_3^+ - CH_3CH_3 + E_{Coul} = {}^+H_3NCH_2COO^- \quad (6\text{-}39)$$
$$CH_3CH_2COO^- + CH_3CH_2NH_3^+ - CH_3CH_2CH_3 + E_{Coul} = {}^+H_3NCH_2COO^- \quad (6\text{-}40)$$

In the above, and in related expressions that follow, unless otherwise said, all gas phase ion data are from the compendium, Reference 2. Furthermore, E_{coul} is taken to be the Coulombic or electrostatic interaction of the positive and negative charges of the zwitterion, and we always explicitly state what we assume to be the dielectric constant that attenuates this attraction. How do we compute these electrostatic interactions? Following from our zwitterionic oxyallyl study in Section 2, we assume:

1. The dielectric constant here is unity: there is no "stuff" between the charged atoms. Furthermore, only point charge interactions is considered.
2. All the unit positive charge is assumed to be located on the nitrogen: other studies have shown this to be a useful,[48] however quantum chemically invalid,[49] assumption in complexation studies of the parent ammonium ion.
3. The unit negative charge is distributed equally between the two oxygens: like the zwitterionic oxyallyl case, no extra anionic self-repulsion between these two like-charged atoms is assumed.
4. The NCC angle has been assumed to be tetrahedral, and both the CCO and OCO angles have been assumed to be exactly 120°. The model is too inexact to want to do any better. Besides, we feel confident that small variations in molecular geometry result in but small effects on the total energy of the zwitterion.
5. Both the C—N and C—C bond lengths have been assumed to be 1.4 Å; the two C—O bonds are assumed to be equivalent and 1.25 Å long. As with the interatomic angles, there is no reason to do better.
6. Two rotamers have been considered: **6,** in which the total five heavy atom skeleton is strictly planar, and **7,** in which the C—N bond is perpendicular to the plane formed by the two carbons and two oxygens.

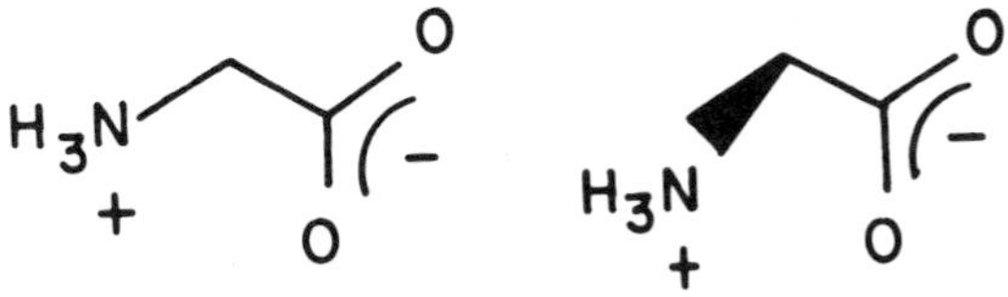

H3N O O − + **6** H3N O O − + **7**

From these geometric and electrostatic assumptions, one finds E_{Coul} to be −103.9 or −107.6 kcal (−438 or −450 kJ), respectively, for rotamers **6** and

7. It would appear that the electrostatics is reasonably independent of the molecular conformation of the zwitterion. As such, we will take the gas phase electrostatic attraction to be about −106 kcal (−445 kJ). Putting together the requisite numbers and assumptions, we thus find from Equations 6-37 to 6-40 that the heats of formation of glycine zwitterion are −63, −66, −70, and −70 kcal (−263, −275, −291, and −291 kJ), respectively. This is a rather wide spread of values. Remember, however, the folkrule that the energetics of ions is much more sensitive to substituent effects than that of neutrals. We conclude that these most simple of macroincrementation reactions are not sensitive to these substituent effects on ions (see Chapters 7 in Volume 2, 5 in Volume 3, and 8 and 9 in Volume 4). For example, implicit in Reactions (6-33) through 6-40 are the assumptions that the proton affinity of methylamine, ethylamine, and glycine are the same as would be that of all primary amines, and that of acetate, propionate, glycinate, and all carboxylate anions are likewise identical. These assumptions, while no doubt reasonable first guesses, are clearly quantitatively wrong if accuracy better than 10 kcal (40 kJ) is desired.

To improve this situation, recall that glycine is both a base and an acid. As such, rather than considering an arbitrary substance with an $—NH_3^+$ group such as methyl or ethyl ammonium ion, why not use glycinium ion? Likewise, rather than considering an arbitrary species containing $—COO^-$ such as acetate or propionate, why not use glycinate? So doing results in the new macroincrementation reaction:

$$NH_2CH_2COO^- + HOOCCH_2NH_3^+ - NH_2CH_2COOH + E_{Coul} = {}^+H_3NCH_2COO^- \qquad (6\text{-}41)$$

where the same six assumptions have been made. The isomerization energy ΔH_{isom} may be directly found if macroincrementation reaction (6-41) is rewritten as follows.

$$NH_2CH_2COO^- + HOOCCH_2NH_3^+ + E_{Coul} = 2H_2NCH_2COOH + \Delta H_{isom} \qquad (6\text{-}42)$$

From this macroincrementation reaction and input ion heats of formation from Reference 2, the heat of formation of the gas phase zwitterion is predicted to be −75 kcal (−315 kJ). Equivalently, from reaction 6-42 the zwitterion lies 18 kcal (75 kJ) higher in energy than the nonpolar form. Although no experimental data are available for comparison, literature quantum chemical calculations give a difference, ΔH_{isom} = 43 kcal (180 kJ).[50] Macroincrementation and quantum chemical theory appear to give rather disparate results. However, as with the zwitterionic oxyallyl study, we feel confident that the stability of glycine zwitterion, like that of most zwitterions, is certainly misdescribed, and probably underestimated. For the same reasons, the quantum chemically calculated isomerization energy is no doubt too high.

These last expressions allow one to easily estimate the isomerization energy in aqueous medium. By analogy to other alkyl ammonium halides and alkali metal carboxylate salts, we anticipate that glycinium chloride and sodium glycinate are strong electrolytes (ie, are completely dissociated. We may thus experiment with Equation 6-41 by adding the implicitly ionized NaCl and "enough" water to both sides result in Equation 6.43.

$$[NH_2CH_2COO^- \; Na^+](aq) + [HOOCCH_2NH_3^+ \; Cl^-](aq) + E_{Coul} = [NaCl](aq) + [^+H_3NCH_2COO^-](aq) + [H_2NCH_2COOH] \, (aq) \quad (6\text{-}43)$$

≪NaCl need not have been added. Instead, one could write simply the alternative macroincrementation reaction

$$[NH_2CH_2COO^-] \, (aq) + [HOOCCH_2NH_3^+] \, (aq) + E_{Coul} = [^+H_3NCH_2COO^-] \, (aq) + [H_2NCH_2COOH](aq)$$

Given correct input numbers, the energies of this reaction and of reactions 6-41 and 6-42 must be equal. Likewise, the heat of solvation of $[NH_2CH_2COO^- \; Na^+](aq) + [HOOCCH_2NH_3^+ \; Cl^-](aq)$ - $[Na^+Cl^-](aq)$ must equal that of $[NH_2CH_2COO^-](aq) + [HOOCCH_2NH_3^+](aq)$. The two quantities when taken from the same archival source, Reference 51, are in fact equal, but since the solvation energies of glycinium and glycinate ions were derived from those of their salts, we have not proved anything. It is much more significant that there is equality to about 4 kcal ($\simeq$ 20 kJ) when the values for the salts were taken from Reference 51 and the values of the bare ions from References 2 and 52. We are not disappointed by this disagreement. Rather, given the quite large variations in the reported sublimation energy of glycine and also in the solvation energies of rather simple ions in the literature, we conclude that our predictions are fundamentally valid even if not quantitatively precise.≫

Alternatively, we may write the right-hand side of Equation 6-43 as $[NaCl](aq) + 2[^+H_3NCH_2COO^-] \, (aq) - \Delta H_{isom}(aq)$, resulting in Equation 6-44. Since we are now considering aqueous media, the dielectric constant is now significant. The value for pure bulk water ($\simeq$ 80) is no doubt too high,[52] perhaps even by a factor of 2 for the water directly solvating the amino acid in solution, but the intramolecular, electrostatic interaction in the zwitterion is severely reduced to about 2 kcal ($\simeq$ 10 kJ). We will thus completely ignore it. It is hardly obvious that reaction 6-43 or 6-44 should be thermoneutral; that is, that $\Delta H_{isom}(aq)$ from this macroincrementation reaction should equal ΔH_r of the isomerization reaction of H_2NCH_2COOH and $^+H_3NCH_2COO^-$ in aqueous medium. Nonetheless let us proceed under the assumption that it is. From our inorganic archive Reference 51 we find data of the thermochemistry on a plethora of NaCl concentrations, one glycinium ion and glycinate in 200 molecules of H_2O, and glycine at a plethora of concentrations. Ideally, one would like to consider infinite dilution for all species. To within better than 1 kcal (4 kJ), the heat of formation of the solution of most nonelectrolytes or 1:1 electrolytes is independent of concentration. Thus taking whatever numbers we can find for NaCl, glycinium ion, glycinate, and glycine, should not result in a severe error. In fact we are encouraged by our accuracy: whereas the literature[53] enthalpy difference of aqueous nonpolar and zwitterionic glycine lies between 10 and 13.5 kcal (42 and 57 kJ), our new theoretical value is 10 kcal (42 kJ).

$$[NH_2CH_2COO^- \, Na^+](aq) + [HOOCCH_2NH_3^+ \, Cl^-](aq) + E_{Coul} = [NaCl](aq) + 2[^+H_3NCH_2COO^-](aq) - \Delta H_{isom}(aq) \qquad (6\text{-}44)$$

We now turn to more complicated amino acids, in particular, those of the structural formula $RCH(NH_2)COOH$. These species, the so called α-amino acids, are the building blocks of proteins and enzymes and so are of considerable biochemical importance. Though occurring naturally as zwitterions, we now consider only the nonpolar form for which numerous macroincrementation reactions may be written. By analogy to those given for glycine, reactions 6-33 to 6-36, we thus write:

$$RCH_2COOH + R'CH_2NH_2 - R'CH_3 = RCH(NH_2)COOH \qquad (6\text{-}45)$$

$$R'CH_2COOH + RCH_2NH_2 - R'CH_3 = RCH(NH_2)COOH \qquad (6\text{-}46)$$

$$RCH(CH_3)COOH + R'CH_2NH_2 - R'CH_2CH_3 = RCH(NH_2)COOH \qquad (6\text{-}47)$$

$$R'CH_2COOH + RCH(CH_3)NH_2 - R'CH_2CH_3 = RCH(NH_2)COOH \qquad (6\text{-}48)$$

$$R'CH(CH_3)COOH + RCH(CH_3)NH_2 - R'CH(CH_3)_2 = RCH(NH_2)COOH \qquad (6\text{-}49)$$

$$RCH(CH_3)COOH + R'CH(CH_3)NH_2 - R'CH(CH_3)_2 = RCH(NH_2)COOH \qquad (6\text{-}50)$$

Quite surprisingly, and certainly regrettably, there is no R′ that allows us to predict the heat of formation of general α-amino acids for *any* of the macroincrementation reactions above. There simply are too few data on branched carboxylic acids and on branched primary amines. How about using data on one amino acid to predict the properties of others?

We hesitate to derive the heat of formation of these species from that of glycine because of possible anomalies arising from the R = H in glycine. That is, regardless of X, macroincrementation reactions of the type

$$RCH(NH_2)COOH = HCH(NH_2)COOH + RX - HX \qquad (6\text{-}51)$$

are suspect because of the unique properties of H, although we have used reactions like this before and we will again in this chapter. A more satisfying macroincrementation reaction is

$$RCH(NH_2)COOH = R_{pd}CH(NH_2)COOH + RX - R_{pd}X \qquad (6\text{-}52)$$

Which amino acid do we choose as our paradigm? Equivalently, which R_{pd} do we choose? We are not inherently limited to the 20 or so different groups corresponding to the 20 natural amino acids found in proteins, but somehow it would be "nice" to use one. Our favorite archive of neutral thermochemistry and more recent journals gave only five additional natural amino acids for which there are thermochemical data on both the solid and gaseous forms.

Proceeding by increasing number of carbon atoms, we start with alanine. Although intuitively we would be hesitant to use CH_3 as R_{pd}, we have done so before and are willing to do so again. The biggest obstacle is that the D and L forms of alanine, as solids, were chronicled[1] to have heats of formation differing by more than 10 kcal (40 kJ)! Since both these measurements were in fact on a single enantiomer (see discussion in Reference 54), we suspect at least one is in error as documented by a new literature measurement.[47] We thus refrain from seriously considering the alanine case; that is, reaction 6-53 will not be used:

$$RCH(NH_2)COOH = CH_3CH(NH_2)COOH + RX - CH_3X \qquad (6\text{-}53)$$

Quite surprisingly, the species with $R_{pd} = C_2H_5$ is *not* a natural amino acid and so the absence of thermochemical information on it is less disappointing. There are, however, data on the amino acid with R = $(CH_3)_2CH$—, valine; $(CH_3)_2CHCH_2$—, isoleucine; $C_6H_5CH_2$—, phenylalanine; the "hetero" amino acid, methionine,[55] with $CH_3SCH_2CH_2$—; and the surprisingly "nonnatural" α-phenylglycine[56] with C_6H_5—. Our "prejudices" suggest using isoleucine for two reasons.

1. Its aliphatic hydrocarbon "tail" should produce no new anomalies, a feature we cannot be sure of a priori for either the aromatic phenylglycine and phenylalanine, or for the sulfur-containing methionine.
2. It contains a primary alkyl group, whereas valine with its isopropyl group might be strained because of vicinal interactions of the CH_3 groups with the NH_2 and/or COOH groups of the remaining part of the molecule.

We thus suggest the reaction

$$RCH(NH_2)COOH = (CH_3)_2CHCH_2CH(NH_2)COOH + RX - (CH_3)_2CHCH_2X \qquad (6\text{-}54)$$

although we have yet to define the X part. Although in principle any X will do, again it is best to choose carefully. For the R groups that constitute amino acids, it probably suffices to find a mimic that is of comparable size—electronic effects are no doubt small. As such, X may be a hydrocarbon group. To ideally mimic—$CH(NH_2)COOH$ by some hydrocarbon moiety, we would thus use —$CH(CH_3)CH(CH_3)_2$; but lacking data on most species containing that group, we opt for those with —$CH(CH_3)_2$. Generally good agreement is found with experiment. [See Table 6-1, where we had to estimate the heat of formation of $(CH_3)_2CHCH_2CH_2SCH_3$ needed for methionine and for completeness we included the results for α-phenylglycine as well.]

TABLE 6-1. Experimental and Estimated Values for the Heat of Formation of Gaseous α-Amino Acids

		Values (kJ)	
Amino acid	R	ΔH_f, Experimental	ΔH_f, Estimated
Glycine	H	−391(5)[a]	−392
Alanine	CH_3	−415(4)[a]	−422
Valine	$(CH_3)_2CH$	−455(1)	−466
α-Phenylglycine	C_6H_5	−281(6)[b]	−283
Phenylalanine	$C_6H_5CH_2$	−313(1)	−309
Methionine	$CH_3SCH_2CH_2$	−414(4)[c]	−417[d]

[a]See Reference 47.
[b]See Reference 56.
[c]See Reference 55.
[d]To calculate the heat of formation of methionine it was necessary to estimate $\Delta H_f[(CH_3)_2CHCH_2CH_2SCH_3]$. The following macroincrementation reactions were used for this thioether.

$$RSCH_3 = C_2H_5SCH_3 + RCH_2CH_3 - C_2H_5CH_2CH_3$$
$$RSCH_3 = C_2H_5SCH_3 + RSH - C_2H_5SH$$

In the particular, for R = CH_3, *n*- and *i*-C_3H_7, and *n*- and *t*-C_4H_9, these two sets of macroincrementation reactions agreed with each other and with experiment to better than 1 kcal (4 kJ). We thus felt confident using the so-derived value of −31 kcal (−130 kJ) where the two individually derived values were −31.5 and −30.8 kcal (−132 and −129 kJ). Interestingly, though perhaps parenthetically, these two reactions also reproduced the heat of formation of $C_6H_5SCH_3$ to the same accuracy. That is, the resonance energy associated with sulfur attached to benzene is negligible—a finding consistent with our later discussion of the rather small resonance energy of thiolesters (see Section 8).

7. THE $n = 5$ SPECIES, PYRROLE, AND THE ENERGETICS ASSOCIATED WITH CONJUGATION IN >NH-CONTAINING HETEROCYLES

The next species we consider is the $n = 5$ compound pyrrole, **8.** Pyrrole is a conjugated ring system with 6π electrons and it is thus not surprising it

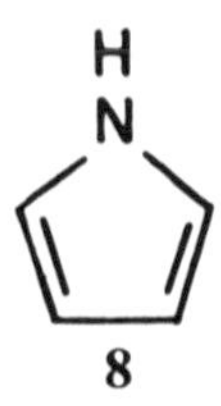

8

has aromatic properties. In this section we estimate the magnitude of the aromaticity of pyrrole and present a qualitative discussion of the consequences of the resulting stabilization. The aromaticity of pyrrole may be expressed in terms of the energy associated with separating the conjugative interactions of the two double bonds from the amino nitrogen. Macroincrementation reaction 6-55 demonstrates pyrrole to be stabilized by 23.4 kcal

8 = **9** + cyclopentadiene − cyclopentane (6-55)

(98 kJ); that is, the left-hand side of this reaction is lower in energy by this quantity than the right-hand side. This is not an unreasonable figure and may be compared with the approximate amount 14 kcal ($\simeq$ 60 kJ) found for both furan and thiophene (**10,** X = $>$ O and $>$ S) and 28.9 kcal (121

10

kJ) found for benzene (**10,** X = —CH=CH—) by the related, but more general, macroincrementation reaction 6-56.

$$\begin{aligned} \mathbf{10} &\equiv \text{cyclo-(CH=CH)}_2\text{X} \\ &= \text{cyclo-(CH}_2\text{CH}_2)_2\text{X } (\mathbf{11}) + \text{cyclo-(CH=CH)}_2\text{CH}_2 \\ &- \text{cyclo-(CH}_2\text{CH}_2)_2\text{CH}_2 \end{aligned} \qquad (6\text{-}56)$$

≪The proposed value for benzene is not the rather customary 36 kcal (151 kJ) because the delocalization energy of a conjugated diene is built into the model compounds used to represent species **10.**≫

Turning briefly to the value of the heat of vaporization of pyrrole, **8** (**10** with X = $>$NH), this quantity is correctly predicted to within 2.4 kcal (10 kJ). This is not a particularly large discrepancy. However, it is disconcerting that pyrrole has a higher vaporization enthalpy, 10.8 kcal (45 kJ), than does the saturated pyrrolidine, **9** (**11** with X = $>$NH) 9.0 kcal (38 kJ), because the latter species would have been expected to have stronger intermolecular

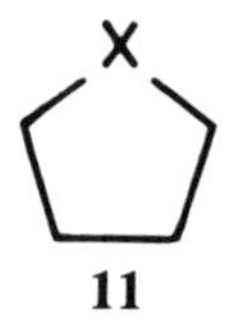

11

hydrogen bonds. Thus the relative heats of vaporization are in the opposite order to our prejudices. Interestingly, the heats of vaporization of the related, but non-hydrogen-bonding pairs cyclopentadiene and cyclopentane, furan and tetrahydrofuran, thiophene and thiolane, and benzene and cyclohexene show no such anomaly: for these species **10** and **11** with X = $> CH_2$, $>$ O, $>$ S, and —CH=CH—, one finds 6.8 and 6.9 kcal (28 and 29 kJ), 6.7 and 7.6 kcal (28 and 32 kJ), 8.4 and 9.3 kcal (35 and 39 kJ), and 8.1 and 7.9 kcal (34 and 33 kJ), respectively. It would appear that to within a tolerable discrepancy of 2 or so kcal ($\simeq$ 10kJ), the relative degree of, and even the absolute presence of, aromaticity has no effect on the heat of vaporization of two otherwise related species. Given this simplifying relationship, we will limit our attention to gas phase properties for the test of this section.

Returning to the aromaticity of pyrrole due to the interaction of the nitrogen lone pair electrons with the double bonds, consider the macroincrementation reactions 6-57 and 6-58, which relate pyrrole with its mono- and dibenzoannelated analogues indole and carbazole, species **12** and **13,** and compare all three species with the corresponding carbocyles, cyclopentadiene, indene, and fluorene. In reaction 6-57, one of the two formal carbon–carbon double bonds in pyrrole is delocalized into an annelated benzene ring as well as with the amine nitrogen, whereas in reaction 6-58 both double bonds are delocalized. This reaction converts a double bond into what may be formally considered to be a 1½ bond

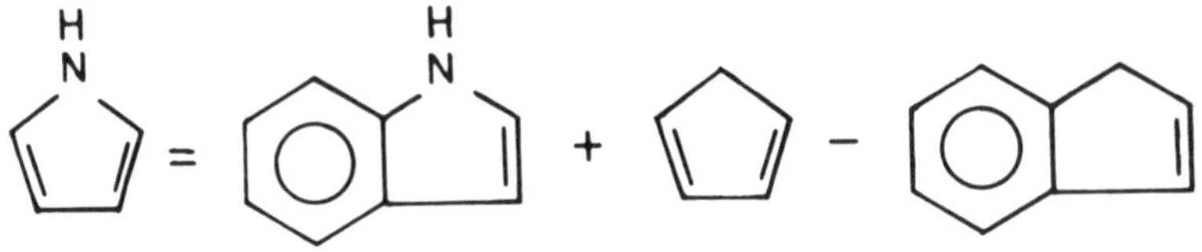

8 = **12** + cyclopentadiene − indene (6-57)

8 = **13** + cyclopentadiene − fluorene (6-58)

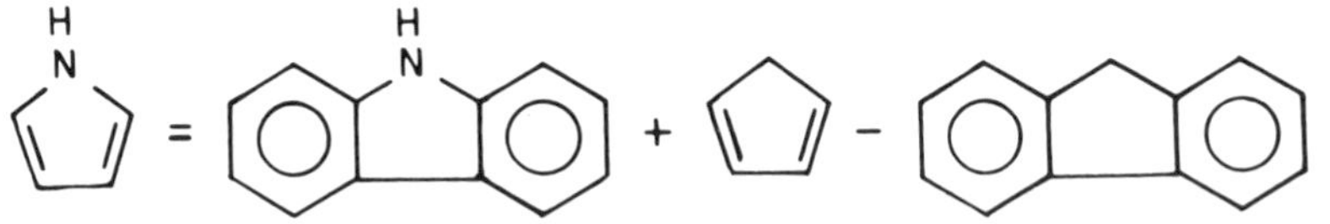

≪This arises from taking half each of a single and a double bond and so corresponds to taking the average contributions from the two well-defined Kékulé structures for the annelating benzene.≫

Therefore we expect a prediction of less resonance stabilization for a singly annelated pyrrole than for the parent species. That is, the heat of formation of pyrrole predicted from the use of reaction 6-57 will be incorrectly high. The value from double annelation (ie, from reaction 6-58) is similarly predicted to be even higher, and naively the discrepancy from this reaction would be expected to be about twice that from reaction 6-57. The first two "predictions" are in fact realized: the discrepancies are 4.1(1.7) kcal (17(7) kJ) and 11.2(1.4) kcal (47(6) kJ), respectively.

≪For the evaluation of this quantity, the heat of formation of fluorene was taken from Reference 58, for all other species our standard archive, Reference 1, was used.≫

Admittedly the factor of 2 is not exact so that our results are to be taken within the allowances of a rather liberal meaning given to the term "about" ($\simeq$). An interesting uniformity is found in the various numbers we have accumulated for the resonance energy of pyrrole: there is a correction of about 5 kcal ($\simeq$ 21 kJ) per half double bond. The first macroincrementation reaction presented in this section (6-54) involved two "whole" double bonds, and so a correction of about 20 kcal ($\simeq$ 84 kJ) might have been expected. This figure is comparable to the much more directly derived value of 23.4 kcal (98 kJ) presented earlier.

In pyrrole, and its mono- and bibenzoannelated species presented above, there is considerable interaction of the amine nitrogen lone pair with the double bonds found in the rest of the molecule. Consider now the case of 1,6-imino-[10]-annulene, **14,** in which this interaction is severely reduced.

NH

14

The following macroincrementation reaction (6-59) may be written, where we opt to add in the total "alphanumeric correction" of about 23 kcal ($\simeq$ 98 kJ) for our prediction of the heat of formation of 1,6-imino[10]annulene to account for the lessened double bond–lone pair interaction. The result is 93(2) kcal (390(6) kJ), in satisfying agreement with the experimental value of 88(2) kcal (367(7) kJ). That the theoretical value is higher than the experimental one is in accord with the expectation of "some" overlap, and accompanying stabilization, of the lone pair on the nonplanar nitrogen with the π electrons of the annulene. It would appear that we have a consistent model for the energetics of >NH-containing heterocycles.

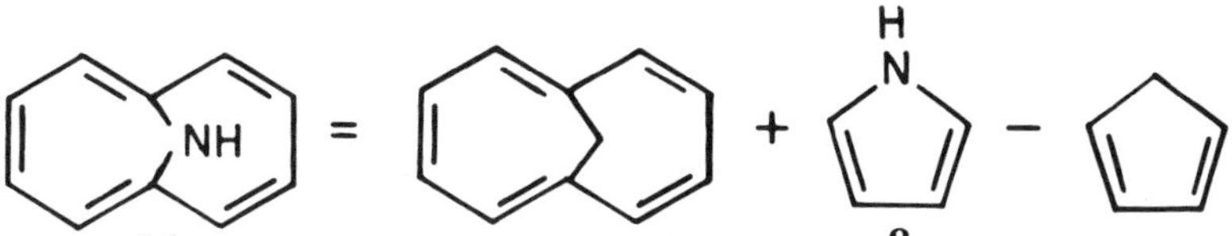

NH = + H N − (6-59)

14 **8**

8. THE $n = 6$ SPECIES, 2- AND 3-KETOTETRAHYDROTHIOPHENES, AND THE ENERGETICS OF THIOPHENE DERIVATIVES

We now turn to the $n = 6$ case of the isomeric 2- and 3-ketotetrahydrothiophenes, **15** and **16**. The reader will soon see that the energetics of these spe-

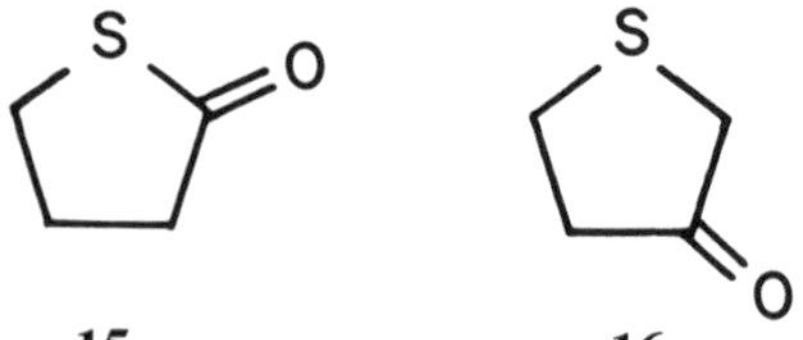

15 **16**

cies is related to the $n = 5$ case of bromoacetone and iodoacetone discussed in Section 5. The experimentally measured heats of formation of the 2- and 3-ketotetrahydrothiophenes (ketothiolanes) are -46.8 and -32.3 kcal (-196 and -135 kJ), and the corresponding heats of vaporization are 9.1 and 12.0 kcal (38 and 50 kJ), respectively. To what do we ascribe the 14.5-kcal (61-kJ) difference in the heats of formation? Which species, which value, if either, may be taken as "normal"? We recognize species **15** as a thiolester, and so it is expected to have some resonance stabilization (see below and Chapter 6 in Volume 4). Likewise species **16** may be recognized as an α-mercapto-ketone, and the near equality of the electronegativities of Br and S suggest some destabilization. Macroincrementation reactions offer the chemist a chance to "synthesize" a ketotetrahydrothiophene that lacks both the stabilizing and destabilizing effects noted above. In particular, consider macroincrementation reaction 6-60.

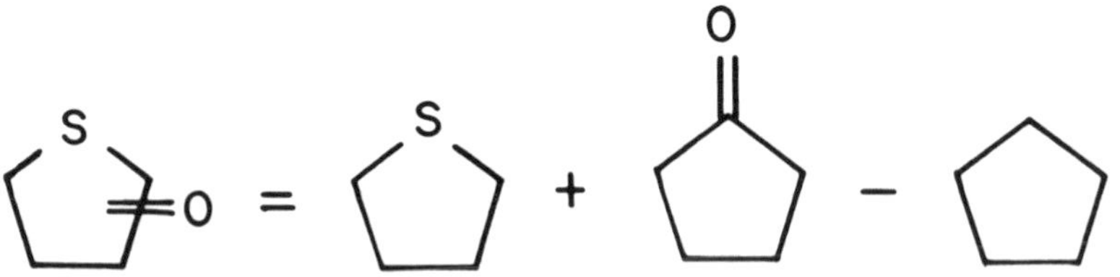

$$\textbf{15, 16} = \text{tetrahydrothiophene} + \text{cyclopentanone} - \text{cyclopentane} \quad (6\text{-}60)$$

This reaction is equally applicable to both ketotetrahydrothiophenes, and so the predictions of any property will be the same: for the heats of formation, -35.9 kcal (150 kJ), and vaporization, 12.7 kcal (53 kJ). Accordingly, the stabilization energy of 2-ketotetrahydrothiophene is 10.9 kcal (46 kJ) and the destabilization energy of the 3-isomer is 3.6 kcal (15 kJ). Likewise, the heat of vaporization of the 2-isomer is 3.6 kcal (15 kJ) low, whereas that of the 3-isomer differs is only 0.7 kcal (3 kJ) too low.

How do these values compare with those derived from related compounds and from whatever sources our intuition can cull? We are not particularly surprised that the 2-isomer is stabilized and the 3-isomer destabilized. However, whereas the 3-isomer is destabilized by a seemingly rather reasonable amount, the stabilization for the 2-isomer is noticeably larger than we would have expected. Being more precise, consider the following acyclic equivalent to the macroincrementation reaction above.

$$CH_3CH_2SC(O)CH_3 = CH_3CH_2SCH_2CH_3 + CH_3CH_2CH_2C(O)CH_3 - CH_3CH_2CH_2CH_2CH_3 \qquad (6\text{-}61)$$

We shall use experimental input from our archive, Reference 1, for most species.

≪In Reference 1, the heat of formation of $CH_3CH_2SCH_2CH_3$ is erroneously written as −93.5 instead of −83.5 kJ, as given in the primary source, Reference 58. How did we know there was an error? First of all, given the reported heats of vaporization (34.2 kJ), of formation of the liquid (−119.4 kJ), and of the gas (−93.5 kJ), the identity $\Delta H_f(\mathrm{l}) = \Delta H_f(\mathrm{g}) + \Delta H_v$ is not fulfilled. Casual observation shows that the heat of vaporization of thioethers and mercaptans depends largely on the number of carbons and is largely independent of structure. The heats of vaporization of almost all other 4-carbon, 1-sulfur species are between 35 and 40 kJ. For the heats of formation to be correct, the heat of vaporization would have had to be 26 kJ. Second, consider the macroincrementation reaction:

$$CH_3CH_2SCH_2CH_3 = CH_3CH_2CH_2SCH_3 + CH_3CH_2XCH_2CH_3 - CH_3CH_2CH_2XCH_3$$

Using any X we choose, —O—, —CHOH—, —C(O)—, —CH_2S—, —CH_2—, the heat of vaporization of both thioethers is nearly the same. In principle, the same macroincrementation reaction can be written to predict the heat of formation of $CH_3CH_2SCH_2CH_3$. From awareness of the "methyl versus ethyl dichotomy," the best X is that which most mimics —S—, and so we choose —CH_2S— and —CH_2—. These result in the prediction that the heat of formation of $CH_3CH_2SCH_2CH_3$ should be nearly identical to that of $CH_3CH_2CH_2SCH_3$. Since the latter value is −82.2(0.9) kJ, the value of −93.5(0.8) kJ for the former is untenable. By contrast, the value −83.5(0.8) kJ is highly reasonable and is adopted here.≫

For the heat of vaporization of $CH_3CH_2SC(O)CH_3$ (taken from Reference 59), we find the acyclic ethyl thiolacetate to be stabilized by 7.9 kcal (33 kJ), a value comparable to—though still meaningfully smaller than—the 10.9 kcal (46 kJ) found for the cyclic 2-ketotetrahydrothiophene. [The first-mentioned quantity, 7.9 kcal (33 kJ) may be labeled the resonance energy of a thiolester, and it is also meaningfully smaller than that found for amides and esters by an analogous macroincrementation reaction.[11]] The energetics of the 3-isomer is reliably reproduced by macroincrementation reaction 6-62.

$$\mathbf{16} \equiv (\text{cyclo-}C_4H_6S){=}O = \text{cyclo-}(CH_2)_4S\ (\mathbf{11},\ X = {>}S) + CH_3COCH_2Br - CH_3CH_2CH_2Br \qquad (6\text{-}62)$$

The predicted value is $-31(2)$ kcal ($-130(8)$ kJ), essentially indistinguishable with the experimental value of $-32.3(0.5)$ kcal (135(2) kJ).

Turning to the heats of vaporization, as mentioned above, there is a 3.6-kcal (15-kJ) discrepancy between the theoretical and experimental heats of vaporization of 2-ketotetrahydrothiophene. Is this a real effect? Again, consider acyclic thiolesters. Generalizing macroincrementation reaction 6-61 by writing Equation 6-63 and choosing a rather wide variety of R groups, R = *n*- and *i*-C_3H_7, and *n*- and *t*-C_4H_9 as well as C_2H_5, which alone had been used earlier,

$$RSC(O)CH_3 = RSCH_2CH_3 + CH_3CH_2CH_2C(O)CH_3 - CH_3CH_2CH_2CH_2CH_3 \qquad (6\text{-}63)$$

we find that thiolesters have heats of vaporization nearly a constant 2 kcal (8 kJ) lower than that predicted by our macroincrementation reactions. [For R = C_2H_5, *n*- and *i*-C_3H_7, and *n*- and *t*-C_4H_9, the experimental values are 9.6, 10.5, 10.1, 11.5, and 10.3 kcal (40, 44, 42, 48, and 43 kJ), whereas the theoretical values are 11.3, 12.3, 12.1, 13.5, and 12.2 kcal (47, 51, 51, 56, and 51 kJ). The discrepancies are thus 1.7, 1.8, 2.0, 2.0, and 1.9 kcal; 7, 7, 9, 8, and 8 kJ.] By analogy to these other thiolesters, the heat of vaporization of the 2-isomer is thus about $3.6 - 1.9 = 1.7$ kcal ($\simeq 8$ kJ) lower than what we would have predicted. The value for the 3-isomer is but 0.7 kcal (3 kJ) lower than we predicted using macroincrementation reaction 6-62. We offer an explanation related to the reduced polarity in α-diketones. It would appear that 3-ketotetrahydrothiophene is a more "reasonable" species than its 2-isomer because its properties are predictable by using macroincrementation reactions, whereas those of the 2-isomer are less predictable.

We are bothered by these seeming anomalies of the 2-isomer. The heats of formation and of vaporization of 2- and 3-methyltetrahydrothiophene **17a** and **17b** are reproduced to better than 0.5 kcal (2 kJ) using macroincrementation reaction 6-64 and the parent tetrahydrothiophene:

$$\mathbf{17} \equiv CH_3\text{-cyclo-}C_4H_7S = \text{cyclo-}(CH_2)_4S\ (\mathbf{11},\ X = {>}S) + CH_3\text{-cyclo-}C_5H_9 - \text{cyclo-}C_5H_{10} \qquad (6\text{-}64)$$

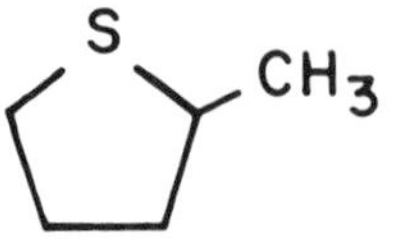

17a 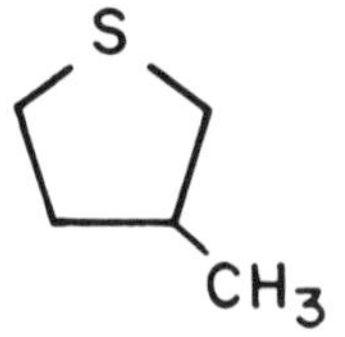 **17b** 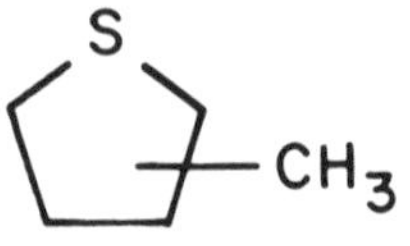 **17**

The same accuracy is found for 2- and 3-methylthiophene **18a** and **18b**, using macroincrementation reaction 6-65 and the parent thiophene:

$$\mathbf{18} \equiv CH_3\text{-cyclo-}C_4H_3S = \text{cyclo-}C_4H_4S\ (\mathbf{10}, X = {>}S) + C_6H_5CH_3 - C_6H_6 \qquad (6\text{-}65)$$

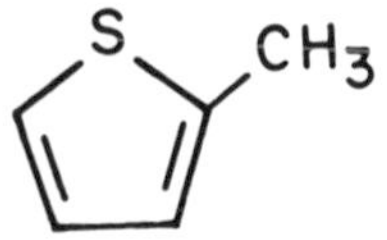

18a 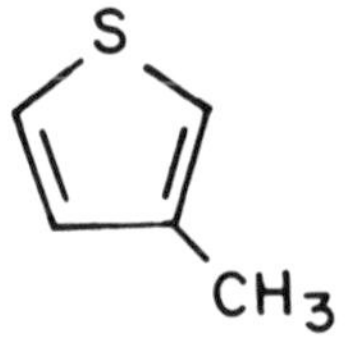 **18b** 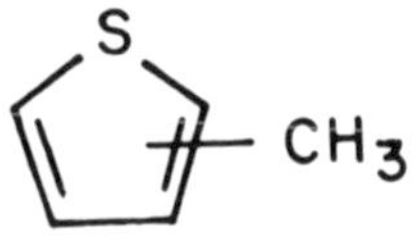 **18**

It would appear that we are generally successful in understanding the thermochemistry of sulfur compounds. The sole discordant measurement is the heat of vaporization of 2-ketotetrahydrothiophene. Let us briefly assume that it was mismeasured, that is, that the doubted experimental value is really the *whole* 3.6 kcal (15 kJ) too low. This means that the heat of formation of the gaseous compound is too low by the same 3.6 kcal (15 kJ) and so the derived resonance energy is too high. We earlier said that the resonance energy for the compound of interest exceeded that of acyclic thiolesters numerically by 10.9 − 7.9 = 3.0 kcal (13 kJ). Thus we see that all the results on ketotetrahydrothiophenes are consistent *if* the heat of vaporization of the 2-isomer is wrong by about 3 kcal ($\simeq$ 14 kJ). In the original paper,[60] the boiling points for the two isomers were given: **15** = 2-, 43°C/1–2 mm and **16** = 3-, 69–71°C/12 mm. These values are sufficiently close to suggest that the heats of vaporization of the isomers are comparable, a result consistent with our suggestion that the value for the heat of vaporization of 2-ketotetrahydrothiophene is significantly in error. We thus suggest remeasuring this quantity.

9. THE *n* = 6 SPECIES, TRIOXANE, AND THE ENERGETICS OF "OXANES" AND "OSES," POLYOXACYCLOALKANES, AND MONOSACCHARIDES

We now turn to another *n* = 6 species, 1,3,5-trioxane, **19**.

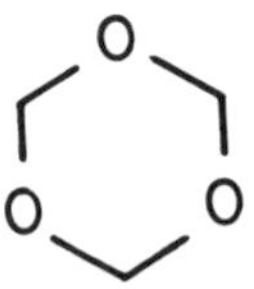

19

The first model we discuss explicitly ignores the 1,3-oxygen, oxygen interactions that we know must exist from our discussion in Section 4 of *gem*-disubstituted methanes and ethanes. (In the special case of geminal interactions involving at least one oxygen, this phenomenon contributes to the "anomeric effect."[34,36,61]) In the particular, consider macroincrementation reaction 6-66 involving "oxane" and cyclohexane.

$$\begin{aligned} \mathbf{19} &\equiv \text{cyclo-}(CH_2O)_3 \\ &= 3\ \text{cyclo-}(CH_2)_5O - 2\ \text{cyclo-}(CH_2)_6 \end{aligned} \qquad (6\text{-}66)$$

The predicted heat of formation of 1,3,5-trioxane is −101.8 kcal (−423 kJ), which is to be compared with the experimental value[62] of −111.4 kcal (−466 kJ). That the theoretical value is not negative enough by about 10 kcal (≃ 40 kJ) is thus directly attributable the stabilizing influence of the three —O—CH_2—O— units found only in the trioxane. This influence may be further thermochemically documented by the related macroincrementation reaction 6-67, for 1,3-dioxane, **20,** in which there is but one —O—CH_2—O— unit.

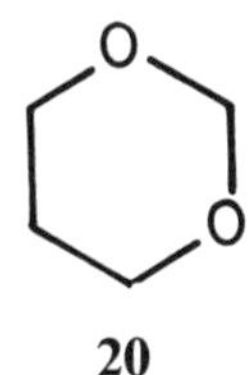

20

$$\begin{aligned} \mathbf{20} &\equiv 1,3\text{-cyclo-}(CH_2)_4O_2 \\ &= 2\ \text{cyclo-}(CH_2)_5O - \text{cyclo-}(CH_2)_6 \end{aligned} \qquad (6\text{-}67)$$

The predicted value, −77.2 kcal (−323 kJ), is 3.6 kcal (15 kJ) less negative than the experimental value,[61] −80.8 kcal (−338 kJ). Taking this 3.6 kcal (15 kJ) as an "alphanumeric" correction, and assuming that the three —O—CH_2—O— units in 1,3,5-trioxane are independent, results in a new predicted heat of formation of −111.9 kcal (−468 kJ). This value is nearly identical to that found experimentally, −111.4 kcal (−466 kJ). (Equivalently, we are assuming that macroincrementation reaction 6-68 is thermoneutral.)

$$\begin{aligned} \mathbf{19} &\equiv \text{cyclo-}(CH_2O)_3 \\ &= \text{cyclo-}(CH_2)_6 + 3[1,3\text{-cyclo-}(CH_2)_4O_2 - \text{cyclo-}(CH_2)_5O] \end{aligned} \qquad (6\text{-}68)$$

Let us try to directly extend our analysis to the related $n = 8$ and 10 species, 1,3,5,7-tetroxane (also called 1,3,5,7-tetraoxacyclooctane and 1,3,5,7-tetroxocane), **21,** and 1,3,5,7,9-pentoxane (also called 1,3,5,7,9-pentaoxadecane and 1,3,5,7,9-pentoxecane), **22.** To do so requires thermochemical data on oxocane, 1,3-dioxocane, oxecane, and 1,3-dioxecane (also called

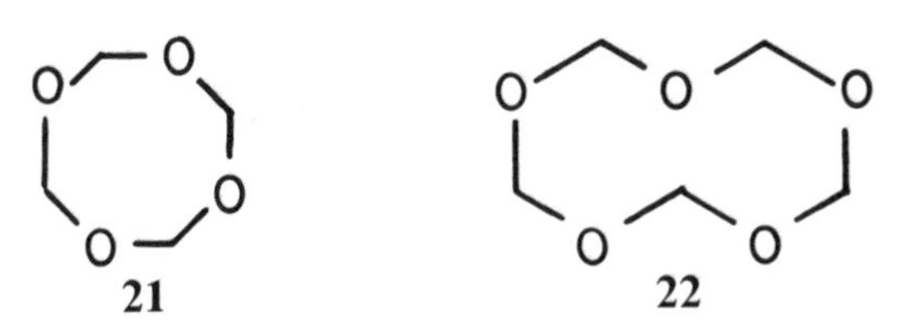

heptamethylene oxide, pentamethylene formal, nonamethylene oxide, and heptamethylene formal, respectively), species **23–26.**

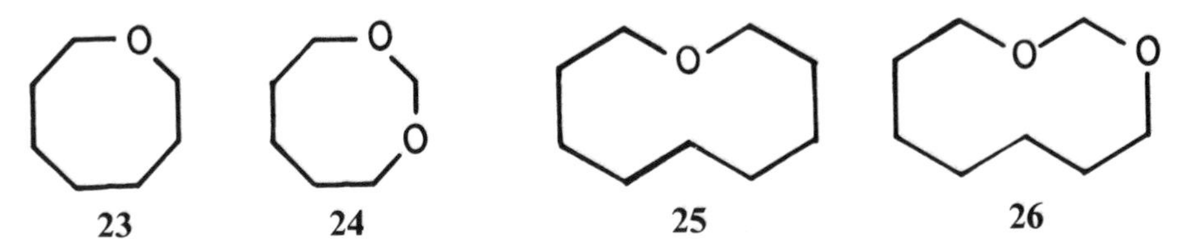

Experimental data are sparse here: there is a directly determined number for **24,** but only indirect information from the polymer literature (Reference 63) exists for **23,** and we know of none at all for either **25** or **26.**

We are not completely thwarted, however, in writing macroincrementation reactions analogous to those above. Let us present two assumptions to be so tested. The first assumption is that 1,3-dioxane is "normal." That is, while the saturated six-membered carbocyclic cyclohexane is known to be as strain free as any acyclic hydrocarbon such as heptane, we will assume that 1,3-dioxane is as strain free as its acyclic counterparts such as $(CH_3CH_2O)_2CH_2$ and check to see whether any inconsistencies arise. The second assumption is that the strain energies of the medium-sized cyclooctane and cyclodecane[64] are unaffected by the presence of oxygen atoms replacing some carbons. Both assumptions may be derived if one tacitly assumes that the inherent stabilizing anomeric interactions and lone pair–lone pair repulsions are independent of the ring size. It is also assumed that transannular repulsion of the C—H bonds of four CH_2 groups in cyclooctane is mimicked by the transannular repulsion of the four oxygens that replace them in tetroxane.

Are these assumptions valid? The simplest test of the first makes use of macroincrementation reactions 6-69 and 6-70.

$$\mathbf{20} \equiv \text{1,3-cyclo-}(CH_2)_4O_2 = \text{cyclo-}(CH_2)_6 + C_2H_5OCH_2OC_2H_5 - C_2H_5CH_2CH_2CH_2C_2H_5 \quad (6\text{-}69)$$

$$\mathbf{20} \equiv \text{1,3-cyclo-}(CH_2)_4O_2 = \text{cyclo-}(CH_2)_6 + C_2H_5OCH_2OCH_3 - C_2H_5CH_2CH_2CH_2CH_3 \quad (6\text{-}70)$$

The first reaction shows 1,3-dioxane to be destabilized by 3 kcal (13 kJ). The second, in which all four species have six heavy (ie, nonhydrogenic atoms), is nearly thermoneutral.

≪For this reaction, the heat of formation of $C_2H_5OCH_2OCH_3$ was estimated by assuming thermoneutrality for the macroincrementation reaction:

$$C_2H_5OCH_2OCH_3 = \tfrac{1}{2}(C_2H_5OCH_2OC_2H_5 + CH_3OCH_2OCH_3)$$

It may appear artificial to worry that $C_2H_5OCH_2OCH_3$ has six heavy atoms while $C_2H_5OCH_2OCH_2CH_3$ has seven. Our choice of diether is perhaps more legitimate when expressed in terms of the total number of carbons bonded to the two oxygens; that is, whereas diethyl formal has a total of five carbons bonded to the two oxygens, both 1,3-dioxane and methylethyl formal have but four. We are trying to incorporate some of the inherent features of the "methyl versus ethyl dichotomy" into the energetics of ring systems.≫

However, whatever value we choose, it is in fact nearly identical to that found in a corresponding treatment of 1,3-dioxocane.

≪That is, using macroincrementation reactions for 1,3-dioxocane that parallel those for 1,3-dioxane, namely:

$$\begin{aligned}\mathbf{24} &\equiv \text{1,3-cyclo-}(CH_2)_6O_2\\ &= \text{cyclo-}(CH_2)_8 + n\text{-}C_4H_9OCH_2O\text{—}n\text{-}C_4H_9\\ &\quad - n\text{-}C_4H_9CH_2CH_2CH_2\text{—}n\text{-}C_4H_9\\ \mathbf{24} &\equiv \text{1,3-cyclo-}(CH_2)_6O_2\\ &= \text{cyclo-}(CH_2)_8 + C_2H_5OCH_2OCH_3 - C_2H_5CH_2CH_2CH_2CH_3\end{aligned}$$

and where *n*-butylmethyl formal was analogously estimated using

$$n\text{-}C_4H_9OCH_2OCH_3 = \tfrac{1}{2}(n\text{-}C_4H_9OCH_2O\text{—}n\text{-}C_4H_9 + CH_3OCH_2OCH_3)$$

result in the prediction that 1,3-dioxocane is destabilized by nearly 4 and 1 kcal (16 and 4 kJ), respectively.≫

This consistency, if not necessarily normalcy, suggests that the second assertion is valid.

Consider the following macroincrementation reactions involving tetroxane and pentoxane:

$$\begin{aligned}\mathbf{21} &\equiv \text{cyclo-}(CH_2O)_4\\ &= \text{cyclo-}(CH_2)_8 + 4[\text{1,3-cyclo-}(CH_2)_4O_2 - \text{cyclo-}(CH_2)_5O] \qquad (6\text{-}71)\\ \mathbf{22} &\equiv \text{cyclo-}(CH_2O)_5\\ &= \text{cyclo-}(CH_2)_{10} + 5[\text{1,3-cyclo-}(CH_2)_4O_2 - \text{cyclo-}(CH_2)_5O] \qquad (6\text{-}72)\end{aligned}$$

Given the seeming validity of our assumptions, we are surprised that whereas the predicted heats of formation of tetroxane and pentoxane are −139.6 and −174.2 kcal (−584 and −729 kJ), the experimental quantities are significantly more negative, −148.2 and −186.2 kcal (−620 and −779 kJ). Alternatively, approximating the heats of formation of tetroxane and pentoxane as 8/6 and 10/6 of that of trioxane results in −148.4 and −185.7 kcal (−621 and −777 kJ), two predicted quantities nearly indistinguishable from experiment. This is in marked contrast to the approximations of the heats of formation of the all-carbon cyclooctane and cyclodecane as 8/6 and 10/6 of that of cyclohexane: the values thus predicted are −39.2 and −49.0 kcal (−164 and −205 kJ) [cf experimental values −29.6 and −36.8 kcal (−124 and −154 kJ)]. This last comparison is quite enlightening. We recognize the difference of −39.2 and −29.6 kcal (−164 and −124 kJ) as the strain energy of cyclooctane, at least when derived from diagonal reference states (see Reference 65 and Chapter 8, Volume 2). Likewise, the difference of −49.0 and −36.8 kcal (−154 and −124 kJ) corresponds to the strain energy of cyclodecane. That cyclooctane and cyclodecane are predicted so poorly while the predictions of tetroxane and pentoxane are so accurate suggests that these oxygen-rich heterocycles are essentially strain free. This corroborates our understanding that the strain energies of medium-sized carbocycles is in large part due to transannular C—H repulsions. For example, compare the "boat–chair" structure that is the most stable conformer of cyclooctane[64] with the C_{4v} "boat–boat" structure that is the most stable conformer of tetroxane.[61]

We close our discussion of the energetics of trioxane, tetroxane, and pentoxane by making a comparison with the gaseous form of the acyclic polymeric species $(CH_2O)_n$. Though this polymer is normally studied as a solid, we may invent a gaseous form by analogy to that recently made in a discussion[66] of the strain energies of various perfluorocycloalkanes in terms of the energetics of the gaseous form of the polymeric species $(CF_2)_n$. Both a casual look and careful statistical analysis[67] show the heat of vaporization of a nonhydroxylic oxygen-containing species of the formula $C_xO_yH_z$ to be comparable to that of the all-carbon analogue $C_{x+y}H_z$. From this, we may employ the same approximate heat of vaporization rule for these species as found earlier for general hydrocarbons[43]:

$$\Delta H_v = 1.12n_c + RT\,(\text{kcal}) = 4.7n_c + 2.5\,(\text{kJ}) \tag{6-73}$$

where n_c is now the total number of carbons and oxygens, $x + y$. Although simple and general rules (see Chapter 3, Volume 2) for estimating heats of sublimation still evade us, it is safe to assume that n_c lies between the heat of vaporization and twice the heat of vaporization. Therefore, the heat of sublimation per CH_2O unit is expected to be more than 2.2 kcal (9 kJ) and less than about 4.5 kcal (≃ 19 kJ). Combining this heat of sublimation with the heat of formation of solid $(CH_2O)_n$ polymer (−42.3 kcal, −177 kJ) results in the value for the gaseous compound of interest, about −39(1) kcal (n−163(4) kJ). Encouragingly, this value is comparable to that found by dividing the heats of formation of trioxane, tetroxane, and pentoxane by 3, 4, and 5, the number of CH_2O units contained therein respectively (−37.1, −37.0, and −37.2 kcal (−155, −155, and −155 kJ) and ignoring the rather small discrepancy due to the strain energies. It is also, in fact, numerically indistinguishable from that found by taking the numerical difference of the heats of formation of gaseous $C_2H_5OCH_2OC_2H_5$ and $C_2H_5OC_2H_5$.

The exercise above was performed not only to test self-consistency of our logic and our numbers. Rather, we recall that there is another and even more important class of species with the same general formula, $(CH_2O)_m$, the simplest of carbohydrates, the monosaccharides. These species, when pure, seemingly occur only as solids—we know of no successful sublimation of any of these compounds. We now endeavor to obtain a relatively simple and admittedly inexact method of estimating the gas phase *and* solid phase heats of formation of these substances. For this we will ignore all the stereochemistry that distinguishes all the various sugars—this is not merely an expression of our ignorance and our laziness. Instead, it arises from the observation that the isomeric solid aldopentoses, α-D-xylose, D-ribose, and D-arabinose, with $m = 5$, have nearly identical heats of formation: −252.8, −251.4, and −252.8 kcal (−1058, −1052, and −1058 kJ). These three species are all cyclic hemiacetals of 2,3,4,5-tetrahydroxypentanal, $CH_2OH(CHOH)_3CHO$, and so we expect their heats of formation to be similar. This similarity is also found for isomeric solid aldohexoses, α-D-glucose, D-mannose, and α-D-galactose, species with $m = 6$, with heats of formation −304.3, −301.8, and −307.3 kcal (−1273, −1263 and −1286 kJ). In this case, all three compounds are cyclic acetals of 2,3,4,5,6-pentahydroxyhexanal, $CH_2OH(CHOH)_4CHO$. Interestingly these values for solid aldohexoses are also nearly identical to those of the isomeric hemiacetals of 1,3,4,5,6-pentahydroxy-2-hexanone, $CH_2OH(CHOH)_3COCH_2OH$, the 2-ketohexoses, L-sorbose and β-D-fructose, −304.0 and −302.6 kcal (−1272 and −1266 kJ). Are these experimental values "reasonable"? We now ask what we would have predicted the heat of formation of these sugars to be.

Let us start with the aldopentoses. Although the textbook literature is confusing as to whether these species are in the furanose or pyranose form, that is, whether they are derivatives of tetrahydrofuran **27** (né **11,** X = $\rangle$O) or of tetrahydropyran, **28** (also called "oxane"), the research literature sug-

27 **28**

gests[68] the latter. The aldopentoses are thus describable as isomeric tetrahydropyran-2,3,4,5-tetraols, **29**.

HO O HO OH OH

29

The energetics of these species may thus be understood using macroincrementation reaction 6-74:

$$\textit{any}\ \text{aldopentose} = \text{cyclo-}(CH_2)_5O + 4(RCHOHR' - RCH_2R') \qquad (6\text{-}74)$$

Since the value of the quantity in parentheses is nearly indistinguishable for $R = R' = CH_3$, $R = R' = C_2H_5$, $R = C_2H_5$, $R' = CH_3$, and $R = R' =$ cyclo-$(CH_2)_5$ (ie, cyclohexanol − cyclohexane), we use this common value of about 40 kcal ($\simeq$ 168 kJ) here and whenever we need to "synthesize" an arbitrary secondary alcohol. We note aldopentoses are also hemiacetals, a class of compounds in equilibrium with generally more stable alcohols and aldehydes. Unfortunately, but not surprisingly, there is thus no direct information on any hemiacetal. However, we can determine the stabilization arising from the hemiacetal structural feature analogously to that noted for 1,3-dioxane earlier in this section and the dihalomethanes in Section 4. In the particular, consider macroincrementation reaction 6-75 for acetals:

$$RCH(OCH_3)_2 = 2RCH(CH_3)(OCH_3) - RCH(CH_3)_2 \qquad (6\text{-}75)$$

For both R = H and CH_3, 5 kcal (21 kJ) of stabilization is found. It seems safe to assume that the same value for hemiacetals. These aldopentoses are also 1,2-diols. Use of the macroincrementation reaction 6-76

$$RCH(OH)CH_2OH = RCH(CH_3)CH_2OH + RCH(OH)CH_2CH_3 - RCH(CH_3)CH_2CH_3 \qquad (6\text{-}76)$$

for ethylene and propylene glycol results in opposite and nearly equal discrepancies of about 1 kcal ($\simeq$ 4 kJ), and so it is safe to assume that no new energy contribution arises from the 1,2-diol feature. Putting all these numbers together results in a predicted heat of formation of *gaseous* aldopentoses of −220 kcal (−920 kJ).

But sugars are solids. What do we predict for the sublimation energy? There are the equivalent of 5 CH_2O units in these species, and so making use of the estimates of sublimation energies in the discussion on "oxanes" results in an initial estimate of $5 \times 3.5 = 17.5$ kcal (75 kJ). These aldopentoses are also intermolecularly hydrogen bonded. With four OH groups, four hydrogen bonds are expected to be associated with each molecule. However, to prevent double counting, we must consider only two per sugar molecule. Since hydrogen bonds for neutral OH groups are "worth" about 5 kcal, there results in an additional $2 \times 5 = 10$ kcal (40 kJ) for the heat of sublimation. The net result for the heat of formation of the solid species is $-220 - 17.5 - 10 = -247.5$ kcal (−1035 kJ). This value is within striking distance of the experimentally observed ones for the three aldopentoses. We are pleased. Indeed, had we taken the heat of sublimation to be the same as the one $(CH_2O)_5$ species we know anything about—namely, pentoxane, 21.3 kcal (89 kJ)—the value would have been −251.3 kcal (−1049 kJ), which is nearly identical to those experimentally found. We are astonished.

Proceeding to the isomeric aldohexoses, we recognize these are also pyranose derivatives.[68] They differ from aldopentoses in that a —CH_2— has been changed into a —CH(—CH_2OH)— group. That is, we have "transformed" a tetrahydropyran-2,3,4,5-tetraol to a 6-hydroxymethyltetrahydropyran-2,3,4,5-tetraol, **30.**

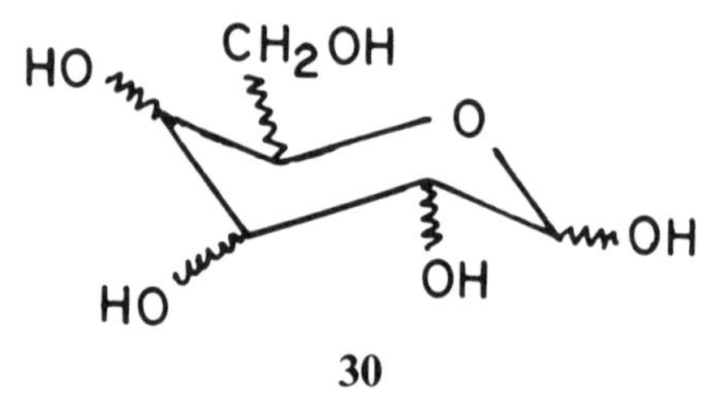

30

Rather than starting de novo, we employ the numbers, either the experimental or the predicted, for aldopentoses. For gas phase numbers we may employ the macroincrementation reaction 6-77:

$$—CH(CH_2OH)— = —CH_2— + RCH(CH_2OH)R' - RCH_2R' \qquad (6\text{-}77)$$

where the —$CH(CH_2OH)$— and —CH_2— are shorthand symbols for the appropriate aldohexose and aldopentose. For $R = R' = CH_3$; $R = CH_3$, $R' = C_2H_5$; and $R = R' =$ —$CH_2CH_2CH_2O$— (ie, tetrahydrofurfuryl alcohol and tetrahydrofuran) there is a nearly constant difference of the heats of formation of the species with and without the affixed CH_2OH of 43(1) kcal (179(6) kJ)). This 43 kcal (179 kJ) is clearly the largest term in the macroincrementation reaction for aldohexoses. However, one should not forget there is also a contribution of about 4 kcal ($\simeq$ 18 kJ) to the sublimation energy because of the presence of one more CH_2O and a 5/2 kcal (22/2 kJ) contribution because of an additional half hydrogen bond. Our prediction is that solid aldohexoses should have a lower heat of formation than aldopentoses by $43 + 4 + 5/2 =$ about 50 kcal ($\simeq$ 210 kJ). The difference experimentally found between the average of the two sets of three sugars is 53 kcal (220 kJ).

Turning to 2-ketohexoses from aldohexoses, we note these species are seemingly also pyranoses[68] (!) and thus may be described as isomeric 2-hydroxymethyltegrahydropyran-2,3,4,5-tetraol, **31.** Consider the conceptual

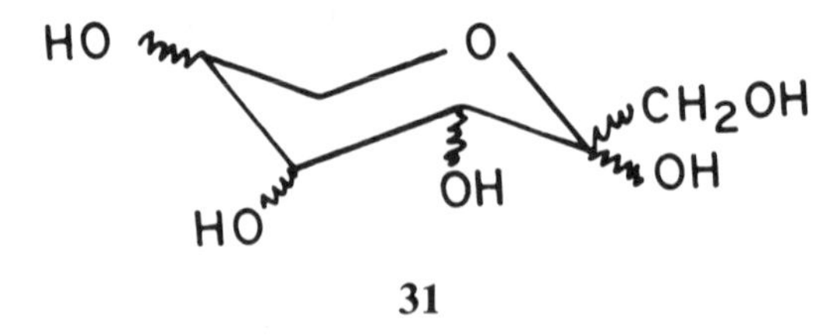

31

isomerization process of an aldohexose into a 2-ketohexose. The 2-carbon of the aldohexose is transformed from a secondary alcohol (hemiacetal) with the substructure —CH_2—CH(OH)—O— into a tertiary alcohol (hemiacetal) containing —CH_2—C(OH)(CH_2OH)—O—, while the 5-carbon is simultaneously transformed from an ether of a secondary alcohol with the substructure —CH(CH_2OH)—O— into that of a primary alcohol containing functionalities into one each of primary and tertiary type is shown below to have comparatively little steric or electronic consequence. That the difference in heats of formation of 1,2- and (either *cis-* or *trans-*) 1,3-dimethylcyclohexane is nearly identical to the differences for 3,3- and 2,4-dimethylpentane suggests that there is little steric effect. Equivalently, macroincrementation reaction 6-78, as written, is thermoneutral to about 1 kcal ($\simeq$3 kJ):

$$1,1\text{-}(CH_3)_2\text{-cyclohexane} - 1,3\text{-}(CH_3)_2\text{-cyclohexane} = CH_3CH_2C(CH_3)_2CH_2CH_3 - (CH_3)_2CHCH_2CH(CH_3)_2 \quad (6\text{-}78)$$

Macroincrementation reaction 6-79 tests the electronic consequences using an acyclic mimic of the change above:

$$(CH_3)_2CHCH_2OH + (CH_3)_3COH - (CH_3)_2CHCH_2CH_3 - (CH_3)_3CCH_3 = 2[CH_3CH(OH)CH_2CH_3 - CH_3CH(CH_3)CH_2CH_3] \quad (6\text{-}79)$$

This reaction is also nearly thermoneutral to about 1 kcal ($\simeq$5 kJ). We thus conclude that the heats of formation of aldohexoses and ketohexoses should be nearly identical, a finding corroborated by experiment.

The sole thermochemical data relating to the difference of the heat of formation of the pyranose and the furanose form of an aldohexose or ketohexose is that derivable from (either α- or β-) methylglucopyranoside, species **32a** and **32b**, and α-methylglucofuranoside, **33**. The two pyranosides

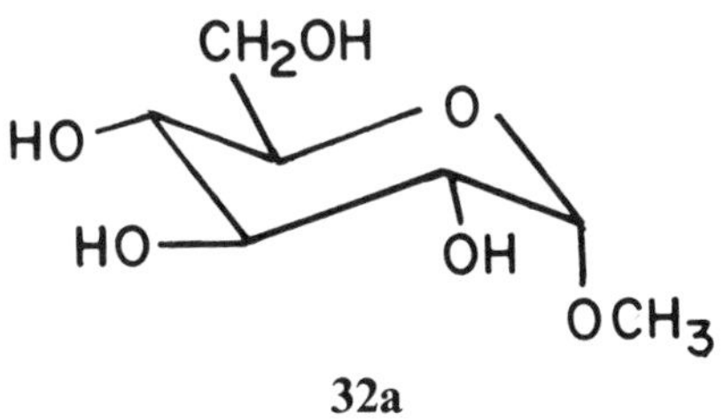

32a 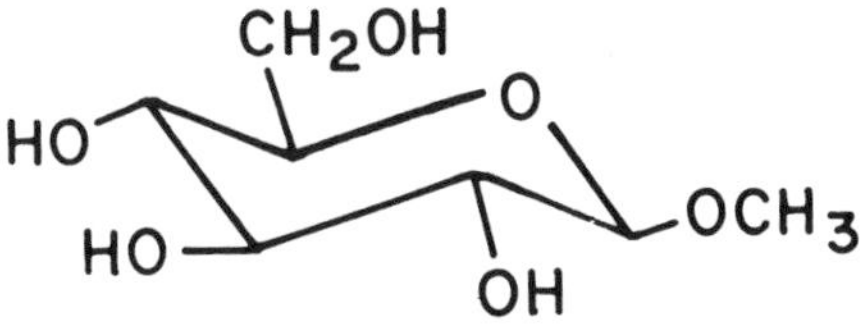 **32b**

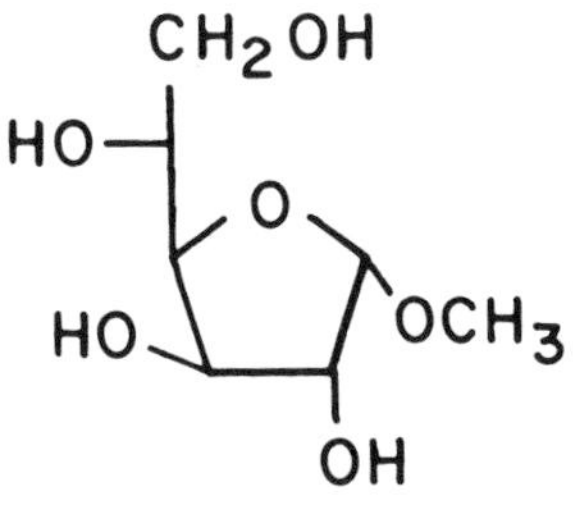

33 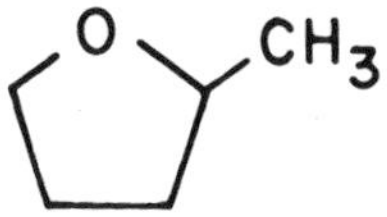 **34**

have the same heat of formation, −295 kcal (−1235 kJ) to within 1 kcal (4 kJ), a finding consonant with the spread of heats of formation within the various classes of monosaccharides discussed above. This experimental value may be theoretically approximated in the following interrelated ways. The first recognizes that the methylation of the free sugar has removed half a hydrogen bond per mole and replaced an —OH group by $—OCH_3$. From the foregoing, we find the first change "costs" about 2.5 kcal ($\simeq$10 kJ). The second change may be approximated by the difference of the heats of formation of $(CH_3)_2CHOH$ and its corresponding methyl ether, 4.8 kcal (20 kJ). Thus the methylpyranoside should have a heat of formation about 7.3 kcal ($\simeq$30 kJ) higher than the free sugar: the experimental value is comparable, 9.6 kcal (40 kJ). To consider furanoses, we must consider the change from tetrahydropyran to 2-methyltetrahydrofuran (**27** to **34**) and accompanying multiple hydroxyl substitution. Despite an absence of experimental thermochemical data on 2-methyltetrahydrofuran, its heat of formation may nonetheless be predicted using macroincrementation reaction 6-80 with X = O.

$$2\text{-}CH_3\text{-cyclo-}[(CH_2)_3CHX] = \text{cyclo-}[(CH_2)_4X] + \tfrac{1}{2}[(CH_3)_2CH—X—CH(CH_3)_2 - CH_3CH_2—X—C_2H_5] \qquad (6\text{-}80)$$

If one additionally incorporates the discrepancy found for the analogous reactions for methylcyclopentane (X = $>$ CH_2, 1 kcal, 4 kJ) and 2-methyltetrahydrothiophene

≪The reader may recall this species as compound **17a.** That macroincrementation reaction 6-80 is so accurate offers another piece of documentation that the value for the heat of formation of $CH_3CH_2SCH_2CH_3$ in Reference 1 is incorrect (see Section 8).≫

(X = $>$ S, 0.3 kcal, 1 kJ) as an "alphanumeric" correction factor, the predicted value is −53.6 kcal (−218 kJ). This derived value for 2-methyltetrahydrofuran is thus about 1 kcal ($\simeq$5 kJ) higher than the experimental value for tetrahydropyran.

The second difference is that the ketohexoses contain a tertiary hemiacetal group whereas the aldohexoses contain a secondary hemiacetal. To mimic this change without introducing any other steric or electronic factors, consider macroincrementation reaction 6-81.

$$(CH_3)_3COH = CH_3CH(OH)CH_2CH_3 + (CH_3)_3CCH_3 - CH_3CH(CH_3)CH_2CH_3 \qquad (6\text{-}81)$$

The "real" *t*-butyl alcohol is more stable than is predicted by this reaction by 1 kcal (4 kJ). This small stabilizing contribution for ketohexoses offsets the destabilizing effects of the greater ring strain found in 2-methyltetrahydrofuran than in tetrahydropyran. Combining these numbers, we expect the furanoside form of ketohexoses to be of comparable stability to the isomeric pyranosides.

The experimental difference for the condensed phase methylglucofuranoside and the isomeric pyranosides is closer to 8.4 kcal (35 kJ), significantly different from that derived using macroincrementation reactions. However, we are not discouraged. Note that the methylfuranoside is a liquid while the pyranosides are both solids. The heat of fusion, by definition, is the difference of the heats of sublimation and of vaporization. From analysis given above on heats of vaporization of compounds containing C and O, we expect the heat of fusion of any of the three methylglucosides (with 13 nonhydrogenic atoms) to be less than 13(1.1) = 14.3 kcal (60 kJ). Sublimation energies are rather idiosyncratic, and thus we hesitate to say how much less. However, heats of fusion of a few hydrocarbons are derivable from the thermochemical data for several hydrocarbons from Reference 1. For the 13 (*all-*) carbon case of diphenylmethane, **35,** the heat of fusion is 4.3 kcal (18 kJ), while the 14 all-carbon cases of 1,4,5,8-tetramethylnaphthalene, 1,2,3,4,5,6,7,8-octahydroanthracene and diamantane, **36–38,** the values are 5.5, 4.5, and 7.2 kcal (23, 19, and 30 kJ).

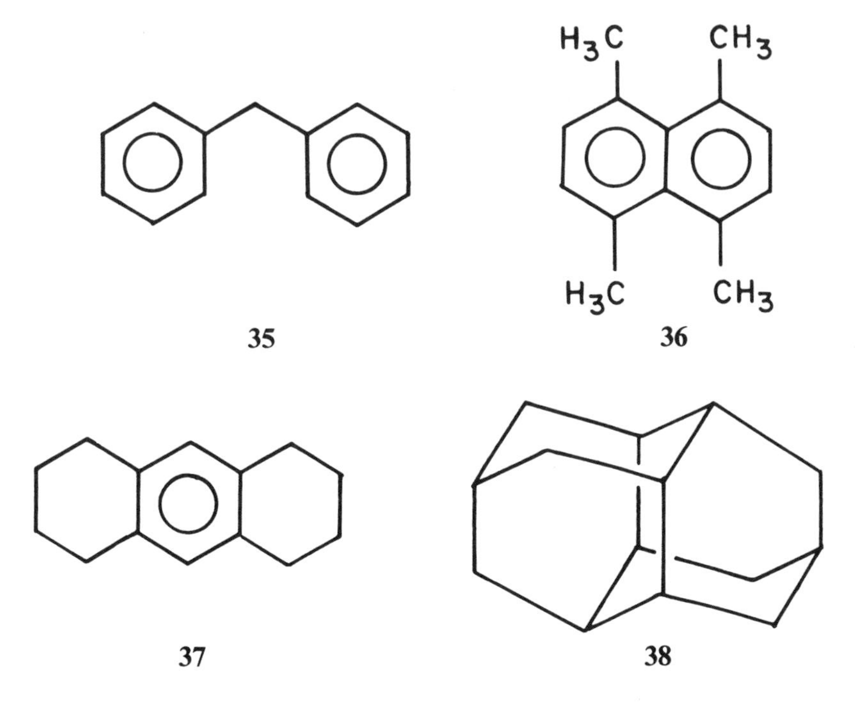

≪We can play this "game" in many ways.[69] For example, knowing the heats of sublimation and of fusion, we can derive the heats of vaporization. For these four hydrocarbons, diphenylmethane, 1,4,5,8-tetramethylnaphthalene, 1,2,3,4,5,6,7,8-octahydroanthracene, and diamantane, species

35–38, the experimental values from Reference 1 are 12.0, 18.1, 15.3, and 20.8 kcal (50, 76, 64, and 66 kJ) whereas the theoretical values using Equation 6-73 (rule 1 from Reference 43) are 15.2, 16.3, 16.3, and 16.3 kcal (64, 68, 68, and 68 kJ). The average discrepancy is about 2.6 kcal ($\simeq$11 kJ), a significant deviation. However, in the interim, a new measurement has been reported for the heat of sublimation of diamantane,[70] 22.9 kcal (96 kJ), resulting in a new derived heat of vaporization, 15.7 kcal (66 kJ). This value is much closer to our "prediction." Encouragingly, for the first species, diphenylmethane, there is an earlier heat of sublimation, 19.6 kcal (82 kJ), that Reference 1 gives (but does not recommend), from which one would derive a heat of vaporization of 15.3 kcal (64 kJ). This value is essentially identical to that from our "prediction," and so we feel it would be desirable for the measurement of the heat of sublimation be remade. If the earlier value for this compound is found to be correct, the average error will be reduced to about 0.6 kcal ($\simeq$3 kJ), comparable to what we had earlier found[43] using heats of vaporization directly measured.≫

This suggests the heat of fusion of any of the three methylglucosides should be about 5 kcal ($\simeq$20) kJ). This last number cancels in large part the discrepancy between theory and experiment for the difference in heats of formation of methylglucofuranoside and the isomeric methylglucopyranosides. We are relieved. It appears that even for the thermochemistry of polyfunctional solids, macroincrementation reactions remain a simple and surprisingly accurate predictive approach.

10. THE $n = 7$ SPECIES, THE METHYLPYRIDINES, AND THE ENERGETICS OF ALKYL-SUBSTITUTED NITROGEN HETEROCYCLES

We now turn to the first set of $n = 7$ species, the isomeric methylpyridines, species **39–41.** Consider the simplest macroincrementation reaction that can be written for these species:

4 5 3 6 N 2 CH_3 CH_3 N CH_3 N

39 **40** **41**

$$\mathbf{39\text{–}41} \equiv CH_3C_5H_4N = C_5H_5N + C_6H_5CH_3 - C_6H_6 \qquad (6\text{-}82)$$

This reaction is equally applicable to all three isomers and indeed does not differentiate between them. The predicted heat of formation, 25.8(0.2) kcal (108(1) kJ) is identical to that observed for the 3-isomer within the error bars, 25.3(0.2) kcal (106(1) kJ), and is somewhat higher than those found for the 2- and 4-isomers, 23.7(0.2) and 24.9(0.2) kcal (99(1) and 104(1) kJ), respectively. These are tolerable discrepancies, but the results are nonetheless in accord with the traditional resonance energy logic for these species. That is, although one can draw dipolar resonance structures involving the methyl group (cf **42–44**), intuitively their importance for the energetics of

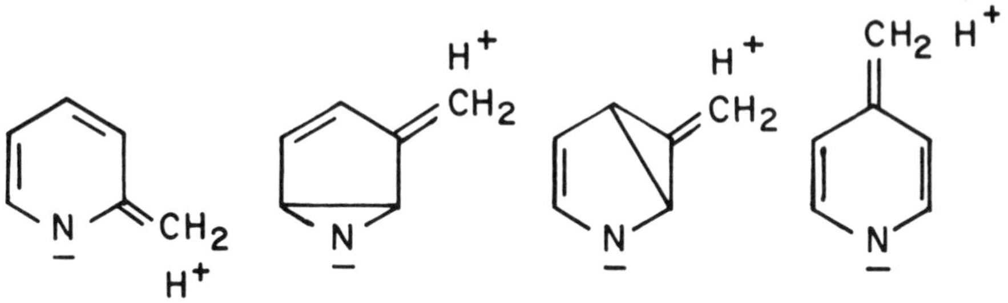

42 **43a** **43b** **44**

x-methylpyridines decreases in the order *x*- = 2- > 4- > 3-, and indeed we might expect the contribution for *x*- = 3- to be negligible.

≪Analogous structures can be drawn for other methylated heterocycles (eg, the isomeric 2- and 3-methylthiophenes, species **18a** and **18b**. In this case of sulfur-containing heterocycles, there is nothing that favors the dipolar resonance structure preferentially for either isomer and, as was shown in Section 8, the heats of formation of 2- and 3-methylthiophene are identical to within 0.2 kcal (1 kJ).≫

Because these dipolar structures are expected to contribute but little to the molecular stability, little effect is expected on the heat of vaporization. Our prediction is realized: the predicted value is 10.6 kcal (44 kJ), and the experimental values are 10.3, 10.6, and 10.8 kcal (43, 44, and 45 kJ).

After a brief discussion that may appear as a digression, we will return to an additional direct use of these numbers for the methylpyridines. Because all three isomers are describable using the same macroincrementation reaction, the differences between theory and experiment in any property of the isomers can be directly compared without discussing the particular macroincrementation reaction. The commonality of a macroincrementation reaction for all three isomeric methylpyridines is paralleled by that for the isomeric dimethylbenzenes, wherein the heats of formation and of vaporization of all three isomers would also be expected to be nearly the same. Indeed they are 4.3 and 10.3 kcal (18 and 43 kJ) for the 1,2-isomer,

4.1 and 10.3 kcal (17 and 43 kJ) for the 1,3-isomer, and 4.3 and 10.0 kcal (18 and 42 kJ) for the 1,4-isomer. Macroincrementation reaction 6-83 reproduces these numbers successfully—the "predictions" are 4.1 and 10.0 kcal (17 and 42 kJ) for the common heats of formation and of vaporization. It would appear that the dimethylbenzenes offer no surprise.

$$C_6H_4(CH_3)_2 = 2C_6H_5CH_3 - C_6H_6 \tag{6-83}$$

Moreover, no surprises are expected for any other simple alkylbenzene. And seemingly none are found. For example, the properties of ethylbenzene may be derived from those of the simpler hydrocarbon, methylbenzene, and from the various dimethylbenzenes, isomeric with each other and with ethylbenzene. In the particular, consider reactions 6-84, 6-85, and 6-86 using benzene, methylbenzene, 1,2-dimethyl-, and 1,3-dimethylbenzene, respectively.

$$C_6H_5C_2H_5 = C_6H_5CH_3 + C_2H_5CHCH_2 - CH_3CHCH_2 \tag{6-84}$$

$$C_6H_5C_2H_5 = 1,2\text{-}C_6H_4(CH_3)_2 + C_2H_5CHCH_2 - (Z)\text{-}CH_3CHCHCH_3 \tag{6-85}$$

$$C_6H_5C_2H_5 = 1,3\text{-}C_6H_4(CH_3)_2 + C_2H_5CHCH_2 - (E)\text{-}CH_3CHCHCH_3 \tag{6-86}$$

The predicted heats of formation are, respectively, 7.2, 6.2, and 6.9 kcal (30, 26, and 29 kJ); the experimental value is 7.0 (29 kJ). The predicted heats of vaporization are, respectively, 9.8, 9.8, and 10.0 kcal (41, 41, and 42 kJ); the experimental value is 10.1 kcal (42 kJ). As said above, no surprises were to be expected for alkylbenzenes, and none were found.

Let us now finally return to alkylpyridines. In the particular, consider the isomeric set composed of six dimethylpyridines, **45**, and the three ethylpyridines, **46**. For the dimethylpyridines, all six heats of formation and heats

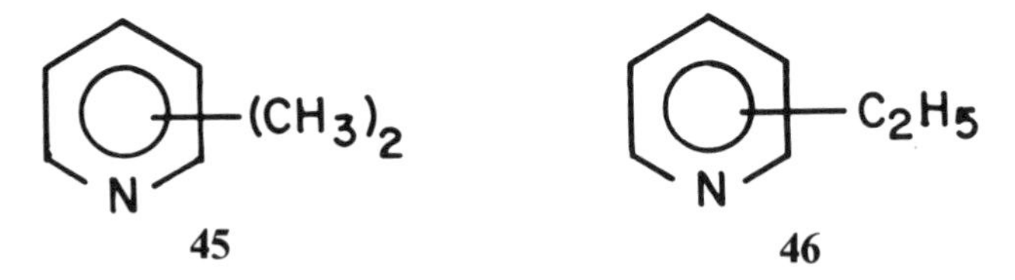

of vaporization are available. Macroincrementation reaction 6-87, analogous to that used for the dimethylbenzenes, may be employed here:

$$\mathbf{45} \leqq i,j\text{-}C_5H_3N(CH_3)_2 = i\text{-}C_5H_4N\text{-}CH_3 + j\text{-}C_5H_4N\text{-}CH_3 - C_5H_5N \tag{6-87}$$

where the i and j are to be taken as the sites of methylation of the pyridine.

≪We say "are to be taken" in that 2- and 6-methylpyridine are identical, as are 3- and 5-, and so, for example, 2,5-dimethylpyridine is to be compared with 2- and 3- (né 5-) methylpyridine.≫

This somewhat fine-tuned expression reproduces all the desired thermochemical properties to better than 1 kcal (4 kJ), suggesting that the methyl groups are weakly coupled with each other. In contrast, the differences among the 2-, 3-, and 4-methylpyridines reflects a stronger coupling of the methyl groups with the pyridine framework.

We now write our final macroincrementation reaction for dimethylpyridines, which builds in explicitly any interaction between the methyl groups.

$$\mathbf{45} \stackrel{\frown}{=} i,j\text{-}(CH_3)_2C_5H_3N$$
$$= i,j\text{-}C_6H_4(CH_3)_2 + i\text{-}CH_3C_5H_4N + j\text{-}CH_3C_5H_4N + C_6H_6 - 2C_6H_5CH_3 - C_5H_5N \qquad (6\text{-}88)$$

Table 6-2 documents that this macroincrementation reaction reproduces all the heats of formation and of vaporization to better than 0.7 kcal (3 kJ).

TABLE 6-2. Heats of Formation of the Six Gaseous Dimethylpyridines and Their Heats of Vaporization[a]

Methyl positions	$\Delta H_f(g)$ (kJ)		ΔH_v (kJ)	
	Experimental	Calculated	Experimental	Calculated
2,3-	68	65	49	48
2,4-	64	62	48	49
2,5-	67	65	48	47
2,6-	59	57	46	47
3,4-	70	70	52	50
3,5-	73	71	50	49

[a] All values are given in kilojoules to "exaggerate" the differences between theory and experiment.

Turning to the ethylpyridines, surprisingly, we have almost no data, even though these are among the simplest of substituted heterocycles. Aware of the "methyl versus ethyl dichotomy," we would like to compare the three isomeric ethylpyridines by such macroincrementation reactions as:

$$\mathbf{46} \stackrel{\frown}{=} C_2H_5C_5H_4N = CH_3C_5H_4N + C_2H_5CHCH_2 - CH_3CHCH_2 \qquad (6\text{-}89)$$
$$\mathbf{46} \stackrel{\frown}{=} C_2H_5C_5H_4N = CH_3C_5H_4N + C_6H_5C_2H_5 - C_6H_5CH_3 \qquad (6\text{-}90)$$

Recall the similarities of the effects of substituents on ethylene and benzene discussed earlier. We may thus expect similar results from these two new equations that utilize methyl- and ethyl-substituted derivatives of ethylene and benzene. This is borne out—consider the heats of formation and of vaporization of the 2-isomer of ethylpyridine, **47**. From reaction 6-89 we

N C2H5

47

find 18.9 and 11.5 kcal (79 and 48 kJ), while from reaction 6-90 we find 18.6 and 11.7 kcal (78 and 49 kJ). What does the experimental literature say? Surprisingly, nothing appears to be known about these two properties! However, they can be subtracted to give the approximate heat of formation of the liquid—7.1(0.4) kcal ($\simeq$30(2) kJ)—to be compared with the literature value of 0.3(1.0) kcal (1(4) kJ)). Clearly something is amiss. No helping insights come from studying the other ethylpyridines because no thermochemical data at all appear to be available for these simple species.

What about invoking our dichotomy? 2-Ethylpyridine is an ethyl derivative of pyridine as much as the methylpyridines are methyl derivatives of pyridine. Saying this allows one to invoke the dichotomy, but this does not provide any insight. Notice however, that both 2-methylpyridine and 2-ethylpyridine contain an alkyl group bound to an sp^2-hybridized carbon, which in turn is bound to an sp^2-hybridized nitrogen. It does not seem unreasonable to interpolate the difference of the heats of formation of methyl and ethyl groups bound to an sp and to an sp^3 to derive the desired difference for sp^2 carbons. Likewise, this sp^2 nitrogen in pyridine should interpolate that of sp nitriles and sp^3 amines. The difference of the experimentally measured heats of formation of gaseous CH_3CN and CH_3CH_2CN is 5.5 kcal (23 kJ) and of $CH_3CH_2NH_2$ and $CH_3CH_2CH_2NH_2$ is 5.3 kcal (22 kJ). [Note that for CH_3CN we are using the new and highly precise heat of formation, 17.7(0.1) kcal (74.0(0.4 kJ)) reported in Reference 71.] Since these species have, respectively, sp- and sp^3-hybridized carbons and nitrogens, the corresponding difference for the species with the sp^2-hybridized carbon and nitrogen may be taken as the average of these two differences.

≪To say "take the average" is a bit sloppy. Since the sp- hybridized atoms utilize half-parts of s and sp^3 utilize quarter-parts of s, a simple average would not result in an atom utilizing one-third parts of s. However, the difference in heats of formation of methyl and ethyl derivatives of nitriles is nearly identical to that of amines and so this discussion brings up an unnecessary complication for the main text.≫

Using this average and using the earlier mentioned and accepted heat of formation of 2-methylpyridine, we find predicted heat of formation, 23.6 − 5.4 = 18.2 kcal (76 kJ), almost identical to that derived earlier, 18.8 kcal (78 kJ).

This result strengthens our concern over the accuracy of the experimental number for the heat of formation of 2-ethylpyridine, but we should still investigate other diagnostic macroincrementation reactions and related "alphanumeric corrections" that might result in different values. These include macroincrementation reaction 6-91 utilizing the sole case other than for the 2-substituted pyridines known to the author where data on both the methyl- and ethyl-substituted nitrogen-containing heterocycles are known. The new results here are nearly the same as we derived before, 19.1 and 11.5 kcal (80 and 48 kJ), where the input data on the alkyloxazolidines **48** and **49** were taken from Reference 72. These findings do not prove, but do strongly suggest, that the experimental value for 2-ethylpyridine is wrong. And we admit that if the experimental value indeed be correct, the result would seriously undermine our confidence in macroincrementation reactions and in any other estimation methodology as well. We strongly urge its remeasurement.

$$\mathbf{47} = \mathbf{39} + \mathbf{48} - \mathbf{49} \qquad (6\text{-}91)$$

11. THE $n = 7$ SPECIES, ANILINE, AND THE ENERGETICS OF SUBSTITUTED BENZENES, NAPHTHALENES, AND HYDRAZINES

The next species we consider is the $n = 7$ case of aniline, $C_6H_5NH_2$. All the macroincrementation reactions discussed in this section employ our knowledge of its experimentally determined heat of formation, 20.8 kcal (87 kJ). As demonstrated in the earlier discussion (Section 3) of the $n = 3$ case of vinyl halides, we expect a near constancy of the difference of the heats of formation of vinyl and phenyl derivatives. We thus predict the heat of formation of vinyl amine using macroincrementation reaction 6-92:

$$CH_2{=}CHNH_2 = C_6H_5NH_2 + CH_2{=}CHR - C_6H_5R \qquad (6\text{-}92)$$

Although while no meaningful quantitative distinction can be made between R = H, CH_3, and C_2H_5 in these cases of vinyl derivatives, we return to our earlier justified preference for C_2H_5 from which the value of 13.6 kcal (57 kJ) is determined. No direct calorimetric measurement of the heat of formation of vinyl amine has been reported. However, combining data from ion energetics measurements on the heat of formation of $(CH_3CHNH_2)^+$, its deprotonation energy to form $CH_3CH{=}NH$, and the isomerization energy of $CH_2CH{=}NH$ and $CH_2{=}CHNH_2$, one finds a value[73] of 11.5(4.1) kcal (48(17) kJ)) compatible with our predicted value.

The second macroincrementation reaction involving aniline, (6-93), is the prediction of the heat of formation of 2- or β-naphthylamine, **50**, a well-known carcinogen.[74]

$$\mathbf{50} \equiv 2\text{-}C_{10}H_7NH_2$$
$$= \beta\text{-}C_{10}H_7NH_2 = C_6H_5NH_2 + 2\text{-}C_{10}H_7X - C_6H_5X \qquad (6\text{-}93)$$

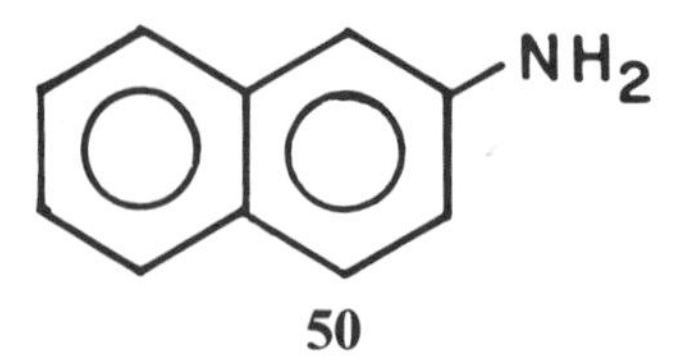

50

Lacking experimental data on 2-ethylnaphthalene,

≪There is a literature value[8] for 2-ethylnaphthalene of 22.9 kcal (96 kJ). This number was obtained from macroincrementation reaction-type logic: since the first macroincrementation reaction below is nearly thermoneutral, it was likewise assumed that so is the second.

$$\text{2-}C_{10}H_7CH_3 - C_{10}H_8 = C_6H_5CH_3 - C_6H_6$$
$$\text{2-}C_{10}H_7C_2H_5 - C_{10}H_8 = C_6H_5C_2H_5 - C_6H_6$$

Equivalently, we note that the following reaction will be thermoneutral as well.

$$\text{2-}C_{10}H_7C_2H_5 - \text{2-}C_{10}H_7CH_3 = C_6H_5C_2H_5 - C_6H_5CH_3$$

However, using the most recent data on 2-methylnaphthalene,[75] the first reaction above shows this species to be less stable than predicted by 1.5 kcal (6 kJ). This strongly suggests that 2-ethylnaphthalene will be likewise less stable than predicted by the second reaction, although the data needed are not available. The third reaction, however, is much more reliably thermoneutral within the tolerances of the "methyl versus ethyl" dichotomy, since hydrogen is generally even more discrepant than any alkyl group. The new, derived value of 30.8 kcal (129 kJ) for the heat of formation of 2-ethylnaphthalene is thus recommended.≫

we must ask what X is best to use. The 2-naphthyl group involves an sp^2 carbon in its bonding much as phenyl and vinyl do, and so it might appear that any X is equally good. However, that possibility is not supported when we examine a collection of possible X. For H, CH_3,[75] OH, and Cl, one finds 36.8, 35.9, 31.8, and 41.3(2.4) kcal (154, 150, 133, and 173(10) kJ). If one interpolates NH_2 between CH_3 and OH, one deduces a value of about 34 kcal ($\simeq$140 kJ). If one considers just polar substituents (ie, OH and Cl), the approximate kcal ($\simeq$40 kJ) difference of 10 predicted by these two choices is rather disconcerting. Which assumption do we make? And if we allow polar substituents, which member do we choose? We know of no way to decide a priori. The literature reports the heat of formation of 2-chloronaphthalene to be more than 4 kcal (17 kJ) higher than the presumably more strained 1-isomer. This seems strange, and so we take the prediction of the heat of formation of 2-naphthylamine from the 2-chloronaphthalene as suspect. Indeed, the literature value for the heat of formation of 2-naphthylamine is 32.5 (2.9) kcal (136(12) kJ), suggesting (a) that the interpolation logic is valid and (b) that the value for 2-chloronaphthalene should be redetermined.

We have just learned some chemistry: substituted naphthalenes are comparatively hard to understand, and the experimental value for the heat of formation of 2-chloronaphthalene is suspect. It seems disconcerting that comparisons between 1- and 2-ring species, derivatives of benzene and naphthalene, are so much less reliable than comparisons between those of 0- and 1-ring species, ethylene and benzene. What does this say about more generally substituted derivatives of the more general many-ring species, the polynuclear aromatic hydrocarbons, discussed in Chapter 1 of Volume 4. Since almost no thermochemical data exist for these species, we conclude that more experiments need to be performed before theory can be adequately predictive for these interesting and important compounds.

We now turn to a brief discussion of the thermochemistry of substituted anilines, reminding the reader that the parent species is of the type $A_mB_{n-m}H_n$ even if most of its derivatives are not. In our first example we compare the isomeric *m*- and *p*-nitroaniline, **51** and **52**, for which all ther-

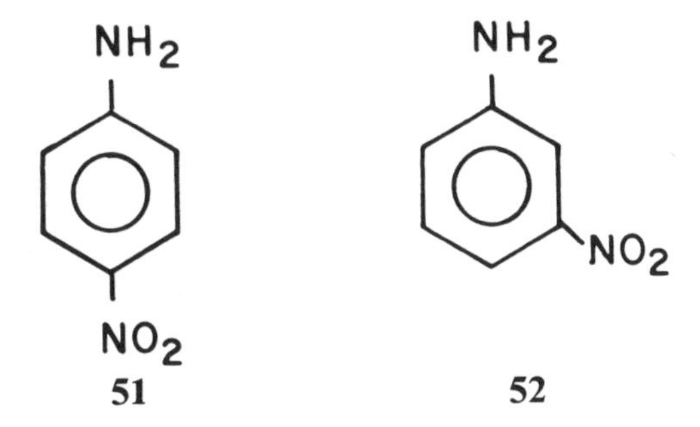

51 **52**

mochemical data was taken from Reference 75. These two species are almost archetypal examples of substituent effects in aromatic compounds and of the often competing roles of inductive and resonance effects on properties such as dipole moment and basicity. The reader may recall the customary explanation in terms of quinonoid resonance structures: whereas both isomers have Kékulé resonance structures that characterize all substituted benzenes, it is only for the *p*-isomer that there is the additionally stabilizing structure, **53**.

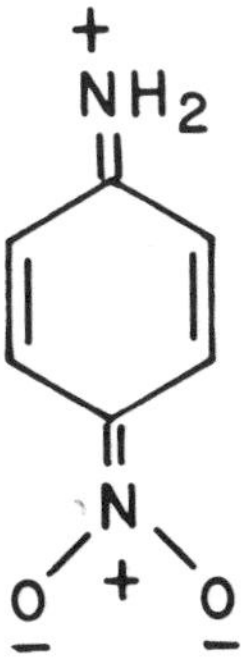

53

Indeed, whereas for the *m*-isomer there are the corresponding two *m*-quinonoid structures **54a** and **54b**, in fact both have been almost totally ignored

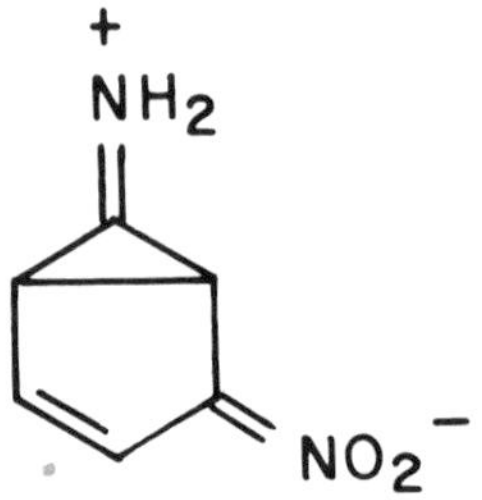

54a 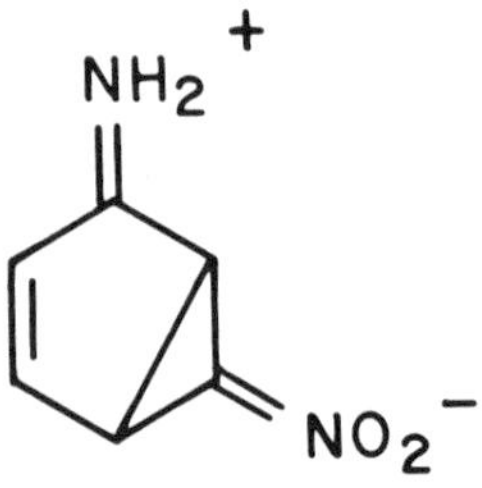 **54b**

in the research and pedagogical literature. We can only surmise that this is because most chemists have presumed them to be nearly irrelevant for the understanding of the observed chemical behavior of this isomer. But is this presumption thermochemically well-founded?

What do macroincrementation reactions say about *m*- and *p*-nitroaniline, and by inference about the two sets of quinonoid resonance structures? For both isomers, there is the logical macroincrementation reaction:

$$x\text{-}O_2NC_6H_4NH_2 = x\text{-}O_2NC_6H_4R + x\text{-}R'C_6H_4NH_2 - x\text{-}RC_6H_4R' \qquad (6\text{-}94)$$

where, of course, *x*- is either *m*- or *p*- on both sides of the equation. Quite surprisingly, there is no set of R and R′ other than the trivial choice of R = R′ = H, such that the thermochemical data are known for all the sustituted nitrobenzenes and anilines and disubstituted benzenes above. It clearly does not behoove us to consider different sets of R and R′ for the two isomers. We note that the thus unavoidable choice of hydrogen as our "substituents" allows us to immediately contrast the two isomers: their difference in heats of formation is precisely the difference in stabilization or resonance energies these species "enjoy" beyond that of aniline and of nitrobenzene. From the most recent experimental calorimetric values of the heats of formation of *m*- and *p*-nitroaniline,[76] we find the difference equals but 15.3(1.0)–13.1(0.5) = 2.2(1.2) kcal (9(5) kJ). This is an amazingly small quantity when one considers the important conceptual roles these two isomers have had in physical organic chemistry in the elucidation of the effects of substituents on relative isomer stability, acidity, basicity, and polarity.

≪The authors of Reference 75 present literature citations to earlier studies of these species from which one could deduce that the difference of the stabilities of *m*- and *p*-nitroaniline was smaller and even that the sign of the difference was reversed.≫

From this finding it is tempting to conclude that not only is the *m*-quinonoid resonance structure irrelevant in understanding *m*-nitroaniline, but likewise the *p*-quinonoid resonance structure contributes little to the chemistry of *p*-nitroaniline.

However, the mere comparability of the resonance energies of the two isomers does *not* mean that the resonance energies are negligible. To find the resonance energies of the two isomers, we look at the energetics of the macroincrementation reactions. For the *p*-isomer, the increment in resonance energy over aniline and nitrobenzene is 4(1) kcal (17(3) kJ), a rather respectable quantity. This conforms with our biases regarding the magnitude of the substituent effects of this species. For the *m*-isomer, the related increment is 2(1) kcal (8(5) kJ). The higher value, $2 + 1 = 3$ kcal (13 kJ), would make the *m*-isomer rather stabilized. The lower value, $2 - 1 = 1$ kcal (4 kJ), would make the *m*-isomer essentially "normal." Our prejudices suggest the latter. However, we must admit this is one case where an uncertainty of ± 1 kcal (4 kJ) is conceptually significant.

Our last example of substituted anilines involves *N*-substitution. The compound of interest is *N*-aminoaniline, more commonly identified as the $n = 8$ species phenylhydrazine, $C_6H_5NHNH_2$. This compound is a historically important reagent in natural product chemistry because it readily reacts with monosaccharides (see Section 9) to form the easily identified 1,2-bishydrazones, more commonly known as osazones.[77] In this reaction, (6-95), two molecules of phenylhydrazine react with the monosaccharide (shown below in the acyclic aldehyde or ketone form, —CHOHCHO or —$COCH_2OH$) to produce new C=N bonds while the N—N bond of one molecule is reductively cleaved.

$$3\, C_6H_5NHNH_2 + \text{either} —COCH_2OH \text{ or } —CHOHCHO \rightarrow$$
$$—C(=NNHC_6H_5)—CH=NNHC_6H_5 + C_6H_5NH_2 + NH_3 \quad (6\text{-}95)$$

From the thermochemical point of view this N—N bond cleavage is interesting because it is not obvious whether to expect phenylhydrazine to be relatively stabilized or destabilized compared to other hydrazines. What does our intuition say? Consider two resonance structures for phenylhydrazine, **55a** and **55b**.

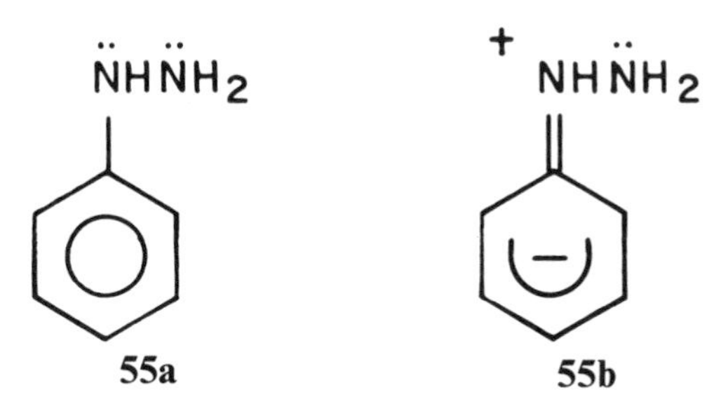

In the latter, dipolar structure, the presence of an —NH_2 group on the positive nitrogen is anticipated to be destabilizing because it is electron withdrawing. However, the same resonance structure lacks the two adjacent lone pairs that destabilize hydrazines in general. This is anticipated to be stabilizing. It is not clear intuitively which effect will be the stronger.

Before commencing with our study of phenylhydrazine using macroincrementation reactions, a brief warning is in order. For all the interest in hydrazine and its derivatives by the chemical community (cf Chapter 1, this volume, and Reference 78), there are almost no thermochemical data. As such, in what is probably the simplest and most natural macroincrementation reaction (6-96)

$$C_6H_5NHNH_2 = C_6H_5NH_2 + RNHNH_2 - RNH_2 \qquad (6\text{-}96)$$

we can imagine for phenylhydrazine, the only organic group R for which the relevant thermochemical data exist for a hydrazine is CH_3. The predicted heat of formation is 49.0 kcal (205 kJ), in almost perfect agreement with experiment, 48.9 kcal (204 kJ). Unfortunately, this may be deceptive because we remember that CH_3 is very often an anomalous substituent. Indeed, this possibility is demonstrated by the formally related reaction 6-97, for which a predicted heat of

$$C_6H_5NHNH_2 = C_6H_5NH_2 + R_2NNH_2 - R_2NH \qquad (6\text{-}97)$$

formation (using again, an R = CH_3 compound) is found to be 45.4 kcal (190 kJ), an error of 3.5 kcal (15 kJ). Again this is using the one dialkylhydrazine for which data exist.

At this stage it is tempting to quit out of frustration at both the lack of data and the inconsistent agreement between theory and experiment. Alternatively, we may quit feeling rather contented, since only the first reaction that is in good agreement is strictly speaking a macroincrementation reaction—in the other the primary and secondary amine groups on the left and two primary, one secondary, and one tertiary amine groups on the right. That is, in the discrepant reaction, not all structural features are preserved. Rather than quitting or ignoring this reaction, we will argue by analogy to the all-carbon versions of both the reactions for phenylhydrazine, the new reactions 6-98 and 6-99:

$$C_6H_5CH_2CH_3 = C_6H_5CH_3 + RCH_2CH_3 - RCH_3 \qquad (6\text{-}98)$$
$$C_6H_5CH_2CH_3 = C_6H_5CH_3 + R_2CHCH_3 - R_2CH_2 \qquad (6\text{-}99)$$

Taking R = CH_3 in both cases, the predicted heats of formation of ethylbenzene are 6.9 and 4.8 kcal (29 and 20 kJ)—the experimental value is 6.9 kcal (29 kJ). It is interesting that the first reaction of this new set is essentially flawless while the second errs in a significant amount. This exactly parallels the situation in macroincrementation reactions 6-96 and 6-97. At the risk of taking the comparison of hydrazines and their all-carbon analogues too far, consider the reactions above for R = H 6-100 and 6-101:

$$C_6H_5NHNH_2 = C_6H_5NH_2 + NH_2NH_2 - NH_3 \qquad (6\text{-}100)$$
$$C_6H_5CH_2CH_3 = C_6H_5CH_3 + CH_3CH_3 - CH_4 \qquad (6\text{-}101)$$

The first reaction predicts a heat of formation of $C_6H_5NHNH_2$ of 54.5 kcal (228 kJ), corresponding to an error of 5.6 kcal (24 kJ), whereas the second predicts a heat of formation of $C_6H_5CH_2CH_3$ of 9.8 kcal (41 kJ), corresponding to an error of 2.8 kcal (12 kJ). Again the discrepancies are large, and again they are larger for the hydrazine than for the all-carbon case.

All these comparisons involving phenylhydrazine and all-carbon analogues inherently make use of isoelectronic species (see Chapter 2,; for example, phenylhydrazine and ethylbenzene are isoelectronic. Also isoelectronic to these two species, and so useful in their comparison, is benzylamine,[79] and this suggests the two new macroincrementation reactions, (6-102) and (6-103):

$$C_6H_5CH_2CH_3 = C_6H_5CH_2NH_2 + CH_3CH_2CH_3 - CH_3CH_2NH_2 \quad (6\text{-}102)$$
$$C_6H_5NHNH_2 = C_6H_5CH_2NH_2 + CH_3NHNH_2 - CH_3CH_2NH_2 \quad (6\text{-}103)$$

The first reaction is nearly flawless in its prediction: the predicted heat of formation of ethylbenzene is 7.4 kcal (31 kJ), and the experimental value is 6.9 kcal (29 kJ). The second reaction, however, shows the much larger discrepancy of 6.4 kcal (27 kJ): the predicted heat of formation of phenylhydrazine is 55.2 kcal (231 kJ), whereas the experimental value is 48.8 kcal (204 kJ). Although we might have anticipated that discrepancies for phenylhydrazine would be larger than for ethylbenzene, we did not anticipate how much larger. Comparing the two reactions more closely, we see that what we omitted in reaction 6-103 was any inclusion of the resonance energy of the aniline substructure, C_6H_5NH—, found in phenylhydrazine. We earlier referred to the appropriate resonance structure for phenylhydrazine, but we did *not* then, nor earlier, discuss the resonance energy of any species with the aniline substructure, including aniline itself. The resonance energy of aniline may be derived by taking it to be the discrepancy of macroincrementation reaction 6-104:

$$C_6H_5NH_2 = C_2H_5NH_2 + C_6H_5CH_3 - C_2H_5CH_3 \quad (6\text{-}104)$$

The theoretical prediction for the heat of formation of aniline is 25.6 kcal (107 kJ),

≪We used $C_6H_5CH_3$ to parallel the other macroincrementation reactions in this section. Had we used $C_6H_5C_2H_5$ (or even C_6H_5H) and $C_2H_5C_2H_5$ (or even C_2H_5H), our conclusions would have changed by no more than 0.3 kcal (1 kJ).≫

the experimental value is 20.8 kcal (87 kJ), resulting in an additional resonance energy stabilization for aniline of 4.8 kcal (20 kJ). Correcting for this, the "real" discrepancy in reaction 6-104 is 6.4 − 4.8 = 1.6 kcal (7 kJ). How does this compare with the other discrepancies for phenylhydrazine when compared with the all-carbon cases 2.8 and 2.1 kcal (12 and 9 kJ)? It would appear that regardless of the macroincrementation reaction we choose, phenylhydrazine is *always* destabilized by about 2 kcal ($\simeq$10 kJ) relative to our expectations based on hydrocarbon chemistry precedents.

12. CONCLUSIONS

The five uses of macroincrementation reactions suggested in this chapter are:

1. To confirm some experimental value that is in doubt.
2. To predict the value of some property of the compound of interest.
3. To compare some property of two species, such as their relative resonance energies or strain energies.
4. To provide a generalization of isodesmic and homodesmotic reactions.
5. To intentionally leave out some interaction and so define a related "alphanumeric" correction term.

Let us proceed through these uses and the various sections of this chapter. Implicitly, they all presented applications of use 1. In that most of our "predictions" corresponded to experimental "reality," we additionally confirmed that macroincrementation reactions are both useful and accurate. Furthermore, this generally high level of success documented that the experimental data we have used are consistent as well as accurate. A small set of experimentally determined numbers were questioned—in particular, we strongly encouraged the remeasuring of the heats of formation of 2-chloronaphthalene, 2-ethylpyridine, and *m*-nitroaniline, the heat of vaporization of 2-ketotetrahydrothiophene, and the heat of sublimation of diphenylmethane.

In the interests of keeping the chapter to a manageable length, few predictions of new numbers were made, and so function 2 was perhaps underutilized. To the author, it was sufficiently educational and rewarding to compare the results of macroincrementation reactions with the published results that had been derived from either calorimetric experiment or quantum chemical calculational theory. As discussed in Section 2, however, the first application of macroincrementation reactions by the author successfully predicted the heat of formation of cyclopropanone.

Implicitly or explicitly, numerous examples of use 3 were made. These include the effects of:

Carbons that were sp^2 hybridized when found in cyclopropane derivatives (Section 2).

Substituents on ethylene and benzene derivatives (Sections 3 and 11).

Geminal and vicinal disubstitution (dihalo, Section 4; dioxy, Section 9).

Adjacent ketosubstitution or other electronegative substitution of ketones and the resulting destabilization (Section 5).

Alkyl substitution on heterocycles (Sections 7,8 and 10).

Solvation on the stability of zwitterions (Section 6).

Delocalization in nitrogen-containing species such as pyrrole and its benzo analogues (Sections 7 and 8) and in sulfur-containing species such as thiolesters (Section 8).

Replacement of —CH_2— in cycloalkanes by —O— and resulting strain energies (Section 9).

Variation in ring sizes in sugars: pyranoses versus furanoses (Section 9).

Hybridization and the "methyl versus ethyl" dichotomy (Section 10).

Phenyl substitution and C—C, C—N, and N—N bond energies (Section 11).

To maximize use of the admittedly sparse experimental data in this chapter, use 4 was largely ignored. Use 5 was often implicit and indirect, but we can now claim to have derived correction terms for:

Carbons that were sp^3 hybridized in cyclopropane derivatives (Section 1).

Zwitterions, both gas phase (Sections 2 and 6) and aqueous (Section 6).

Substituted ethylene and benzene derivatives (Sections 3 and 11).

Geminal and vicinal dihalo substitution (Section 4), and dioxy substitution (Section 9).

α-Diketones and α-haloketones (Section 5).

Thiolesters (Section 8), pyrrole and its benzo analogues indole and carbazole (Section 7).

Alkyl thiophenes and pyridines (Sections 8 and 10).

Cycloalkanes and their polyoxa analogues (Section 9).

Substituted anilines and hydrazines (Section 11).

In summary, macroincrementation reactions have been demonstrated to be a generally powerful yet conceptually simple method for the prediction and understanding of chemical properties. The author hopes that the reader shares his enthusiasm for this approach.

ACKNOWLEDGMENTS

Thanks are extended to Deborah Van Vechten for her editorial suggestions and Eugene S. Domalski, Rhoda D. Levin, and Sharon G. Lias for their hospitality at the National Bureau of Standards, where the author was on sabbatical during much of the research of this manuscript.

REFERENCES

1. Pedley, J.B.; Rylance, J. "Sussex-N.P.L. Computer-Analysed Thermochemical Data: Organic and Organometallic Compounds." University of Sussex: Brighton, UK, 1977. Professor Pedley has informed the author that the second edition of this compendium is in press and will be published by Methuen Chapman and Hall, Ltd.
2. All heats of formation of ions in this chapter and additional heats of formation of neutral species not in Reference 1, first edition will be found in the compendium: Lias, S.G.; Liebman, J.F.; Bartmess, J.E.; Holmes, J.L.; Lossing, F.P.; Levin, R.D.; Motevalli-Aliabadi, M. "Gas Phase Ion Thermochemistry." *J. Phys. Chem. Ref. Data.* To be submitted.

3. Benson, S.W. "Thermochemical Kinetics: Methods for the Estimation of Thermochemical Data and Rate Parameters," 2nd ed.; Wiley: New York, 1976.
4. Bondi, A. "Physical Properties of Molecular Crystals, Liquids and Glasses," Wiley: New York, 1968.
5. Cox, J.D.; Pilcher, G. "Thermochemistry of Organic and Organometallic Compounds," Academic Press: New York, 1970.
6. Janz, G.J. "Thermodynamic Properties of Organic Compounds: Estimation Methods, Principles and Practices," Academic Press: New York, 1967.
7. Reid, R.C.; Prausnitz, J.M.; Sherwood, T.K. "The Properties of Gases and Liquids: Their Estimation and Correlation," 3rd ed.; McGraw-Hill: New York, 1977.
8. Stull, D.R.; Westrum, E.F., Jr.; Sinke, G.C. "The Chemical Thermodynamics of Organic Compounds," Wiley: New York, 1969.
9. Rosenstock, H.M.; Dannacher, J.; Liebman, J.F. *Radiat. Phys. Chem.* **1982,** *20,* 7. [Although this is the first time the word "macroincrementation" has been used, it is not the first time the author has used the concept. Two early "informal" or "implicit" applications of his were to the heat of formation of cyclopropanone[10] and for application to the resonance energies of amides and esters.[11] (Also see Section 2 of this chapter for an "update" and thus expansion of the former)].
10. Liebman, J.F.; Greenberg, A. *J. Org. Chem.* **1974,** *39,* 123.
11. Liebman, J.F.; Greenberg, A. *Biophys. Chem.* **1974,** *1,* 222.
12. (a) Montgomery, R.L. Rossini, F.D. *J. Chem. Thermodyn.* **1978,** *10,* 471. (b) Sellers, P.; Stridh, G.; Sunner, S. *J. Chem. Eng. Data* **1978,** 23, 2150. (c) Sunner, S.; Wulff, C. A. *J. Chem. Thermodyn.* **1983,** *12,* 505. (d) See the mathematical analysis in Chapter 8 of Volume 2 of this series. This chapter also presents mathematical analysis that compares the fundamental similarities and differences of group and macroincrement methods.
13. Hehre, W.J.; Ditchfield, R.; Radom, L.; Pople, J.A. *J. Am. Chem. Soc.* **1970,** *92,* 4796. Many of the subsequent articles by these authors have made use of this approach.
14. George, P.; Trachtman, M.; Bock, C.W.; Brett, A.M. *Theor. Chim. Acta* **1975,** *38,* 121. Many of the subsequent articles by these authors have made use of this approach (eg, Chapter 6, Volume 4, this series).
15. Dill, J.D.; Greenberg, A.; Liebman, J.F. *J. Am. Chem. Soc.* **1979,** *101,* 6814.
16. Greenberg, A.; Liebman, J.F. "Strained Organic Molecules." Academic Press: New York, 1978, pp. 66 and 94.
17. Rodrigues, H.J.; Chang, J.-C.; Thomas, T.E. *J. Am. Chem. Soc.* **1976,** *98* 2027.
18. Bock, H.; Mohmand, S.; Hirabayashi, T.; Semkow, A. *Chem. Ber.* **1982,** *115,* 339.
19. Greenberg, A.; Liebman, J.F., "Strained Organic Molecules." Academic Press: New York, 1978, p. 11.
20. Lias, S.G.; Liebman, J.F.; Levin, R.D. *J. Phys. Chem. Ref. Data* **1984,** *13,* 695.
21. Radom, L.; Hariharan, P.C.; Pople, J.A.; Schleyer, P.v.R. *J. Am. Chem. Soc.* **1973,** *95,* 6531.
22. Raghavachari, K.; Whiteside, R.A.; Pople, J.A.; Schleyer, P.v.R. *J. Am. Chem. Soc.* **1981,** *103,* 5649.
23. Greenberg, A; Liebman, J.F.; Dolbier, W.R., Jr.; Medinger, K.S.; Skancke, A. *Tetrahedron* **1983,** *38,* 1533.
24. See the Chapters 3 and 5, this volume, and the numerous references cited therein.
25. Liberles, A.; Greenberg, A.; Lesk, A. *J. Am. Chem. Soc.* **1972,** *94,* 8685.
26. Franklin, J.L.; Dillard, J.G.; Rosenstock, H.M.; Herron, J.T.; Draxl, K.; Field, F. H. "Ionization Potentials, Appearance Potentials and Heats of Formation of Gaseous Positive Ions," *Nat. Stand. Ref. Data Ser.* U.S. National Bureau of Standards NSRDS-NBS 26, 1969. This work may be considered the intellectual great-grandfather of References 2 and 20.
27. Osamura, Y.; Borden, W.T.; Morokuma, K. *J. Am. Chem. Soc.* **1984,** *106,* 5112.
28. The reader is advised of the existence of two topics of anions: those that are "closed shell" and may simply be spoken of as the conjugate base of some acid, and those that are "open shell" and may simply be spoken of as the radical anion of some molecule. Both types have computational difficulties. See Radom, L. In "Applications of Electronic Structure Theory," Schaefer, H. F., III, Ed.; Plenum Press: New York, 1977, and references cited therein.
29. Sunil, K.K.; Jordan, K.D.; Shepard, R. *Chem. Phys.,* **1984,** *88,* 65.
30. The varying charges given in Reference 27, and Schaad, L. J.; Hess, B. A. Jr.; Zahradnik, R. (*J. Org. Chem.* **1981,** *461,* 1909) for the carbon and oxygen atoms in zwitterionic oxyallyl

amply document that the well-known Mulliken population analysis method does not give quantitatively unambiguous results. Discussion of this species is further complicated by the existence of another state of oxyallyl, a singlet biradical.
31. Fuchs, R. *J. Chem. Educ.* **1984,** *61,* 113. Strictly speaking, Fuchs might "disapprove" of some of what we are doing in this section because C_6H_5X is to be compared with $(CH_3)_3CX$ in what he calls "methathetical isodesmic reactions." These reactions may be seen to represent a special subclass of macroincrementation reactions.
32. See the discussion in Kolesov, V.P.; Papina, T.S. *Russ. Chem. Rev.* (Engl. transl.) **1983,** *52,* 425.
33. Epiotis, N.D.; Larson, J.R.; Yates, R.L; Cherry, W.R.; Shaik, S.; Bernardi, F. *J. Am. Chem. Soc.* **1977,** *99,* 7460.
34. Radom, L.; Hehre, W.J.; Pople, J.A. *J. Am. Chem. Soc.* **1971,** *93,* 289.
35. Kudchadker, S.A.; Kudchadker, A.P. *J. Phys. Chem. Ref. Data* **1978,** *7,* 1285.
36. (a) Kirby, A.J. "The Anomeric Effect and Related Stereoelectronic Effects at Oxygen." Springer-Verlag: Berlin, 1983. (b) Delongschamp, P. "Stereoelectronic Effects in Organic Chemistry." Pergamon Press: New York, 1983.
37. Greenberg, A.; Sprouse, S.D.; Liebman, J.F. unpublished observations.
38. Conant, J.B.; Kirner, W.R. *J. Am. Chem. Soc.* **1924,** *46,* 223.
39. Streitwieser, A., Jr. *Chem. Rev.* **1956,** *56,* 571; "Solvolytic Displacement Reactions." McGraw-Hill: New York, 1962, p. 28.
40. Carrion, F.; Dewar, M.J.S. *J. Am. Chem. Soc.* **1984,** *106,* 3531.
41. Kende, A.S. *Org. React.* **1960,** *11,* 261.
42(a). Schwarzenbach, G.; Wittwer, C. *Helv. Chim. Acta* **1947,** *30,* 663, 669.
42(b). Selman, S.; Eastham, J.F. *Q. Rev.* **1962,** p. 221.
43. Chickos, J.S.; Hyman, A.S.; Ladon, L.H.; Liebman, J.F. *J. Org. Chem.* **1981,** *46,* 4294.
44. (a) Suenram, R.D.; Lovas, F.J. *J. Mol. Spectrosc.* **1978,** *72,* 372. (b) Brown, R.D.; Godfrey, P.D.; Stoery, J.W.V.; Bassez, M.-P. *J. Chem. Soc. Chem. Commun.* **1978,** p. 547. (c) Schafer, L.; Sellers, H. J. C.; Suenram, R.D.; Lovas, F.J. *J. Am. Chem. Soc.* **1980,** *102,* 7180.
45. Locke, M.J.; McIver, R.T., Jr. *J. Am. Chem. Soc.* **1983,** *105,* 4226.
46. This is by now a standard statement in biochemistry. For extensive discussions of this on the biochemical/physical chemical interface, see: (a) Edsall, J.T.; Wyman, H. "Biophysical Chemistry," Vol. 1; Academic Press: New York, 1958, Chapter 5. (b) Greenstein, J.P.; Winitz, M. "Chemistry of the Amino Acids," Vol. 1. Wiley: New York, 1961, Chapter 4.
47. Ngauv, S.N.; Sabbah, R.; Laffitte, M. *Thermochim. Acta* **1977,** *20,* 371. The reader should be alerted to the fact that there is a spread of nearly 14 kcal (60 kJ) in the four values of the heat of sublimation of glycine reported in the literature. This should indicate some of the experimental complications in studying the thermochemistry of zwitterions. Indeed, Locke and McIver[45] chose a different value for the heat of sublimation of glycine, and so there are quantitative differences of results from this source and those derived from other papers. Admittedly, we have not tried to reconcile the differences, since they seem to disagree primarily in what seemingly remains a subjective choice of which sublimation energy to take for glycine.
48. Castora, F.J.; Meot-Ner (Mautner), M.; Liebman, J.F. Unpublished results.
49. Greenberg, A.; Winkler, R.; Smith, B.M.; Liebman, J.F. *J. Chem. Educ.* **1982,** *59,* 367.
50. Wright, L.R.; Borkman, R.F., *J. Am. Chem. Soc.* **1980,** *102,* 6207. This paper compiles the results of numerous other quantum chemical calculations on the zwitterionic and nonpolar forms of glycine from the literature. The wide spread of energy differences reported amply documents the assertion of the sensitivity of calculations on zwitterions made in Reference 29.
51. Wagman, D.D.; Evans, W.H.; Parker, V.B.; Schumm, R.H.; Halow, I.; Bailey, S.B.; Churney, K. L.; Nuttall, R. L. "The NBS tables of chemical thermodynamics properties. Selected values for inorganic and C_1 and C_2 organic substances in SI units." *J. Phys. Chem. Ref. Data 11* **1982,** Suppl. 2.
52. This, too, by now is a standard statement in biochemistry (eg, see Edsall and Wyman[46a]).
53. Haberfield, P. *J. Chem. Educ.* **1980,** *57,* 346, gives a value of 10 kcal (42 kJ) for this quantity, whereas Locke and McIver (cf Reference 45, their footnote 38) present a reevaluated value of 13.5 kcal (57 kJ).
54. Domalski, E.S. *J. Phys. Chem. Ref. Data* **1972,** *1,* 221, in particular, p. 248.
55. Sabbah, R.; Skoulika, S. *Thermochim. Acta* **1980,** *36,* 179.
56. Sabbah, R.; Minadakis, S. *Thermochim. Acta* **1978,** *36,* 179.

57. Kudchadker, A.P. Kudchadker, S.A.; Wilhoit, R.C.; Gupta, S.K. "Acenaphthylene, Acenaphthene, Fluorene and Fluoranthene," API Monograph; 715-81. American Petroleum Institute: Washington, D.C., 1981.
58. Hubbard, W.N.; Good, W.D.; Waddington, G. *J. Phys. Chem.* **1958,** *62,* 614.
59. Wadso, I. *Acta Chem. Scand.* **1966,** *20,* 544.
60. Geiseler, G.; Sawistowsky, *Z. Phys. Chem. (Leipzig)* **1972,** *250, 43.*
61. Norskov-Lauritsen, L.; Allinger, N.L. *J. Comput. Chem.* **1984,** *5,* 326.
62. Bystrom, K.; Mansson, M. *J. Chem. Soc. Perkin Trans. 2* **1982,** 565.
63. Penczek, S.; Kubisa, P.; Matyjaszweski, K. "Cationic ring polymerization." *(Adv. Polym. Sci.* **1980,** *37);* Springer-Verlag: Berlin. This study reports that the heat of polymerization of cyclooctane and of its oxa- and 1,3-dioxa derivatives are comparable.
64. Greenberg, A.; Liebman, J.F. "Strained Organic Molecules." Academic Press: New York, 1978, pp. 65–70.
65. Van Vechten, D.; Liebman, J.F. *Isr. J. Chem.* **1981,** *21,* 105.
66. Liebman, J.F.; Dolbier, W.R., Jr.; Greenberg, A. *J. Phys. Chem.,* in press.
67. Ladon, L.H.; Liebman, J.F. Unpublished observations.
68. Schaffer, R. In "The Carbohydrates," Pigman, W.; and Horton, D., Eds.; Academic Press: New York, 1972.
69. Chickos, J.S.; Annunziata, R.; Hyman, A.S.; Ladon, L.H.; Liebman, J. F. Unpublished observation.
70. (a) Clark, T.; Knox, T. McO.; Mackle, H.; Kervey, M. A.; Rooney, J.J. *J. Am. Chem. Soc.* **1975,** *97,* 3835. (b) Clark, T.; Knox, T. McO; Mackle, H.; Kervey, M.A.; Rooney, J.J. *J. Am. Chem. Soc.* **1979,** *101,* 2404.
71. An, X.-W.; Mansson, M. *J. Chem. Thermodyn.* **1983,** *15,* 287.
72. Hamilton, W.S.; Thompson, P.; Pustejovsky, S. *J. Chem. Eng. Data* **1976,** *21,* 428.
73. Ellenberger, M.R.; Eades, R.A.; Thomson, M.W.; Farneth, W.E.; Dixon, D.A. *J. Am. Chem. Soc.* **1979,** *101,* 7151.
74. Clayton, D.B.; Garner, R.C. In "Chemical Carcinogenesis," Searle, C.E., Ed.; American Chemical Society: Washington, D. C., 1976, Chapter 8.
75. Sabbah, R.; Chastel, P.; Laffitte, M. *Thermochim. Acta* **1974,** *10,* 353.
76. Nishiyama, A.; Sakiyama, N.; Seki, S.; Horita, H.; Otsubo, T.; Misumi, S. *Bull. Chem. Soc. Japan* **1983,** *53,* 889.
77. See, for example, Cheronis, N.D.; Entrikin, J.B. "Semimicro Qualitative Organic Analysis." T.Y. Crowell: New York, 1947, pp. 297–304.
78. (a) Smith, P.A.S. "Derivativers of Hydrazine and Other Hydronitrogens Having N—N Bonds." Benjamin/Cummings: Reading, Mass., 1983. (b) Schmidt, E.W. "Hydrazine and Its Derivatives: Preparation, Properties, Applications." Wiley: New York, 1984.
79. The thermochemical data on this species were obtained from Carson, A. S.; Laye, P.G.; Yurekli, M. *J. Chem. Thermodyn.* **1977,** *9,* 827.

CHAPTER 7

Generation of Long Carbon–Carbon Single Bonds in Strained Molecules by Through-Bond Interaction*

Eiji Ōsawa

Hokkaido University, Sapporo, Japan

Ken Kanematsu

Kyushu University, Fukuoka, Japan

CONTENTS

Statements in angular brackets such as this, ≪ ≫, are intended to supplement the text by offering details of arguments, additional examples, and extensions of the ideas presented. They may be omitted in an initial reading without loss of continuity.

*Part 26 of the Series "Application of Potential Energy Calculations to Organic Chemistry". Part 25: Jaime, C.; Ōsawa, E. *J. Mol. Struct.* **1985,** *126,* 363.

MOLECULAR STRUCTURE AND ENERGETICS, Vol. 3

1. INTRODUCTION

The length of a chemical bond usually remains constant for the same type of bond among different molecules. In fact, internuclear distances are used as primary criteria for finding bonds in diffraction analyses of molecular structure. A closer look at the reported bond lengths reveals, however, considerable variations for each type of bond. For example, the standard[1] length of the sp^3-hybridized C—C bond, with which this chapter is concerned, is 1.54 Å, but actually it ranges between 1.4 and 1.8 Å, reflecting diversity in the kinds and magnitudes of forces that influence equilibrium bond distances. Since bond lengths can be determined with a precision of a few thousandths of an angstrom unit by modern diffraction and spectroscopic techniques, this basic structural parameter will become one of the most useful probes of intramolecular environments, if the factors governing bond lengths are known in sufficient detail.

Classical interpretations for the changes in the bond length in such terms as covalent radius,[2] electronegativity,[3] bond order,[4] and the number of adjacent atoms[5,6] are being replaced by more "microscopic" measures such as hybridization,[7,8] overlap integral,[8] steric repulsion,[9,10] and orbital interactions.[11] Advanced computational methods now afford estimates of bond lengths with remarkably high precision. For instance, errors in reproducing unstrained C—C bond lengths by the ab initio gradient method using the 4-21G basis set[12] and the MM2 force field[13] are 0.006–0.010 Å. Nonetheless, calculations give only the result of balancing among intra- and intermolecular forces, and it is the conceptual insight into the nature of these microscopic interactions that offers perspectives on how to control this fundamental property of a molecule—its bond lengths.

In this chapter we mention a method of elongating the C—C bond under circumstances having considerable generality. The method, an extension of Mislow's[14] discovery on bond lengthening by through-bond orbital interaction, is not entirely new but has never been treated systematically before. Long bonds can be readily cleaved; hence their reactivities are also discussed. We limit the scope of our interests here exclusively to C—C single bonds in neutral hydrocarbons.

≪We realize that this limitation imposes omission of such interesting topics as the theoretical calculations of extremely long one-electron C—C bonds ($\simeq$ 1.9 Å) in organic cation radicals (Bellville, D. J.; Bauld, N. L. *J. Am. Chem. Soc.* **1982,** *104,* 5700; Lathan, W. A.; Hehre, W. J.; Pople, J. A. ibid, **1971,** *93,* 808), very strong through-bond interaction across Si—Si bonds (Sakurai, H.; Nakadaira, Y.; Hosomi, A.; Eriyama, Y.; Kabuto, C. ibid, **1983,** *105,* 3359), and the relation between C—O bond length and reactivity in ethers and esters (Allen, F. H.; Kirby, A. J. ibid, **1984,** *106,* 6197; Briggs, A. J.; Glenn, R.; Jones, P. G.; Kirby, A. J.; Ramaswamy, P. ibid, **1984,** 106, 6200; Jones, P. G.; Kirby, A. J. ibid, **1984,** *106,* 6207).≫

≪Regarding a perspective on the bond lengths in inorganic crystals, see I. D. Brown, *Acta Crystollogr.* **1977,** *B33,* 1305.≫

2. MAJOR FACTORS AFFECTING BOND LENGTHS

Bond length varies not only by intramolecular circumstances but also by its *definition.* There are a number of definitions depending primarily on the inclusion or exclusion of vibrational effects and on whether the experimental method employed gives the average internuclear *distances* or the distances between average nuclear *positions.*[13,15,16] Bond lengths as they are obtained from experimental determinations are supposed to be subjected to refinements that involve removal of the influence of vibrations so that they approach r_e, corresponding to the minimum of potential energy which mimics a hypothetical motionless state at 0 K. Different experimental measures of bond length are discussed in Chapter 1, Volume 2 of this series. Ab initio molecular orbital (MO) calculations give r_e.[17] On the other hand, molecular mechanics (MM)[13,18] and semiempirical MO methods like modified neglect of differential overlap, (MNDO),[19] both used extensively in our work, are parameterized to reproduce thermally averaged structures.

A. Elongation

Longer bonds[20] are seen more often than shorter ones, probably because internuclear repulsion increases more rapidly than the destabilization due to thinning out of bonding electrons. Let us start with factors that tend to elongate C—C bonds. Repulsion between vicinal substituents is an obvious cause, as demonstrated in Rüchardt and Beckhaus'[10] extensive studies on the effects of various substituents on the properties of ethane C—C bonds. However, as is well known,[13,18] stretching of a C—C bond is the most "expensive" way of relieving intramolecular strain, compared to other deformation modes involving smaller force constants such as angle bending and bond rotation. For this reason, elongation beyond 1.6 Å by repulsion alone occurs only under extreme congestion as in **1**,[21] **2**,[22] and **3**.[23] [*Note:* Numbers on the bonds in structures **1–100** denote lengths in angstrom units.]

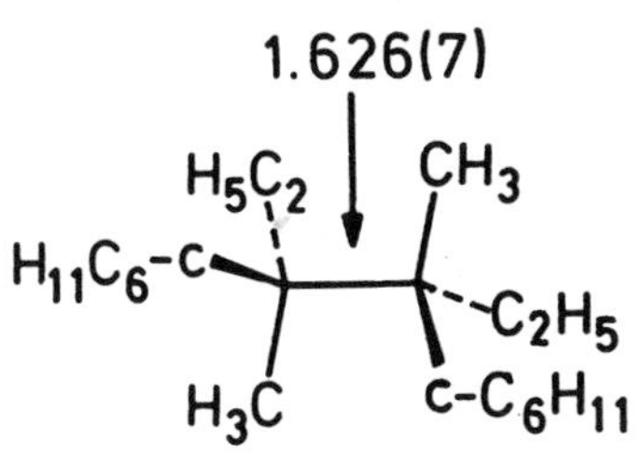

1.626(7)

1

1.641

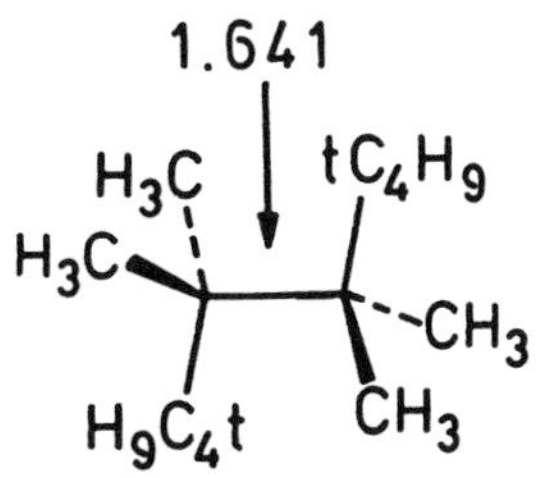

2

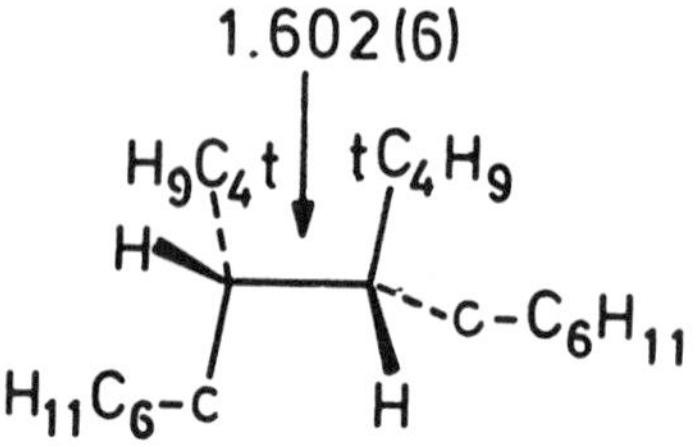

1.602(6)

3

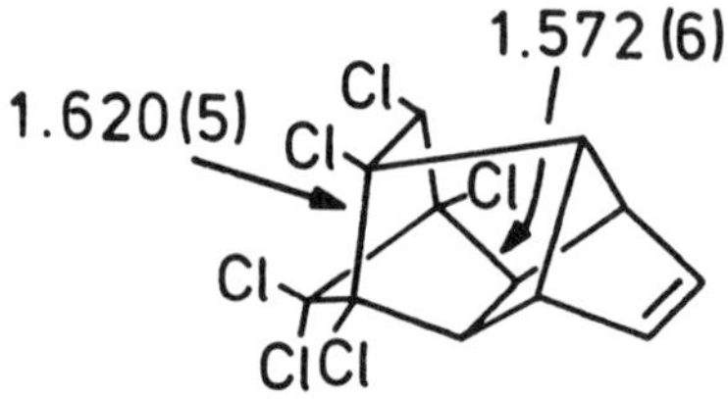

1.572(6)

1.620(5)

4

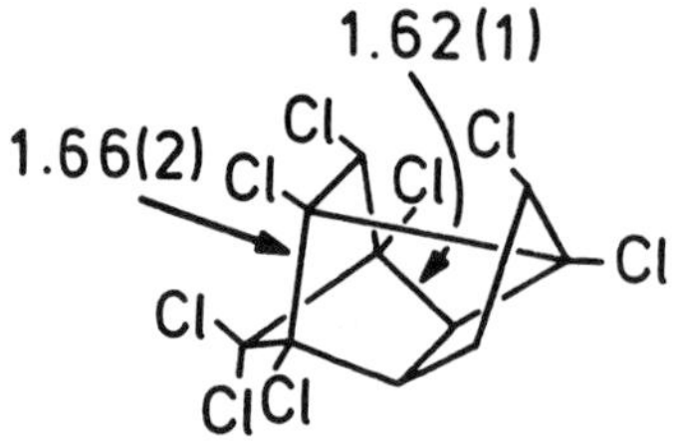

1.62(1)

1.66(2)

5

1.61(1)

1.60(1)

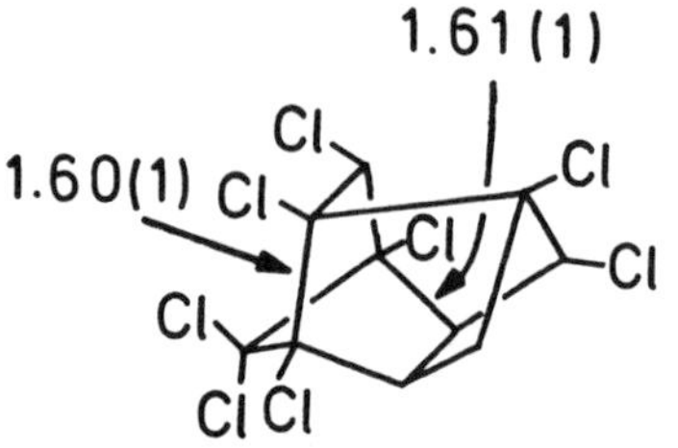

6

≪The gauche form (**3**) is highly deformed (X-ray, *t*-Bu—C—C—*t*-Bu dihedral angle 98°), and dominates in solution, gas, and solid phases. Molecular mechanics indicates this conformer to be 10 kcal/mol more stable than the anti form. The reason for the unusual behavior of this highly crowded molecule is the destabilization of the anti form, where there is no effective way of releasing nonbonded repulsion across the central bond due to molecular symmetry. For similar examples, see Ōsawa, E.; Shirahama, H.; Matsumoto, T. *J. Am. Chem. Soc.* **1979,** *101,* 4824.≫

Long C—C bonds have often been observed in some polycyclic and bridged molecules, like polychlorinated insecticides (**4**,[24] **5**,[25] and **6**[25]). The fact that only bonds in certain locations of these cage structures are elongated suggests that some van der Waals or Coulombic repulsion is concentrated at these positions for some "mechanical" reason, in addition to the withdrawal of electrons by the electronegative chlorines from bonding MOs.

The longest known C—C bonds occur in derivatives of bisnorcaradiene (7) carrying at least one methyl group at C_{11} (Table 7-1).[26]

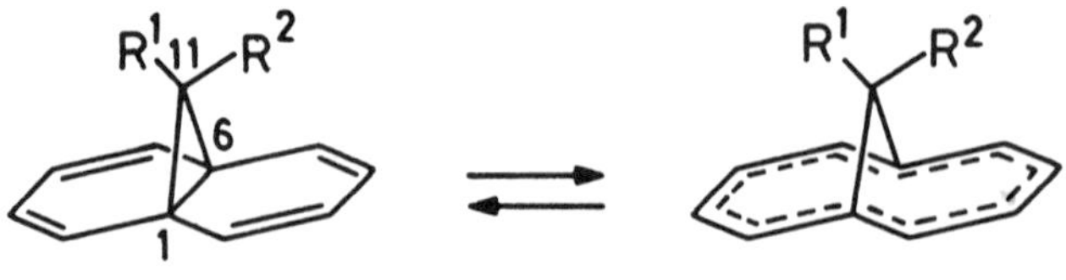

7

8

≪The longest record is 1.851 Å for **7** (R^2 = Me, R^2 = CN).[26]≫

Table 7-1. C_1-C_6 Bond Lengths, r, in Structure **7**, Determined by X-Ray Analysis

R^1	R^2		r (Å)[a]	Ref.
Me	Me		1.799[b]	26a
Me	CN	α-Form	1.817[b]	26b
		β-Form	1.712[c]	26c

[a]Standard deviation less than 0.004 Å.
[b]Average of two different molecules in the unit cell.
[c]Shrinks to 1.622 Å at −100°C.

The extremely long C_1—C_6 bond of about 1.8 Å in **7** has aroused great interest.[23] The central question is whether this bond can be called a real chemical bond. Rapid equilibrium with 1,6-methano[10]annulene (**8**) through valence isomerization is likely to occur with a small barrier height according to ab initio calculations.[27a,b] This possibility was, however, excluded in crystals on the basis of careful evaluation of the effects of thermal motion.[26] This bond length in **7** is sensitive to temperature change,[26c] and this property is attributed to the angle bending oscillation at C_1—C_{11}—C_6. For this reason, Simonetta[26] proposes that this molecule be considered fluxional in the vicinity of the C_1—C_6 bond. Such variability in the length may be a common characteristic of a very long and weak bond. It should be emphasized that this picture does not mean that an average structure between **7** and **8** is observed; it indicates only nuclear oscillation within the structure **7**.

What is the situation about the C_1—C_6 bond of norcaradiene (**9**)? Those bonds in chromium tricarbonyl complexes of bridged norcaradiene (**10**[28] and **11**[29]) are very long.

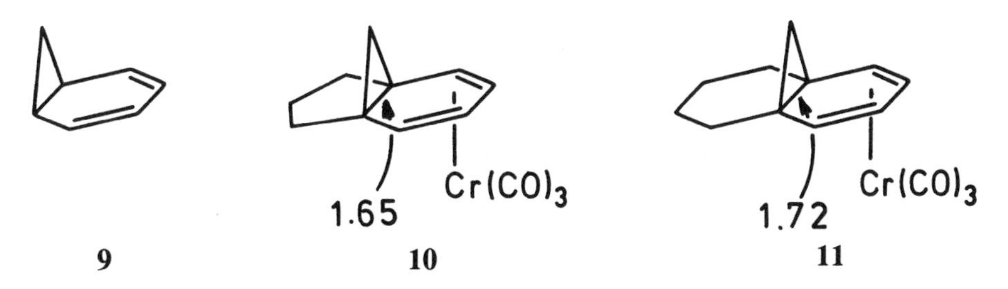

≪In contrast, two cobalt complexes of bisnorcaradiene, **12** and **13**, have normal C_1—C_6 bond lengths (Mues, P.; Benn, R.; Krüger, C.; Tsay, Y.-H.; Vogel, E.; Wilke, G. *Angew. Chem.* **1982,** *94,* 879). The contrast between **10** and **11** and **12** and **13** has been explained as follows by Professor J. Liebman: Cobalt (Co) needs only 4 electrons to fill its valence shell; thus it interacts only with the butadienes in **12** and **13**. In contrast, Cr $(CO)_3$ needs 6 electrons to fill its valence shell and therefore interacts with the cyclopropane ring also.≫

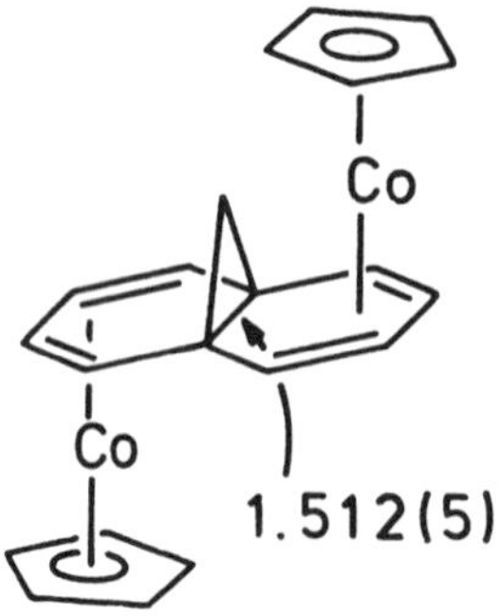

Co Co 1.512(5)

12

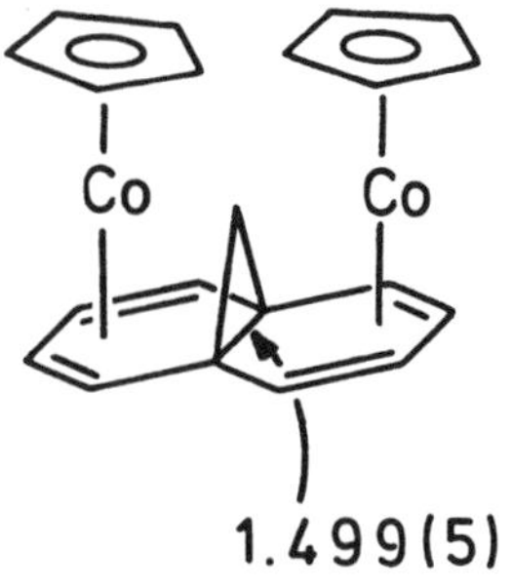

Co Co 1.499(5)

13

The molecular structure of norcaradiene itself (**9**) has not been studied experimentally since the tropylidene valence isomer was observed, but STO-3G calculations including gradient geometry optimization suggest considerable elongation at the C_1—C_6 bond to 1.58 Å.[30]

In connection with the last-mentioned calculations, Cremer and co-workers[31] used a novel method for characterizing the long bond. The method, originally proposed by Bader and associates,[32] is based on the calculated electronic charge density distribution, and the bond is defined by the presence of a "bond critical point" where the gradient of charge density attains a minimum value along the path of maximum density linking two neighboring nuclei in a molecule. According to this method, it is also possible to determine the point of bond breaking in the course of bond elongation by the vanishing of the "bond critical point." The Bader criteria have been applied to the problem of deciding whether 1*H*,3*H*-1,3-diborete (**14**) and bicyclobutane-2,4-dione (**15**) exist in monocyclic (**14a, 15a**) or bicyclic (**14b, 15b**) structures.[33]

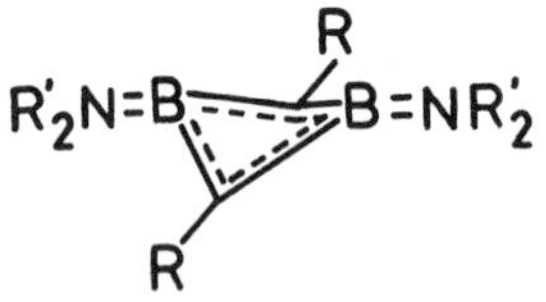

$R'_2N{=}B$ R $B{=}NR'_2$ R

14a

$R'_2N{=}B$ R $B{=}NR'_2$ R

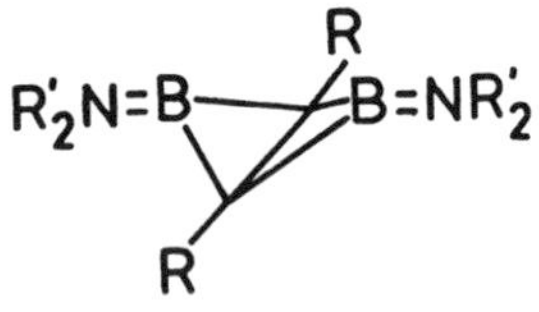

14b

≪Aromatic monocyclic planar structures were eliminated on the basis of high calculated total energies.[33]≫

X-Ray analysis of **14** (R = *t*-Bu, R′ = Me) gave 1.814 Å as the $C_1 \ldots C_3$ distance, a number that seems to rival the $C_1 \ldots C_6$ distance in 11- cyano-11-methyl-1,6-methano[10]annulene (Table 7-1). An ab initio 3–21G/gradient calculation of **14** (R = R′ = H) gave a similar value of 1.794 Å for this distance, but the Bader criteria for a bond were not satisfied in this region. Hence the bicyclic structure **14b** was excluded in favor of the delocalized structure **14a**. The same applies to **15:** calculations gave a C_1—C_3 distance of 1.762 Å, but the bond critical point did not exist between C_1 and C_3; hence the correct structure must be **15a**.[33] It would be interesting to see how the Bader criteria apply to the cyclobutadiene dication (which is also nonplanar)[34] and to **7**.[35]

The dominant factor of the C_1—C_6 bond lengthening in **7** has been considered to be the π-donor property of the methyl group, which weakens the C_1—C_6 bond by increasing its antibonding character at the filled Walsh_{as} orbital (**16**).[36] Although this mechanism is limited to cyclopropane deriva-

16

tives, an orbital interaction effect that generally elongates a C—C bond was recently brought to our attention by Mislow.[14] When a pair of p-type orbitals separated by three bonds are appropriately aligned, as illustrated in Figure 7-1 with 1,2-diphenylethane (**17**) in syn–parallel conformation, they interact through intervening bonds. Symmetry restriction allows only two types of interactions, π_{-}–σ^* and σ–π^*_{+}. Both interactions work to destabilize the intervening σ bonds, the former by populating the antibonding character there, and the latter by transferring bonding electrons into a virtual orbital. The former type predominates over the latter, according to quantitative perturbational MO treatment using *p,p′*-dibenzene (**18**) as a model.[14] (Weakening of a mediating σ bond in the through-bond interaction has also been mentioned by Gleiter.[11b])

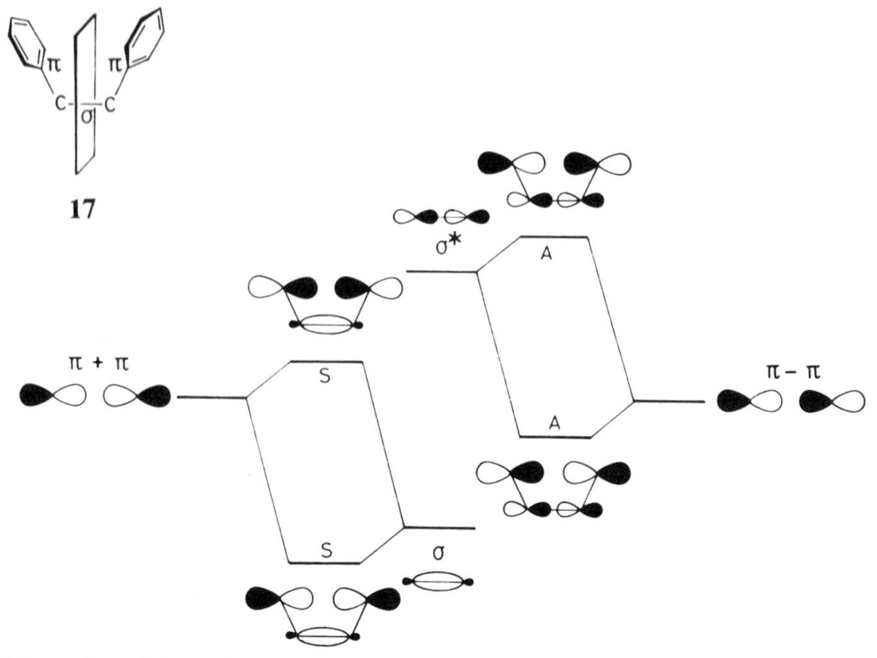

Figure 7-1. Schematic illustration of through-bond interactions between bonding and antibonding orbitals of 1,2-diphenylethane (**17**).

A "longitudinal" C—C bond of **18** interacts with four favorably aligned π orbitals and both MNDO and STO-3G calculations indicate that this bond is as long as 1.6 Å.[14b,37] Although **18** is unknown, anthracene photodimer **19**[38] and a number of related molecules[39] have the same structural fea-

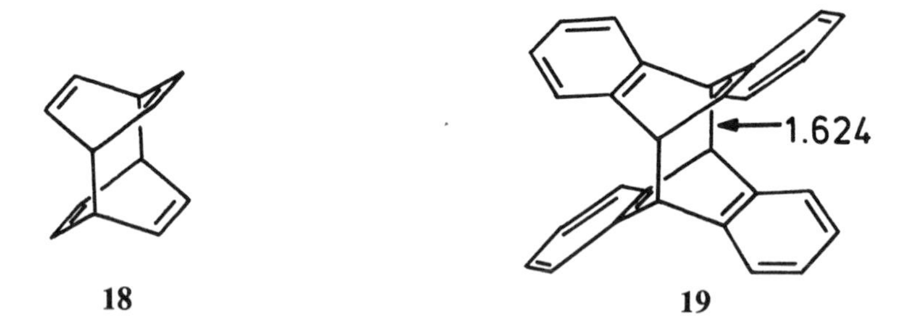

ture regarding the through-bond interactions as **18**, and the mediating "longitudinal" bonds in these molecules have been found by X-ray analysis to exceed 1.6 Å. An interesting point here is that these long bonds had once been attributed unanimously to nonbonded repulsion between the two anthracene rings, which are forced to take close, face-to-face disposition.[38,39] However, this repulsion is not that strong: the central ring of anthracene is greatly puckered upon dimerization, which means that the facing benzo rings are actually far apart. MM calculations of **19** and related structures give normal lengths for all the longitudinal bonds.[14a,39] MO calculations at the levels of MNDO and STO-3G approximations satisfactorily reproduce X-ray structural features, including these bond lengths. Comparison of MO and MM results turned out to be a convenient way of detecting the bond elongation by the through-bond mechanism.[14a]

More recently, an extraordinarily long C—C bond (1.781(15) Å) has been found in an organosilicon compound (**20**),[40] and in this case the elongation

Me_3Si $SiMe_3$ 1 3 Me_2Si $SiMe_2$

20

seems to be caused by a mechanism different from those mentioned above.[41] Some time ago, Stohrer and Hoffmann[42] predicted that a carbon atom would be stabilized in trigonal pyramidal valence if the three atoms forming the basal plane of a trigonal pyramid had a low-lying unoccupied MO suitable for accepting a π electron from the carbon atom, and they cited Si as the potential ligand atom for this purpose. The X-ray geometry[40] at the bridge-head carbon (**20**) exactly corresponds to this situation, and because of the donation of a π electron to the ligand Si atom, the "vertical" C—C bond in **21**, which corresponds to the C_1—C_3 bond in **20**, is elongated. Note that the

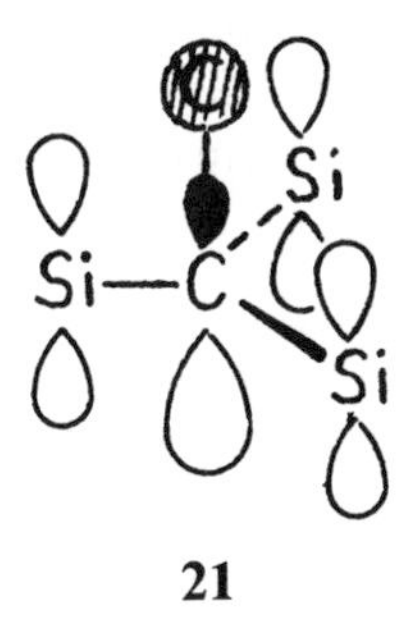

21

distance of the C_1—C_3 bent bond in the parent bicyclo[1.1.0]butane **22** (1.498(4) Å)[43] is shorter than normal; hence the elongation of the C_1—C_3 bond in **20** relative to that of the parent molecule amounts to 0.28 Å or 19%!

B. Contraction

A brief mention of the occurrence of bond shortening is now in order. The well-known effect of high s character on decreasing C—C bond length as

22 23

proposed by Dewar[44] and later generalized by Maksić[8] does not appear to lead to great changes in bond lengths. Even in 1,1′-binorbornane (**23**), the pivot bond (1.515(5) Å) is only marginally shorter than normal.[45] The rather surprisingly short central bond length of 1.558(3) Å observed for 9,9′-bitriptycyl (**24**) has also been explained by high s character. In this case, MM gave a 0.031-Å longer bond, and this was taken as indirect evidence for the rehybridization of the central carbon atoms.[46]

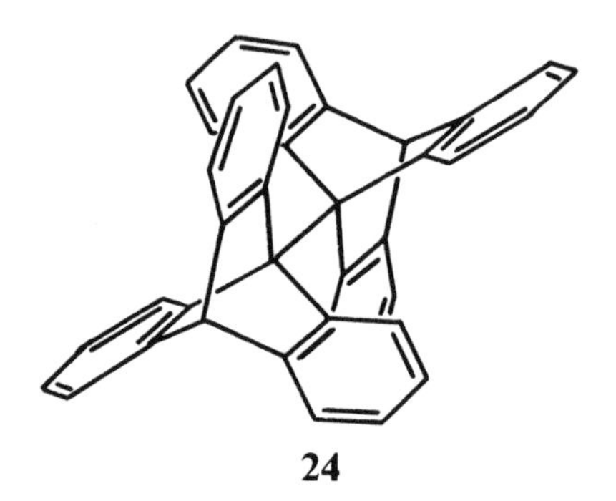

24

Perez and Brisse describe in their series of papers on the X-ray analysis of alkylene glycol dibenzoate derivatives and bis(2-hydroxyethyl)terephthalate, that C_{sp3}—C_{sp3} distances in these molecules are invariably short, close to 1.50 Å (Table 7-2).[47] Valence angles OCC that involve the short C—C bond also are abnormal ($\simeq$105°). Rigid motion correction

TABLE 7-2. $C_{SP3}-C_{SP3}$ Bond Lengths, *r*, in (**A**) Alkylene Glycol Dibenzoate and Derivatives and (**B**) Bis(2-hydroxyethyl)terephthalate Determined by X-Ray Analysis

Specimen[a]	*n*	Ar	*r* (Å)	Ref.
A	2	C_6H_5	1.499(2)	47a
	2	C_6H_4-*p*-Cl	1.493(2)	47b
	3	C_6H_4-*p*-Cl	1.504(3)	47c
	6	C_6H_5	1.485(4)[b]	47d
			1.523(4)	
			1.496(4)	
B			1.499	47e

[a] **A** = $ArCOO(CH_2)_nOCOAr$; **B** = $HOCH_2CH_2OCO-C_6H_4(p)-COOCH_2CH_2OH$.
[b] In order from outer to inner C—C bond.

in the case of ethylene glycol dibenzoate explained only about 0.007 Å of the decrease. The bond contraction observed here is by no means remarkable, but the tendency seems persistent. It should be worthwhile to study other compounds having vicinally oxygenated C—C bonds.

More striking bond compression has been reported for a hexaphenylethane derivative (**25**), the central bond length of which was found by X-ray analysis to be 1.47(3) Å.[48]

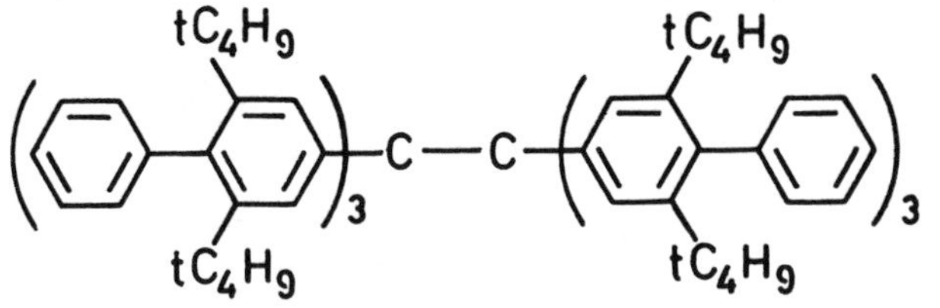

25

≪Hexaphenylethane itself does not exist: McBride, J.M. *Tetrahedron,* **1974,** *30,* 2009.≫

This short bond is in conflict with other examples of highly substituted ethanes (whose central bond never contracted but always lengthened)[10] and with theoretical calculations.[20a] One of us has addressed this problem in some detail, concluding that either the reported value is grossly in error or a novel shrinkage effect may be operative.[20] At least one other paper that appeared thereafter expresses a critical view of the X-ray results.[49] Winter and co-workers[50] continue intensive X-ray studies concerning this problem. One interesting fact that emerged from Winter's efforts is that the central bond of 1,2-diphenylethane itself (**26**, 1.517(3) Å) is slightly shorter than

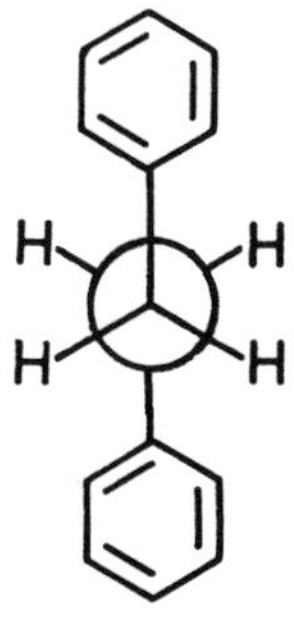

26

normal.[50] All other known central bond distances of diphenylethane derivatives are also shorter than the standard C_{sp3}—C_{sp3} bond length. 1,2-

Bis(2,4,6-tri-*tert*-butyl)ethane (**27**) is in an eclipsed anti–clinal conformation in the crystal, but the observed central bond length (1.567(4) Å)[49] is certainly short for the crowded situation. There seems to be some tendency for the central bond in the open-chain 1,2-diarylethanes to take short distances.

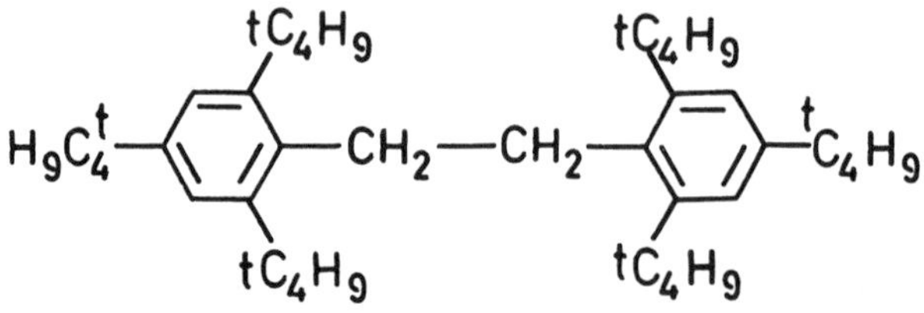

27

3. ELONGATION BY ENHANCED THROUGH-BOND INTERACTIONS

We discuss in this section on how to increase the effect of bond weakening by the through-bond mechanism of Mislow[14] mentioned briefly above. Recent works on the through-bond interaction, pioneered by Gleiter,[51] Verhoeven,[52] Paddon-Row,[53] and Imamura[54] and their associates, as well as by others,[55] have revealed general characteristics of this type of orbital interaction. According to basic rules of through-bond interaction, two requirements to increase the interactions are: (a) to increase the resonance integral and (b) to decrease the energy gap between the interacting orbitals. The first can be fulfilled by designing suitable molecular geometries; parallel (or anti–parallel) planar alignment with correct phase relation provides the ideal situation. The second requirement will be met, in the case of the dominant π–σ^* interaction, by raising the π level and/or by lowering the σ^* level. The following discussion is classified according to the second requirement. Since our own entry in this field was marked by a chance observation of a through-bond interaction between vicinal phenyl groups,[56] we first consider the effect of energy levels associated with C—C bond upon mediating the interaction between two vicinal phenyl groups. Then interactions between other types of orbital are discussed.

A. Orbital Interactions Between Vicinal Phenyl Groups

a. Through a Cyclobutane Bond

In the course of studying valence isomerization systems suitable for solar energy utilization, we encountered an abnormally long C—C bond (1.657(5) Å) in a pentacyclic cage ketone (**28**).[56] We first suspected, as other people did with anthracene photodimers (**19**: see above), that the elongation might have resulted from repulsion between two vicinal phenyl rings disposed in

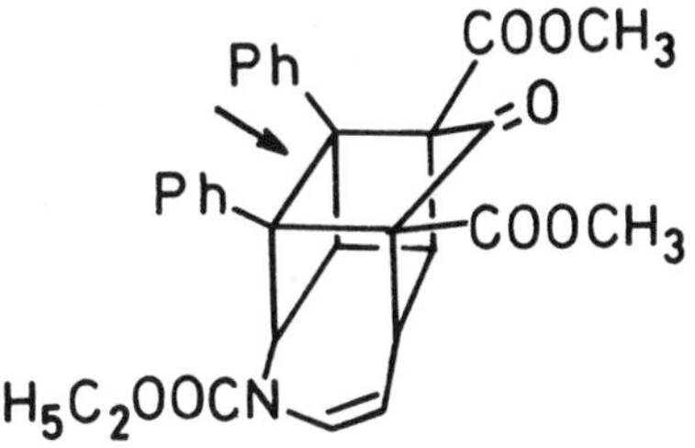

28

the face-to-face orientation. However, molecular mechanics pi electron (MMPI)[13] calculations of model structures of **28** gave normal bond lengths of 1.56–1.58 Å for the C—C bond carrying vicinal phenyl groups in the perpendicular orientation. Hence steric forces are not the major cause of the observed bond elongation. Operation of through-bond interaction was inferred from two pieces of evidence. First, MNDO calculation of **28** reproduced the X-ray structure well, including the abnormal bond length (1.64 Å). Second, replacement of a phenyl group with methyl followed by complete relaxation except for the long C—C bond length brought about significant increase in Mulliken electron population at this bond.

≪A tactic first employed by Mislow.[14]≫

If we take the MM bond lengths as reference points, the elongation of the C_{Ph}—C_{Ph} bond amounts to 0.08–0.1 Å. Such a remarkable elongation is likely to have been assisted by an angularly strained cyclobutane bond, which probably has a low σ^* level.

≪Note that this does not apply to σ^* of a cyclopropane. See the next section.≫

A similarly substituted pentacyclic compound, **29**, also turned out to have an abnormally long C_{Ph}—C_{Ph} bond (1.635(7) Å).[57] MM and MNDO calculations of this molecule gave exactly parallel responses regarding the distance and electron population of this bond, as in **28**.[57,58]

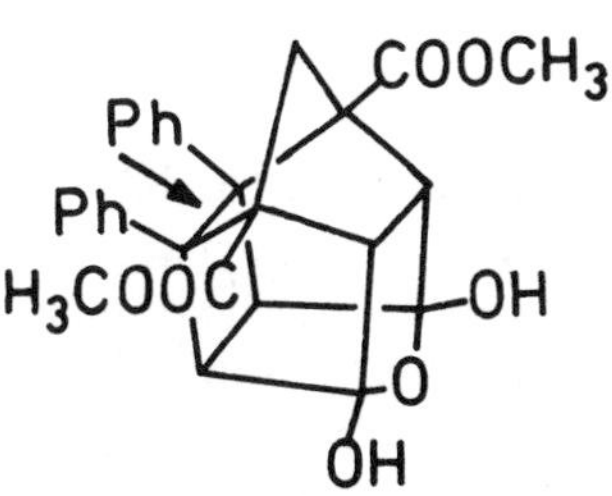

29

At this point it was considered desirable to confirm these interpretations on simple models.

≪In complex cage structures, it is not always clear how strain will be distributed over the entire molecule. For example, highly strained **30** appears

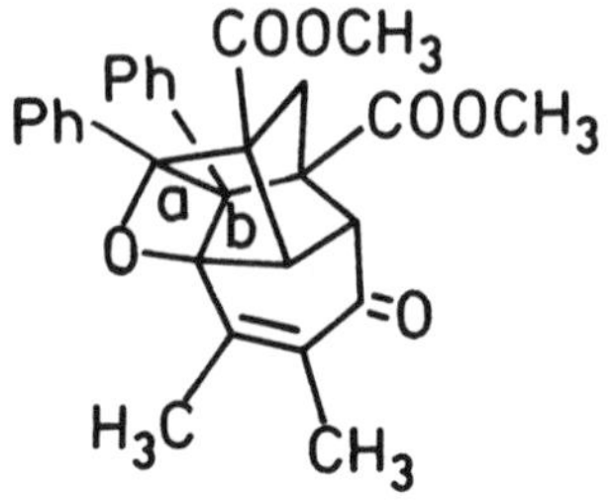

30

to have several possibilities of enhanced through-bond coupling between phenyl, α,β-unsaturated carbonyl, and the oxygen lone pair in the oxetane ring. However, bonds *a* and *b* are normal according to X-ray analysis.[59]≫

1,4-Diarylbicyclo[2.2.0] hexanes (**31**) appear to be good models for two reasons. First, the C_1—C_4 bond orbital of the parent hydrocarbon is bent,[60]

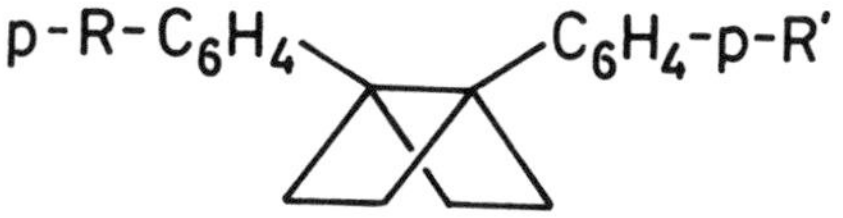

31

hence is likely to have high p character with its vacant orbital at a low energy level. Second, the two aryl ring planes are oriented face to face and bisected by the C_1—C_4 bond vector in the MM2- and MNDO-optimized structures,[58] giving the geometry appropriate for through-bond interaction.

≪No experimental information is available on the structure of **31**. For bicyclo[2.2.0]hexane, the observed C_1—C_4 bond length of 1.577(17) Å[60] is well reproduced by MM2 (1.572 Å),[56] MNDO (1.573 Å),[58] and ab initio 4-31G/gradient (1.573Å)[61] calculations.≫

Preliminary MNDO calculations of **31** (R, R′ = NH_2, O^-, CH_2^-, NO_2, CN) indeed predict C_1—C_4 bond lengths longer than 1.6 Å.[58]

Vicinally-arylated cyclobutanes are still simpler (albeit less strained) models, but no work has been done on the through-bond interaction. In a more complex system, [2.2](9,10)anthracenophene photoisomer (**32**), one of the cyclobutane bonds (indicated by arrow) seems to be ideally disposed for mediating coupling between π orbitals of the benzo groups. This bond has

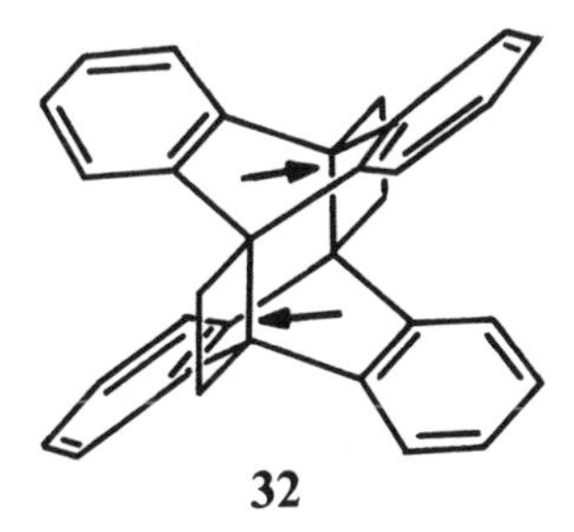

32

been shown by early X-ray work to have the extraordinarily long distance of 1.77 Å,[62] which might appear to indicate remarkably strong through-bond interaction.[14a] This distance is, however, in the disputable range according to the Bader criteria (see above).[63] Preliminary neutron diffraction analysis of a giant single crystal of **32** revealed a much shorter value of 1.64(1) Å for this bond.[39] MNDO calculations gave values close to the latter.[39] This bond length is still significantly longer than the corresponding bond in anthracene photodimer **19**,[38] and indicates some positive effect of the cyclobutane ring.[39]

b. Through a Cyclopropane Bond

Cyclopropane itself does not necessarily appear to be highly suitable for mediating the through-bond coupling by the π–σ^* mechanism, since the vacant Walsh orbital **33** is very high in energy despite its ideal geometry.[65–68]

33

As in cyclobutanes, fused cyclopropanes like bicyclo[1.1.0]butane (**22**) are more attractive for our thesis. This highly strained system has been the target of intensive studies.[41,69,70] It is likely that one of the low-lying vacant MOs of **22** is predominantly localized at the C_1—C_3 bond with appropriate symmetry to allow interaction with one of the high-lying occupied phenyl π orbitals in 1,3-diphenyl derivatives. The situation here is somewhat complicated because the highly bent C_1—C_3 bond has an equilibrium internuclear distance r_{13} shorter than normal and because this distance is influenced by such other geometrical factors as the valence angle at the bridgehead carbon, the flap angle θ between two cyclopropane rings (Figure 7-2), and the substituents at C_2 and C_4. Recently F. H. Allen[41] demonstrated abnormal elongation of the C_1—C_3 bond by the through-bond interaction between phenyl groups. Figure 7-2 is a reconstruction of the relation between r_{13} and θ[71] for **22** and its derivatives **34–41** based on geometrical data collected by Allen.[41] Although most of the points fall between the theoretically predicted range (dashed curves),[71] two points, **34a** and **41**, show significant deviations,

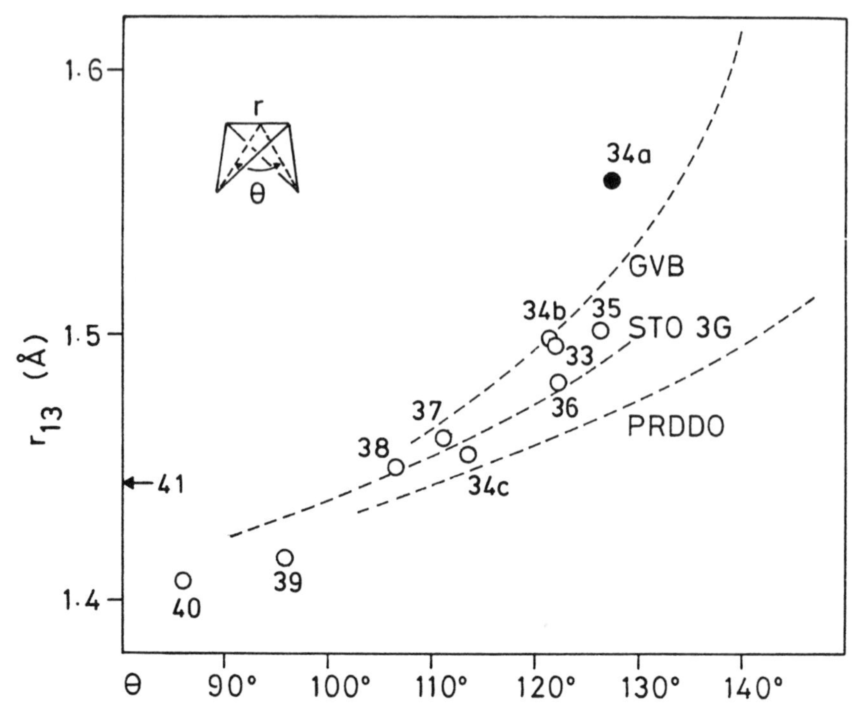

Figure 7-2. Relation between the length of bridgehead bond (r_{13}) and flap angle θ between three-membered rings of bicyclo[1.1.0]butane **22** and derivatives **34–41.** See Reference 41 for the sources of data points. Dashed curves are given by theoretical calculations [generalized valence bond (GVB), STO−3G, and partial retention of diatomic and differential overlap (PRDDO) methods] of Paddon-Row and coworkers.[71]

indicating abnormal elongation for the corresponding θ value. Only in these two compounds are the phenyl ring planes at right angles with the C_1—C_3 bond.[41] In **34b** and **34c**, phenyl groups are out of the bisecting orientation. Less pronounced elongation of r_{13} in **41** compared to **34a** is due to the presence of the 1,3-carbonyl bridge, which works to shorten the r_{13} bridge.[36,72]

≪Allen[41] terms this interaction simply "conjugation" and regards the r_{13}–θ relation as linear. Bicyclobutane geometries are discussed in greater detail in Chapter 5 of this volume.≫

Ph Ph Ph Ph Ph Ph

R R R R R R

a **b** **c**

34 $R = COOCH_3$

NC CN $COOCH_3$ N N $COOCH_3$ S

35 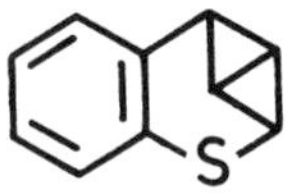 **36** 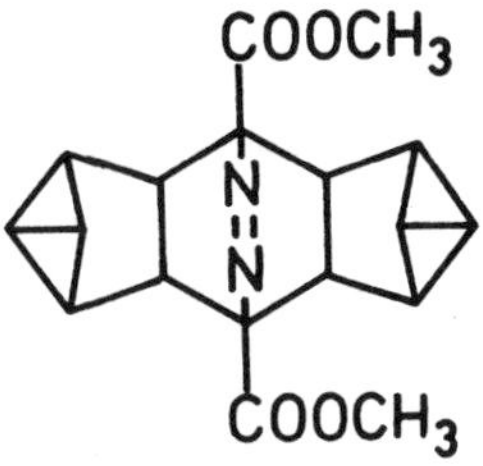 **37**

CH_3COO $OCOCH_3$ O

38 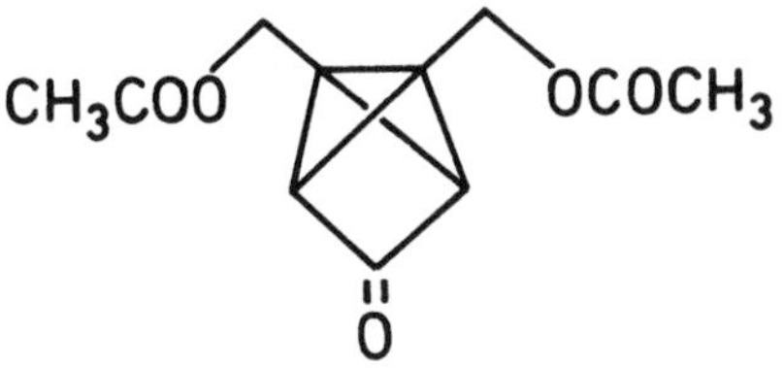 **39**

H_3C CH_3 O Ph Ph O

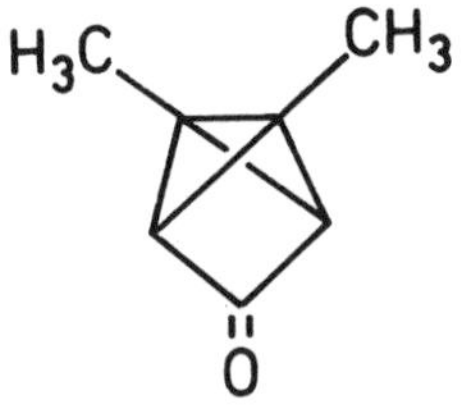

40 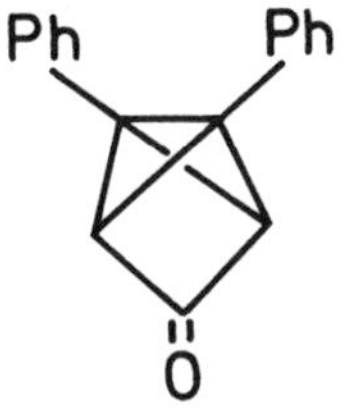 **41**

c. Through Acyclic Bonds: Polyarylethanes

Mislow and co-workers have discussed the possibility of elongating the central bond of polyphenylethanes by the through-bond coupling of π orbitals.[73,74] The simplest among polyphenylethanes, 1,2-diphenylethane, takes, in the crystal, the anti–periplanar conformation **26**,[75] with both phenyl planes almost perpendicular (72°) to the plane made by the three central bonds,[76] thus meeting at least the first requirement for the through-bond interaction.

≪The preference for perpendicular conformation (nomenclature given by Anderson and Pearson[77]) is rather surprising for the following reasons. First, theoretical calculations (MM, CNDO) predict that syn–clinal conformation is only slightly more stable than anti–periplanar.[76,78] Second, a parallel conformation like **42** must fit well in close crystal packing. Finally, in ethylben-

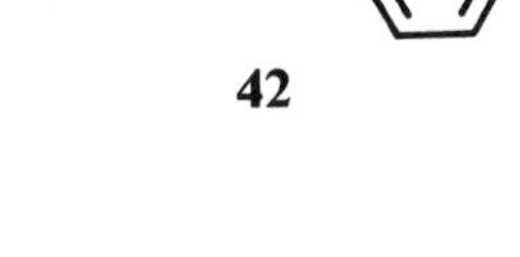

42

TABLE 7-3. Comparison Between Observed and Calculated Central C—C Bond Lengths, r, in Polyphenylethanes and Related Molecules

Compounds	ϕ_r[a]	X-ray analysis r (Å)	Ref.	MM approach[b] r (Å)	Ref.
26	72[c]	1.517(3)	50	1.534(CFF)	82
				1.535(MM2)[d]	
1,1,1,2-Tetraphenylethane	−6.3[e]	1.567(2)[e]		1.553 (CCF)	82
	40.3			1.571 (A71)	83
	−73.8			1.555 (MM2)[d]	
	89.6				
1,1,2,2-Tetraphenylethane	25.7	1.540	74	1.545 (CFF)	82
	23.4	(1.555)[f]		1.545 (MM2)[d]	
	−72.5			1.556 (A71)	74
	−69.7			1.547 (MM2′)[d]	
Pentaphenylethane					
X-ray conformation[g]	2.0[h]	1.606[i]		1.585 (MM1)[g]	
	50.5	1.612(3)[h]			
	−76.0				
	82.9				
	−47.3				
Mislow conformation[g]				1.561 (CFF)	82
				1.571 (MM2)[d,j]	
				1.595 (A71)	73
				1.602 (MM1)[g]	
1,2-Bis(t-butyl)-1,2-diphenylethane					
meso	—[k]	1.573(8)[l]		1.561 (MM2)[j]	
DL	—[k]	1.589(6)[l]		1.558 (MM2)[j]	
1,1,2,2-Tetra-n-butyl-1,2-diphenylethane, **45**					
	86	1.638(6)	85	1.612 (MM2)[j]	
				1.614 (MM2)	58
9,9′-Bifluorenyl		1.542[m]		1.545 (A71)	82
				1.543 (A71)	73
9-t-Butyl-9-(9′-fluorenyl)fluorene, **48**					
	−72.9	1.585(2)	89	1.572 (A71)	82
	60.7				
10,10′-Bianthronyl, **49**		1.60[90]			
Anthralin dimer, **50**		1.612(7)	91		

[a]Dihedral angle C_{ethane}–C_{ethane}–C_{ipso}–C_{ortho} in degrees. Only the values observed by X-ray analysis are given. Calculated values are close to those observed.

[b]Sources for MM programs are as follows:

A71 N. L. Allinger's 1971 force field with Mislow's parameters for the phenyl group: Allinger, N. L.; Tribble, M. T.; Miller, M. A.; Wertz, D. W. *J. Am. Chem. Soc.* **1971,** *93,* 1637. Andose, J. D.; Mislow, K. ibid, **1974,** *96,* 2168.

CFF Quantum Consistent Force Field/Pi Electron: Warshel, A. *Comput. Chem.* **1977,** *1,* 195,

MM1 Used with "harder" van der Waals parameters: Reference 84.

MM2 Used with special constants for the phenyl group: Allinger, N. L. *J. Am. Chem. Soc.* **1977,** *99,* 8127. Allinger, N. L.; Yuh, Y. H. *Quantum Chem. Program Exch.* **1980,** *11,* 395. Allinger, N. L. *Quantum Chem. Program Exch. Bull.* **1983,** *3,* 32. See also note *j*.

MM2′ Jaime, C.; Ōsawa, E. *Tetrahedron* **1983,** *39,* 2769.

[c]See text.

[d]Unpublished work.

[e]Destro, R.; Pilati, T.; Simonetta, M. *Acta Crystallogr.* **1980,** *B36,* 2497.

[f]1,1,2,2-Tetrakis(2-methoxyphenyl)ethane: Daly, J. J.; Sanz, F.; Sneeden, R. P. A.; Zeiss, H. H. *J. Chem. Soc. Perkin Trans. 2* **1972,** 1614.

zene, the perpendicular (**43a**) and parallel (**43b,c**) conformations have almost the same energies, although there is some disagreement as to which is the more stable.[79–81] Available information is, however, not adequate to judge whether the through-bond interaction is responsible for the observed conformation **26**.≫

a **b** **c**

43

The second condition should be fulfilled by increasing congestion. However, as shown in Table 7-3,

≪Hexaphenylethane is not included here (see above).≫

at least for unclamped molecules of this table, from the top down to bis(*tert*-butyl)diphenylethane, the ethane C—C bond lengths are normal despite suitable ϕ_r angles and are actually reproduced by MM calculations.[73,74,82,83]

≪Even if ϕ_r's are close to 90° (Table 7-3), C_{aryl}–C_{ethane}–C_{ethane} angles do not close in to the favored range near 90° (see **44**), because geminal aryl rings tend to stack on top of each other to obviate geminal repulsions.[74]≫

44

Hence it might appear that there are no remarkable through-bond effects in polyphenylethanes.

Nonetheless, this conclusion may not be totally safe, if the natural central bond length of polyphenylethanes is shorter than the usual value for the C_{sp3}—C_{sp3} bond as mentioned in Section 2. Then the MM-calculated lengths

---

[g]Similar, but discrete energy minima. See Reference 84.
[h]Acetone solvate: Bernardinelli, G.; Gerdil, R. *Helv. Chim. Acta* **1981,** *64,* 1365.
[i]THF solvate: Destro, R.; Pilati, T.; Simonetta, M. *J. Am. Chem. Soc.* **1978,** *100,* 6507.
[j]Beckhaus, H.-D. *Chem. Ber.* **1983,** *116,* 86.
[k]Not calculated but stereopicture indicates near-perpendicular orientation.
[l]Beckhaus, H.-D.; McCullough, K. J.; Fritz, H.; Rüchardt, C.; Kitschke, B.; Lindner, H.-J.; Dougherty, D. A.; Mislow, K. *Chem. Ber.* **1980,** *113,* 1867.
[m]Dougherty, D. A.; Llort, F. M.; Mislow, K.; Blount, J. F. *Tetrahedron* **1978,** *34,* 1301.

in Table 7-3 will become considerably shorter, and the observed lengths will look longer.

Setting this somewhat "elongation-favored" view aside for the moment, the observed central bond length in pentaphenylethane nevertheless attests considerable strain across this bond. Recent recalculation of this conformation using modified MM parameters gave 1.585 Å for this bond,[84] still considerably *shorter* than the X-ray value. The difference between the calculated and observed values probably corresponds to the elongation by the through-bond interaction.

An extremely long central C—C bond observed in **45** (1.638(6) Å)[85] also could not be reproduced by MM calculations, and this seems to be the first

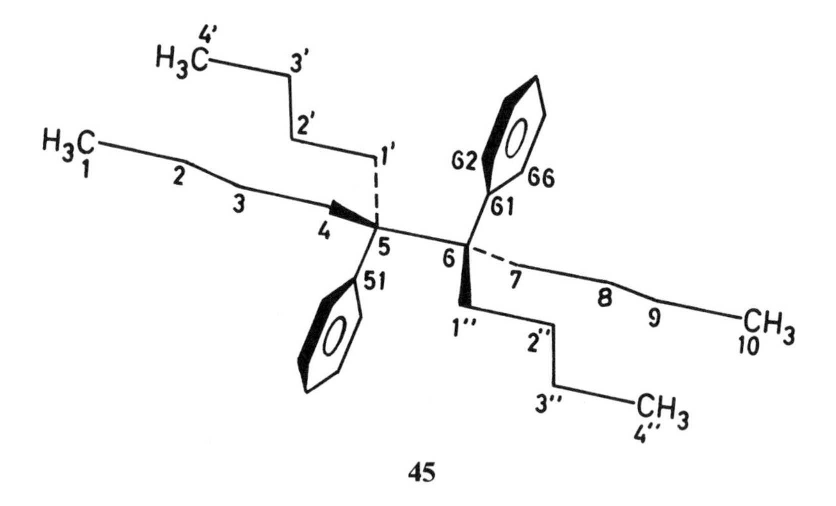

45

example of through-bond interaction enhanced by prestretching of the C—C bond, which probably lowered its σ* level.[58] The structure involving two all-trans *n*-decane chains as revealed by X-ray analysis might appear to be strainless but is actually highly strained.[85] There are a pair of g^Pg^M arrangements[86] along C_3—C_4—C_5—$C_{1'}$—$C_{2'}$ and $C_{2''}$—$C_{1''}$—C_6—C_7—C_8, where $C_{2'}$–C_3 and $C_{2''}$–C_8 interactions are strongly repulsive. Any deformation to relieve these nonbonded interactions leads to further straining. For example, if the $C_{1''}$–C_6–C_7 angle is opened to increase the $C_{2''}$ . . . C_8 distance, four sets of gauche, $C_{1''}$–C_4, $C_{1''}$–C_{51}, C_7–$C_{1'}$, and C_7–C_{51}, as well as two sets of progauche interactions, $C_{1''}$–C_{62} and C_7–C_{66}, increase.

≪Progauche interaction is of the 1,4-type resembling gauche and involving a C_{sp2}—C_{sp2} bond at the one end.[88]≫

Thus, the observed conformation **45** is inescapably locked into a high-energy minimum where stretching of this bond is the only way to decrease the overall strain. Hence "expensive" stretching of the central bond took place, thereby promoting the through-bond coupling of two phenyl groups in the proper orientation.[58] This interpretation has recently been confirmed by X-ray analysis of the lower homologues **46** and **47**.[87] All the features responsible for straining the central bond in **45** exist in **47a**, which in its C_{2h} conformation

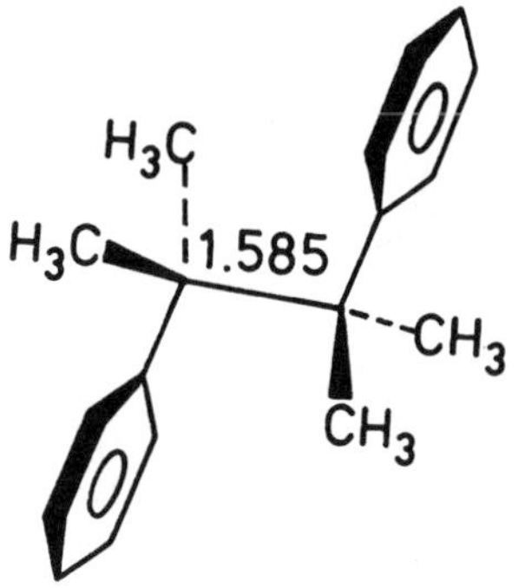

46

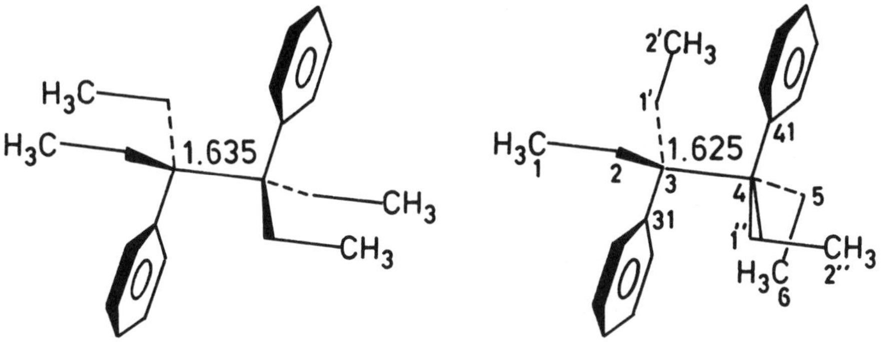

a **b**

47

≪Idealized point group: the observed molecular symmetry in crystal is stated to be C_i.[87]≫

has a central bond length comparable to that of **45**. The crystal of **47a** contains another conformation (**47b**) wherein two of the *endo*-methyls ($C_{2'}$ and C_6) are rotated into the gauche position. Whereas this conformer still contains a pair of g^Pg^M arrangements, $C_{2'}$—$C_{1'}$—C_3—C_4—C_{41} and C_6—C_5—C_4—C_3—C_{31}, these involve CH_3–C_{sp2} end interactions, which are less demanding than CH_3–CH_3 (in **47a**) or CH_2–CH_2 (in **45**) interactions.[88] In accordance with this expectation, the central bond of **47b** is 0.01 Å shorter than that in **47a**. The central bond of **46** is *much* shorter than those in **45** and **47a,b**, reflecting the smaller strain across the bond in **46**.

Clamped polyphenylethanes are considered more apt to involve through-bond coupling.[73] Although 9,9′-bifluorenyl is unremarkable in this respect, introduction of a *tert*-butyl group at C_9 (**48**) significantly elongates the cen-

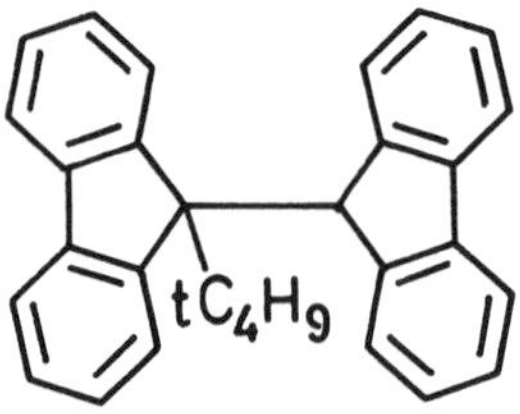

tC4H9

48

tral bond (Table 7-3). Simonetta and co-workers[83] have noted that **48** provides the only case where the MM-calculated central bond length did not agree with the observed value within 0.01 Å. This difference probably reflects additional lengthening by the enhanced through-bond coupling. Abnormally long central bond lengths in 10,10′-bianthronyl (**49**)[73,90] and bianthralin (**50**),[91] as revealed by X-ray analysis (Table 7-3), appear to have

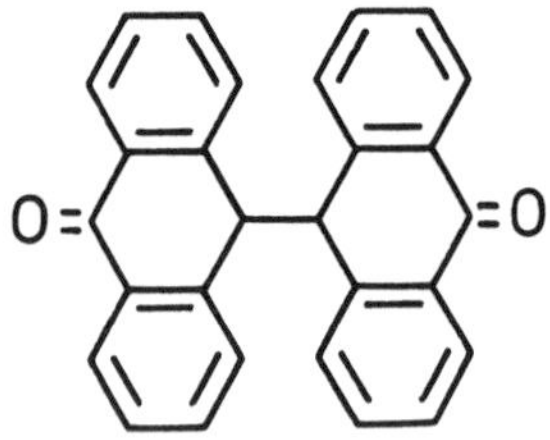

O= =O HO— —OH O= =O HO— —OH

49 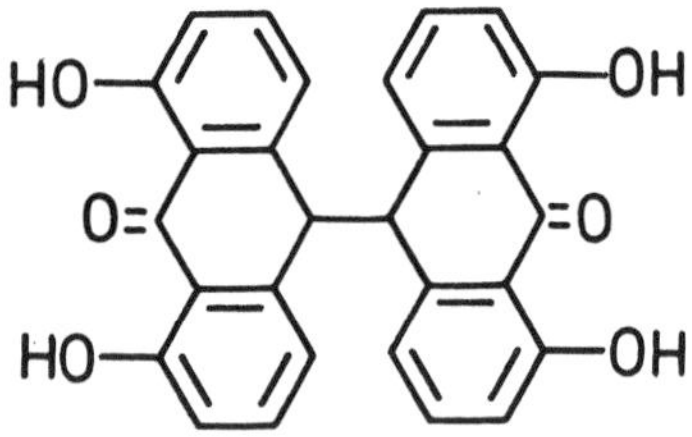 **50**

the same origin. Again, repulsion between two anthralin rings having synclinal overlap was at first considered to be responsible for the bond elongation in **50**.

≪Three more crowded bisfluorenyl derivatives have been suggested to show lengthening of the pivot bond by the through-bond coupling mechanism.[73]≫

d. In Cyclophanes

The possibility of through-bond interaction involving the C_2H_4 bridge in [2.2]paracyclophane (**51**), first suggested by Gleiter[92] in 1969, has since been

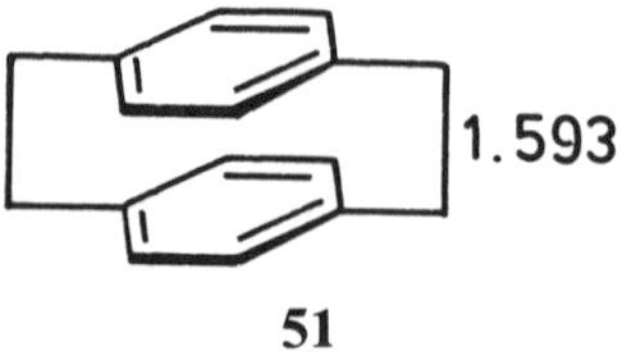

1.593

51

verified by photoelectron spectroscopy.[51c,93] The idea is now usually taken into account in theoretical treatments of [2_n]cyclophanes.[94,95] Although the through-bond interaction in **51** between two benzene rings is overridden by the through-space interaction,[95] near eclipsing in bridge methylene groups,[96] which contributes to the high strain energy of **51** (29.6 kcal/mol),[97,98] must be a contributing factor for preelongation of the CH_2—CH_2 bond and enhanced through-bond interaction across this bond. According to X-ray analysis, the bridge bond is certainly long (1.593 Å), even though the accuracy of this value is limited by the libration problem.[96]

Recent X-ray analysis revealed that bridge bonds in other [2_n]paracyclophanes (**52**[99] and **54–57**[100]) are universally lengthened. The only

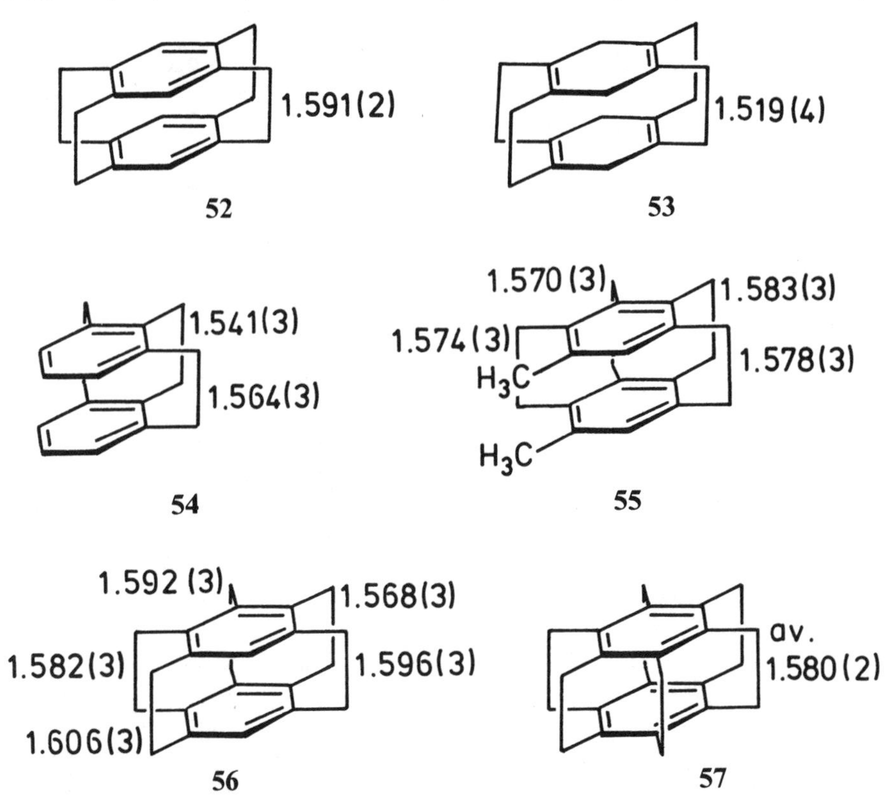

exception is the middle bridge bond of **54**, which is of normal length because of the high tilt of the benzene ring (dihedral angle 42°).[100] Interestingly, the less strained **53**, a Birch reduction product of **52**, has bridge bond length of (1.519(4) Å),[99] almost identical with that of 1,2-diphenylethane (**25**). Separate evaluation of the through-bond effect in the overall lengthening of bridge bonds may be a difficult task, but it can be approached by modifying the bridge. In **58**, for example, analysis of the photoelectron spectrum indicates enhanced through-bond interaction compared to **51**.[51c] It would be

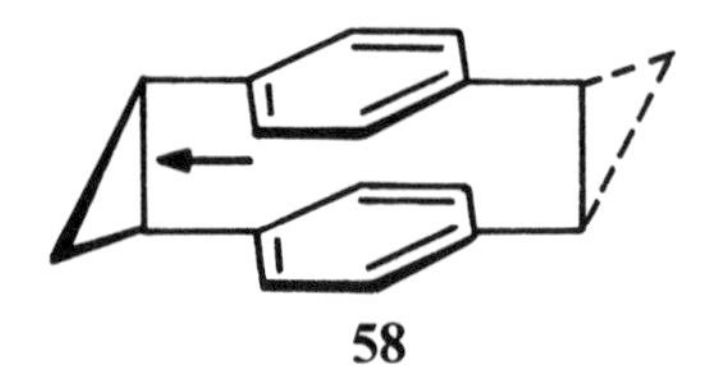

58

interesting to determine lengths of the C_t—C_t cyclopropane bond of this molecule.

B. Interactions Between Orbitals of Other Types

a. Olefinic Double Bonds

Pentacyclic tetraene (**59**) gives a useful illustration of the effect of a strained σ bond in mediating orbital interactions between two isolated double bonds.

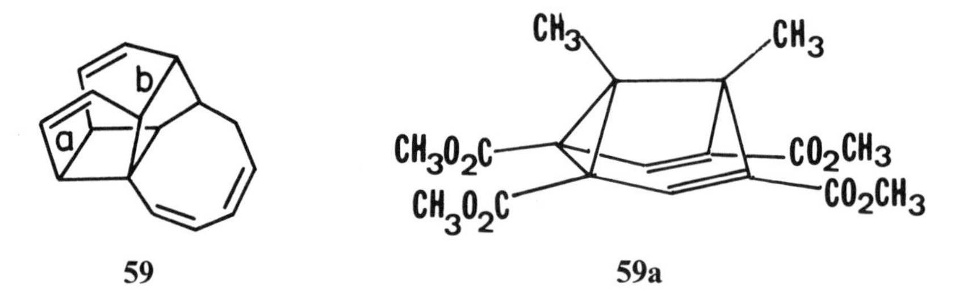

59 **59a**

According to X-ray analysis,[101] bond *a* is 1.617(3) Å, whereas *b* is 1.578(3) Å; according to our theory, the more strained bond *a* must be a better mediator of the through-bond interaction, hence more weakened than bond *b*.[58,102]

Two double bonds in semibullvalene should be able to interact through C_2—C_8, a cyclopropane bond. This bond is indeed very long (1.600 Å).[103] In one derivative (**59A**), this bond is remarkably lengthened (1.782(5) Å).[104] The additional lengthening compared to parent hydrocarbon may be interpreted as arising from the π-orbital interaction through this bond, which had been "preweakened" by the π-donating group R' attached to it; one of the two carbomethoxy groups is in an ideal geometry to cause this effect.[104]

Wiberg's "mini-superphane" (**60**, tricyclo[4.2.2.2^{2,5}]dodeca-1,5-diene)[105a]

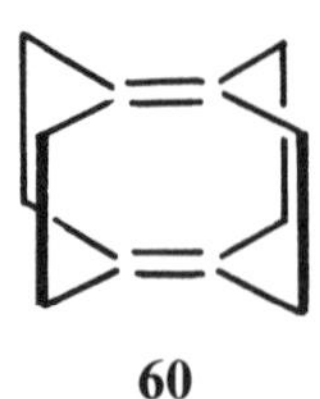

60

is an extremely interesting system from our point of view. The long C_{sp3}—C_{sp3} bond, as revealed by X-ray analysis (1.596(1) Å), may be interpreted as arising from the coupling of a pair of strained double bonds (first ionization potential 8.6 eV)[105b] through the ideally oriented C_{sp3}—C_{sp3} bond, which probably is prestretched by repulsion between closely constrained double bonds (at a distance of 2.395 Å) and also by repulsion arising in the eclipsed conformation (D_{2h}, not D_2).

≪Professor Wiberg holds an alternative view regarding this point. See Reference 105a.≫

If this explanation is true, **60** will be the first instance of out-of-plane deformation (35.6°) and van der Waals repulsion between C=C bonds playing significant roles in the enhanced through-bond interaction.

A potentially interesting case for the enhanced through-bond intereaction between olefinic double bonds will be the C_1—C_8 bond of strained ansaradiene (**61**).[106] Hexacyclopropylethane (**62**) provides another possibly favorable case of enhanced through-bond interaction between Walsh orbitals. Although one would certainly expect a rather long central bond for this fully substituted ethane (MM2, 1.615 Å), X-ray analysis revealed a still longer C—C distance of 1.636(5) Å.[107] The planes of the cyclopropyl rings are all nearly parallel with the central bond.

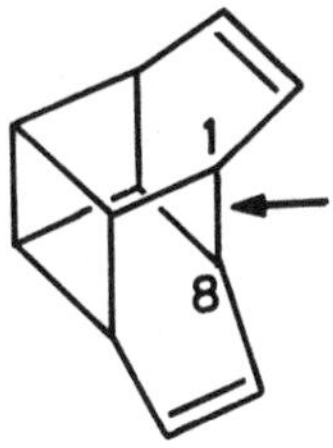

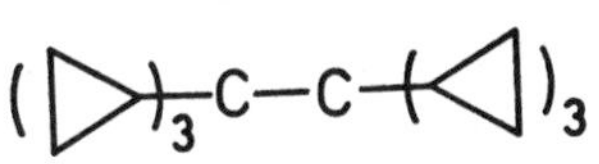

61 **62**

b. Carbonyl Groups

It has been experimentally recognized for some time that α,β-dicarbonyl cyclopropanes (**63**) and cyclobutanes (**64**) undergo reductive cleavage at the

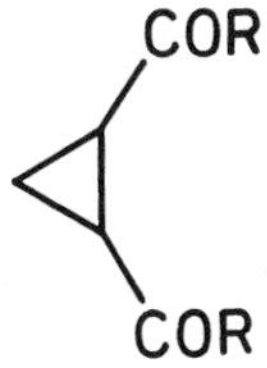

63 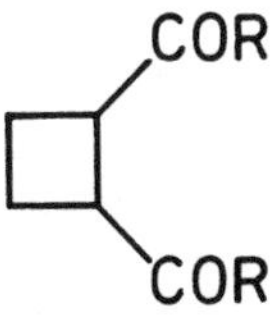 **64**

vicinally substituted C—C bond.[108] These facts led us to suspect that π orbitals of carbonyl groups in these strained molecules interact to elongate the intervening, angularly strained C—C σ bond already in the ground state. This possibility appears to have been implied by Mislow[14b] when he mentioned the significant bond lengthening in **65** and **66** in terms of through-bond coupling.

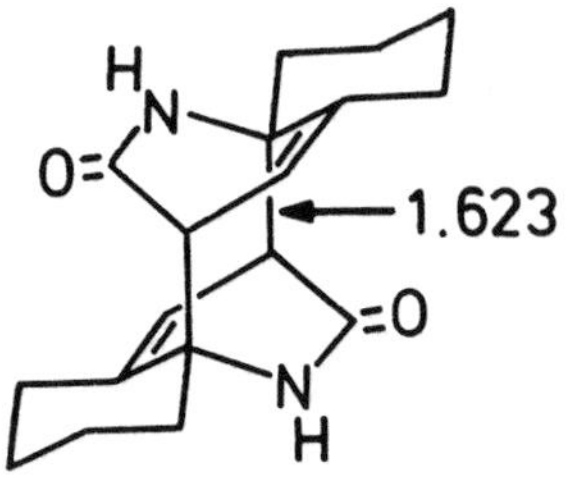

1.623

65

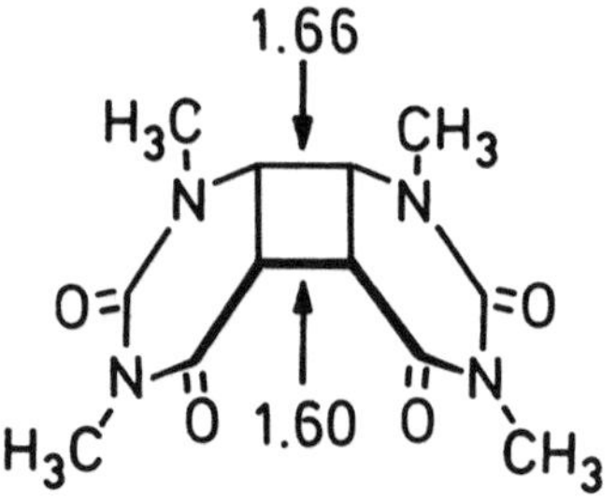

1.66

1.60

66

The first negative answer to this hypothesis came from an X-ray analysis of **67**,[109] wherein the C_8–C_9 cyclobutane bond is of normal length (1.563 Å).[110] Then we tested pentacyclic triketone (**68**) as a model for a computational check. The calculated cyclobutane bonds (arrows) were normal by MM2 (1.5521 Å) and slightly longer by MNDO 1.5592 Å). It seems likely that the effect of through-bond interaction between vicinal carbonyl groups on the length of the mediating C—C bond must be small, if indeed it exists.[58]

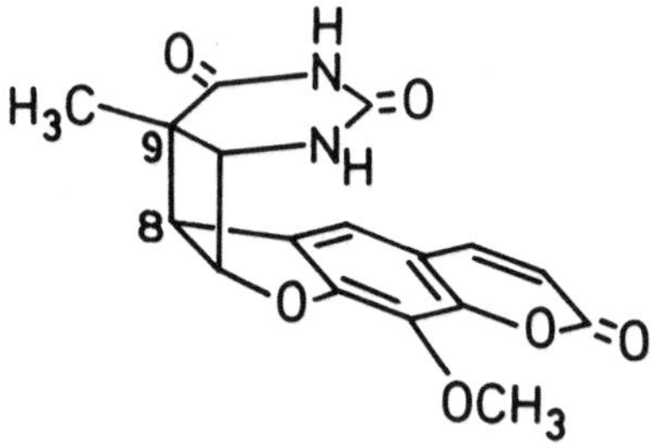

67

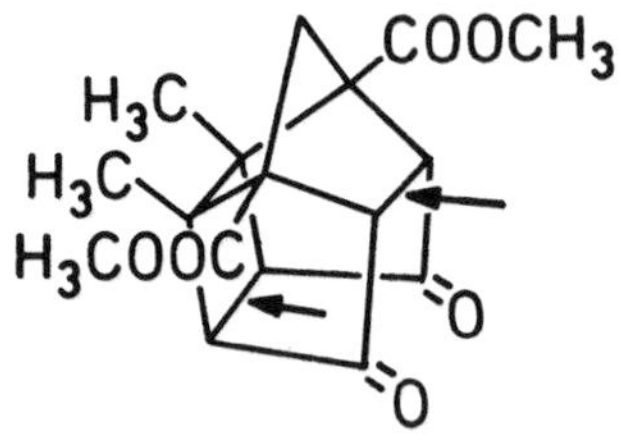

68

Notwithstanding, recent X-ray analysis of related compounds **69**[111] and **70**[112,113] revealed rather long C_{sp3}—C_{sp3} bond lengths in the O=CC—CC=O sequence, and the lengthening was explained in terms of the steric repulsion between carbonyl groups alone.[111] It was then felt desirable to do a careful comparison of bond lengths in the parent skeleton, Cookson's half-winged birdcage diketone, **71**. Table 7-4 summarizes calculated lengths of the key C—C bonds in terms of difference from the respective mean lengths and

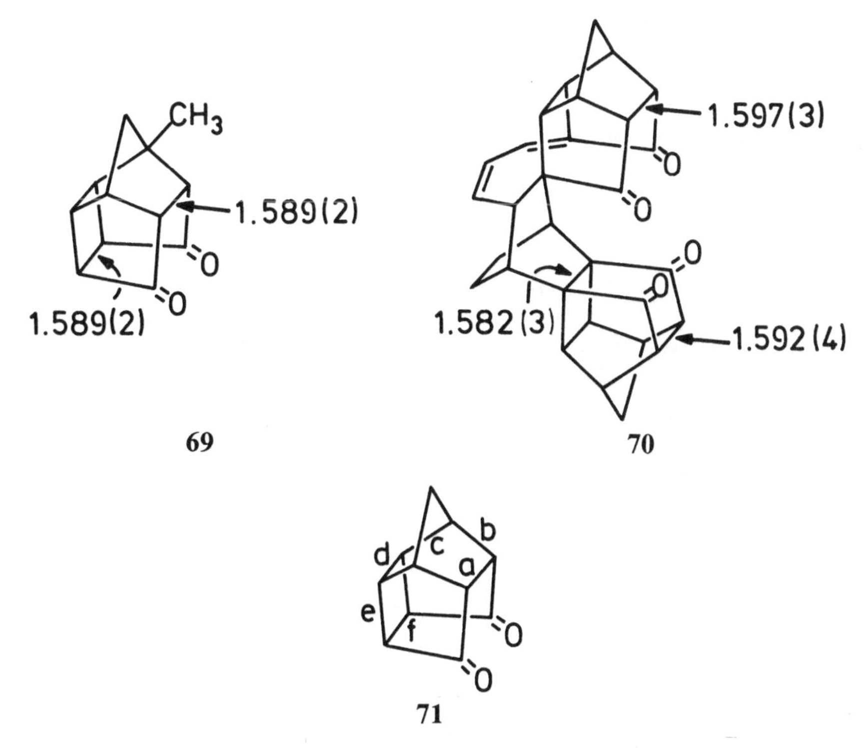

TABLE 7-4. Key C_{sp3}—C_{sp3} Bond Lengths in Cookson's Half-Winged Birdcage Diketone **71** and in the Corresponding Partial Skeletons of **69** and **70**

	Bond lengths (Å)[a]			
	X-ray		Calculated: **71**	
Bond[b]	**69**[c]	**70**[d]	MM2'[e]	MNDO[e]
[Average[f]	1.564	1.557	1.5510	1.5743]
a	0.025	0.035	0.0065	0.0137
b	−0.007	−0.011	−0.0018	0.0054
	−0.013	−0.018	−0.0018	0.0054
c	−0.002	−0.023	−0.0052	0.0024
	−0.007	−0.017	−0.0052	0.0024
d	−0.011	−0.013	0.0015	0.0011
e	−0.001	0.006	0.0007	−0.0071
	−0.005	0.015	0.0007	−0.0071
f	0.025	0.025	0.0042	−0.0065

[a]Numbers are relative to average length for each compound.
[b]See structure drawing **71**.
[c]Reference 111.
[d]Bond lengths courtesy of Professor G. Mehta.
[e]Bond lengths from this work.
[f]$(a + 2b + 2c + d + 2e + f)/9$.

compares these with corresponding variations of observed bond lengths in **69** and **70**. MNDO gave considerable lengthening for bond *a* but no special feature for the more strained *f*. MM2′ agrees better with the X-ray trend in that both bonds *a* and *f* are lengthened, but the magnitudes of lengthening are only one-sixth of those expected. These results may be taken as indicating the inadequacy of the present computational techniques, and the steric explanation of bond lengthening of Marchand and Suri[111] probably holds.

The failure to detect through-bond interaction between carbonyl π orbitals even under ideal circumstances is not surprising for two reasons. First, the highest occupied π level of carbonyl is much lower than that of phenyl.[58] Second, the electron density of this orbital is shifted toward oxygen, causing less effective overlap with the mediating orbitals.[114]

Perhaps more promising as the orbital for the through-bond interaction is one of the lone pair orbitals of carbonyl oxygen (p_y, **72**), which is the high-

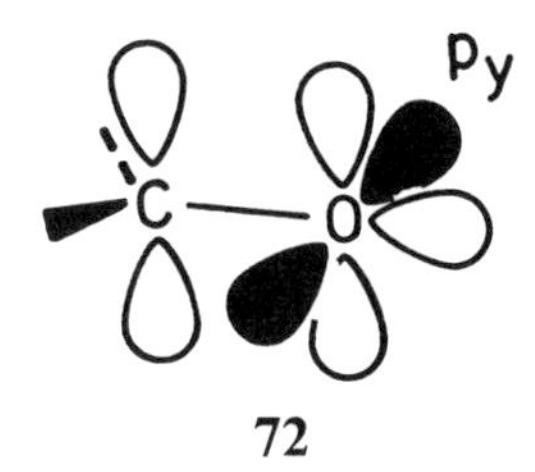

72

est occupied molecular orbital (HOMO) and is coplanar with the bonds extending from carbonyl carbon.[115] Although this particular possibility has not been studied yet, lone pairs are certainly potential orbitals for the through-bond interaction, as discussed below.

C. Lone Pairs

Principally, the high-energy lone pairs should readily engage in the n-σ^* and n^*-σ interactions. Naturally enough, both Hoffman[11a] and Gleiter[11b] used the lone pair as a typical orbital for illustrating through-bond interaction. To our knowledge, the first example of the *enhanced* through-bond coupling between lone pairs was given by Dewey, Miller, and Michl,[116] who noted a large red shift in the n → π^* transition of a pyrazine derivative (**73**). We have also demonstrated the effective coupling of lone pair orbitals through the central C—C bond of bicyclo[2.2.0]hexane (**74**) by computation.[58] When

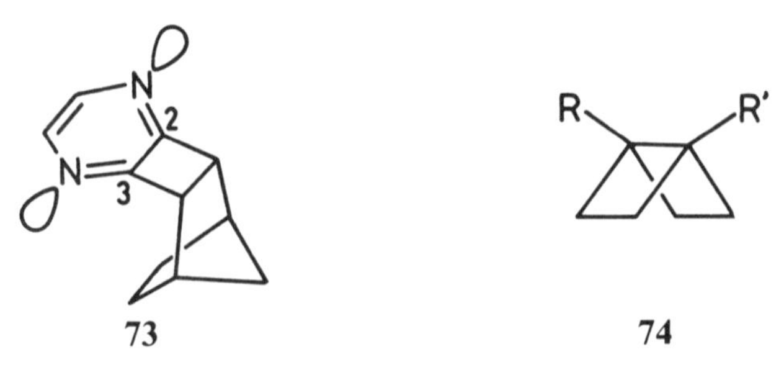

73 **74**

R and R′ are both O^- or both CH_2^-, the calculated MNDO C_1—C_4 bond lengths are 1.638 and 1.657 Å, respectively.[58] Interestingly, when a push-pull combination involving lone pairs is used for R, R′ in **74**, the through-bond interaction appears to be strengthened. When R = O^- and R′ = CO_2^-, for example, the MNDO C_1—C_4 bond length is 1.669 Å. When R′ is changed to more powerful electron-withdrawing substituents like NO_2 and CN, the C_1—C_4 bond *dissociates* during geometry optimization![58]

Highly complicated situations arise when two pairs of push-pull substituents are attached on C—C bonds (**75**).[117]

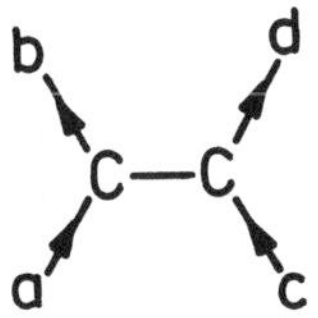

75

≪Such systems have been studied by Viehe[118] and others[119] to produce stable "capto-dative" radicals upon cleavage of the C—C bond.[120]≫

In addition to the through-bond interaction between lone pairs of "pushing" substituents (*a*/*c* in **75**), push-pull interactions through the bond (*b*/*c* and *a*/*d*), and possibly, mixing of a stabilized radical nature in the ground state may all work to elongate the intervening bond. MNDO calculations of such double push-pull systems (**76, 77**) indeed predict highly elongated cyclobutane bonds.[117]

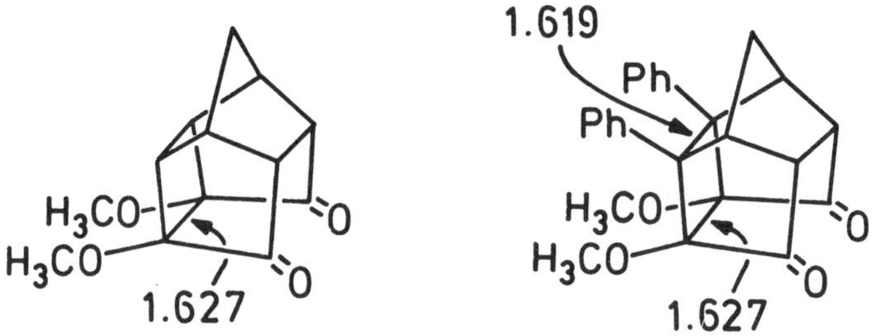

76 **77**

4. CHEMICAL BEHAVIOR OF LONG BONDS

A C—C bond longer than about 1.6 Å is generally prone to cleave; hence its generation or even the recognition of a potentially weak bond provides a new strategy in organic synthesis or novel aspects for studies on reactivities. This section illustrates instances wherein cleavage of a C—C bond during reaction can be interpreted as being caused or assisted by *enhanced* through-bond interaction.

A. Simple Thermolysis

Admirably systematic and extensive work by Rüchardt and Beckhaus and their colleagues on the thermolysis of substituted ethanes demonstrated large dependence of the rate on the length of the bond being cleaved, among other factors.[10,21,23,87,107,120,121] Some of their molecules are vicinally substituted with aryl groups like **46** and **47,** and these turned out to be among the most labile in thermolysis reactions. Notwithstanding, no abnormality has been noticed for polyarylethanes in the linear relationship between the measured free enthalpy of activation of C—C bond homolysis and the calculated total strain energy in the ground state.[10] This happened probably because the strain in polyarylethanes is overestimated by MM calculations, and this error is compensated by the decreased activation energy due to the through-bond interaction.

For structures of the *p,p'*-dibenzene (**18**) type, simultaneous thermal cleavage of the two longitudinal C—C bonds is disallowed by the orbital symmetry rule.[122] Thus, in the case of strained **78**[123] (Scheme 1), only one bond cleaves upon heating, initiating a clean [3,3]sigmatropic shift, giving pure **79.**

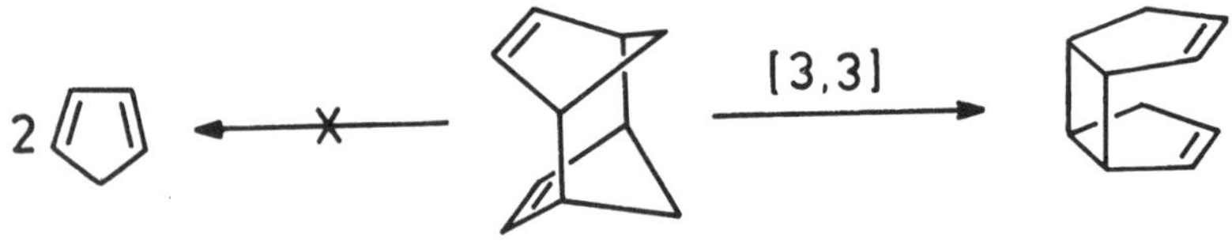

2 [3,3]

78 **79**

Scheme I.

≪Lengths of longitudinal bonds are unknown.≫

If the sigmatropic bond shift is disfavored by high instability of the product, as in the case of **80** (Scheme 2), no bond cleavage takes place at all despite the obvious weakening of the longitudinal bonds. Actually **80** is stable up to about 700°C![124]

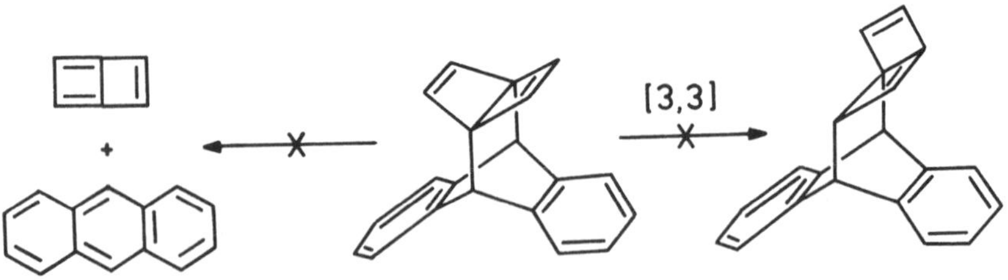

+ [3,3]

80

Scheme II.

≪In an article[125] citing this work, the structure of **80** is incorrectly drawn.≫

A remarkable effect of the through-bond interaction on the cleavage of the mediating bond has been observed with pentacyclic cage compound **81** (Scheme 3). When R = Ph, the decarbonylative ring-opening reaction gives **82** in quantitative yield (k_1 at 230°C in *o*-dichlorobenzene, 2.6×10^{-5} s^{-1}), whereas no such reaction occurs when R = H or Cl.[126–128]

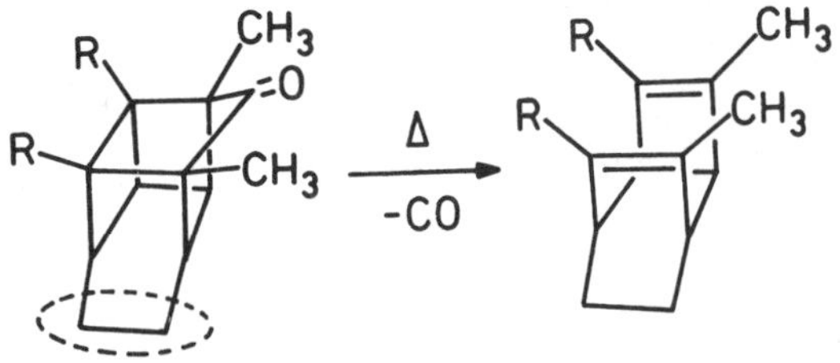

R R CH3 =O CH3 Δ -CO R R CH3 CH3

81 **82**

Scheme III.

≪The rate of reaction is sensitive to the length of the alkylene bridge (ethano in the case of **81**) located remote from the reaction center.[126,127] Preliminary MM studies on the strain distribution did not give any clear picture on the cause of rate variation.[128] For thermolyses of **28,** a compound related to **81,** see Reference 56.≫

Similarly, pentacyclic diketone (**83,** Scheme 4) undergoes cycloreversion as fast as **81** (k_1 at 230°C in diphenylether, 43×10^{-5} s^{-1}) when R = Ph,[117] but much more severe conditions are required to effect the same reaction when R = H.[117,129]

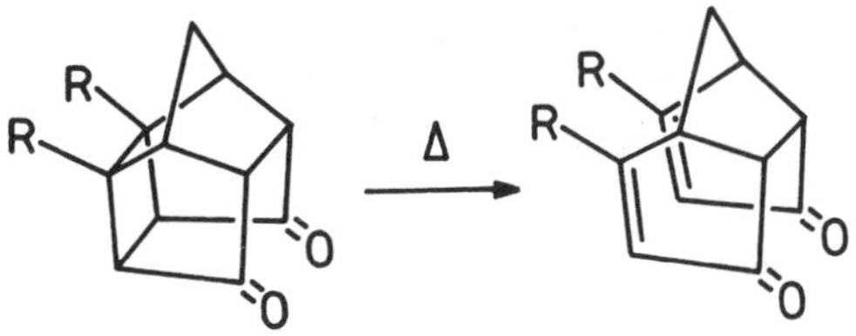

R R O O Δ R R O O

83 **84**

Scheme IV.

The thermolysis is greatly accelerated by vicinal methoxy substituents capable of through-bond interaction through their lone electron pairs. Both **76** and **77** undergo cycloreversion at 90°C in quantitative yield. Rate enhancements cannot be fully accounted for by the known stabilization energy of the methoxy group on a carbon radical.[117] Results of MNDO bond length calculations of **76** and **77** have been mentioned above.

B. Heterolysis

Cycloreversion of **76** and **77** proceeds much faster than simple thermolysis under catalysis by a Lewis acid such as $AlCl_3$ or BF_3 etherate. The solution reaction is complete in a few seconds at O°C to give quantitative yields of ring-opened products![130,131]

The long C—C bond (1.588(4) Å, R = CH_3) of **85** (Scheme 5) dissociates spontaneously in a polar solvent at ambient temperature to give a stable carbocation and carbanion ($\Delta H°$ 5.5 ± 0.5 kcal/mol, R = Ph). Upon cooling, these ions coordinate to give back covalent **85.**[132]

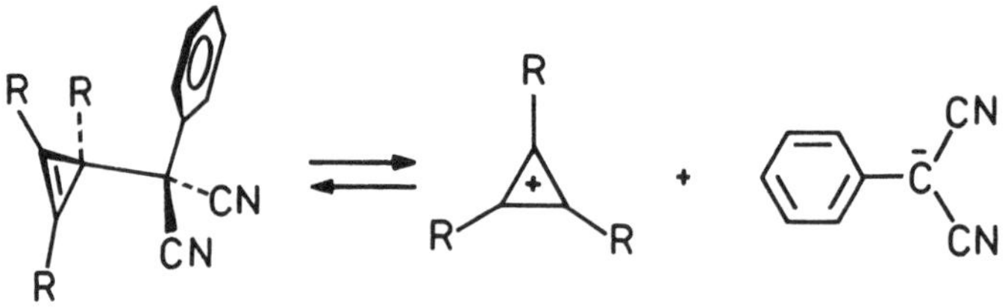

85

Scheme V.

In hypostrophene **86** (Scheme 6), available experimental as well as theoretical evidence[133] indicates that the two longitudinal C—C bonds of unknown length (arrowed) mediate strong interaction between π orbitals of the two double bonds. It is hence very likely that these bonds have a high propensity to heterolytic cleavage under ionic conditions. This premise was proved by the elegant experimental work of Paquette and associates.[133] Ionic bromination of **86** gives tricyclic bromide (**87**). Since all attempts to intercept possible intermediates in this reaction have failed, the conversion from the initial adduct **88** to the final product **87** is likely to be a concerted electronic reorganization as shown.[133a] One of the "mediator" bonds *a* and the most strained central bond *b* of the bicyclo[2.2.0]hexane partial structure migrate in this process.

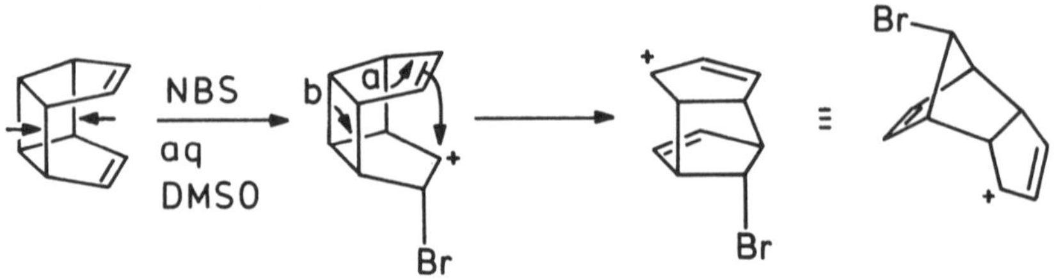

NBS aq DMSO

86 **88** **87**

Scheme VI.

≪Paquette attributes the driving force of the migration of bond *a* to its electron richness.[1331] However, this bond actually is electron poor (see above), and its heightened energy level must be the driving force.≫

Acetolysis of tosylates (**89, 90,** Scheme 7) further illustrates the high mobility of a "mediator" bond. *endo*-tosylate **90** solvolyzes by the S_N1 mechanism to give only **91,** and in this respect this reaction resembles ionic bromination of the parent hydrocarbon **86.** On the other hand, product distribution and the rate in the solvolysis of *exo*-tosylate **89** are best explained

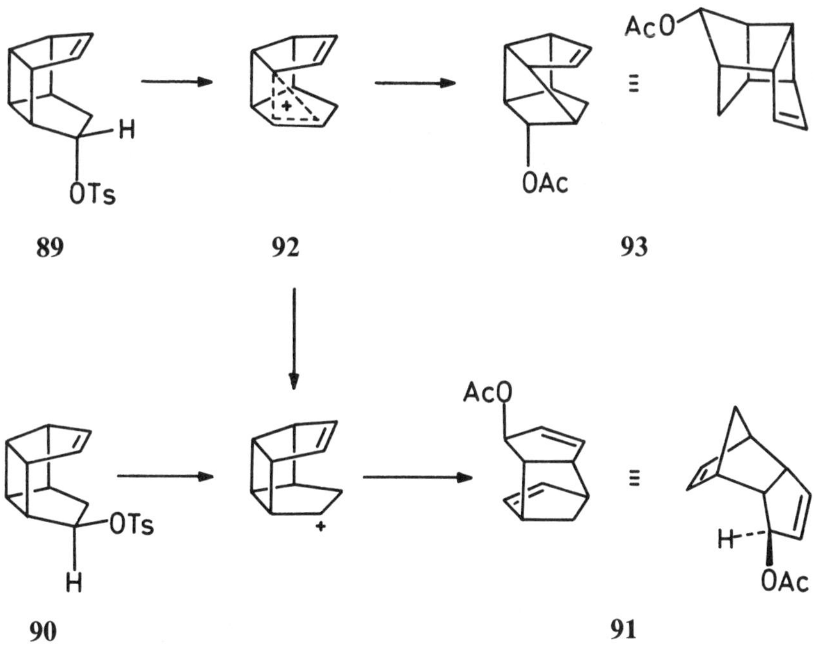

Scheme VII.

by postulating a σ-delocalized, nonclassical transition state **92.**[133b] In addition to **91,** tetracyclic acetate **93** is formed as the result of backside attack of acetate ion at **92.** A large exo/endo rate ratio of 51 is consistent with the anchimeric assistance by the "mediator" σ bond.[133b]

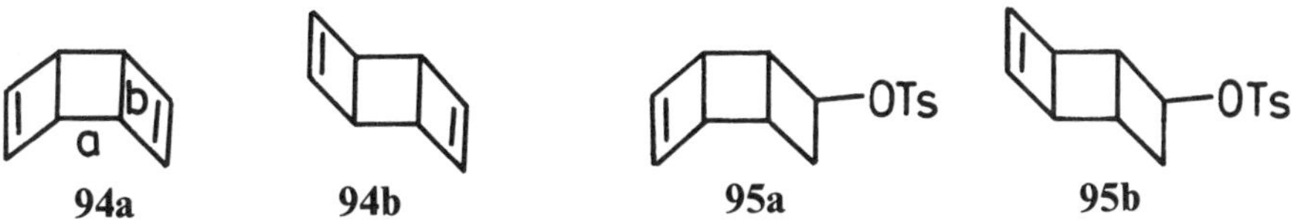

Tricyclo[4.2.0.0^{2,5}]octa-3,7-dienes (**94**) and their tosylates (**95**) proved disappointing with regard to the expected lability of the mediator bond *a*: bromination did not initiate any skeletal rearrangements, and solvolysis gave only cyclooctane derivatives. Clearly, the more strained "ladder" bond *b* competes favorably with the less strained mediator bond *a*.[133c] Only when the ladder bond is removed by inserting a methylene bridge, as in **96**, could cleavage (by photolysis) of the C_1—C_2 bond be realized.[134]

2 1

96

C. Cleavage of Cyclobutane Bonds by Photosensitized Oxidation–Reduction

The orbital interaction between the in-phase combination of π (or n) orbitals and the mediating σ orbital (the left-hand side of Figure 7-1) *increases* the energy level of one of the high-lying π orbitals. A consequence of such per-

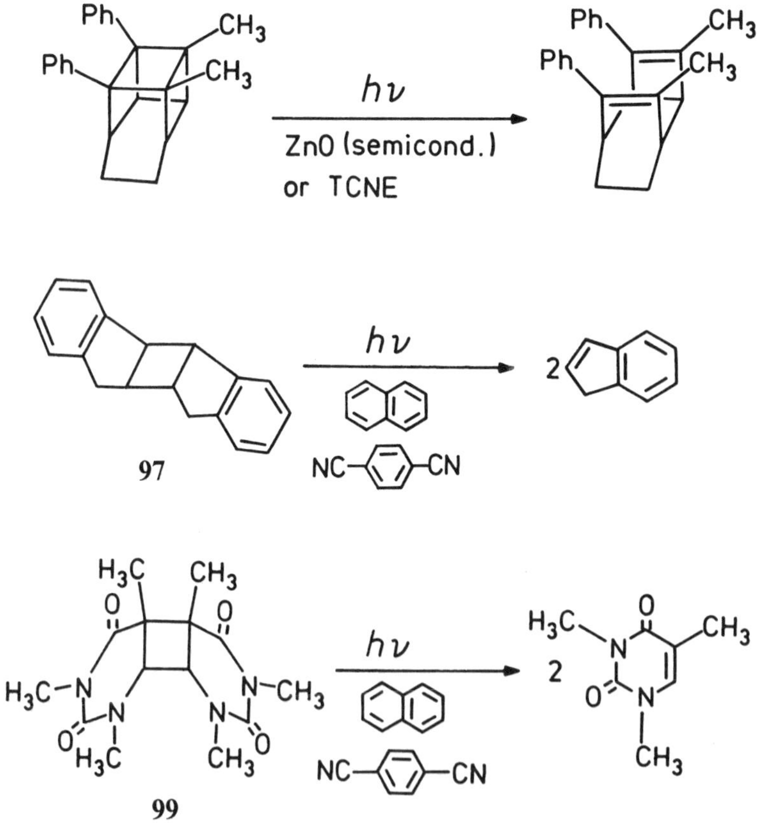

Ph Ph CH3 CH3 → hν, ZnO (semicond.) or TCNE → Ph Ph CH3 CH3 — Ref 135

97 → hν, naphthalene, NC–C6H4–CN → 2 indene — 136

99 (H3C, CH3, O, N–CH3) → hν, naphthalene, NC–C6H4–CN → 2 (H3C, O, CH3, N, O, N, CH3) — 137

Scheme VIII.

turbation is the increased HOMO level, which leads to higher ability for such a molecule to transfer an electron to an acceptor. Partial electron transfer, like π complexing with a triplet sensitizer or charge-transfer complexing with tetracyanoethylene, triggers further elongation of the mediator bond and finally leads to the opening of the cyclobutane ring. Several examples of this type of reaction are given in Scheme 8.

The importance of through-bond interaction in these reactions was first noticed by the total inertness of *trans,anti*-indene dimer **98** under the reac-

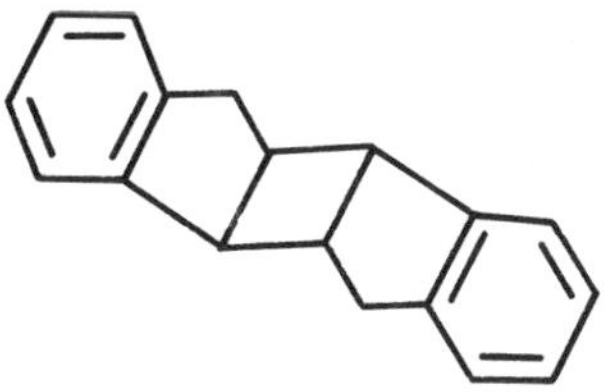

98

tion conditions that effected rapid and quantitative generation of indene from *trans,syn*-isomer **97.**[136] Failure to effect the "photorepair" dissociation of *cis,anti*-dimethylthymine dimer **100,** in contrast to *cis,syn*-dimer **99,**[137] demonstrates that the through-bond interaction occurs not in NC—CC=O but in either the NC—CN or the O=CC—CC=O system. Our previous observations suggest that both bonds *a* and *b* of **99** must have undergone considerable prestretching, by steric effect for bond *a* (note the *syn*-butane type interaction between two methyls), and by the through-bond interaction for bond *b*.

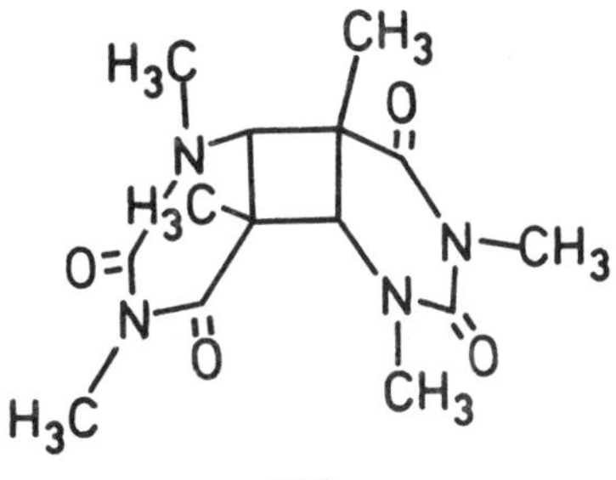

H₃C CH₃ O N H₃C O= N N—CH₃ N O O H₃C CH₃

100

5. CONCLUSIONS

Previous work on through-bond interaction has been concerned mainly with such novel structures as **18, 86,** and **94.** Two independent research groups, Sakurai's[136] and ours,[56] realized that such conventional groups as phenyl and nitrogen or oxygen lone pairs can be brought into strong orbital interactions when attached vicinally to a *strained* C—C bond. The interaction in such systems is so strong that the "mediator" bond is significantly elongated and is cleaved preferentially under various conditions. The latter finding provides a new opportunity to control regional reactivity in a molecule.

Although available examples are still rather scarce and the exact scope and limitations of the technique of controlling bond length are yet to be explored, it is fortunate that computational chemistry has matured to the point that typical consequences of through-bond interaction—namely, the change in ordering of MO levels and the variation of bond lengths—can be calculated with considerable confidence. It is also possible to evaluate the stretching force constant and the dissociation energy of a "mediator" bond. Steric contributions can be readily estimated by the MM method. It is not always necessary to synthesize every molecule to be tested, but promising candidates can be effectively screened by these computational techniques.

6. SUMMARY

Major factors affecting C—C bond lengths have been discussed in some detail. Elongation is caused by steric repulsion among bulky substituents, by localization of strain in some polycyclic structures, by introducing electropositive substituents on cyclopropane rings, and by through-bond interaction. Bond contraction occurs by increasing s character, and probably by introducing special substituents like ester and phenyl in vicinal positions. Among these factors, orbital interaction through strained bonds, which sometimes elongates the intervening bond to a remarkable extent, was discussed in detail. Vicinal phenyl groups are effective in lengthening the C—C bonds in fused cyclopropanes, in cyclobutanes, and in highly strained acyclic compounds like **45.** Ideal substrate skeletons are bicyclo[1.1.0]butanes (**34a**) and bicyclo[2.2.0]hexanes (**31**). All known polyarylethanes can be considered to have elongated central C—C bonds if the standard length in this type of molecule is shorter than has heretofore been believed. [2_n]Cyclophanes are another class of polyarylethanes having elongated C—C bonds. A cage polyene (**59**) was mentioned as a good example for comparing the effects of strain in four- and five-membered rings on the through-bond interaction between two olefinic double bonds. "Mini-superphane" (**60**) is cited as a potential example of enhanced through-bond interaction in which interacting orbitals are all strained. That carbonyl groups do not effectively interact through a strained C—C bond is illustrated by model calculations and confirmed by X-ray analysis. Lone pair electrons on nitrogen and oxygen atoms attached vicinally to a strained C—C bond often cause significant elongation of the bond. The most remarkable bond elongation occurred (computationally) between push-pull substituents on bridgehead positions of bicyclo[2.2.0]hexane, leading to spontaneous breaking of the central bond.

Chemical consequences of through-bond interactions were briefly mentioned. Long bonds are cleaved preferentially under thermolysis conditions. Heterolysis of long bonds in ionizing conditions occurs much more rapidly. The following examples were discussed in the light of the enhanced through-bond coupling mechanism: ring opening of cage molecules **81** and **83,** ionic bromination and solvolysis of hypostrophene (**86**) and derivatives, cleavage of cyclobutane bonds in polycyclic cage compounds by photosensitized oxidation–reduction, in indene dimers (**97**), and in model photorepair of DNA base dimers (**99**).

ACKNOWLEDGMENTS

We thank Professor G. Mehta for information on the X-ray analysis of structure **70.** Ms. Teruyo Fujiyoshi kindly read the entire manuscript and gave invaluable comments and criticism. Our works described herein were partially supported by the Japanese Ministry of Education through Grants-in-Aid for Scientific Research.

REFERENCES

1. A widely used standard value is 1.537(5) Å: Sutton, L.E. "Tables of Interatomic Distances and Configuration in Molecules and Ions", Supplement 1956–1959; The Chemical Society: London, 1965.
2. Pauling, L. "The Nature of the Chemical Bond", 3rd ed.; Cornell University Press: Ithaca, N.Y., 1960.
3. Iijima, T. *J. Cryst. Soc. Japan* **1975,** *17,* 21.
4. Coulson, C.A. *Proc. R. Soc. London* **1939,** *A169,* 413.
5. Kuchitsu, K. In "MTP International Review of Science", Physical Chemistry Series 1, Vol. 2; Allen, G., Ed.; Medical and Technical Publishing Co.: Oxford, 1972, Chapter 6.
6. Furka, A. *Croat. Chem. Acta* **1983,** *56,* 191.
7. Mislow, K. *Tetrahedron Lett.* **1964,** 1415.
8. Maksić, Z.B.; Randić, M. *J. Am. Chem. Soc.* **1970,** *92,* 424.
9. Bartell, L.S. *J. Chem. Phys.* **1960,** *32,* 827; *Tetrahedron* **1962,** *17,* 177; ibid, **1978,** *34,* 2891.
10. Rüchardt, C.; Beckhaus, H.-D. *Angew. Chem. Int. Ed. Engl.* **1980,** *19,* 429.
11. (a) Hoffmann, R. *Acc. Chem. Res.* **1971,** *4,* 1. (b) Gleiter, R. *Angew. Chem. Int. Ed. Engl.* **1974,** *13,* 696.
12. Schäfer, L.; Van Alsenoy, C.; Scarsdale, J.N. *J. Mol. Struct. (Theochem)* **1982,** *86,* 349.
13. Burkert, U.; Allinger, N.L. "Molecular Mechanics". American Chemical Society: Washington, D.C., 1982.
14. (a) Dougherty, D.A.; Hounshell, W.D.; Schlegel, H.B.; Bell, R.A.; Mislow, K. *Tetrahedron Lett.* **1976,** 3479. (b) Dougherty, D.A.; Schlegel, H.B.; Mislow, K. *Tetrahedron* **1978,** *34,* 1441.
15. Kuchitsu, K.; Oyanagi, K. *Faraday Discuss. Chem. Soc.* **1977,** *62,* 20.
16. Dunitz, J.D. "X-Ray Analysis and the Structure of Organic Molecules". Cornell University Press: Ithaca, N.Y., 1979, pp. 248–249.
17. Van Alsenoy, C.; Scarsdale, J.N.; Schäfer, L. *J. Comput. Chem.* **1982,** *3,* 53.
18. (a) Ermer, O. "Aspekte von Kraftfeldrechnungen", Wolfgang Bauer Verlag: Munich, 1981. (b) Ōsawa, E.; Musso, H. *Angew. Chem. Int. Ed. Engl.* **1983,** *22,* 1; *Top Stereochem.* **1982,** *13,* 117.
19. (a) Dewar, M.J.S.; Thiel, W. *J. Am. Chem. Soc.* **1977,** *99,* 4899, 4907. (b) Dewar, M. J. S.

J. Mol. Struct. **1983,** *100,* 41. (c) Stewart, J.J.P.; Csaszar, P.; Pulay, P. *J. Comput. Chem.* **1982,** *3,* 227.
20. For collections of literature on long bonds, see: (a) Ōsawa, E.; Onuki, Y.; Mislow, K. *J. Am. Chem. Soc.* **1981,** *103,* 7475. (b) Hounshell, W. D.; Dougherty, D.A.; Hummel, J.P.; Mislow, K. ibid, **1977,** *99,* 1916.
21. Beckhaus, H.-D.; Kratt, G.; Lay, K.; Geiselmann, J.; Rüchardt, C.; Kitschke, B.; Lindner, H.J. *Chem. Ber.* **1980,** *113,* 3441.
22. Winiker, R.; Beckhaus, H.-D.; Rüchardt, C. *Chem. Ber.* **1980,** *113,* 3456.
23. Beckhaus, H.-D.; Hellmann, G.; Rüchardt, C. *Chem. Ber.* **1978,** *111,* 3764.
24. Khan, A.A.; Baur, W.H.; Khan, M.A.Q. *Acta Crystallogr.* **1972,** *B28,* 2060.
25. Knox, J.R.; Raston, C.L.; White, A.H. *Aust. J. Chem.* **1979,** *32,* 553.
26. (a) Bianchi, R.; Morosi, G.; Mugnoli, A.; Simonetta, M. *Acta Crystallogr.* **1973,** *B29,* 1196. (b) Bianchi, R.; Pilati, T.; Simonetta, M. ibid, **1978,** *B34,* 2157. (c) Bianchi, R.; Pilati, T.; Simonetta, M. *J. Am. Chem. Soc.* **1981,** *103,* 6426.
27. (a) Farnell, L.; Radom, L. *J. Am. Chem. Soc.* **1982,** *104,* 7650. (b) Cremer, D.; Dick, B. *Angew. Chem.* **1982,** *94,* 877. (c) Bürgi, H.B.; Shefter, E.; Dunitz, J.D. *Acta Crystallogr.* **1975,** *31,* 3089.
28. Beddoes, R.L.; Lidley, P.F.; Mills, O.S. *Angew. Chem.* **1970,** *82,* 293.
29. Barrow, M.J.; Mills, O.S. *J. Chem. Soc. A* **1971,** 1982.
30. Haddon, R.C. Quoted in Reference 31.
31. Cremer, D.; Kraka, E.; Slee, T.S.; Bader, R.F.W.; Lan, C.D.H.; Nguyen-Dang, T.T.; MacDougall, P.J. *J. Am. Chem. Soc.* **1983,** *105,* 5069.
32. (a) Bader, R.F.W.; Anderson, S.G.; Duke, A.J. *J. Am. Chem. Soc.* **1979,** *101,* 1389. (b) Bader, R.F.W.; Nguyen-Dang, T.T. *Adv. Quantum Chem.* **1981,** *14,* 63. (c) Bader, R.F.W.; Slee, T.S.; Cremer, D.; Kraka, E. *J. Am. Chem. Soc.* **1983,** *105,* 5061.
33. Schleyer, P.v.R.; Budzelaar, P.H.M.; Cremer, D.; Kraka, E. *Angew. Chem.* **1984,** *96,* 374.
34. Chandrasekhar, J.; Schleyer, P.v.R.; Krogh-Jespersen, K. *J. Comput. Chem.* **1981,** *2,* 356.
35. See, however, Cremer, D.; Kraka, E. *Angew. Chem. Int. Ed. Engl.* **1984,** *23,* 627.
36. Hoffmann, R. *Tetrahedron Lett.* **1970,** *2907.*
37. Engelke, R.; Hay, P.J.; Kleier, D.A.; Wadt, W.R. *J. Am. Chem. Soc.* **1984,** *106,* 5439.
38. Choi, C.S.; Marinkas, P.L. *Acta Crystallogr.* **1980,** *B36,* 2491.
39. Dougherty, D.A.; Choi, C.S.; Kaupp, G.; Ōsawa, E.; Buda, A.B.; Rudzinski, J.M., submitted to *J. Chem. Soc. Perkin Trans. 2.*
40. Fritz, G.; Wartanessian, S.; Matern, E.; Hönle, W.; Schnering, H.G.v. *Z. Anorg. Allg. Chem.* **1981,** *475,* 87.
41. Allen, F. H. *Acta Crystallogr.* **1984,** *B40,* 306.
42. Stohrer, W.-D.; Hoffmann, R. *J. Am. Chem. Soc.* **1972,** *94,* 779.
43. Cox, K.W.; Harmony, M.D.; Nelson, G.; Wiberg, K.B. *J. Chem. Phys.* **1969,** *50,* 1976.
44. Dewar, M.J.S. "Hyperconjugation". Ronald Press: New York, 1962.
45. Alden, R.A.; Kraut, J.; Traylor, T.G. *J. Am. Chem. Soc.* **1968,** *90,* 74.
46. Ardebili, M.H.P.; Dougherty, D.A.; Mislow, K.; Schwartz, L.H.; White, J.G. *J. Am. Chem. Soc.* **1978,** *100,* 7994.
47. (a) Perez, S.; Brisse, F. *Acta Crystallogr.* **1976,** *B32,* 470. (b) Perez, S.; Brisse, F. *Can. J. Chem.* **1975,** *53,* 3551. (c) Perez, S.; Brisse, F. *Acta Crystallogr.* **1976,** *B32,* 1518. (d) Perez, S.; Brisse, F. ibid, **1977,** *B33,* 1673. (e) Kashino, S.; Haisa, M. ibid, **1975,** *B31,* 1819.
48. (a) Stein, M.; Winter, W.; Rieker, A. *Angew. Chem.* **1978,** *90,* 737. (b) Winter, M. *Fresenius Z. Anal. Chem.* **1980,** *304,* 279.
49. Olah, G.; Field, L.D.; Watkins, M.I.; Malhotra, R. *J. Org. Chem.* **1981,** *46,* 1761.
50. Winter, M.; Butters, T.; Rieker, A.; Butsugan, Y. *Z. Naturforsch.* **1982,** *37b,* 855.
51. (a) Gleiter, R.; Haider, R.; Spanget-Larsen, J.; Bischof, P. *Tetrahedron Lett.* **1983,** *24,* 1149. (b) Jähne, G.; Gleiter, R. *Angew. Chem. Int. Ed. Engl.* **1983,** *22,* 488. (c) Gleiter, R.; Eckert-Maksić, M.; Schäfer, W.; Truesdale, E. A. *Chem. Ber.* **1982,** *115,* 2009. (d) Gleiter, R.; Gubernator, K. ibid, **1982,** *115,* 3811. (e) Spanget-Larsen, J.; de Korswagen, C.; Eckert-Maksić, M.; Gleiter, R. *Helv. Chim. Acta* **1982,** *65,* 968. (f) Gleiter, R.; Gubernator, K.; Grimme, W. *J. Org. Chem.* **1981,** *46,* 1247. (g) Spanget-Larsen, J.; Gleiter, R.; Paquette, L.A.; Carmody, M.J.; Degenhardt, C.R. *Theor. Chim. Acta* **1978,** *50,* 145. (h) McMurray, J.E.; Haley, G.J.; Matz, J.R.; Clardy, J.C.; Van Duyne, G.; Gleiter, R.; Schäfer, W.; White, D.H. *J. Am. Chem. Soc.* **1984,** *106,* 5018.
52. (a) Pasman, P.; Rob, F.; Verhoeven, J.W. *J. Am. Chem. Soc.* **1982,** *104,* 5127. (b) Verhoeven, J.W.; Pasman, P. *Tetrahedron* **1981,** *37,* 943. (c) Verhoeven, J.W. *Rec. Trav.*

Chim. Pays-Bas **1980,** *99,* 369. (d) Dekkers, A.W.J.D.; Verhoeven, J.W.; Speckamp, W.N. *Tetrahedron* **1973,** *29,* 1691.

53. (a) Paddon-Row, M.N. *Acc. Chem. Res.* **1982,** *15,* 245. (b) Paddon-Row, M.N.; Patney, H.K.; Brown, R.S. *Aust. J. Chem.* **1982,** *35,* 293. (c) Schipper, P.E.; Paddon-Row, M.N. ibid, **1982,** *35,* 1755. (d) Balaji, V.; Jordan, K.D.; Burrow, P.D.; Paddon-Row, M.N.; Patney, H.K. *J. Am. Chem. Soc.* **1982,** *104,* 6849. (e) Paddon-Row, M.N.; Patney, H.K.; Brown, R.S.; Houk, K.N. ibid, **1981,** *103,* 5575.
54. (a) Imamura, A.; Tachibana, A.; Ohsaku, M. *Tetrahedron* **1981,** *37,* 2793. (b) Imamura, A.; Ohsaku, A. ibid, **1981,** *37,* 2191. (c) Ohsaku, M.; Shigemi, S.; Murata, H.; Imamura, A. ibid, **1981,** *37,* 297. (d) Imamura, A.; Ohsaku, M.; Akagi, K. ibid, **1983,** *39,* 1291.
55. (a) Greenberg, A.; Liebman, J.F. *J. Am. Chem. Soc.* **1981,** *103,* 44. (b) Baker, A.D.; Scharfman, R.; Stein, C.A. *Tetrahedron Lett.* **1983,** *24,* 2957. (c) Duddeck, H.; Feuerhelm, H.-T. *Tetrahedron* **1980,** *36,* 3009. (d) Surjan, P.R.; Mayer, I.; Kertesz, M. *J. Chem. Phys.* **1982,** *77,* 2454. (e) Borden, W.T.; Davidson, E.R. *J. Am. Chem. Soc.* **1980,** *102,* 5409. (f) Lee, I.; Cheun, Y.G.; Yang, K. *J. Comput. Chem.* **1982,** *3,* 565. (g) Goldberg, A.H.; Dougherty, D.A. *J. Am. Chem. Soc.* **1983,** *105,* 284. (h) Martin, H.-D.; Eisenmann, E.; Kunze, M.; Bonacic-Koutecky, V. *Chem. Ber.* **1980,** *113,* 1153.
56. Harano, K.; Ban, T.; Yasuda, M.; Ōsawa, E.; Kanematsu, K. *J. Am. Chem. Soc.* **1981,** *103,* 2310.
57. Okamoto, Y.; Harano, K.; Yasuda, M.; Ōsawa, E.; Kanematsu, K. *Chem. Pharm. Bull.* **1983,** *31,* 2526.
58. Ōsawa, E.; Ivanov, P.M.; Jaime, C. *J. Org. Chem.* **1983,** *48,* 3990.
59. Harano, K.; Okamoto, Y.; Yasuda, M.; Ueyama, K.; Kanematsu, K. *J. Org. Chem.* **1983,** *48,* 2728.
60. Andersen, B.; Srinivasan, R. *Acta Chem. Scand.* **1972,** *26,* 3468.
61. Wiberg, K.B.; Wendoloski, J.J. *J. Am. Chem. Soc.* **1982,** *104,* 5679.
62. Ehrenberg, M. *Acta Crystallogr.* **1966,** *20,* 183.
63. See also Reference 16, p. 339.
64. Jorgensen, W.L.; Salem, L. "The Organic Chemist's Book of Orbitals". Academic Press: New York, 1973.
65. Reference 64, pp. 22–23, 154–155.
66. Jason, M.E.; Ibers, J.A. *J. Am. Chem. Soc.* **1977,** *99,* 6012.
67. Tinant, B.; De Block, I.; Declercq, J.P.; Germain, G.; Van Meersche, M.; Leroy, G.; Weiler, J. *Bull. Soc. Chim. Belg.* **1982,** *91,* 629.
68. Mancini, V.; Passini, P.; Santini, S. *J. Chem. Soc. Chem. Commun.* **1978,** 100.
69. Greenberg, A.; Liebman, J.F. "Strained Organic Molecules". Academic Press: New York, 1978, pp. 70.
70. Dill, J.D.; Greenberg, A.; Liebman, J.F. *J. Am. Chem. Soc.* **1979,** *101,* 6814.
71. Paddon-Row, M.N.; Houk, K.N.; Daud, P.; Garner, P.; Schappert, R. *Tetrahedron Lett.* **1981,** *22,* 4799.
72. See also: Gleiter, R.; Haider, R.; Bischof, P.; Zefirov, N.S.; Boganov, A.M. *J. Org. Chem.* **1984,** *49,* 375.
73. Hounshell, W.D.; Dougherty, D.A.; Hummel, J.P.; Mislow, K. *J. Am. Chem. Soc.* **1977,** *99,* 1916.
74. Dougherty, D.A.; Mislow, K.; Blount, J.F.; Wooten, J.B.; Jacobus, J. *J. Am. Chem. Soc.* **1977,** *99,* 6149.
75. Cruickshank, D.W.J. *Acta Crystallogr.* **1949,** *2,* 65.
76. Ivanov, P.; Pojarlieff, I.; Tyutyulkov, N. *Tetrahedron Lett.* **1976,** 775.
77. Anderson, J.E.; Pearson, H. *J. Chem. Soc. Perkin Trans. 2* **1977,** 699.
78. Ivanov, P.; Pojarlieff, I.; Tyutyulkov, N. *Commun. Dept. Chem. Bulg. Acad. Sci.* **1976,** *9,* 516. For extensive computational studies of 1,2-diphenylethane derivatives, see: Pojarlieff, I.G.; Ivanov, P.M.; Berova, N.D. *J. Mol. Struct.* **1983,** *91,* 283, and preceding papers of this series.
79. Perpendicular conformation **43a** more stable: Schaefer, T.; Kruczynski, L.J.; Krawchuk, B.; Sebastian, R.; Charlton, J.L.; McKinnon, D.M. *Can. J. Chem.* **1980,** *58,* 2452, and references cited therein.
80. Methyl-parallel more stable: Stokr, J.; Pivcova, H.; Schneider, B.; Dirlikov, S. *J. Mol. Struct.* **1972,** *12,* 45.
81. H-parallel more stable: Camail, M.; Proutiere, A.; Verlaque, P. *J. Phys. Chem.* **1975,** *79,* 1962.

82. Watkins, M.I.; Olah, G.A. *J. Am. Chem. Soc.* **1981,** *103,* 6566.
83. Favini, G.; Simonetta, M.; Todeschini, R. *J. Am. Chem. Soc.* **1981,** *103,* 3679.
84. Bernardinelli, G.; Gerdil, R. *Helv. Chim. Acta* **1981,** *64,* 1372.
85. Littke, W.; Drück, U. *Angew. Chem. Int. Ed. Engl.* **1979,** *18,* 406.
86. Jaime, C.; Ōsawa, E. (a) *J. Chem. Soc. Chem. Commun.* **1983,** 708; (b) *J. Chem. Soc. Perkin Trans. 2* **1984,** 995; (c) *J. Mol. Struct.* **1985,** *126,* 363.
87. Kratt, G.; Beckhaus, H.-D.; Lindner, H.J.; Rüchardt, C. *Chem. Ber.* **1983,** *116,* 3235.
88. Kane, V.V.; Martin, A.R.; Jaime, C.; Ōsawa, E. *Tetrahedron* **1984,** *40,* 2919.
89. Destro, R.; Pilati, T.; Simonetta, M. *Acta Crystallogr.* **1980,** *B36,* 2550.
90. Ehrenberg, M. *Acta Crystallogr.* **1967,** *22,* 482.
91. Ahmed, F.R.; Neville, G. A. *Acta Crystallogr.* **1982,** *B38,* 2930.
92. Gleiter, R. *Tetrahedron Lett.* **1969,** 4453.
93. Kovac, B.; Allan, M.; Heilbronner, E. *Helv. Chim. Acta* **1981,** *64,* 430, and references cited therein.
94. (a) Spanget-Larsen, J. *Theor. Chim. Acta (Berl.)* **1983,** *64,* 187. (b) Doris, K.A.; Ellis, D.E.; Ratner, M.A.; Marks, T.J. *J. Am. Chem. Soc.* **1984,** *196,* 2491.
95. Review: Heilbronner, E.; Yang, Z.-Z. *Top. Curr. Chem.* **1983,** *115,* 1.
96. Hope, H.; Bernstein, J.; Trueblood, K.N. *Acta Crystallogr.* **1972,** *B28,* 1733.
97. Nishiyama, K.; Sakiyama, M.; Seki, S.; Horita, H.; Otsubo, T.; Misumi, S. *Tetrahedron Lett.* **1977,** 3739.
98. See also Reference 69, pp. 156 ff.
99. Hanson, A.W. *Acta Crystallogr.* **1977,** *B33,* 2003.
100. Hanson, A.W.; Cameron, T.S. *J. Chem. Res. (S)* **1980,** 336; ibid, **1980,** 4201.
101. Ōsawa, E.; Henke, H.; Schröder, G. *Tetrahedron Lett.* **1976,** 847.
102. Reference 13, p. 156.
103. Wang, Y.C.; Bauer, S.H. *J. Am. Chem. Soc.* **1972,** *94,* 5651.
104. Miller, L.S.; Grohmann, K.; Dannenberg, J.J.; Todaro, L. *J. Am. Chem. Soc.* **1981,** *103,* 6249.
105. (a) Wiberg, K.B.; Adams, R.D.; Okarma, P.J.; Matturo, M.G.; Segmuller, B. *J. Am. Chem. Soc.* **1984,** *106,* 2200. (b) Wiberg, K.B.; Matturo, M.G.; Okarma, P.J.; Jason, M.E. ibid, **1984,** *106,* 2194.
106. Martin, H.-D.; Pföhler, P. *Angew. Chem.* **1978,** *90,* 901.
107. Bernlöhr, W.; Beckhaus, H.-D.; Peters, K.; von Schnering, H.-G.; Rüchardt, C. *Chem. Ber.* **1984,** *117,* 1013.
108. (a) Wenkert, E.; Yoder, J.E. *J. Org. Chem.* **1970,** *35,* 2986. (b) Dekker, J.; Martins, F.J.C.; Kruger, J.A.; Goosen, A.J. *Tetrahedron Lett.* **1974,** 3721. (c) Dekker, J.; Martins, F.J.C.; Kruger, J.A. ibid, **1975,** 2489. (d) Hamner, E.R.; Kemmitt, R.D.W.; Smith, M.A.R. *J. Chem. Soc. Chem. Commun.* **1974,** *841.*
109. Land, E.J.; Rushton, F.A.P.; Beddoes, R.L.; Bruce, J.M.; Cernik, R.J.; Dawson, S.C.; Mills, O.S. *J. Chem. Soc. Chem. Commun.* **1982,** 22.
110. Calculated using atomic coordinates supplied from the Cambridge Crystallograpic Data Centre.
111. Marchand, A.P.; Suri, S.C. *J. Org. Chem.* **1984,** *49,* 670.
112. Mehta, G.; Singh, V.; Srikrishna, A.; Cameron, T.S. *Tetrahedron Lett.* **1979,** 4595.
113. Bond lengths supplied by Professor Mehta.
114. See, eg, Reference 64, p. 78.
115. Raber, D.J.; Janks, C.M.; Johnston, M.D., Jr.; Raber, N.K. *J. Am. Chem. Soc.* **1980,** *102,* 6591. See also Reference 55, p. 84.
116. Dewey, H.J.; Miller, R.D.; Michl, J. *J. Am. Chem. Soc.* **1982,** *104,* 5298.
117. Okamoto, Y.; Kanematsu, K.; Fujiyoshi, T.; Ōsawa, E. *Tetrahedron Lett.* **1983,** *24,* 5645.
118. (a) Viehe, H.G.; Merenyi, R.; Stella, L.; Janousek, Z. *Angew. Chem. Int. Ed. Engl.* **1979,** *18,* 917. (b) Merenyi, R.; de Mesmaeker, A.; Viehe, H.G. *Tetrahedron Lett.* **1983,** *24,* 2765.
119. (a) Hart, D. J. *Science* **1984,** *223,* 883. (b) Korth, H.-G.; Lommes, P.; Sustmann, R. *J. Am. Chem. Soc.* **1984,** *106,* 663.
120. See, however, Zamkanei, M.; Kaiser, J.H.; Birkhofer, H.; Beckhaus, H.-D.; Rüchardt, C. *Chem. Ber.* **1983,** *116,* 3216.
121. Some recent papers: (a) Kratt, G.; Beckhaus, H.-D.; Rüchardt, C. *Chem. Ber.* **1984,** *117,* 1748. (b) Bernlöhr, W.; Beckhaus, H.-D.; Rüchardt, C. ibid, **1984,** *117,* 1026. (c) Hellmann, S.; Beckhaus, H.-D.; Rüchardt, C. ibid, **1983,** *116,* 2238.

122. Woodward, R.B.; Hoffmann, R. "The Conservation of Orbital Symmetry". Verlag Chemie: Weinheim, 1970.
123. Paquette, L.A.; Doecke, C.W.; Klein, G. *J. Am. Chem. Soc.* **1979,** *101,* 7599.
124. Johnson, G.C.; Bergman, R.G. *Tetrahedron Lett.* **1979,** 2093.
125. Butler, D.N.; Russell, R.A.; Waring, R. B.; Warrener, R. N. *Aust. J. Chem.* **1984,** *37,* 1293.
126. Tezuka, T.; Yamashita, Y.; Mukai, T. *J. Am. Chem. Soc.* **1976,** *98,* 6051.
127. See also Reference 56 and (a) Yamashita, Y.; Mukai, T.; Tezuka, T. *J. Chem. Soc. Chem. Commun.* **1977,** 532. (b) Mukai, T.; Yamashita, Y. *Tetrahedron Lett.* **1978,** 357.
128. Ōsawa, E.; Aigami, K.; Inamoto, Y. *J. Chem. Soc. Perkin Trans. 2* **1979,** 181.
129. Mehta, G.; Reddy, A. V.; Srikrishna, A. *Tetrahedron Lett.* **1979,** 4863.
130. Mehta, G.; Reddy, D.S.; Reddy, A.V. *Tetrahedron Lett.* **1984,** *25,* 2275.
131. Okamoto, Y.; Senokuchi, K.; Kanematsu, K. *Chem. Pharm. Bull.* **1984,** *32,* 4593.
132. Arnett, E. M.; Troughton, E.B.; McPhail, A.T.; Molter, K.E. *J. Am. Chem. Soc.* **1983,** *105,* 6172.
133. (a) Paquette, L.A.; James, D.R.; Klein, G. *J. Org. Chem.* **1978,** *43,* 1287. (b) Klein, G.; Paquette, L. A. ibid, **1978,** *43,* 1293. (c) Pquette, L. A.; Carmody, M.J. ibid, **1978,** *43,* 1299.
134. Haselbach, E.; Eberbach, W. *Helv. Chim. Acta* **1973,** *56,* 1944.
135. (a) Okada, K.; Hisamitsu, K.; Mukai, T. *J. Chem. Soc. Chem. Commun.* **1980,** 941. (b) Mukai, T.; Satō, K.; Yamashita, Y. *J. Am. Chem. Soc.* **1981,** *103,* 670.
136. (a) Majima, T.; Pac, C.; Sakurai, H. *J. Am. Chem. Soc.* **1980,** *102,* 5265. (b) Majima, T.; Pac, C.; Takamuku, S.; Sakurai, H. *Chem. Lett.* **1979,** 1149. (c) Majima, T.; Pac, C.; Sakurai, H. *J. Chem. Soc. Perkin Trans. 1* **1980,** 2705.
137. Majima, T.; Pac, C.; Kubo, J.; Sakurai, H. *Tetrahedron Lett.* **1980,** 377.

Addendum

CHAPTER 1

Recent unpublished work has given more quantitative information on the exceedingly slow electron transfer rate between hydrazine radical cations and hydrazines, both in solution and in the gas phase.

In contrast to the anti bent radical cations from **10, 42,** and **53,** some syn bent cations show rapid enough electron exchange with their neutral forms to allow convenient measurement of the second order rate constant for electron transfer (k_2 in Equation 1-4). For the unsaturated analogue of **38** with a C_4, C_5 double bond, **53,** k_2 is $5.4 \times 10^4\ M^{-1}s^{-1}$ in CD_2Cl_2 (see Ref. 67, Chapter 1). Counterion and solvent effects prove to be quite small. Variable temperature studies (-22 to $+23$°C) show that the 10.9 kcal/mol free energy barrier for electron transfer between $\mathbf{53}^+NO_3^-$ and **53** in CD_2Cl_2 is composed of $\Delta H^\ddagger = 7.4$ kcal/mol, $\Delta S^\ddagger$ about -12 cal/deg-mol. The x-ray structures of **53** and $\mathbf{53}^+NO_3^-$ document the magnitude of the geometry change for this electron transfer. The NN bond distance shortens from 1.49_7 to 1.34_0 Å (10.5%), and the nitrogens flatten from α(av) 112.0 to 117.5° (52% of the 10.5° between tetrahedral and planar nitrogens). Interestingly, MNDO calculations (carried out by the method used for H_4N_2) estimate $\Delta H^\ddagger$ for this electron transfer to be about 8.8 kcal/mol, surprisingly close to the observed value.

Variable temperature HPMS studies[1] have allowed separation of the $\Delta G°$ (550 K) values into enthalpy and entropy terms, allowing measurement of the enthalpy of relaxation, ΔH_r, observed to vary from 1.4_0 eV for Me_4N_2 to 1.0_0 eV for **19.** Gas phase electron transfer between $R_4N_2^+$ and R_4N_2 is also very slow. For example, k_2 for $Me_4N_2^+ + (EtMeN)_2$ (electron transfer is exothermic by 3 kcal/mol) is 2.2×10^{-12} cm^{3}molec^{-1}s^{-1}, corresponding to a collision efficiency of 0.0017. Such low collision efficiencies have previously only been observed for heavy atom transfer reactions. The activation energy for this reaction was found to be 2.7 kcal/mol, showing that the electron transfer reaction has a barrier higher than the energy well for complexation of the radical cation with the neutral species. The depth of this well for $Me_4N_2^+$ complexing with Me_4N_2 has been measured at 13 kcal/mol[1], showing that the electron transfer barrier for these gauche hydrazines in the gas phase is even higher than that measured in solution for highly branched, protected hydrazines which have θ held near 180 and 0°.

See Ogawa and coworkers[2] for a review of rotational and inversion barriers in acyclic hydrazines, including very recent work by Lunazzi and Macciantelli[3] on barriers in compounds with simple alkyl groups, including a rotational barrier of 6.0 kcal/mol (−137°C) in Me_4N_2 and inversion barriers of 7.5 kcal/mol (−112°C) and 5.1 (−160°C) at the Me_2N and iPr_2N nitrogens, respectively, of Me_2NNiPr_2.

CHAPTER 4

A general definition of the intrinsic thermodynamic stability of a chemical species has been advanced and reviewed by Leroy.[4] The stabilization energy (SE°) for the species is defined as the difference between its heat of atomization and the sum of its bond energy terms. The bond energy scheme and the use of reference compounds have been described in detail.[5] Thus, for example, CH_2F_2 and CHF_3 have SE° values of 7.8 and 15.2 kcal/mol, respectively, relative to CH_3F (SE° = 0 by definition). Similarly, the SE° values for CH_3CHF_2 and CH_3CF_3 are 12.5 and 24.6 kcal/mol, respectively. Although this treatment of substituent effects on thermodynamic stability usually agrees qualitatively with the incremental stabilization concept that was emphasized in this chapter, the two approaches often give markedly different quantitative results.

A new summary of ab initio calculations with a 4-31G basis set on a wide variety of mono-substituted ethylenes, cyclopropanes, and benzenes is available.[6] Isodesmic stabilization energies for substituents, including F and CF_3, have been derived and employed in linear-free energy relationships to analyze the relative importance of σ- and π-substituent effects. Notably, a comparison of acetylene derivatives with the corresponding ethyl derivatives indicated F and CF_3 destabilize a triple bond by 19.2 and 11.4 kcal/mol, respectively, which agree well with the estimates given in this chapter.

The equilibrium geometry of CF_3OCF_3 has been calculated using a 4-31G basis set.[7] The computed bond lengths and bond angles agree well with experiment, except for the $\angle COC$ which is ca. 6° larger than experiment. The STO-3G basis set gave less satisfactory results. New values for bond dissociation energies of perfluoroalkyl iodides are reported and compared with previously published results.[8] The following values are recommended: $D°(CF_3{-}I)$ = 53.5 kcal/mol, $D°(C_2F_5{-}I)$ = 52.3 kcal/mol, $D°(i{-}C_3F_7{-}I)$ = 51.4 kcal/mol.

A detailed study of the conformational isomers of $n{-}C_4F_{10}$, $n{-}C_6F_{14}$, and $n{-}C_8F_{18}$ in different physical states has appeared.[9] From an analysis of IR, Raman, and approximate force field data, both $n{-}C_4F_{10}$ and $n{-}C_6F_{14}$ exist in their all-trans forms in the solid state. In the gas or liquid states, a less stable gauche isomer of $n{-}C_4F_{10}$ with a dihedral angle of 120 ± 15° is present, and there is evidence for at least three non-trans forms of $n{-}C_6F_{14}$.[9]

Several reports on the relative effects of fluorination on the structure and stability of oxiranes versus cyclopropanes have appeared. Microwave structures of cis- and trans-1,2-difluoroethylene oxide[10,11] and tetrafluoroethylene oxide[12] have been determined and an electron diffraction structure of hexafluoropropylene oxide (HFPO) has been reported.[13] In contrast to what is observed for cyclopropanes, all ring bond lengths nearly linearly change (shorten) from ethylene oxide to cis-1,2-difluoroethylene oxide to tetrafluoroethylene oxide. A theoretical analysis of the interaction between fluorine and occupied ring orbitals has accounted for this trend.[14] The E_a for pyrolysis of tetrafluoroethylene oxide (31.6 kcal/mol) is 25.8 kcal/mol lower than that for ethylene oxide itself, and HFPO thermally decomposes exclusively into CF_2: and CF_3COF (E_a = 36.3 kcal/mol) with a half-life of about 6 hr at 165°C.[15] The chemistry of HFPO has been reviewed recently.[16]

CHAPTER 5

A very important contribution to the concept of substituent effects on strain in the cyclopropane series has been made by Cremer and Kraka.[17] They have employed an electron density model which indicates the importance of the substituent in enhancing or decreasing surface delocalization (sigma aromaticity). An accompanying article uses these principles to assess strain in other three-membered rings.[18] The field effect as seen in 4-substituted bicyclo(2.2.2)octane-1-carboxylic acids has been modeled using ab initio calculations.[19] A very thorough study of the conjugation in vinylcyclopropane has appeared.[20] A study of enthalpies of hydrogenation of phenyl-substituted alkynes includes data for diphenylacetylene which indirectly provides a value of 22 kcal/mol for the resonance stabilization in diphenylcyclopropenone,[21] in virtually perfect agreement with our own value. Fuchs and coworkers[22] have calculated substituent effects on allenes and have found the LFER approach to be useful.

CHAPTER 6

The addendum of this chapter consists solely of new experimental numbers that directly relate to the thermochemical quantities discussed earlier in the main body of the text.

In section 2 we used a semiempirical heat of formation of cyclopropyl cation in macroincrementation reaction (6-15) to estimate the heat of formation of the cation which has been directly determined[23] via photoionization of the radical. This quantity, 25.5 ± 0.9 kcal/mol (107 ± 4 kJ/mol), is comparable to that used earlier. As such, all of our previous findings and logic remain unchanged.

In section 9 the need for a new heat of sublimation of diphenylmethane was indicated. In the interim this quantity has been measured[24] and found to be 19.9 kcal/mol (83 kJ/mol). Together with a new heat of fusion[24] (4.6 kcal/mol, 19 kJ/mol), it is seen that our prediction for the heat of vaporization from Equation 6-73 of 15.2 kcal/mol (64 kJ/mol) is essentially indistinguishable from that derived from experiment, 15.3 kcal/mol (64 kJ/mol). It would thus appear that our calculations of hydrocarbon heats of phase change are accurate to better than ± 1 kcal/mol (4 kJ/mol).

In section 10, the heats of formation and of vaporization of the methyl and dimethylpyridines were discussed. New values for the heats of vaporization have been reported.[25] Macroincrementation reaction (6-88) combined with these new numbers results in even closer agreement between experiment and calculation than before: for the heat of vaporization, the average error has been reduced to 0.2 kJ/mol, while for the heat of formation, the average error is now 1.2 kJ/mol—the units of kJ/mol chosen again to "exaggerate" the discrepancies. We also note the heats of vaporization recently determined[26] for two of the five isomeric trimethylpyridines, the 2,3,6- and the 2,4,6-. The new macroincrementation reaction:

$$\text{i,j,k}-(CH_3)_3C_5H_2N = \text{i,j,k}-C_6H_3(CH_3)_3 + \text{i}-CH_3C_5H_4N + \text{j}-CH_3C_5H_4N + \text{k}-CH_3C_5H_4N + 2C_6H_6 - 3C_6H_5CH_3 - 2C_5H_5N$$

results in predicted values for the heats of vaporization for the 2,3,6- and 2,4,6-isomers of 50.8 and 50.6 kJ/mol to be compared to the experimental 50.6 and 50.3. There remain no experimental heats of formation with which to compare theoretical predictions. Likewise, there are no new direct experimental numbers for the heat of formation of any ethylpyridine. However, we have recently become aware of indirect kinetics experiments[27] on these species that show the earlier macroincrementation reaction (6-90) to be nearly thermoneutral for all three isomers. This corroborates our derived number for the 2-isomer and supports our contention that a directly measured thermochemical heat of formation of 2-ethylpyridine needs to be obtained.

Finally, we note the appearance of a newly evaluated value[28] for the heat of formation of vinylamine which would be useful in Section 11. This new quantity, 10 (± 5) kcal/mol (42 (± 21) kJ/mol), is still compatible with the prediction of macroincrementation reaction (6-92), though smaller than that suggested before by a generally significant 1.5 kcal/mol (6 kJ/mol).

ADDENDUM REFERENCES

1. Meot-Ner(Mautner), M., Rumack, D.T. Unpublished work.
2. Ogawa, K.; Takeuchi, Y.; Suzuki, H.; Yoshida, H. *J. Mol. Str.*, **1985,** *126,* 445.
3. Lunazzi, L.; Macciantelli, D. *Tetrahedron,* **1985,** *41,* 1991.
4. Leroy, G. *Adv. Quantum Chem.*, **1985,** *17,* 1.
5. Leroy, G. *Int. J. Quantum Chem.*, **1983,** *23,* 277.
6. Greenberg, A.; Stevenson, T.A. *J. Am. Chem. Soc.*, **1985,** *107,* 3488.
7. Pacansky, J.; Liu B. *J. Phys. Chem.*, **1985,** *89,* 1883.
8. Ahonkhai, S.I.; Whittle, E. *Int. J. Chem. Kinetics,* **1984,** *16,* 543.
9. Campos-Vallette, M.; Rey-Lafon, M. *J. Mol. Struct.*, **1984,** *118,* 245.
10. Gilles, C.W. *J. Mol. Spectrosc.*, **1978,** *71,* 85.

11. LaBrecque, G.; Gilles, C.W.; Raw, T.T.; Agopovich, J.W. *J. Am. Chem. Soc.*, **1984,** *106,* 6171.
12. Agopovich, J.W.; Alexander, J.; Gilles, C.W.; Raw, T.T. *J. Am. Chem. Soc.*, **1984,** *106,* 2250.
13. Beagley, B.; Pritchard, R.G.; Banks, R.E. *J. Fluorine Chem.*, **1981,** *18,* 159.
14. Deakyne, C.A.; Cravero, J.P.; Hobson, W.S. *J. Phys. Chem.*, **1984,** *88,* 5975.
15. Kennedy, R.C.; Levy, J.B. *J. Fluorine Chem.*, **1976,** *7,* 101.
16. Millauer, H.; Schwertfeger, W.; Siegemund, G. *Angew. Chem. Int. Ed. Engl.*, **1985,** *24,* 161.
17. Cremer, D.; Kraka, E. *J. Am. Chem. Soc.*, **1985,** *107,* 3811.
18. Cremer, D.; Kraka, E. *J. Am. Chem. Soc.*, **1985,** *107,* 3800.
19. Inamoto, N.; Masuda, S.; Niwa, J. *Bull. Chem. Soc. Japan,* **1985,** *58,* 158.
20. Klahn, B.; Dyczmons, V. *J. Mol. Struct.*, **1985,** *122,* 75.
21. Davis, E.H.; Allinger, N.L.; Rogers, D.W. *J. Org. Chem.*, **1985,** *50,* 3601.
22. Furet, P.; Matcha, R.L.; Fuchs, R. Unpublished results courtesy of R. Fuchs.
23. Dyke, J.; Ellis, A.; Johnathan, N.; Morris, A. *J. Chem. Soc. Farad. Trans.* In press; and personal communication from J. Dyke.
24. Chickos, J.S.; Annunziata, R.; Ladon, L.H.; Hyman, A.S.; Liebman, J.F. Unpublished results.
25. Monomethyl: Majer, V.; Svoboda, V.; Lencka, M. *J. Chem. Thermodyn.*, **1984,** *16,* 1019.
26. Di and trimethyl: Majer, V.; Svoboda, V.; Lencka, M. *J. Chem. Thermodyn.*, **1985,** *17,* 365.
27. Barton, B.D.; Stein, S.E. *J. Chem. Soc. Farad Trans. 1,* **1981,** *77,* 1755.
28. Lias, S.G. Personal communication.

CUMULATIVE CONTENTS: VOLUMES 1 TO 3

(Article title followed by volume number with chapter in parentheses)

GENERAL INDEX

A

B

C

D

H

I

L

M

N

O

P

Q

R

S